ADVANCES IN

Biodegradation

AND BIOREMEDIATION OF

Industrial Waste

ADVANCES IN
Biodegradation
AND BIOREMEDIATION OF
Industrial Waste

RAM CHANDRA

CRC Press
Taylor & Francis Group
Boca Raton London New York

CRC Press is an imprint of the
Taylor & Francis Group, an **informa** business

CRC Press
Taylor & Francis Group
6000 Broken Sound Parkway NW, Suite 300
Boca Raton, FL 33487-2742

First issued in paperback 2020

Version Date: 20150202

ISBN 13: 978-0-367-57583-0 (pbk)
ISBN 13: 978-1-4987-0054-2 (hbk)

Visit the Taylor & Francis Web site at
http://www.taylorandfrancis.com

and the CRC Press Web site at
http://www.crcpress.com

Contents

Preface

Bioremediation and detoxification of environmental pollutants due to industrial activities is a global challenge in the current scenario for sustainable development of human society. The detailed knowledge of pollutants and their metabolic mineralisation is prerequisite for the monitoring of environmental pollutants. Although the diverse metabolic capabilities of microorganisms and their interactions with hazardous organic and inorganic compounds have been revealed in the recent past, the knowledge explored in the areas of bioremediation and biodegradation during the recent past is scattered and not easily accessible to readers. Therefore, the present book has compiled the available advanced knowledge of biodegradation and bioremediation of various environmental pollutants, which are a real challenge to environmental researchers in the current scenario. In general, the bioremediation and biodegradation processes are typically implemented in a relatively cheaper manner and are applicable on a large scale. Besides, only a few bioremediation techniques have even been successfully implemented to clean up the polluted soil, oily sludge and groundwater contaminated by petroleum hydrocarbons, solvents, pesticides and other chemicals. Still, some pollutants released from tanneries, distilleries and the pulp paper industry are a challenge to scientists due to lack of proper knowledge regarding the persistent organic pollutants discharged from these industries and the process of their detoxification. Similarly, the safe disposal and biodegradation of hospital waste is also a real challenge worldwide for human health.

For this book, a number of experts from universities, government research laboratories and industry have shared their specialised knowledge in environmental microbiology and biotechnology. Chapters dealing with microbiological, biochemical and molecular aspects of biodegradation and bioremediation have covered numerous topics, including microbial genomics and proteomics for the bioremediation of industrial waste. The roles of siderophores and the rhizosphere bacterial community for phytoremediation of heavy metals have been also described in detail with their mechanisms. The mechanism of phytoremediation of soil polluted with heavy metals is still not very clear to all researchers. Therefore, the current advances in phytoremediation have been included in this book. The relationship of metagenomes with persistent organic pollutants present in the sugarcane molasses–based distillery waste and pulp paper mill wastewater after secondary treatment has been also described. The role of biosurfactants for bioremediation and biodegradation of various pollutants discharged from industrial waste has been described as they are tools of biotechnology. In the bioremediation process, the role of potential microbial enzymatic processes has been described; these are very important tools for understanding bioremediation and biodegradation. The book has also described the latest knowledge regarding the biodegradation of tannery and textile waste. The role of microbes in plastic degradation bioremediation and recycling of urban waste is highlighted properly. Although the microbial degradation of hexachlorocyclohexane and other pesticides has been emphasised earlier in detail, the recent development of bioremediation of various xenobiotics is still not well documented and circulated; hence, this book has described the latest information. The biodegradation of complex industrial waste is a major challenge for sustainable development in the current scenario. Therefore, this book has given emphasis on the role of different bioreactors for treatment of complex industrial waste. Thus, this book will facilitate to the environmental engineering student also. This book has also given special emphasis to phytoremediation and the role of wetland plant rhizosphere bacterial ecology and the bioremediation of industrial wastewater. Therefore, this book will provide an opportunity for a wide range of readers, including students, researchers and consulting professionals in biotechnology, microbiology, biochemistry, molecular biology and environmental sciences. We gratefully acknowledge the cooperation and support of all the contributing authors for the publication of this book.

Editor

Ram Chandra is a professor and founder head of the Department of Environmental Microbiology at Babasaheb Bhimrao Ambedkar Central University in Lucknow, India. He obtained his BSc (Hons) in 1984 and MSc in 1987 from Banaras Hindu University in Uttar Pradesh, India. Subsequently, a PhD was awarded in 1994. He started his career as Scientist B at the Industrial Toxicology Research Centre Lucknow in the area of biotechnology in 1989. Finally he became a senior principal scientist (Scientist F) in 2009 in the area of environmental microbiology at the Indian Institute of Toxicology Research (IITR) in Lucknow.

He subsequently joined as a professor and head of the Department of Environmental Microbiology (2011) at Babasaheb Bhimrao Ambedkar Central University in Lucknow. He has done leading work on bacterial degradation of lignin from pulp paper mill waste and molasses melanoidin from distillery waste. Consequently, he has authored or coauthored more than 90 original research articles in national and international peer-reviewed journals of high impact published by Springer, Elsevier and John Wiley and Sons, Inc. In addition, he has published 18 book chapters and 1 book. He has vast experience in strategic R & D management preparation of scientific reports. He has also been awarded for writing 25 popular scientific articles in Hindi. He also attended and presented more than 65 national and international conference papers on microbiology, biotechnology and environmental biology. He is a life member of various scientific societies. He also offered different scientists for training under the International Programme from Germany and Nigeria. He has chaired various scientific sessions of different scientific conferences. He is also a guest reviewer for various national and international journals in his discipline. He was also a member of a delegation team that visited Japan for the study of environmental protection from industrial waste. He is a member of the American Society for Microbiology, USA, and a life member of the National Academy of Sciences, Allahabad, India. Based upon this outstanding contribution in the areas of environmental microbiology and environmental biotechnology, he has been awarded a Fellow of the Academy of Environmental Biology, the Association of Microbiologist India and Biotech Research Society of India.

Contributors

Munees Ahemad
Department of Agricultural
   Microbiology
Faculty of Agricultural Sciences
Aligarh Muslim University
Aligarh, U.P., India

Zulfiqar Ahmad
Department of Environmental
   Sciences
University of California
Riverside, California

Nadeem Akhtar
Department of Biotechnology
Thapar University
Patiala, Punjab, India

Muhammad Arshad
Department of Soil and
   Environmental Sciences
University of Agriculture
Faisalabad, Pakistan

Neena Capalash
Department of Biotechnology
Panjab University
Chandigarh, India

Ram Chandra
Department of Environmental
   Microbiology
School for Environmental Sciences
Babasaheb Bhimrao Ambedkar
   Central University
Lucknow, India

Hao Chen
Department of Agricultural and
   Biological Engineering
University of Florida
Gainesville, Florida

David Crowley
Department of Environmental
   Sciences
University of California
Riverside, California

June Fang
Department of Agricultural and
   Biological Engineering
University of Florida
Gainesville, Florida

Bin Gao
Department of Agricultural and
   Biological Engineering
University of Florida
Gainesville, Florida

Boutheina Gargouri
Laboratoire d'Electrochimie et
   Environnement, Ecole nationale
   d'ingénieurs de Sfax
Université de Sfax
Sfax, Tunisia

Arun Goyal
Department of Biotechnology
Indian Institute of Technology
Guwahati, Assam, India

Dinesh Goyal
Department of Biotechnology
Thapar University
Patiala, Punjab, India

Rasna Gupta
Department of Biochemistry
Dr. RML Avadh University
Faizabad, India

Vijaya Gupta
Department of Microbiology
Panjab University
Chandigarh, India

Bee Hameeda
Department of Microbiology
Osmania University
Hyderabad, India

Muhammad Imran
Department of Environmental
   Sciences
University of Gujrat
Gujrat, Pakistan

Jawed Iqbal
Department of Microbiology and
   Immunology
H. M. Bligh Cancer Research
   Laboratories
Rosalind Franklin University of
   Medicine and Science
Chicago Medical School
North Chicago, Illinois

Rohini Karandikar
Department of Biosciences and
   Bioengineering
Indian Institute of
   Technology-Bombay
Powai, India

Mohamed Yahya Khan
Department of Microbiology
Osmania University
Hyderabad, India

Vineet Kumar
Department of Environmental
   Microbiology
School for Environmental Sciences
Babasaheb Bhimrao Ambedkar
   Central University
Lucknow, India

Monika Mishra
School of Environmental Sciences
Jawaharlal Nehru University
New Delhi, India

Prashant S. Phale
Department of Biosciences and
   Bioengineering
Indian Institute of
   Technology-Bombay
Powai, India

Kandasamy Ramani
Department of Biotechnology
School of Bioengineering
SRM University
Kattankulathur, Chennai, India

Gopal Reddy
Department of Microbiology
Osmania University
Hyderabad, India

Gaurav Saxena
Department of Environmental
   Microbiology
School for Environmental Sciences
Babasaheb Bhimrao Ambedkar
   Central University
Lucknow, India

Ganesan Sekaran
Environmental Technology Division
CSIR-Central Leather Research
  Institute
Adyar, Chennai, India

Prince Sharma
Department of Microbiology
Panjab University
Chandigarh, India

Rajat Pratap Singh
Department of Biochemistry
Dr. RML Avadh University
Faizabad, India

Ram Lakhan Singh
Department of Biochemistry
Dr. RML Avadh University
Faizabad, India

Randhir Singh
Department of Biosciences and
  Bioengineering
Indian Institute of
  Technology-Bombay
Powai, India

T. H. Swapna
Department of Microbiology
Osmania University
Hyderabad, India

Indu Shekhar Thakur
School of Environmental Sciences
Jawaharlal Nehru University
New Delhi, India

Adeline Su Yien Ting
School of Science
Monash University Malaysia
Jalan Lagoon Selatan
Selangor Darul Ehsan, Malaysia

Shengsen Wang
Department of Agricultural and
  Biological Engineering
University of Florida
Gainesville, Florida

1

Phytoremediation of Environmental Pollutants: An Eco-Sustainable Green Technology to Environmental Management

Ram Chandra, Gaurav Saxena and Vineet Kumar

CONTENTS

1.1 Introduction

A pollution-free environment is one of the major challenges of the 21st century. Most conventional remedial technologies are expensive and cause the pollution of the environment. To avoid this global problem, bioremediation, typically referring to microbe-based cleanup, and phytoremediation, or plant-based cleanup, have gained much attention as effective low-cost and eco-sustainable alternatives to conventional remedial technologies for the cleanup of a broad spectrum of hazardous pollutants (Pilon-Smits 2005).

Phytoremediation is a green technology that makes use of green plants with their associated microbiota for the in situ remediation of environmental pollutants that can be organic and inorganic. Organic pollutants include trichloroethylene (TCE), trinitrotoluene (TNT), atrazine, oil, gasoline, benzene, toluene, polycyclic aromatic hydrocarbons (PAHs), methyl tertiary butyl-ether (MTBE) and polychlorinated biphenyls (PCBs). On the other hand, inorganic contaminants occur as natural elements in the Earth's crust, including plant macronutrients such as nitrates and phosphates; micronutrients such as Zn, Cr, Fe, Ni, Mo, Mn and Cu; nonessential elements such as V, Cd, Co, Se, Hg, F, Pb, As and W; and radionuclides such as ^{238}U, ^{137}Cs and ^{90}Sr. Environmental pollutants, whether organic or inorganic, severely affect human health and environments (Bridge 2004).

Some plants that absorb toxic metals and help to clean them from soils are termed hyperaccumulators, which have been shown to be resistant to heavy metals and are capable of accumulating those metals into their roots and leaves and transporting these soil pollutants in high concentrations. There is a need to identify suitable plants, which can colonise the polluted site and remove, degrade or immobilise the pollutant of environmental interest.

Phytoremediation is the only eco-friendly alternative for developing nations, such as India, where funding is lacking. It can also be an income-generating technology, especially if metals removed from soil can be used as bio-ore to extract utilisable metal, that is, phytomining (Angle et al. 2001), and energy can be generated through biomass burning (Li et al. 2003). The overall result of carefully and well-planned phytoremediation–phytomining would be a commercially and economically viable metal product (i.e., metal-enriched bio-ore) and land better suited for agricultural operations or general habitation (Boominathan et al. 2004). Substantial research efforts are currently underway to realise the economic potential of these technologies (Ghosh and Singh 2005) with several plant species now recognised as suitable for the phytoremediation.

In this chapter, several aspects of phytoremediation are discussed: phytoextraction, phytodegradation, rhizofiltration, phytostabilisation and phytovolatisation. Combining these technologies offers the greatest potential to effectively phytoremediate the polluted environment. An appropriate application of plant growth–promoting rhizobacteria (PGPRs) is one of the most useful and eco-friendly techniques that is currently considered a useful process in phytoremediation.

Moreover, the growing interest in molecular genetics has increased our understanding of the mechanism of heavy metal tolerance in plants, and many transgenic plants have displayed increased heavy metal tolerance. Further, improvement in plants by genetic engineering, that is, by modifying properties such as metal uptake, transport, accumulation and tolerance to plants, will open up the endless possibilities of phytoremediation.

1.2 Phytoremediation and Associated Phytotechnologies

The concept of phytoremediation arose in the 1980s from the inherent ability that some plant species displayed accumulating high levels of toxic metal concentration in their tissues or organs. Along the years, a number of related technologies were developed that enabled the practical application of higher plants to decontaminate soil and water, and then 'phytoremediation' started to be used in the scientific literature around 1993. The definition later evolved into 'phytotechnologies' (ITRC 2001), meaning a wide range of technologies that can be applied to remediate pollutants through (1) stabilisation; (2) volatilisation; (3) metabolism, including rhizosphere degradation; and (4) accumulation and sequestration. A comprehensive treatise on phytotechnologies can be found in McCutcheon and Schnoor (2003).

Phytoremediation is an eco-sustainable, noninvasive, promising green technology for in situ treatment of environmental pollutants, accomplished by the use of plants and their associated microbiota for the uptake, sequestration, detoxification or volatilisation of pollutants from soils, water, sediments and possibly air. This technology can be applied to both organic and inorganic pollutants present in soil (solid substrate), water (liquid substrate) or air (Salt et al. 1998). The concept of phytoremediation is presented in Figure 1.1.

However, the application of phytoremediation technology has been reviewed by many researchers (Table 1.1).

The major and overall objective of this technique was to collect the pollutants from the media and turn them into an easily extractable form (plant tissues). It accomplished the growth of plants in a polluted matrix, either artificially (constructed wetlands) or naturally, for a required growth period to remove pollutants from the matrix or facilitate immobilisation (binding/containment) or degradation (detoxification) of the pollutants.

Phytotechnologies are the set of techniques that make use of plants to achieve environmental goals. These techniques use plants to extract, degrade, contain or immobilise pollutants in soil, groundwater, surface water and other polluted media. These remediate a wide range of pollutants

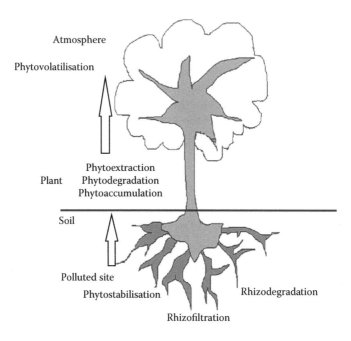

FIGURE 1.1
(See color insert.) Concept of phytoremediation technologies.

TABLE 1.1

Some of the Applications of Phytoremediation

Mechanism	Pollutant	Media	Plant	Status	Reference
Phytoextraction	Zn, Cd and As	Soil	*Datura stramonium* and *Chenopodium murale*	Applied	Varun et al. (2012)
Phytodegradation	As	Soil	*Cassia fistula*	Applied	Preeti et al. (2011)
Phytostabilisation	Mn	Soil	*Chondrila juncea* and *Chenopodium botrys*	Soil	Cheraghi et al. (2011)
Phytoextraction	^{137}Cs	Soil	*Catharanthus roseus*	Applied	Fulekar et al. (2010)
Phytoextraction	Cr	Soil	*Anogeissus latifolia*	Applied	Mathur et al. (2010)
Phytodegradation	Zn and Cd	Soil	*Vetiveria, sesbania, Viola, sedum, Rumex*	Field demo	Mukhopadhyay and Maiti (2010)
Phytodegradation	Pb and Cd	Soil	*Jatropha curcas* L.	Applied	Mangkoedihardjo and Surahmaida (2008)
Phytostabilisation	Cd	Soil	Sunflower	Applied	Zadeh et al. (2008)
Phytodegradation	U	Soil	*Brassica juncea*	Field demo	Huhle et al. (2008)

using several different mechanisms dependent on the application, although not all mechanisms are applicable to all pollutants or all matrices.

Thus, phytotechnologies may potentially (1) clean up moderate to low levels of selected elemental and organic pollutants over large areas, (2) maintain sites by treating residual pollution after cleanup is achieved, (3) act as a buffer against potential waste release, (4) aid voluntary clean-up efforts, (5) facilitate nonpoint source pollution control and (6) offer an effective form of monitored natural attenuation (McCutcheon and Schnoor 2003).

Several types of phytoremediation can be defined as follows.

1.2.1 Phytoextraction

This is also known as phytoaccumulation, phytoabsorption and phytosequestration because it uses pollutant-accumulating plants to remove pollutants, such as metals or organics, from soil via root absorption and concentrates them in above-ground harvestable plant parts. Unlike the destructive degradation mechanisms, this technique yields a mass of plant and pollutant (typically metals) that must be transported for disposal or recycling. It is also a concentration technology that generates a much smaller mass to be disposed of when compared to excavation and landfilling. It is being evaluated in a Superfund Innovative Technology Evaluation (SITE) demonstration and may also be used for pollutant recovery and recycling. It also has environmental benefits because it is a low-impact technology. Furthermore, during phytoextraction, plants cover the soil, and thus erosion and leaching will be reduced.

It involves (1) cultivation of the suitable plant/crop species on the polluted site, (2) removal of harvestable plant parts containing metal from the site and (3) post-harvest treatments (including composting, compacting and thermal treatments) to reduce the biomass volume and/or weight for disposal as a hazardous waste or for its recycling to recover valuable metals.

Two types of phytoextraction have been suggested: continuous or natural phytoextraction and induced, enhanced or chemically assisted phytoextraction (Lombi et al. 2001a). Continuous phytoextraction is the use of plants, usually hyperaccumulators, that accumulate particularly high levels of the toxic pollutants throughout their lifetime, and induced phytoextraction enhances toxin accumulation at a single time point by the addition of accelerants or chelators to the soil. After the plants have been allowed to grow for some time period, they were harvested and either incinerated or composted to recycle the metals. This procedure may be repeated as necessary to bring soil pollutant levels down to permissible limits. If plants are incinerated, the ash must be disposed of in a hazardous waste landfill, but the ash volume will be less than 10% of the volume that would be created if polluted soil itself were dug up for treatment. In some cases, it is possible to recover metals through a process known as phytomining, which is usually reserved for precious metals.

Metals such as Cu, Ni and Zn are the most suitable candidates for phyto-extraction because the majority of the approximately 400 known plants that absorb unusually large amounts of metals have a high affinity for accumulating these metals.

The main factors limiting phytoextraction efficiency are (1) soil metal phytoavailability and (2) metal translocation to above-ground plant parts. To increase these, the use of soil amendments has been suggested and tested by several authors (Blaylock and Huang 2000). Ethylene diamine tetraacetic acid (EDTA) is a complex agent that has been used in agriculture since the 1950s as an additive in micronutrient fertilisers (Bucheli-Witschel and Egli 2001). Recently, an experiment was conducted showing the effect of EDTA on the phytoextraction ability of *Eleusine indica* (grass). Results revealed that the grass showed a relatively good response to EDTA application, and higher levels of Cu and Cr concentration in the root suggested that the grass may be a good metal excluder with the possibility of extracting Pb from polluted soils (Garba et al. 2012). EDTA also assists in mobilisation and subsequent accumulation of soil pollutants such as Zn, Cd, Ni, Cu, Cr and Pb in *Brassica juncea* (Indian mustard) and *Helianthus anuus* (sunflower). The ability of other metal chelators, such as CDTA, DTPA, EGTA, EDDHA and NTA, to enhance metal accumulation has also been assessed in various plant species (Lombi et al. 2001b). However, there may be risks associated with using certain chelators considering the high water solubility of some chelator–toxin complexes, which could result in movement of the complexes to deeper soil layers (Lombi et al. 2001b) and potential groundwater and estuary contamination.

There are two important factors that should be considered when evaluating the potential of a plant as a phytoextractor: bioconcentration and biomass production. The former is defined as the ratio between the concentration of the pollutant in the shoot and in the soil. It serves as an indicator of the capacity of a plant to accumulate toxic compounds. Biomass production is also critical in order for phytoextraction to be commercially viable because it decreases the number of crops required to complete the remediation of a given site (McGrath and Zhao 2003).

1.2.2 Rhizofiltration

Rhizofiltration is the use of plants, both terrestrial and aquatic, to absorb, concentrate and precipitate pollutants in aqueous sources with low pollutant concentration in their roots (Jadia and Fulekar 2009). It is used for cleaning polluted surface waters or wastewaters, such as industrial discharge, agricultural runoff or acid mine drainage, by adsorption or precipitation of metals onto roots or absorption by roots or other submerged organs of metal-tolerant aquatic plants. It is similar to phytoextraction but is mainly concerned with the remediation of contaminated groundwater rather than polluted soils. The advantages of rhizofiltration are (1) the ability to use both

terrestrial and aquatic plants for either in situ or ex situ applications and
(2) the pollutants do not have to be translocated to the shoots. Thus, species
other than hyperaccumulators may be used. It remediates metals such as As,
Cu, Cd, Pb, Cr, V and Ni and radionuclides (U and Cs) (USEPA 2000; Jadia
and Fulekar 2009). For this, plants must not only be metal-resistant but also
have a high absorption surface and must tolerate low oxygen concentration
(Dushenkov et al. 1995). The ideal plants for rhizofiltration should produce
significant amounts of root biomass or root surface area, be able to accumulate and tolerate considerable amounts of target metals, involve easy han-
dling and a low maintenance cost and have a minimum of secondary waste
that requires disposal. Terrestrial plants are more suitable for rhizofiltration
because they produce longer, more substantial and often fibrous root systems with large surface areas for metal adsorption (Raskin and Ensley 2000).
Pteris vittata, commonly known as Chinese brake fern, is the first known
hyperaccumulator (Ma et al. 2001). Several aquatic plant species that possess the ability to remove heavy metals from water have been reviewed by
several authors (Dierberg et al. 1987; Mo et al. 1989; Zhu et al. 1999). Indian
mustard (*Brassica juncea*) and sunflower (*Helianthus annuus*) are most promising for metal removal from water. Indian mustard effectively removes Cd,
Cr, Cu, Ni, Pb and Zn (Dushenkov et al. 1995), whereas sunflower absorbs Pb
(Dushenkov et al. 1995) and U (Dushenkov et al. 1997) from hydroponic solutions. Indian mustard could effectively remove a wide range (4 to 500 mg/L)
of Pb concentration (Raskin and Ensley 2000).

1.2.3 Phytostabilisation

This is also known as in-place inactivation or phytoimmobilisation of pollutants by plant roots and is primarily used for the remediation of soil,
sediment and sludges (USEPA 2000). It is used to limit pollutant mobility,
preventing migration into groundwater or air and phytoavailability in soil
and water, thus preventing its spread throughout the food chain because
pollutants are absorbed and accumulated by roots, adsorbed onto roots or
precipitated in the rhizosphere. It can also be used to reestablish a plant community on sites that have been denuded due to the high metal pollution levels. Once a community of tolerant species has been established, the potential
for wind erosion (and thus spread of the pollutant) is reduced, and leaching of the soil contaminants is also reduced. It can occur through sorption,
precipitation, complexation or metal valence reduction (Ghosh and Singh
2005). It is useful in the treatment of Pb as well as As, Cu, Cr, Zn and Cd
(Jadia and Fulekar 2009). The advantages of phytostabilisation are that the
disposal of hazardous material/biomass is not required (USEPA 2000) and
it is very effective when rapid immobilisation is needed to preserve ground
and surface waters. However, the major disadvantage is that the pollutants
always remain in the soil, and therefore, it requires regular monitoring. It

has also been used to treat polluted land areas affected by mining activities and superfund sites. Plants with high transpiration rates, such as grasses, sedges, forage plants and reeds, are useful in phytostabilisation because they decrease the amount of groundwater migrating away from the pollutant-carrying site. Combining these plants with hardy, perennial, densely rooted or deep rooting trees (poplar, cottonwood) can be an effective approach (Berti and Cunningham 2000).

1.2.4 Phytovolatilisation

Phytovolatilisation refers to the uptake and transpiration of organic pollutants and primary organic compounds by the plants, comparatively at low concentrations. Here, water-soluble pollutants are taken up by the plant roots, pass through the plant or are modified by the plant, are transported to the leaves and are volatilised into the atmosphere through the stomata. It is also reported that the use of phytoextraction and phytovolatilisation of metals by plants offers a viable remediation on commercial projects (Sakakibara et al. 2007). It has been primarily used for mercury removal wherein the mercuric ion is transformed into the less toxic elemental Hg (Ghosh and Singh 2005). It has also been successful in radioactive tritium (3H), an isotope of hydrogen; it is decayed to stable helium with a half-life of about 12 years. It is the most controversial of all phytoremediation technologies because some metals, such as Se, Hg and As, may exist in a gaseous state in the environment. It is also reported that some naturally occurring or genetically modified plants, such as *Chara canescens* (muskgrass), *B. juncea* (Indian mustard) and *Arabidopsis thaliana*, possess the capability to absorb heavy metals and convert them to a gaseous state within the plant and subsequently release them into the atmosphere (Ghosh and Singh 2005).

1.2.5 Phytodegradation

It is the breakdown or conversion of highly toxic organic pollutants into less toxic forms via the action of enzymes secreted within plant tissue (Suresh and Ravishankar 2004). It is an enzyme-catalysed metabolism of pollutants. Plants produce some enzymes, such as dehalogenase and oxygenase, which help in degradation of the organic pollutant. It is independent of the activity of rhizosphere microorganisms. Some plant enzymes have been identified that are involved in the breakdown of ammunition wastes; chlorinated solvents, such as TCE (trichloroethylene); and others that degrade organic herbicides (Newman et al. 1997). Plant enzymes that metabolise contaminants may be released into the vicinity of the rhizosphere, where they may participate in pollutant transformation. Enzymes, such as nitro-reductase, dehalogenase, peroxidase, nitrilase and laccase, have been discovered in plant sediments and soils (Suresh and Ravishankar 2004).

1.2.6 Rhizodegradation

Like phytodegradation, rhizosphere degradation involves the enzymatic breakdown of organic pollutants but through microbial enzymatic activity. These breakdown products are either volatilised or incorporated into the microorganisms and the soil matrix of the rhizosphere. The types of plants growing in the contaminated area influence the amount, diversity and activity of microbial populations (Jones et al. 2004; Kirk et al. 2005). Grasses with high root density, legumes that fix nitrogen and alfalfa that fixes nitrogen and has high evapotranspiration rates are associated with different microbial populations. These plants create a more aerobic environment in the soil, which stimulates microbial activity that enhances oxidation of organic chemical residues (Jones et al. 2004; Kirk et al. 2005). Secondary metabolites and other components of the root exudates also stimulate microbial activity, a by-product of which may be degradation of organic pollutants (Pieper et al. 2004).

1.3 Mechanism of Metal Hyperaccumulation in Plants

The process of metal hyperaccumulation in plants is accomplished in several steps (Figure 1.2).

Solubilisation of the metal from the soil matrix

Most of the metals in soil occur in insoluble forms; thereby they are not available for plant uptake. To overcome these problems, plants use two methods to desorb metals from the soil matrix: (1) rhizosphere acidification through the action of plasma membrane proton pumps and (2) secretion of ligands capable of chelating the metal. Plants have evolved these processes to solubilise essential metals from the soil, but soils containing high concentrations of toxic metals will release both essential and toxic metals to solution (Lasat 2000).

Uptake into the root

There are two available mechanisms by which soluble metals can enter into the root: (1) symplast by crossing the plasma membrane of the root endodermal cells – or they can enter the root – and (2) apoplast through the space between cells. Although it is possible for solutes to travel up through the plant by apoplastic flow, the more efficient method of moving up the plant is through the vasculature of the plant, called the xylem. To enter the xylem, solutes must cross the Casparian strip, a waxy coating, which is impermeable to solutes unless they pass through the cells of the endodermis. Therefore, to enter the xylem, metals must cross a membrane, probably through the action

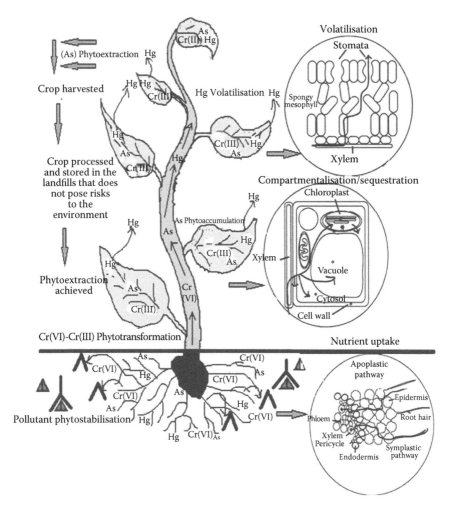

FIGURE 1.2
(See color insert.) Different mechanism of phytotechnologies: phytoextraction of As from the soil to aerial parts of plant (leaves and stems), phytotransformation of Cr(VI) from the soil to Cr(III) in the aerial parts of the plant, phytostabilisation of metal contaminants in soil and phytovolatisation of Hg from the soil.

of a membrane pump or channel. Most toxic metals are thought to cross these membranes through pumps and channels intended to transport essential elements. Excluder plants survive by enhancing specificity for the essential element or pumping the toxic metal back out of the plant (Hall 2002).

Transport to the leaves

Once the solutes are loaded into the xylem, the flow of the xylem sap will transport the metal to the leaves, where it must be loaded into the cells of the leaf, again crossing a membrane. The cell types in which the metals are deposited vary between hyperaccumulator species.

Detoxification and/or chelation

At any point along the pathway, the metal could be converted to a less toxic form through the process of chemical conversion or by complexation. Various oxidation states of toxic elements have very different uptake, transport and sequestration or toxicity characteristics in plants. Toxin chelation by endogenous plant compounds can have similar effects on all of these properties as well. As many chelators use thiol groups as ligands, the sulphur (S) biosynthetic pathways have been shown to be crucial for hyperaccumulator function (van Huysen et al. 2004) and for possible phytoremediation strategies.

Sequestration and volatilisation

The final step for most metal accumulation is the metal sequestration away from any cellular processes it might disrupt. It usually occurs in the plant vacuole, where the metal/metal ligand must be transported across the vacuolar membrane. Metals may also remain in the cell wall instead of crossing the plasma membrane into the cell as the negative charge sites on the cell walls may interact with polyvalent cations (Wang and Evangelou 1994). Selenium may also be volatilised through the stomata.

Metallothioneins (MTs)

Metallothioneins are cysteine-rich, low molecular weight proteins synthesized on ribosomes according to the mRNA information. Four categories of these proteins, class-I MTs from mammalian cells and class II from yeast MTs, occur in plants, which are encoded by at least seven genes in *A. thaliana* (Cobbett and Goldsbrough 2002). When the MT gene of *Pisum sativum* (PsMTA) was expressed in *A. thaliana*, more Cu (several-fold) accumulated in the roots of the transformed than of the control plants. Similarly, the *A. thaliana* metallothionein proteins AtMT2a and AtMT3 were introduced as fluorescent protein-fused forms into the guard cells of *Vicia faba*. The MTs protected guard cell chloroplasts from degradation upon exposure to Cd by reducing the presence of reactive oxygen species. It was concluded that the Cd stays bound to the MT in the cytoplasm and is not sequestered into the vacuole as occurs when Cd is detoxified by phytochelatins (PCs) (Lee et al. 2004).

Transporters

Transporters are required for toxic metal (ion) exclusion, transporting the metal into the apoplastic space and vacuole where it would be less likely to exert a toxic effect (Tong et al. 2004). Overexpressing lines exposed to lethal concentrations of Zn or Cd translocated these metals at a greater extent to the shoot; in contrast, the metal level was found to be rather similar in roots, indicating that the metal uptake by the roots compensated for the increased metal translocation to the shoot (Verret et al. 2004). The vacuole is considered to be the main metal storage site in yeast and plant cells; thus phytochelation–metal complexes are pumped into the vacuole. YCF1 from *Saccharomyces cerevisiae* is one of the best-known vacuolar transporters. It is a Mg ATP-energised glutathione S-conjugate transporter (Song et al. 2003). Other valuable transporter proteins include the *A. thaliana* antiporter CAX2 (Hirschi et al. 2000); LCT1, a nonspecific transporter for Ca^{2+}, Cd^{2+}, Na^+ and K^+ (Antosiewicz and Hennig 2004); the *Thlaspi caerulescens* heavy metal ATPase; TcHMA4 (Papoyan and Kochian 2004), a novel family of cysteine-rich membrane proteins that mediate Cd resistance in *A. thaliana*; and AtMRP3, an ABC transporter (Bovet et al. 2005).

1.4 Plant Response to Environmental Pollutants

Plants have three basic strategies for growth in metal-contaminated soil (Raskin et al. 1994):

Metal excluders

Plant species that prevent metal from entering their aerial parts or maintain low and constant metal concentration over a broad range of metal concentration in soil mainly restrict metal in their roots by altering their membrane permeability, changing the metal-binding capacity of cell walls or exuding more chelating substances.

Metal indicators

Plant species that actively accumulate metal in their aerial tissues and generally reflect the metal level in the soil tolerate the existing concentration level of metals by producing intracellular metal-binding compounds (chelators) or alter the metal compartmentalisation pattern by storing metals in nonsensitive parts.

Metal accumulator plant species

They can concentrate metal in their aerial parts to levels far exceeding that in the soil. Hyperaccumulators are plants that can absorb high levels of contaminants concentrated either in their roots, shoots and/or leaves.

1.5 Hyperaccumulators for Phytoremediation

Successful phytoremediation depends on those plants (woody or herbaceous) that can accumulate desired levels of heavy metal concentration in their shoots (a hundred- to a thousandfold) without visible symptoms, termed hyperaccumulators, and the phenomenon is termed hyperaccumulation (i.e. the ability to accumulate at least 0.1% of the leaf dry weight in a heavy metal), which is only exhibited by <0.2% of angiosperms (Baker and Whiting 2002), making the selection of native plant species for phytoremediation a tough task. The ideal plants for phytoremediation should have the ability to hyperaccumulate heavy metals, have a fast growth rate, have the ability to tolerate high salt concentration and pH, have high biomass and be easily harvestable and must uptake and translocate metals to aerial parts efficiently (Sharma 2011). The main criteria for hyperaccumulator plants are given in Table 1.2.

The hyperaccumulators that have been most extensively studied by scientists include *Thlaspi* sp., *Arabidopsis* sp. and *Sedum alfredii* sp. (both genera belong to the family of Brassicaceae and Alyssum). However, Baker et al. (2000) found many species that can be classified as hyperaccumulators based on their capacity to tolerate toxic concentration as given in Table 1.3.

TABLE 1.2

Criteria for Metal Accumulation in Plants

Factor Influencing	Reference
Accumulating ability	Zhou and Song (2004)
Tolerance ability	Sun et al. (2009)
Removal efficiency (RE)	Soleimani et al. (2010)
Bioconcentration factor (BCF)	Yoon et al. (2006)
Bioaccumulation coefficient (BAC)	McGrath and Zhao (2003)
Transfer factor (TF)	Liu et al. (2010)

TABLE 1.3

Some Metal Hyperaccumulator Plant Species

Species	Metal	Reference
Biden spilosa	Cd	Sun et al. (2009)
Brassica junceae	Ni and Cr	Saraswat and Rai (2009)
Thlaspi caerluescen	Cd, Zn and Pb	Banasova et al. (2008)
Solanum nigrum L.	Cd	Sun et al. (2008)
Sedum alferedii	Cd	Sun et al. (2007)
Helicotylenchus indicus	Pb	Sekara et al. (2005)
Alyssum lesbiacum	Ni	Cluis (2004)
Pistia stratiotes	Zn, Pb, Ni, Hg, Cu, Cd and Cr	Odjegba and Fasidi (2004)
Pityrogramma calomelanos	As	Dembitsky and Rezanka (2003)
Thordisa villosa	Cu	Rajakaruna and Bohm (2002)
Croton bonplandianus	Cu	Rajakaruna and Bohm (2002)

Currently, there are about 420 species belonging to about 45 plant families reported as metal hyperaccumulators (Cobbett 2003). Although new hyperaccumulators continue to be discovered from field collections (Kramer 2003), only a few species have been tested in the laboratory to confirm their hyperaccumulating behaviours. However, a problem associated with most of the hyperaccumulators is the insufficient biomass and growth rate (Kramer and Chardonnens 2001). Many researchers consider the best way to transfer the appropriate characteristics of hyperaccumulators into high biomass plants (Kramer and Chardonnens 2001). To do so, it is required to understand how these plants tolerate and accumulate such high heavy metal concentrations.

1.6 Plant Growth–Promoting Rhizobacteria in Environmental Restoration

Microorganisms that are present in the rhizosphere of a plant are known as rhizobacteria (also called plant growth–promoting rhizobacteria or PGPR). Various species of bacteria, such as *Pseudomonas, Azospirillum, Azotobacter, Klebsiella, Enterobacter, Alcaligenes, Arthrobacter, Burkholderia, Bacillus* and *Serratia,* have been reported to enhance plant growth (Kloepper et al. 1989; Glick et al. 1995; Joseph et al. 2007). PGPR have been initially used in agriculture and forestry to increase plant yield as well as growth and tolerance to disease. It has been recently reported that PGPR plays a critical role in environmental remediation, particularly to overcome plant stress under flooded, high-temperature and acidic conditions (Lucy et al. 2004). The metal-resistant plant growth–promoting bacteria (PGPB) can serve as an effective metal sequestering and growth–promoting bioinoculant for plants in metal-stressed soil (Kloepper et al. 1989). The deleterious effects on plants of heavy metals taken up from the environment can be lessened with the use of PGP bacteria or mycorrhizal fungi (Joseph et al. 2007). The soil microbes, PGPR, phosphate solubilising bacteria, mycorrhizal-helping bacteria (MHB) and arbuscular mycorrhizal fungi (AMF) in the rhizosphere of plants growing on trace metal–contaminated soils play an important role in phytoremediation (Kloepper et al. 1989). PGPR include a diverse group of free-living soil bacteria that can improve host plant growth and play an important role in mitigating the toxic effects of heavy metals on the plants (Belimov et al. 2004). A high metal concentration can even be toxic for metal-hyperaccumulating and metal-tolerant plants. This is partly attributable to iron deficiency in a range of different plant species (Wallace et al. 1992) in heavy metal–contaminated soil. Moreover, the low iron content of plants that are grown in the presence of high levels of heavy metals generally results in these plants becoming chlorotic because iron deficiency inhibits both chloroplast development and chlorophyll bio-synthesis (Imsande 1998). However, microbial iron-siderophore complexes can be taken up by plants and thereby serve as an iron source for plants (Wang et al. 1993). It was therefore reasoned that the best way to prevent plants from becoming chlorotic in the presence of high levels of heavy metals was to provide them with an associated siderophore-producing bacterium. This suggests that some plant growth–promoting bacteria can significantly increase the growth of plants in the presence of heavy metals, including nickel, lead and zinc (Burd et al. 2000), thus allowing plants to develop longer roots and get better established during early stages of growth (Glick et al. 1995). Once the seedling is established, the bacterium can also help the plant acquire sufficient iron for optimal plant growth. It is crucial to highlight that because the efficient use of PGPR is limited to slight and moderately polluted sites, the most important limiting factor for the application of PGPR is their tolerance to the heavy metal concentration. Based on the amount and type of organic compounds, which are mostly exuded from plant roots (Myers et al. 2001), as well as the amount and the type of heavy metals (Sandaa et al. 1999), the PGPR population between plants could be different among the same species in the polluted soils or even between the different growing stages of an individual plant. A number of new research studies carried out in relation to the effects of PGPR on plant growth and/or heavy metal concentration in polluted soil are given in Table 1.4.

1.6.1 Plant Growth–Promoting Rhizobacteria in Terrestrial Plants

The alteration of the rhizospheric microbial complex in the uptake of essential elements, such as Mn^{+2} and Fe^{+3} (Barber and Lee 1974), and the efficiency of phytoremediation (O'Connell et al. 1996) have been well documented. Hasnain and Sabri (1997) showed an improvement in the growth of *Triticum aestivum* seedlings in different Pb concentrations when their seeds were inoculated with two *Pseudomonas* strains as compared to the uninoculated control. The safety of their usage is one of the most important considerations that should be taken into account before deciding on whether to use PGPR for phytoremediation purposes. For example, *Burkholderia cepacia* is a multi-drug-resistant PGPR with health risk potentials (Lee et al. 2008), but at the same time, it has been shown to have special abilities in increasing the efficiency of phytoremediation.

1.6.2 Plant Growth–Promoting Rhizobacteria in Aquatic Plants

Aquatic plants are relatively newly approved organisms for remediation purposes; these include rhizofiltration, phytofiltration and constructed wet-lands (Zurayk et al. 2001; Bennicelli et al. 2004; Abou-Shanab et al. 2008). These aspects of phytoremediation have attracted more attention because of

TABLE 1.4

Some Current Research in Relation to the Effects of PGPR on the Plants in Heavy Metal–Contaminated Soils

PGPR	Plant	Heavy Metal	Effect(s)	Reference
Bradyrhizobium sp., *Pseudomonas* sp., *Ochrobactrum cytisi*	*Lupinus luteus*	Pb, Cu and Cd	Decreased the metal accumulation; however, plant biomass increased	Dary et al. (2010)
Bacillus subtilis, Bacillus cereus, Pseudomonas aeruginosa, Flavobacterium sp.	*Orychophragmus violaceus*	Zn	Increased shoot biomass and Zn accumulation	He et al. (2010)
Pseudomonas aeruginosa	Maize	Cr, Pb	Increased the uptake by shoot by a factor of 5.4 and 3.4, respectively	Braud et al. (2009)
Ralstonia metalidurans	Maize	Cr	Increased the accumulation of Cr in shoots by a factor of 5.2	
Achomobacter xylosoxidans strain Ax10	*Brassica juncea*	Cu	Increased the length of root and shoot, fresh and dry weight significantly and extensively improved the Cu uptake of *B. juncea* plants as compared to the control	Ma et al. (2009)
Microbacterium sp. G16, *Pseudomonas fluorescens* G10	Rape	Pb	Increased root elongation of inoculated rape seedlings and total Pb accumulation as compared to the control plants	Sheng et al. (2008)
Pseudomonas aeruginosa	Black gram plants	Cd	Lessened the accumulation of Cd in plants; showed extensive rooting and enhanced plant growth	Ganesan (2008)
Burkholderia sp. J62	Maize and tomato	Cd, Pb	Increased the biomass of maize and tomato plant significantly; the increased Pb and Cd content in tissue varied from 38% to 192% and from 5% to 191%, respectively	Jiang et al. (2008)
Bradyrhizoium sp. RM8	Green gram var. K851	Ni, Zn	Increased plant growth and decreased uptake of heavy metals by plant	Wani et al. (2007)

the increase in water pollution. Due to the new approach, most of the current research still focuses on wetland hyperaccumulator species. Nonetheless, the availability of information on the effects of rhizospheric or rhizoplanic bacteria on the uptake of metal by plants rooted in aquatic systems is rather scarce. So et al. (2003) demonstrated that bacterial species resistant to Cu^{2+} or Zn^2, isolated from water hyacinths (*Eichhornia crassipes*), had led to an increase in the Cu^{2+} removal capacity of this plant species. Xiong et al. (2008), who worked on *Sedum alfredii* (a terrestrial plant) in an aqueous medium with rhizospheric bacteria, suggested that rhizospheric bacteria appeared to protect the roots against heavy metal toxicity. The number of bacteria on the root surface of terrestrial plants is approximately 107 cell/cm^2 (Kennedy 1998), but this was found to decrease to 106 cell/cm^2 in aquatic plants (Fry and Humphrey 1978). The difference in the population of bacteria could be attributed to several factors, such as the variability of oxygen flux around the roots of aquatic plants, which might change the equations of phytoremediation in the different media.

1.7 Transgenic Approach to Phytoremediation

A plant's phytoremediation efficiency can be substantially improved using genetic engineering technologies. Most of the current transgenic research is focused on understanding the genomics behind the ability of some plants and bacteria to modify or remove pollutants (Doty 2008). Transgenic research on a variety of applications is occurring for constructed treatment wetlands, field crops and tree plantations for several contaminants. Before that date, no full-scale applications of transgenic, or genetically modified, plants for polluted site remediation are known. A few laboratory and pilot studies have shown promising results in using transgenic plants for phytoremediation and are given in Table 1.5.

1.8 Technological Development

Phytoremediation is a new cleaning concept, potentially applicable to a variety of environmental pollutants. Major limitations are the lack of research data related to the metal mass balance. It is not easy to estimate phytoremediation cost due to the absence of economic data. Recently, a group of scientists categorised a variety of metals related to research status of phytoextraction, readiness for commercialisation and regulatory acceptance of phytoremediation (Lasat 2000) (Table 1.6).

TABLE 1.5

Applications of Some Recently Developed Genetically Engineered Plants for Phytoremediation with Other Additional Benefits

Transgenic Plant	Targeted Pollutant	Proposed Additional Benefits	Reference
Brassica juncea with ATP sulfurylase from *Arabidopsis thaliana* and Se Cysmethyltransferase (SMT) from *Astragalus bisulcatus*	Enhance Se accumulation, tolerance and volatilisation	Biodiesel production, carbon sequestration	Dhankher et al. (2012)
Hybrid poplar (*Populus sciboldii* × *Populus grandidentata*) with manganese perodixase (*MnP*) gene from *Trametes versicolor*	Increased degradation of bisphenol A	Biomass for bioenergy, pulp, charcoal, carbon sequestration	Iimura et al. (2007)
Populus deltoids with bacterial mercuric ion reductase (*merA*) gene	Enhanced mercuric ion reduction and resistance	Biomass for bioenergy, pulp, charcoal, carbon sequestration	Che et al. (2003)
Populus canescens overexpressing γ-glutamylcysteine synthetase	Tolerance to Zn stress	Biomass for bioenergy, pulp, charcoal, carbon sequestration	Bittsanszkya et al. (2005)
Hybrid aspen (*Populus tremula* × *Populus tremuloides*) expressing bacterial nitroreductase (*pnrA*)	Enhanced bioremediation of TNT	Biomass for bioenergy, pulp, charcoal, carbon sequestration	van Dillewijn et al. (2008)
Hybrid poplar (*Populus tremula* × *Populus alba*)	Removal of TCE, vinyl chloride, CCl$_4$, benzene and chloroform	Biomass for bioenergy, pulp, charcoal, carbon sequestration	Doty et al. (2007)
Populus trichocarpa overexpressing γ-glutamylcysteine synthetase from poplar	Increased tolerance to chloroacetanilide herbicides	Biomass for bioenergy, pulp, charcoal, carbon sequestration	Gullner et al. (2001)

TABLE 1.6

Recent Research Status, Readiness for Commercialisation and Regulatory Acceptance of Phytoremediation for Some Metal and Metalloid Pollutants

	Recent Research Status	
Metal Pollutants	Commercial Readiness[a]	Regulatory Acceptance[b]
Ni	4	Y
Co	4	Y
Se	4	N
Pb	4	Y
Hg	3	N
Cd	2	Y
Zn	3	Y
As	1	N

Source: Adapted from Mukhopadhyay, S., and Maiti, S.K., *Applied Ecology and Environmental Research* 8, 3, 207–222, 2010.

[a] Rating: 1 – basic research underway; 2 – laboratory stage; 3 – field deployment; 4 – under commercialisation.

[b] Y – yes; N – no.

1.9 Advantages and Disadvantages

Phytoremediation is a natural process that uses potential plants to clean contaminants from a polluted site. However, through biotechnological methods, these potential plants can be used for environmental protection and human health welfare. Due to the specific nutrient and ecophysiological properties, phytoremediation may also be an effective method for concentrating and harvesting valuable metals that are thinly dispersed in the ground, and simultaneously, it offers an interesting option for the remediation of contaminated sites. But due to limited knowledge on phytoremediation among the scientific communities, it is still a new developing technology; moreover, the intrinsic characteristic of phytoremediation limits the size of niche that it occupies in the contaminated site undergoing remediation. The main advantages and disadvantages of phytoremediation technology are summarized in Table 1.7.

TABLE 1.7

Advantages and Disadvantages of Phytoremediation

Advantages	Limitations
Cheap and aesthetically pleasing (no excavation required)	The plant must be able to grow in the polluted media.
Soil stabilisation and reduced water leaching and transport of inorganics in the soil	The plant can accumulate inorganics that it can reach through root growth and is soluble in soil.
Generation of a recyclable metal-rich plant residue	Time-consuming process can take years for pollutant concentrations to reach regulatory levels (long-term commitment).
Applicability to a wide range of toxic metals and radionuclides	The pollutant must be within or drawn toward the root zones of plants that are actively growing.
Minimal environmental disturbance as compared to conventional remedial methods	It must not pose harm to human health or further environmental problems.
Removal of secondary air or water-borne wastes	Climatic conditions are the limiting factor.
Enhanced regulatory and public acceptance	Introduction of exotic plant species may affect biodiversity.

1.10 Future Outlook

Today, phytoremediation is still being researched, and much of the current research is laboratory based, where plants grown in a hydroponic setting are fed with heavy-metal diets. Although these results are promising, scientists are ready to admit that solution culture is quite different from that of soil because, in soil, most of the metals occur as insoluble forms and are less available, and that is the biggest problem. There are several technical impediments that need to be caught up. Both agronomic practices and plant genetic abilities need to be optimised to develop commercially and economically viable practices. Many hyperaccumulators remain to be discovered, and there is a need to know more about their eco-physiology. Optimisation of the process, proper understanding of plant heavy metal uptake and proper biomass disposal are still required. Future research is required to develop plants with high growth rates, high biomass, improved metal uptake, translocation and tolerance via genetic engineering for effective phytoremediation. For better acceptance in the remediation industry, it is important that transgenic science continues to be tested in the field. In that context, it will be helpful if regulatory restrictions can be regularly reevaluated to make the use of transgenics for the phytoremediation less cumbersome. Moreover, the selection and testing of multiple hyperaccumulators could enhance the phytoremediation rate, making this process successful for a pollution-free environment (Suresh and Ravishankar 2004).

1.11 Regulatory Considerations

A range of existing federal and state regulatory programs may pertain to site-specific decisions regarding the use of this technology. These programs include those established under the Resource Conservation and Recovery Act (RCRA), which deals with specific waste management activities; the Comprehensive Environmental Response, Compensation, and Liability Act (CERCLA) referred to as 'Superfund' to attain a general cleanup standard assuring human health and environment protection; the Clean Air Act (CAA) to regulate hazardous air pollutant emissions from source categories; the Toxic Substances Control Act (TSCA) to regulate the use of plants intended for commercial bioremediation; the Federal Insecticide, Fungicide, and Rodenticide Act (FIFRA) to regulate the use of pesticides; the Federal Food, Drug, and Cosmetic Act (FFDCA) to regulate the use of phytoremediation plants as food; and statutes enforced by the U.S. Department of Agriculture (USEPA 2000).

1.12 Research Needs

1. Further exploration of plants suitable for phytoremediation is required.
2. Continued field demonstration is required to determine the extent of pollutant removal by selected plant species.
3. High-resolution microanalyses of hyperaccumulator plants by SEM or TEM are required to determine the discrete site of metal sequestration and bioaccumulation in specific plant organs, tissues, cells and organelles.
4. There is a need to evaluate the procedure for disposal, processing and volume reduction of polluted biomass.
5. Studies on root and other plant biomass decomposition in soil are required to understand the kinetics and cycling of contaminants.
6. There is a need to extend the investigation of the most promising research on phytoremediation, which also includes the following:
   a. Mechanism of pollutant uptake, transport and accumulation in plant tissues
   b. Better understanding of rhizosphere interaction among plant roots, microorganisms and other biota
   c. Role of both natural and artificial metal chelators and their metal complexes, their dynamics and decomposition in rhizosphere and plant tissues

    d. Development of fertilisers and other soil amendments to enhance the phytoremediation efficiency of hyperaccumulators

    e. Development of transgenic plants for efficient phytoremediation of environmental pollutants

    f. Breeding of genetically altered pollutant accumulators and degraders

1.13 Concluding Remarks

Environmental pollution is a global concern; hence, phytoremediation is a new evolving cleanup science that has the potential to be low cost, low impact and eco-friendly because this technology relies on green plants to remediate the polluted sites. It will be the most suitable alternative for developing nations, such as India, where this technology is at its nascent stage, knowledge of suitable phytoremediation plants is particularly limited or still being searched for and funding is a major problem. Financial resources should be devoted to a better understanding of the ecology and behaviour of green plants in polluted environments. Testing and controls in field research are still needed in order to fully understand the movement and final fate of pollutants using phytoremediation. In each case, particular attention is paid to the nature of the pollutants, the physiography of the environment polluted and the mix of pollutants present. A basic knowledge of the mechanism by which plants take up trace and toxic elements is also required. Efforts should be made toward the conservation of the remaining and establishment of more mangrove plant species, including other types of vegetation in their ecological zone in such a way that will assist in exploiting phytoremediation. It is also important that public awareness about this technology is considered, and clear and more precise information is made available to the public to raise its worldwide acceptability as a global sustainable green technology.

Acknowledgements

Financial assistance from the Department of Science & Technology (DST), Science and Engineering Research Board (SERB), Government of India, to Prof. Ram Chandra and the Rajeev Gandhi National Fellowship (RGNF) from the University Grant Commission (UGC) to Mr. Vineet Kumar, PhD, is highly acknowledged.

References

Abou-Shanab, R.A.I., Ghanem, K., Ghanem, N., and Al-Kolaibe, A. 2008. The role of bacteria on heavy-metal extraction and uptake by plants growing on multi-metal-contaminated soils. *World Journal of Microbiology and Biotechnology* 24: 253–262.

Angle, J.S., Chaney, R.L., Baker, A.J.M, Li, Y., Reeves, R., Volk, V., Roseberg, R., Brewer, E., Burke, S., and Nelkin, J. 2001. Developing commercial phytoextraction technologies: Practical consideration. *South African Journal of Science* 97: 619–623.

Antosiewicz, D.M., and Hennig, J. 2004. Overexpression of LCT1 in tobacco enhances the protective action of calcium against cadmium toxicity. *Environmental Pollution* 129: 237–245.

Baker, A.J.M., and Whiting, S.M. 2002. In search for the Holy Grail – Another step in understanding metal hyperaccumulation? *New Phytologist* 155: 1–7.

Baker, A.J.M., McGrath, S.P., Reeves, R.D., and Smith, J.A.C. 2000. Metal hyperaccumulator plants: A review of the ecology and physiology of a biological resource for phytoremediation of metal-polluted soil. In Terry, N., and Banuelos, G. (eds.): Lewis Publishers, Boca Raton, FL, pp. 5–107.

Banasova, V., Horak, O., Nadubinska, M., and Ciamporova, M. 2008. Heavy metal content in *Thlaspi caerulescens* J. et C. Presl growing on metalliferous and non-metalliferous soils in central Slovakia. *International Journal of Environmental Pollution* 33(2): 133–145.

Barber, D.A., and Lee, R.B. 1974. The effect of microorganisms on the absorption of manganese by plants. *New Phytology* 73: 97–106.

Belimov, A.A., Kunakova, A.M., Safronova, V.I., Stepanok, V.V., Yudkin, L.Y., Alekseev, Y.V., and Kozhemyakov, A.P. 2004. Employment of rhizobacteria for the inoculation of barley plants cultivated in soil contaminated with lead and cadmium. *Microbiology (Moscow)* 73(1): 99–106.

Bennicelli, R., Stepniewska, Z., Banach, A., Szajnocha, K., and Ostrowski, J. 2004. The ability of *Azolla caroliniana* to remove heavy metals Hg(II), Cr(III), Cr(VI) from municipal waste water. *Chemosphere* 55: 141–146.

Berti, W.R., and Cunningham, S.D. 2000. Phytostabilizaton of metals. In *Phytoremediation of toxic metals: Using plants to clean up the environment*, Raskin, I., and Ensley, B. (eds.). Wiley Inter Science, NY, pp. 71–88.

Bittsanszkya, A. et al. 2005. Ability of transgenic poplars with elevated glutathione content to tolerate Zinc (2$^+$) stress. *Environmental International* 31: 251–254.

Blaylock, M.J., and Huang, J.W. 2000. Phytoextraction of heavy metals. In *Phytoremediation of toxic metals: Using plants to clean up the environment*, Raskin, I., and Ensley, B.D. (eds). John Wiley & Sons, NY, pp. 53–69.

Boominathan, R., Saha-Chaudhary, N.M., Sahajwala, V., and Doran, P.M. 2004. Production of nickel bio-ore from hyperaccumulator plant biomass: Applications in phytomining. *Biotechnology and Bioengineering* 86: 243–250.

Bovet, L., Feller, U., and Martinoia, E. 2005. Possible involvement of plant ABC transporters in cadmium detoxification: A cDNA sub-library approach. *Environmental International* 31: 263–267.

Braud, A., Jezequel, K., Bazot, S., and Lebeau, T. 2009. Enhanced phytoextraction of an agricultural Cr and Pb contaminated soil by bioaugmentation with siderophore producing bacteria. *Chemosphere* 74: 280–286.

Bridge, G. 2004. Contested terrain: Mining and the environment. *Annual Review of Environment and Resources* 29: 205–259.

Bucheli-Witschel, M., and Egli, T. 2001. Environmental fate and microbial degradation of aminopolycarboxylic acids. *FEMS Microbiology Review* 25: 69–106.

Burd, G.I., Dixon, D.G., and Glick, B.R. 2000. Plant growth promoting bacteria that decrease heavy metal toxicity in plants. *Canadian Journal of Microbiology* 46(3): 237–245.

Che, D.S., Meagher, R.B., Heaton, A.C.P., Lima, A., Rugh, C.L., and Merkle, S.A. 2003. Expression of mercuric ion reductase in eastern cottonwood (*Populus deltoides*) confers mercuric iron reduction and resistance. *Plant Biotechnology Journal* 1: 311–319.

Cheraghi, M., Lorestani, B., Khorasani, N., Yousefi, N., and Karami, M. 2011. Findings on the phytoextraction and phytostabilization of soils contaminated with heavy metals. *Biology of Trace Elements and Research* 144: 1133–1141.

Cluis, C. 2004. Junk-greedy Greens: Phytoremediation as a new option for soil decontamination. *Biotechnology Journal* 2: 61–67.

Cobbett, C. 2003. Heavy metals and plants – Model systems and hyperaccumulators. *New Phytologist* 159: 289–293.

Cobett, C., and Goldsbrough, P. 2002. Phytochelatins and metallothioneins: Roles in heavy metal detoxification and homeostasis. *Annual Reviews in Plant Biology* 53: 159–182.

Dary, M., Chamber-Perez, M.A., Palomares, A.J., and Pajuelo, E. 2010. In situ phytostabilisation of heavy metal polluted soils using *Lupinus luteus* inoculated with metal resistant plant-growth promoting rhizobacteria. *Journal of Hazardous Material* 177: 323–330.

Dembitsky, V.M., and Rezanka, T. 2003. Natural occurrence of arseno-compounds in plants, lichens, fungi, algal species, and microorganisms. *Plant Science* 165(6): 1177–1192.

Dhankher, O.P. et al. 2012. Biotechnological approaches for phytoremediation. In *Plant biotechnology and agriculture*, Altman, A., and Hasegawa, P.M. (eds.). Academic Press, pp. 309–328.

Dierberg, F.E., Debus, T.A., and Goulet, N.A. 1987. Removal of copper and lead using a thin-film technique. In *Aquatic plants for water treatment and resource recovery*, Reddy, K.R., and Smith, W.H. (eds.). Magnolia Publishing, NY, pp. 497–504.

Doty, S.L. 2008. Enhancing phytoremediation through the use of transgenics and endophytes. *New Phytologist* 179: 318–333.

Doty, S.L. et al. 2007. Enhanced phytoremediation of volatile environmental pollutants with transgenic trees. *Proceedings of the National Academy of Sciences of the United States of America* 104(43): 16816–16821.

Dushenkov, S., Vasudev, D., Kapulnik, Y., Gleba, D., Fleisher, D., Ting, K.C., and Ensley, B. 1997. Removal of uranium from water using terrestrial plants. *Environmental Science and Technology* 31: 3468–3474.

Dushenkov, V., Nanda Kumar, P.B.A., Motto, H., and Raskin, I. 1995. Rhizofiltration – The use of plants to remove heavy metals from aqueous streams. *Environmental Science and Technology* 29: 1239–1245.

Fry, J.C., and Humphrey, N.C.B. (1978). Techniques for the study of bacteria epiphytic on aquatic macrophytes. In *Techniques for the study of mixed populations*, Lovelock, D.W., and Davies, R. (eds.). Academic Press, London, pp. 1–29.

Fulekar, M.H., Singh, A., Thorat, V., Kaushik, C.P., and Eapen, S. 2010. Phytoremediation of 137 Cs from low level nuclear waste using *Catharanthus roseus*. *Indian Journal of Pure and Applied Physics*. 48: 516–519.

Ganesan, V. 2008. Rhizoremediation of cadmium soil using a cadmium-resistant plant growth-promoting rhizo-pseudomonad. *Current Microbiology* 56: 403–407.

Garba, S.T., Osemeahon, A.S., Maina, H.M., and Barminas, J.T. 2012. Ethylene-diaaminetetraacetate (EDTA)-assisted phytoremediation of heavy metal contaminated soil by *Eleusine indica* L. Gearth. *Journal of Environmental and Chemical Ecotoxicology* 4(5): 103–109.

Ghosh, M., and Singh, S.P. 2005. A review on phytoremediation of heavy metals and utilization of its byproducts. *Applied Ecology and Environmental Research* 3: 1–18.

Glick, B., Karaturovic, D., and Newell, P. 1995. A novel procedure for rapid isolation of plant growth promoting Pseudomonas. *Canadian Journal of Microbiology* 41: 533–536.

Gullner, G., Komives, T., and Rennenberg, H. 2001. Enhanced tolerance of poplar plants overexpressing gamma-glutamycylcysteinesynthetase towards chloroacetanilide herbicides. *Journal Experimental Botany* 52: 971–979.

Hall, J.L. 2002. Cellular mechanism for heavy metal detoxification and tolerance. *Journal of Experimental Botany* 53: 1–11.

Hasnain, S., and Sabri, A.N. 1997. Growth stimulation of *Triticum aestivum* seedlings under Cr-stresses by non-rhizospheric pseudomonad strains. *Environmental Pollution* 97: 265–273.

He, C.Q., Tan, G.E., Liang, X., Du, W., Chen, Y.L., Zhi, G.Y., and Zhu, Y. 2010. Effect of Zn-tolerant bacterial strains on growth and Zn accumulation in *Orychophragmus violaceus*, *Applied Soil Ecology* 44: 1–5.

Hirschi, E.D., Korenkey, V.D., Wilganewski, N.I., and Wagner, G.I. 2000. Expression of Arabidopsis CAX2 in tobacco. Altered metal accumulation and increased manganese tolerance. *Plant Physiology* 124: 128–133.

Huhle, B., Heilmeier, H., and Merkel, B. 2008. Potential of *Brassica juncea* and *Helianthus annuus* in phytoremediation for uranium. In Uranium. *Mine Hydrogeology*. pp. 307–318.

Iimura, Y., Yoshizumi, M., Sonoki, T., Uesugi, M., Tatsumi, K., Horiuchi, K.I., Kajita, S., and Katayama, Y. 2007. Hybrid aspen with a transgene for fungal manganese peroxidase is a potential contributer to phytoremediation of the environment contaminated with bisphenol A. *Journal of Wood Science* 53: 541–544.

Imsande, J. 1998. Iron, sulfur, and chlorophy II deficiencies: A need for an integrative approach in plant physiology. *Physiologia Plantarum* 103(1): 139–144.

Interstate Technology and Regulatory Cooperation (ITRC). 2001. Phytotechnology Technical and Regulatory Guidance Document. Available at http://www.state .nj.us/dep/dsr/bscit/Phytoremediation.pdf (accessed August 16, 2014).

Jadia, C.D., and Fulekar, M.H. 2009. Phytoremediation of heavy metals: Recent techniques. *African Journal of Biotechnology* 8(6): 921–928.

Jiang, C.Y., Sheng, X.F., Qian, M., and Wang, Q.Y. 2008. Isolation and characterization of a heavy metal-resistant Burkholderia sp. from heavy metal-contaminated paddy field soil and its potential in promoting plant growth and heavy metal accumulation in metal polluted soil. *Chemosphere* 72: 157–164.

Jones, R., Sun, W., Tang, C.S., and Robert, F.M. 2004. Phytoremediation of petroleum hydrocarbons in tropical coastal soils. II. Microbial response to plant roots and contaminant. *Environmental Science and Pollution Research* 11: 340–346.

Joseph, B., Patra, R.R., and Lawrence, R. 2007. Characterization of plant growth promoting Rhizobacteria associated with chickpea (*Cicer arietinum* L). *International Journal of Plant Production* 1(2): 141–152.

Kennedy, A.C. 1998. The rhizosphere and spermosphere. In *Principles and applications of soil microbiology*, Sylvia, D.M., Fuhrmann, J.J., Hartel, P.G., Zuberer, D.A. (eds.). Prentice-Hall, Upper Saddle, NJ.

Kirk, J., Klironomos, J., Lee, H., and Trevors, J.T. 2005. The effects of perennial ryegrass and alfalfa on microbial abundance and diversity in petroleum contaminated soil. *Environmental Pollution* 133: 455–465.

Kloepper, J.W., Lifshitz, R., and Zablotowicz, R.M. 1989. Free-living bacterial inocula for enhancing crop productivity. *Trends in Biotechnology* 7(2): 39–43.

Kramer, U., and Chardonnens, A.N. 2001. The use of transgenic plants in the bioremediation of soils contaminated with trace elements. *Applied Microbiology and Biotechnology* 55: 661–672.

Kramer, U. 2003. Phytoremediation to phytochelatin – Plant trace metal homeostasis. *New Phytologist* 158: 4–6.

Lasat, M.M. 2000. Phytoextraction of metals from contaminated soil: A review of plant/soil/metal interaction and assessment of pertinent agronomic issues. *Journal of Hazardous Substance Research* 2: 1–25.

Lee, C.S., Lee, H.B., Cho, Y.G., Park, J.H., and Lee, H.S. 2008. Hospital-acquired *Burkholderia cepacia* infection related to contaminated benzalkonium chloride. *Journal of Hospital Infection* 68: 280–282.

Lee, J., Shim, D., Song, W.Y., Hwang, I., and Lee, Y. 2004. Arabidopsis metallothioneins 2a and 3 enhance resistance to cadmium when expressed in *Vicia faba* guard cells. *Plant Molecular Biology* 54: 805–815.

Li, Y.M., Chaney, R., Brewer, E., Rosenberg, R., Angle, S.J., Baker, A.J.M., Reeves, R.D., and Nelkin, J. 2003. Development of technology for commercial phytoextraction of nickel: Economic and technical consideration. *Plant and Soil* 249: 107–115.

Liu, W., Zhou, Q., An, J., Sun, Y., and Liu, R. 2010. Variations in cadmium accumulation among Chinese cabbage cultivars and screening for Cd-safe cultivars. *Journal of Hazardous Material* 173(1–3): 737–743.

Lombi, E., Zhao, F., McGrath, S., Young, S., and Sacchi, G. 2001a. Physiological evidence for a high-affinity cadmium transporter highly expressed in a *Thlaspi caerulescens* ecotype. *New Phytol* 149: 53–60.

Lombi, E., Zhao, F.J., Dunham, S.J., and McGrath, S.P. 2001b. Phytoremediation of heavy metal contaminated soils: Natural hyperaccumulation versus chemically enhanced phytoextraction. *Journal of Environmental Quality* 30: 1919–1926.

Lucy, M., Reed, E., and Glick, B.R. 2004. Applications of free living plant growth-promoting rhizobacteria. *Antonie Van Leeuwenhoek* 86: 1–25.

Ma, L.Q., Komar, K.M., Tu, C., Zhang, W.H., Cai, Y., and Kennelley, E.D. 2001. A fern that hyperaccumulates arsenic: A hardy, versatile, fast-growing plant helps to remove arsenic from contaminated soils. *Nature* 409: 579.

Ma, Y., Rajkumar, M., and Freitas, H. 2009. Inoculation of plant growth promoting bacterium *Achromobacter xylosoxidans* strain Ax10 for the improvement of copper phytoextraction by *Brassica juncea*. *Journal of Environmental Management* 90: 831–837.

Mangkoedihardjo, S., and Surahmaida. 2008. *Jatropha curcas* L. for phytoremediation of lead and cadmium polluted soil. *World Applied Science Journal* 4: 519–522.

Mathur, N., Singh, J., Bohra, S., Bohra, A., Mehboob Vyas, M., and Vyas, A. 2010. Phytoremediation potential of some multipurpose tree species of Indian Thar Desert in oil contaminated soil. *Advances in Environment Biology* 4(2): 131–137.

McCutcheon, S.C., and Schnoor, J.L. 2003. Phytoremediation: Transformation and control of contaminants. John Wiley & Sons, Inc., Hoboken, NJ, p. 987.

McGrath, S.P., and Zhao, F.J. 2003. Phytoextraction of metals and metalloids from contaminated soils. *Current Opinion in Biotechnology* 14: 277–282.

Mo, S., Chol, D.S., and Robinson, J.W. 1989. Uptake of mercury from aqueous solution by Duckweed: The effect of pH, copper and humic acid. *Environmental Health* 24: 135–146.

Mukhopadhyay, S., and Maiti, S.K. 2010. Phytoremediation of metal mine waste. *Applied Ecology and Environmental Research* 8(3): 207–222.

Myers, R.T., Zak, D.R., White, D.C., and Peacock, A. 2001. Landscape level patterns of microbial community composition and substrate use in upland forest ecosystems. *Soil Science Society of America Journal* 65: 359–367.

Newman, L.A., Strand, S.E., Choe, N., Duffy, J., Ekuan, G., Ruszaj, M., Shurtleff, B.B., Wilmoth, J., Heilman, P., and Gordon, M.P. 1997. Uptake and biotransformation of trichloroethylene by hybrid poplars. *Environmental Science and Technology* 31: 1062–1067.

O'Connell, K.P., Goodman, R.M., and Handelsman, J. 1996. Engineering the rhizosphere expressing a bias. *Trends in Biotechnology* 14: 83–88.

Odjegba, V.J., and Fasidi, I.O. 2004. Accumulation of trace elements by Pistia stratiotes: Implications for phytoremediation. *Ecotoxicology* 13(7): 637–646.

Papoyan, A., and Kochian, L.V. 2004. Identification of *Thlaspi caerulescenns* genes that may be involved in the heavy metal hyperaccumulation and tolerance. Characterization of a novel heavy metal transporting ATPase. *Plant Physiology* 139: 3814–3823.

Pieper, D.H., Martins dos Santos, V.A.P., and Golyshin, P.N. 2004. Genomic and mechanistic insights into the biodegredation of organic pollutants. *Current Opinion in Biotechnology* 15: 215–224.

Pilon-Smits, E.A.H. 2005. Phytoremediation. *Annual Review of Plant Biology* 56:15–39.

Preeti, P.P., Tripathi, A.K., and Shikha, G. 2011. Phytoremediation of arsenic using *Cassia fistula* Linn. seedling. *International Journal of Research and Chemistry of Environment* 1: 24–28.

Rajakaruna, N., and Bohm, B.A., 2002. Serpentine and its vegetation: A preliminary study from Sri Lanka. *Journal of Applied Botany – Angewandte Botanik* 76: 20–28.

Raskin, I., and Ensley, B.D. 2000. Phytoremediation of toxic metals: Using plants to clean up the environment. John Wiley & Sons, Inc., NY.

Raskin, I., Kumar, P.B.A.N., Dushenkov, S., and Salt, D. 1994. Bioconcentration of heavy metals by plants. *Current Opinion in Biotechnology* 5: 285–290.

Sakakibara, M., Aya, W., Masahiro, I., Sakae, S., and Toshikazu, K. 2007. Phytoextraction and phytovolatilization of arsenic from As-contaminated soil by *Pteris vittata*. *Proceedings of the Annual International Conference on Soils, Sediments, Water and Energy* 12: 26.

Salt, D.E., Smith, R.D., and Raskin, I. 1998. Phytoremediation. *Annual Reviews of Plant Physiology, Plant Molecular Biology* 49: 643–668.

Sandaa, R.A., Torsvik, V., and Enger, Ø. 1999. Analysis of bacterial communities in heavy metal-contaminated soils at different levels of resolution. *FEMS Microbiology, Ecology and Engineering* 30: 237–251.

Saraswat, S., and Rai, J.P.N. 2009. Phytoextraction potential of six plant species grown in multi-metal contaminated soil. *Chemical Ecology* 25(1): 1–11.

Sekara, A., Poniedzialeek, M., Ciura, J., and Jedrszczyk, E. 2005. Cadmium and lead accumulation and distribution in the organs of nine crops: Implications for phytoremediation. *Polish Journal of Environmental Studies* 14(4): 509–516.

Sharma, H. 2011. Metal hyperaccumulation in plants: A review focusing on phytoremediation technology. *Journal of Environmental Science and Technology* 4: 118–138.

Sheng, X.F., Xia, J.J., Jiang, C.Y., He, L.Y., and Qian, M. 2008. Characterization of heavy metal-resistant endophytic bacteria from rape (*Brassica napus*) roots and their potential in promoting the growth and lead accumulation of rape. *Environmental Pollution* 156: 1164–1170.

So, L.M., Chu, L.M., and Wong. P.K. 2003. Microbial enhancement of Cu^{2+} removal capacity of *Eichhornia crassipes* (Mart.). *Chemosphere* 52: 1499–1503.

Soleimani, M., Hajabbasi, M.A., Afyuni, M., Mirlohi, A., Borggaard, O.K., and Holm, P.E. 2010. Effect of endophytic fungi on cadmium tolerance and bioaccumulation by *Festuca arundinacea* and *Festuca pratensis*. *International Journal of Phytoremediation* 12(6): 535–549.

Song, W.Y., Sohn, E.J., Martinoia, E., Lee, Y.J., Yang, Y.Y., Jasinski, M., Forestier, C., Hwang, I., and Lee, Y. 2003. Engineering tolerance and accumulation of lead and cadmium in transgenic plants. *Nature Biotechnology* 21: 914–919.

Sun, Q., Ye, Z.H., Wang, X.R., and Wong, M.H. 2007. Cadmium hyperaccumulation leads to an increase of glutathione rather than phytochelatins in the cadmium hyperaccumulator *Sedum alfredii*. *Journal of Plant Physiology* 164(11): 1489–1498.

Sun, Y., Zhou, Q., and Diao, C. 2008. Effects of cadmium and arsenic on growth and metal accumulation of Cd-hyperaccumulator *Solanum nigrum* L. *Bioresource Technology* 99(5): 1103–1110.

Sun, Y., Zhou, Q., Wang, L., and Liu, W. 2009. Cadmium tolerance and accumulation characteristics of *Bidenspilosa* L. as a potential Cd-hyperaccumulator. *Journal of Hazardous Material* 161(2–3): 808–814.

Suresh, B., and Ravishankar, G. 2004. Phytoremediation – A novel and promising approach for environmental clean-up. *Critical Reviews in Biotechnology* 24: 97–124.

Tong, Y.P., Kneer, R., and Zhu, Y.G. 2004. Vacuolar compartmentalization: A second-generation approach to engineering plants for phytoremediation. *Trends in Plant Science* 9: 7–9.

United States Environmental Protection Agency (USEPA). 2000. Introduction to Phytoremediation. EPA 600/R-99/107. U.S. Environmental Protection Agency, Office of Research and Development, Cincinnati, OH.

van Dillewijin, P. et al. 2008. Bioremediation of 2,4,6-trinitrotoluene by bacterial nitroreductase expressing transgenic aspen. *Environmental Science and Technology* 42: 7405–7410.

van Huysen, T., Terry, N., and Pilon-Smits, E.A.H. 2004. Exploring the selenium phytoremediation potential of transgenic Indian mustard overexpressing ATP sulfurylase or cystathionine-y-synthase. *International Journal of Phytoremediation* 6: 1–8.

Varun, M., DSouza, R., Pratas, J., and Paul, M.S. 2012. Metal contamination of soils and plants associated with the glass industry in North Central India: Prospects of phytoremediation. *Environment Science and Pollution Research* 19: 269–281.

Verret, F., Gravot, A., Auroy, P., Leonhardt, N., David, P., Nussaume, L., Vavasseur, A., and Richaud, P. 2004. Overexpreassion of AtHMA4 enhances root-to-shoot translocation of zinc and cadmium and plant metal tolerance. *FEBS Letter* 576: 306–312.

Wallace, A., Wallace, G.A., and Cha, J.W. 1992. Some modifications in trace elements toxicities and deficiencies in plants resulting from interactions with other elements and chelating agents. The special case of iron. *Journal of Plant Nutrition* 15(2): 1589–1598.

Wang, J., and Evangelou, V.P. 1994. Metal tolerance aspects of plant cell walls and vacoules. In *Handbook of plant and crop physiology*, Pessaraki, M. (ed.). Marcel Dekker, Inc., NY, pp. 695–717.

Wang, Y., Brown, H.N., Crowley, D.E., and Szaniszlo, P.J. 1993. Evidence for direct utilization of a siderophore, ferroxamine B, in axenically grown cucumber. *Plant Cell and Environment* 16(5): 579–585.

Wani, P.A., Khan, M.S., and Zaidi, A. 2007. Effect of metal tolerant plant growth promoting *Bradyrhizobium* sp. (vigna) on growth, symbiosis, seed yield and metal uptake by green gram plants. *Chemosphere* 70: 36–45.

Xiong, J., He, Z., Liu, D., Mahmood, Q., and Yang, X. 2008. The role of bacteria in the heavy metals removal and growth of *Sedum alfredii Hance* in an aqueous medium. *Chemosphere* 70: 489–494.

Yoon, J., Cao, X., Zhou, Q., and Ma, L.Q. 2006. Accumulation of Pb, Cu, and Zn in native plants growing on a contaminated Florida site. *Science of Total Environment* 368(2–3): 456–464.

Zadeh, B.M., Savaghebi-Firozabadi, G.R., Alikhani, H.A., and Hosseini, H.M. 2008. Effect of sunflower and amaranthus culture and application of inoculants on phytoremediation of the soils contaminated with cadmium. *Americam-Eurasian Journal of Agriculture and Environment Science* 4: 93–103.

Zhou, Q.X., and Song, Y.F. 2004. Principles and methods of contaminated soil remediation. Science Press, Beijing, p. 568.

Zhu, Y.L., Zayed, A.M., Qian, J.H., Desouza, M., and Terry, N. 1999. Phytoaccumulation of trace elements by wetland plants: II, *Water Hyacinth. Journal of Environmental Quality* 28: 339–344.

Zurayk, R., Sukkariyah, B., Baalbaki, R., and Ghanem, D.A. 2001. Chromium phytoaccumulation from solution by selected hydrophytes. *International Journal of Phytoremediation* 3: 335–350.

2

Microbial Cells Dead or Alive: Prospect, Potential and Innovations for Heavy Metal Removal

Adeline Su Yien Ting

CONTENTS

2.1 Introduction

Heavy metals in trace quantities are important to all living organisms to regulate physiological developments (Park et al. 2006). However, metal concentrations in the environment often exceed permissible levels due to the rampant discharge of high loads of metals from vigorous urbanisation, industrialisation and anthropogenic activities. The metal-laden waste effluents come primarily from the following industries (but are not confined to them): the metallurgy industry, surface-finishing industry, energy and fuel production industry and fertilizer and pesticide industry. Effluents from each of these industries vary in the type and concentration of heavy metals discharged. Heavy metals of major concern include toxic metals (Hg, Cr, Pb, Zn, Cu, Ni, Cd, As, Co and Sn), radionuclides (U, Th, Ra and Am) and precious metals (Pd, Pt, Ag, Au and Ru) (Wang and Chen 2006). Among these metals, lead (Pb), copper (Cu), mercury (Hg), cadmium (Cd) and chromium (Cr) are key environmental pollutants (Gadd and Fomima 2011). Metal toxicity is the result of frequent discharge and accumulation (and, in some cases, the bio-magnification) of these metals in the environment. Their persistence in the environment is enhanced by their ability to exist as various chemical species attributed to their interaction with biotic and abiotic environmental factors. Metals exist as cations or anions, in oxidised or reduced forms or in complex or hydroxylated metal forms (Gadd 2009). These bioavailable forms cause toxicity upon uptake by living organisms (microbes, plants, animals and humans) posing serious health and environmental hazards (Manasi et al. 2014). Excessive Cu, Cd and nickel (Ni) are known to cause liver and kidney failure as well as chronic asthma (Dal Bosco et al. 2006).

Numerous methods and technologies have been employed to treat metal-laden waste effluents. In the early days, physicochemical approaches were adopted to remove metals, relying primarily on chemical precipitation, filtration, flotation, electrochemical treatment, ion exchange, membrane-related process and evaporation (Figure 2.1). Over the years, these methods were discovered to have several limitations (Wang and Chen 2009). Chemical precipitation and electrochemical treatments are ineffective for treatment of effluents low in metal ion concentrations (1 to 100 mg/L) (Satapathy and Natarajan 2006). Ion exchange and membrane technologies can only treat small volumes of effluents as treating large volumes is expensive. These methods also demand high usage of chemicals and incur additional cost to manage the disposal of toxic sludge generated. The introduction of adsorbents resolved some of the issues from the use of physicochemical approaches. However, adsorbents too were gradually found to be costly, particularly the

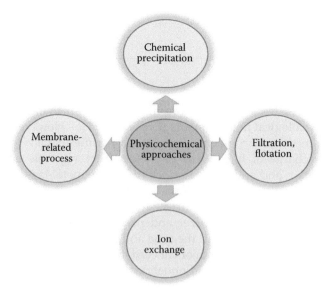

FIGURE 2.1
Conventional metal-removal techniques are primarily based on physicochemical approaches.

use of activated carbon (Park et al. 2006). This prompted explorations on low-cost adsorbents as an alternative, leading to the introduction of biosorbents. Over the years, a variety of biosorbents have been discovered and investigated for their efficacy in removing metals. These include the use of agricultural wastes (corn core) (Vijayaraghavan and Yun 2008), industrial wastes (fermentation wastes) (Agarwal et al. 2006), algae biomass and microbial biomass (*Bacillus subtilis, Rhizopusarrhizus Saccharomyces cerevisiae*) (Wang and Chen 2009), which have all demonstrated encouraging results in absorbing metals.

Among the many types of biosorbents, microbial cells are found to be highly useful. Microbes are abundant and ubiquitous in nature, relatively easy and cheap to culture and able to generate sufficient biomass for use (Wang and Chen 2006). Their potential in detoxifying pollutants was first identified when microbes in reed beds and wetlands demonstrated bioremediation activities. Subsequent investigations revealed that microbial tolerance to toxic metals is dependent on their ability to adapt and regulate metals to a level sufficient for cell function while avoiding toxicity. Microbes have been found to effectively decrease concentrations of metals in solutions from ppm to ppb, mostly through biosorption (Wang and Chen 2006). Biosorption is a process in which metals bind to the cell surface and thus occur in both dead and live cells. Over the years, both dead and live microbial cells have been extensively explored to identify and harness potential microbes as biosorbents for treatment of metal-laden waste effluents. Innovations have also been incorporated to realise the maximum sorption efficacy of the microbial biosorbents and to improve desorption and reusability potential. The following discuss the use of both dead and live microbial cells as well as approaches in introducing innovations to these biosorbents for improved biosorption activities.

2.2 Microbes for Heavy Metal Removal

Microbial cells of various origins have been investigated for metal removal with bacteria and fungi dominating most of the literature. These microbial cells, in the form of dead or live cells or even their derivatives (polysaccharides), have demonstrated significant capacity for metal removal. Common environmental species, such as *Bacillus subtilis* (Wang and Chen 2009), *B. thuringiensis* strain OSM29 (Oves et al. 2013), *B. circulans* (Sahoo et al. 1992), *Pseudomonas fluorescens* (Choudhary and Sar 2009), *P. putida* (Pardo et al. 2003) and *P. aeruginosa* ASU 6a (Gabr et al. 2008), have been studied extensively for their metal biosorption activities. In recent years, diverse groups of microbes with metal biosorption potential were revealed. These include a variety of rhizobacteria (*Achromobacter, Arthrobacter, Azotobacter, Azospirillum, Enterobacter, Serratia*) (Gray and Smith 2005), members of actinobacteria

(*Streptomyces* spp., *S. mirabilis* P16B-1, *S. zinciresistens*) (Ma et al. 2011; Lin et al. 2012; Schütze et al. 2014) and lactic acid bacteria (*Bifidobacterium longum* 46, *B. lactis* Bb12, *Lactobacillus fermentum* ME3) (Halttunen et al. 2007). Novel isolates have also been attempted to evaluate their potential in removing metals. They include enriched consortium of sulphate-reducing bacteria (SRB) (Kieu et al. 2011) and the commercial mixture of Effective Microorganisms (EM-1™ Inoculant) (Ting et al. 2013).

A similar trend was observed in studies related to fungal biosorbents with which ubiquitous microfungi (including yeast) and macrofungi have been extensively explored. Common isolates with established metal removal capacity include *Aspergillus niger, Mucor* spp., *Rhizopus nigricans, R. arrhizus, R. javanicus, Penicillium chrysogenum, Trichoderma atroviride, T. asperellum, Saccharomyces* spp., *Candida albicans, C. utilis, C. tropicalis, Lentinus edodes* and *Termitomyces clypeatus* (Kapoor and Viraraghavan 1995; Lopez-Errasquin and Vazquez 2003; Ahluwalia and Goyal 2007; Aksu and Balibek 2007; Zafar et al. 2007; Bayramoglu and Arica 2008; Baysal et al. 2009; Das and Guha 2009; Ting and Choong 2009a; Wang and Chen 2009; Tan and Ting 2012). The use of unicellular yeasts, particularly *S. cerevisiae*, as biosorbents has grown increasingly important despite their fairly recent discovery. Although *S. cerevisiae* demonstrated only mediocre capacity for metal uptake when compared with other fungi, usage of yeast is an economically feasible and favourable approach as yeast is often a by-product of the fermentation industry (Wang and Chen 2006). Yeasts as biosorbents are therefore abundant, and the recycling of fermentation wastes could also generate additional income for the fermentation industry.

The conventional approach in biosourcing microbial biosorbents is primarily via isolation from contaminated environmental samples. Industrial waste effluents of the steel industries, iron foundries, electroplating industries, polluted water/soil and mine spoils are examples of contaminated sites where isolates have been recovered (Chatterjee et al. 2010; Viraraghavan and Srinivasan 2011; Oves et al. 2013). These polluted sites are popular as microbes exposed to polluted sources typically exhibit high metal removal rates. This was demonstrated by *Halomonas* BVR1 isolated from the waste of electronic industries, which removed 80 mg/L of Cd (Manasi et al. 2014). Industrial waste–derived *Rhizopus arrhizus* and *Saccharomyces cerevisiae* also showed similar high metal-removal rates (Wang and Chen 2009). Tan and Ting (2012) and Ting and Choong (2009a,b) have also successfully isolated *T. asperellum* and *Stenotrophomonas maltophilia* from a river sediment sample (Penchala River). Isolates were first screened for their metal tolerance in plate assays (with increasing metal concentrations), gradually cultured to obtain sufficient biomass to perform biosorption and bioaccumulation tests (Figure 2.2). Their isolates demonstrated both biosorption and bioaccumulation activities toward Cu, supporting many other earlier studies using isolates from polluted sites. Therefore, for many years, microbial isolates were sourced from polluted environments.

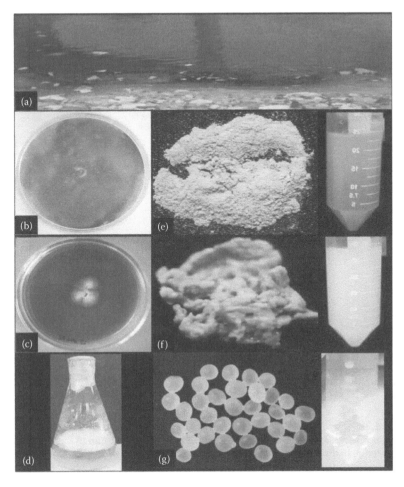

FIGURE 2.2
(See color insert.) Typical process of (a) sampling (b) isolating and (c) screening from polluted environments to determine metal tolerance. Potential isolates are subsequently (d) cultured to generate sufficient biomass for metal biosorption tests. The biomass can be prepared as (e) non-viable, (f) viable or (g) immobilised forms to determine their biosorption potential.

Interestingly, in a novel study, Ting et al. (2011) observed that polluted indoor waste could also harbour microbes with metal-tolerant attributes. In their preliminary assessment on metal tolerance and biosorption activities of filamentous fungi recovered from analytical wastewater (water residues from atomic absorption spectroscopy [AAS]) in the laboratory, they found nine fungal isolates (members of *Penicillium* sp., *Aspergillus* sp., *Trichoderma* sp. and *Fusarium* sp.) with metal-tolerance potential (Figure 2.3). Of these, three isolates (*Penicillium* sp., *Fusarium* sp. and *Aspergillus* sp.) showed tolerance to mostly Al, Cr and Zn (up to 50 mM), followed by Cu and Pb (up

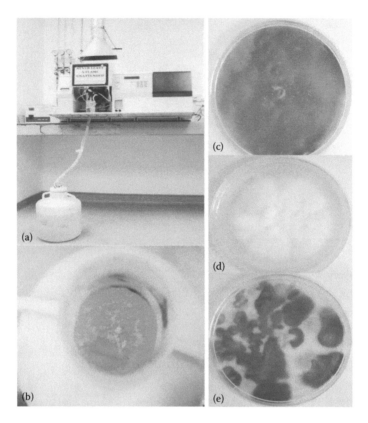

FIGURE 2.3
(See color insert.) (a) Wastewater from AAS where (b) growth of fungi in the wastewater can be detected visually. Isolates (c) 6 (*Penicillium* sp.), (d) 9 (*Fusarium* sp.) and (e) 10 (*Aspergillus* sp.) showed the most potential in removing various metals via biosorption. (Compiled and modified from Ting ASY et al., *Proceedings of the International Congress of the Malaysian Society for Microbiology.* 8–11 December 2011, Penang, Malaysia, pp. 110–113, 2011.)

to 20 mM) and the least tolerance to Cd (2–4 mM). Subsequent biosorption tests revealed isolates removed between 13.25 and 15.78 mg/g metal. A comparison with the literature revealed that although their biosorption activities were inferior to environmental isolates, their metal-tolerance potential was commendable. From this preliminary observation, it was concluded that microbes derived from waste, irrespective of indoor or environmental origin, have the ability to tolerate metal concentrations and demonstrate metal removal potential (albeit to a lesser extent for indoor isolates) via biosorption.

On the contrary, metal-tolerant isolates have also been recovered from natural unpolluted environments. Sahoo et al. (1992) were among the first few who discovered that pure isolates from laboratories could be used. They identified and documented the tolerance mechanisms of the polysaccharide-producing bacterium *Bacillus circulans,* a pure isolate from the National Chemical Laboratory, Pune, in removing Cu and Cd via metal-binding complexation. Puyen et al. (2012) further concur by isolating *Micrococcus luteus* DE2008 from a microbial mat community to absorb Pb and Cu. These isolates from unpolluted environments have efficient metal biosorption activities. Thus, it was evident that metal tolerance and biosorption capacity are not exclusive to isolates from polluted environments. Nevertheless, not all isolates from unpolluted sources were beneficial. The novel attempt by Ting et al. (2013) on the use of the commercial concoction of Effective Microorganisms (EM) (EM-1 Inoculant) in alginate-immobilised and free cell forms revealed that EM on its own was not specifically beneficial in removing metals. In their study, removal of Cr, Cu and Pb was attributed mainly to the role of alginate. In the absence of alginate, free cells removed only 0.160, 0.859 and 0.755 mg/mL or Cr, Cu and Pb compared to 0.940, 2.695 and 4.011 mg/g by alginate-immobilised EM, respectively (Figure 2.4). Thus, immobilisation was beneficial and contributed to metal biosorption rather than the role of EM.

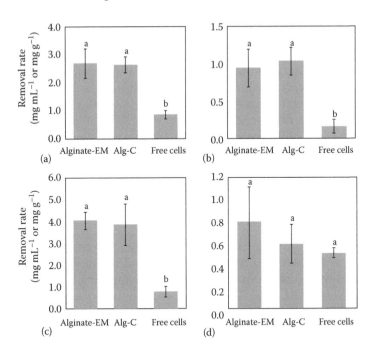

FIGURE 2.4
Mean of (a) Cu, (b) Cr, (c) Pb and (d) Zn removed by alginate-immobilised EM (Alginate-EM), plain alginate beads (Alg-C) and free cells of EM (free cells). Amount of metal removed by alginate-based biosorbents are expressed as mg/g, and for free cell, EM is expressed as mg/L. Means with the same letters are not significantly different ($HSD_{(0.05)}$). Bars indicate standard error of means. (Compiled and modified from Ting ASY et al., *Bioresource Technology,* 147, 636–639, 2013; Ting ASY et al., *Proceedings of the International Congress of the Malaysian Society for Microbiology.* 8–11 December 2011, Penang, Malaysia, pp. 110–113, 2011.)

Thus, comparisons on the biosorption activities between isolates originating from polluted and unpolluted sources is highly suggestive that exposure to high metal concentrations may be beneficial to render the isolates tolerant to metals. However, it is not the sole determinative factor of successful metal removal by microbes because high tolerance to metals does not necessarily translate to efficient biosorption capacity (Ting et al. 2011). In fact, there is little, if any, correlation between metal tolerance and biosorption properties (Zafar et al. 2007). Instead, tolerance and biosorption appears to be species-dependent, irrespective of the origin of the isolates.

Metal tolerance and sorption activities of the microbial species are dependent on the type of cells used (dead or live cells) and their cell wall structure and composition (Halttunen et al. 2007). For bacteria, their cell walls are composed of various peptidoglycan carboxyl groups or phosphate groups, which are the primary sites for metal binding and sorption for Gram-positive and negative bacteria, respectively (Halttunen et al. 2007; Gadd and Fomina 2011).

The functional groups on the cell surface differ in various species, and as a result, strain-specific characteristics among bacteria in adsorbing metals are demonstrated. For example, metal preference is observed between the two Gram-positive bacteria *B. thuringiensis* OSM29 and *M. luteus,* and the former prefers Ni and Cu (Oves et al. 2013) while the latter uptakes more Pb and Cu (Puyen et al. 2012). Similarly, metal affinity varies among Gram-negative bacteria. *P. fluorescens, P. putida* and *G. thermodenitrificans* demonstrated varying degrees of metal preference and uptake of Cu, Ni, Co, Cd (Choudhary and Sar 2009), Cu, Zn, Cd and Pb (Pardo et al. 2003) and Fe, Cr, Co and Cu (Chatterjee et al. 2010), respectively. For fungal cells, metal sorption occurs typically via binding to chitins, glucans, mannans, phenolic polymers (carboxyl, phenolic and methoxyl groups) and, to a certain extent, to lipids and pigments (melanin) found in the fungal cell wall. Because these compositions vary between species, their biosorption capacity differs from one species to another as well.

In recent years, a new emerging group of microbes was studied for their role in metal removal (Rajkumar et al. 2009; Ma et al. 2011). These microbes are called endophytes. Endophytes are defined as microbes that colonise the internal tissues of plants without causing symptoms, infections or negative effects on their host. The role of endophytes in metal removal was first highlighted in the literature when endophyte–host plant association (often with a phytoremediator or a hyperaccumulator) had been observed to improve plant growth, ameliorate toxicity and promote phytoextraction efficiency in controlled sites (Rajkumar et al. 2009; Babu et al. 2013). It is, however, not clear as to the exact role of endophytes in these plant–endophyte associations. For example, it is not established if phytoextraction is attributed to the plant itself or to the endophyte–host plant association. Furthermore, how endophytes survive and tolerate high concentrations of heavy metals is also poorly understood. One of the earliest and most extensively studied endophyte communities was the endophytes associated with the hyperaccumulator *Solanum nigrum* L. Endophytes from *S. nigrum* were identified to include members of actinobacteria (43%), proteobacteria (23%), bacteroidetes (27%) and firmicutes (7%) (Luo et al. 2011). The most common species recovered were the endophytic *Bacillus* sp. (Guo et al. 2010), which are ubiquitously found in most plants (Babu et al. 2013). The hyperaccumulator *S. nigrum* L. also harboured the fungal endophyte *Microsphaeropsis* sp. (LSE10), which may have a role in enhancing phytoremediation as the endophyte showed good biosorption efficacy for Cd. Other endophytic fungi from various plants have also demonstrated potential to improve phytoremediation. They include *Mucor* sp. isolate CBRF59 (Deng et al. 2011), *Trichoderma* sp., *Aspergillus* sp., the arbuscular mycorrhizal fungi (AMF) and species of *Phoma, Alternaria* and *Peyronellaea* (Li et al. 2012).

Sim (2013) conducted preliminary isolation and screening of endophytes from *Phragmites* in a landfill to determine their tolerance, adaptive tolerance behaviour and biosorption potential. This study was distinctive as the endophytes were studied independent of the host plant with the aim of examining their potential as biosorbents. In the study, a total of 21 fungal endophytes with tolerance to Cd, Cu, Cr, Pb and Zn were isolated. Three of the endophytic isolates with the most potential were identified as *T. asperellum* (isolate 11), *Phomopsis* sp. (isolate 9) and *Saccharicola bicolour* (isolate 22) (Figure 2.5). Of the three, *T. asperellum* has the most potential as the isolate showed tolerance to 2000 ppm of Al, Cd and Cr with no significant change to radial growth. *Phomopsis* sp. and *S. bicolour* were, however, implicated by increasing metal concentrations (2000 ppm) with notable changes to cultural morphology and pigmentation (Figure 2.5). All isolates have similar biosorption capacities; more Cd was removed with the total adsorbed 19.78, 20.05 and 19.57 mg/g for isolates *T. asperellum, Phomopsis* sp. and *S. bicolour*, respectively. They were least effective in removing Cr with only 16.75, 16.90 and 16.14 mg/g, respectively.

Results from Sim (2013) agree with Rajkumar et al. (2009) in which endophytes isolated from host plants exposed to high metal concentrations conferred similar metal tolerance to the endophytes. This presumably is due to adaptations by endophytes as a consequence of exposure of host plants to high metal concentrations. Nevertheless, there was no obvious evidence or significant difference in the superiority of metal tolerance of endophytes from plants found in areas polluted with heavy metal from endophytes from plants in unpolluted areas as reported by Shen et al. (2013). This suggests that although endophytes in host plants from polluted sites may 'acquire' metal-tolerance traits due to continuous exposure to metal stress, the endophytes are not necessarily advantaged by these circumstances.

From these discussions, it is evident that the use of microbial cells as biosorbents has numerous benefits that outweigh the use of other biosorbents. Microbes are ubiquitous, are resilient to a wide range of environmental conditions and can be sourced easily for use. Microbial biomass is also easy to culture using unsophisticated fermentation techniques and inexpensive

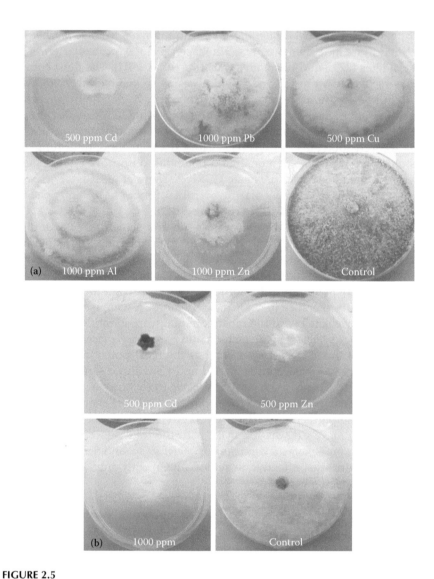

FIGURE 2.5
(See color insert.) Metal tolerance of endophytic (a) *T. asperellum* (isolate 11) and (b) *Saccharicola bicolour* (isolate 22) from *Phragmites* toward various metals at their maximum tolerable concentrations. Changes in pigmentation and colony diameter are the two most common responses of the endophytes to high metal concentrations.

(Continued)

growth media (under controlled conditions) to generate sufficient biomass (Kapoor and Viraraghavan 1995). In addition, microbial biosorbents can also be derived from waste or spent biomass from industries, thereby presenting an economically feasible solution (generating side incomes) to remove wastes (spent biomass) (Ahluwalia and Goyal 2007). Lastly, microbial biosorbents

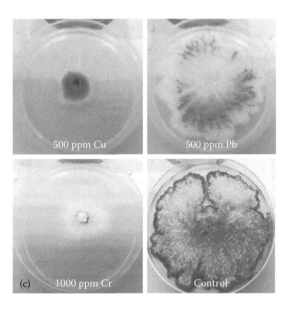

FIGURE 2.5 (CONTINUED)
(See color insert.) Metal tolerance of endophytic (c) *Phomopsis* sp. (isolate 9) from *Phragmites* toward various metals at their maximum tolerable concentrations. Changes in pigmentation and colony diameter are the two most common responses of the endophytes to high metal concentrations. (Compiled and modified from the study by Sim CSF, Metal tolerance and biosorption potential of endophytic fungi from Phragmites. Project Report, Monash University Malaysia, p. 47, 2013.)

are also amenable to genetic and morphological manipulations, which may result in the innovation of biosorbents.

2.3 Dead Microbial Cells for Metal Removal

Both live and dead cells have been used to remove metals, although comparisons will generally indicate that dead cells have better removal efficacy than live cells, as illustrated in Figure 2.6 (Ting and Choong 2009a,b; Huang et al. 2013). Dead cells are derived either by growing them until the lag phase, by autoclaving or by boiling (Puranik and Panikar 1999; Tan and Ting 2012). The dead cells remove metals via biosorption, a passive process that is independent of metabolic processes. In other words, metal biosorption of the dead cells occurs solely via metal binding to the surface of the cell biomass (Gadd 2009). As such, this process is no longer biological in nature but a physicochemical process. The biosorption efficacy is influenced by the structural components (functional groups) found on the surface of the cells, which include the carboxyl, phosphate, hydroxyl and amino groups

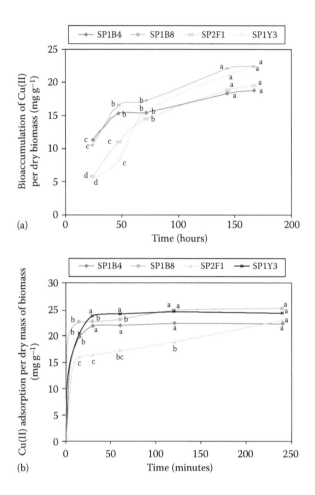

FIGURE 2.6
Removal efficacy of Cu by (a) live and (b) dead cells. Isolates SP1B4 and SP1B8 are bacterial isolates, and SP2F1 and SP1Y3 are fungal isolates. Means with the same letters for each isolate are not significantly different ($HSD_{(0.05)}$). (Compiled and modified from studies by Ting and Choong 2008.)

(Choudhary and Sar 2009; Gadd 2009). These structural components differ among the microbial cells with the peptidoglycan carboxyl groups serving as the primary binding sites for cations in Gram-positive bacteria while Gram-negative bacteria rely on the phosphate groups and chitins for fungi. The functional groups each have a different capacity to bind to the various metals and are dependent on biosorption conditions (pH, temperature, initial metal concentrations, agitation speed) (Avery et al. 1993; Vijayaraghavan and Yun 2008; Febrianto et al. 2009; Park et al. 2010).

Typically, the influence of pH is one of the main factors with poor metal biosorption observed in both low (acidic conditions) and high (alkaline conditions) pH conditions (Tan and Ting 2012). Under acidic conditions (pH < 6), the functional groups on the cell surface have positive charges, which are repulsive to the cationic metals. In alkaline conditions (pH > 7), most metals form precipitates (Gadd 2009; Tan and Ting 2012). Variations in metal preference and optimum uptake within species have been reported as a consequence of the response of functional groups to different pH conditions. For example, adsorption of Pb, Fe, Cr, Co, Cu, Zn and Cd by *Geobacillus thermodenitrificans* was optimum at pH 4.5, 6.5, 7.0, 4.5, 7.5, 5.0 and 6.0, respectively (Chatterjee et al. 2010). The metal-binding efficacy also differs under single and mixed metal conditions. The heterogeneous ions either elicit interactions between metal ions or compete for binding sites to the surface groups (Avery et al. 1993).

The use of dead cells as biosorbents is highly advantageous. They are more easily and cheaply produced, have broader application range and are more amenable to treatments to improve metal biosorption and recovery (Figure 2.7). Generation of dead biomass also removes the need for a continuous supply of costly nutrient feed for growth. In fact, some dead biomass can be procured from industries (fermentation wastes), providing alternatives for the industries to be rid of the waste. Dead cells can be used in all conditions (various pH, temperature and metal concentrations) for rapid metal removal (few minutes to hours) as biosorption occurs as a growth-independent process, not governed by physiological constraints nor

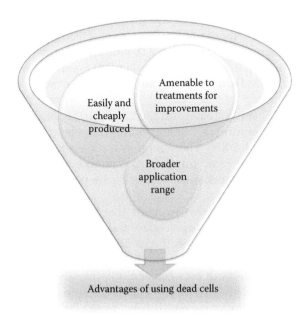

FIGURE 2.7
Highlights of the advantages of using dead cells as biosorbents.

limited by metal toxicity. In addition, dead cells are in innate forms, allowing a higher metal load leading to more efficient metal uptake (Ahluwalia and Goyal 2007). Dead cells are also more amenable to treatment of harsh chemicals for metal recovery (desorption) and for the regeneration and reuse of the biomass. Immobilisation of dead cells in various matrices also generates improved biosorbents that do not clog the water systems (Tan and Ting 2012). The broad application of dead cells as biosorbents, however, does have some limitations. For dead cells, no improvements on biological processes are possible, only chemical modifications to recover metals (Ahluwalia and Goyal 2007). Biosorbents from dead cells may also require desorption as functional sites on the surface of the biosorbents are rapidly exhausted in the metal binding processes. This early saturation phenomenon is a rate-limiting step unless desorption is introduced.

2.4 Isotherm and Kinetic Models for Biosorption Processes and Mechanisms

The efficacies of dead cells in adsorbing metals are also easier to predict as they can be fitted into equilibrium and kinetic models. Many models have been used to illustrate how metals are adsorbed onto biosorbents and removed, defining the adsorption equilibrium with biosorption. Modelling also enables the understanding of process mechanisms and how to optimise processes. The Freundlich (Equation 2.1), Langmuir (Equation 2.2), Redlich-Paterson (Equation 2.3) and Sips (Equation 2.4) equations are some examples (Table 2.1). Freundlich and Langmuir are the two more commonly used models as the former can describe the equilibrium relationship for most heterogeneous sorbent systems, and the latter further exemplifies the equilibrium model by defining that adsorption is monolayer and has saturation limits (Febrianto et al. 2009). These models are therefore useful to describe the capacity of sorbent to accumulate sorbate to equilibrium (Gadd 2009), identify the adsorption mechanism and compare pollutant uptake capacities of the various biosorbents.

The Freundlich model further explains that biosorption is influenced by pH, temperature, pore and particle size and their distribution, functional groups on the surface of the biosorbents and cation exchange capacity; the Langmuir model reiterates that the number of sites, accessibility, availability of sites and binding strength between sites and metals are factors determining the biosorption efficacy of biosorbents (Febrianto et al. 2009). However, both models have limitations. Freudlich is limited by the absence of saturation limit and thus is not applicable over a wide range of concentrations, and Langmuir is strictly based on monolayer assumptions and thus cannot be used

TABLE 2.1

Summary of Equilibrium and Kinetic Equations Used in Biosorption Studies

Equations	Models	Equations	References
Equation 2.1	Freundlich	$q_e = K_F C_e^{1/n}$	Aksu and Donmez (2006)
Equation 2.2	Langmuir	$q_e = q_{max} \dfrac{K_L C_e}{1 + K_L C_e}$	Aksu and Donmez (2006)
Equation 2.3	Redlich-Paterson	$q_e = \dfrac{K_{RP} C_e}{1 + a_{RP} C_e^{\beta}}$	Dursun (2006)
Equation 2.4	Sips	$q_e = q_{max} \dfrac{(K_S C_e)^{\gamma}}{\left(1 + (K_S C_e)^{\gamma}\right)}$	Vijayaraghavan et al. (2006)
Equation 2.5	Brunauer-Emmer-Teller (BET)	$q_e = q_{max} \dfrac{B C_e}{(C_e - C_s^*)\left[1 + (B-1)(C_e - C_s^*)\right]}$	Kiran and Kaushik (2008)
Equation 2.6	Pseudo first-order kinetic	$\ln(q_e - q) = \ln q_e - k_1 t$	Mukhopadhyay et al. (2007)
Equation 2.7	Pseudo second order kinetic	$t/q = (t/q_e) + (1/k_2 q_e^2)$	Ho (2006)

for heterogeneous surfaces. Therefore, in instances in which an estimation of biosorption of multimetals is required, modified multicomponent Langmuir and multicomponent Freundlich models are used (Aksu et al. 2002). This is to account for interference and competition between the various metals, which is a rather common occurrence in wastewater (Aksu et al. 2002; Aksu and Donmex 2006). Both the Langmuir and Freundlich models are, however, invalid for multimetal systems; instead the Brunauer-Emmett-Teller (BET) isotherm (Equation 2.5) (Table 2.1), derived from nonbiological systems, is more appropriately used to describe multilayer adsorption (Gadd 2009).

In addition to equilibria models, kinetic models are also used to determine the experimental data and the rate-limiting steps, which influence the mass transport and chemical reaction processes. Prediction of the rate of adsorption is crucial to the adsorption system design. Kinetic models are calculated from experimental data derived from batch studies evaluating the biosorption efficacy of various biosorbents and sorbate types. Linear regressions are then used to determine the best-fitting kinetic rate equation for the particular biosorbents–sorbate tested in response to varying initial concentrations as well as pH, temperature and agitation speeds among others (Ho 2006). Among the many models available, pseudo-first (Equation 2.6) and pseudo-second (Equation 2.7) order kinetic models are the more popular equations for heavy metal biosorption (Table 2.1). Pseudo-first order and second order compare the calculated adsorption value (q_e) to predicted q_e value with the former often having predicted q_e values lower than the experimental q_e values. On the contrary, pseudo-second-order models have calculated q_e values similar to experimental values and also have higher correlation coefficients greater than 0.98 (Mukhopadhyay et al. 2007). This is attributed to the fact that this model takes into account the interaction of valency forces between adsorbent–adsorbate, which is crucial to biosorption. Hence, in most kinetic studies in biosorption systems, compliance to the pseudo-second-order model is typically found.

2.5 Live Microbial Cells for Metal Removal

Live cells have several mechanisms of metal resistance and metal uptake. The uptake of metals by live cells is via the active mode known as bioaccumulation. To facilitate metal uptake, live cells excrete metal-binding metabolites, such as the polysaccharide-based extracellular polymeric substances (EPS) in various forms, which include capsules, slimes, sheaths and biofilms (Comte et al. 2008). Other beneficial substances expressed include bioactive molecules (methallothioneins, siderophores) and organic acids and biosurfactants, which all aid in metal binding processes. In addition, live cells detoxify metals via regulation of metal efflux systems and metal valence transformation and volatilisation, which all contribute to a more complete removal process (Figure 2.8). Although live microbial cells as biosorbents for metal removal have been attempted, reports often indicate that their metal removal efficacies are often inferior to those by dead cells. Nevertheless, the use of live cells does have benefits, particularly in achieving a more complete metal removal process via methylation and valence transformation, and live cells are also amenable to improvements on biological processes via genetic engineering.

Briefly, extracellular polymeric substances (EPS), which consist of protein, uronic acid and carbohydrate, are produced in the presence of high concentrations of metals. The increase in the production often coincided with the increase in the cellular diameter of metal-exposed cells, reflecting cellular response to the toxic effect of the metals (Puyen et al. 2012). The EPS can cohesively entrap, precipitate and adsorb metal particulates for metal removal due to their anionic nature, which attracts and binds easily to cationic metals (Comte et al. 2008). These negatively charged exopolymers are particularly efficient in binding metals such as Pb, Cd and U (Schiewer and Volesky 2000). This has been demonstrated in *Staphylococcus aureus*, *Micrococcus luteus* and *Azotobacter* spp. In addition, some bacteria produce siderophores, which bind metals, particularly metals similar to iron (Fe), such as aluminium (Al), gallium (Ga), Cr and Cu (Schiewer and Volesky 2000). Siderophores elevate toxicity by solubilising unavailable forms of heavy metal bearing Fe to form complexes that are mobile for assimilation. This is a profound method of ameliorating toxicity in hyperaccumulator plants in which siderophore-forming bacteria aid in root-mediated processes for uptake of complexed metals by plants (Rajkumar et

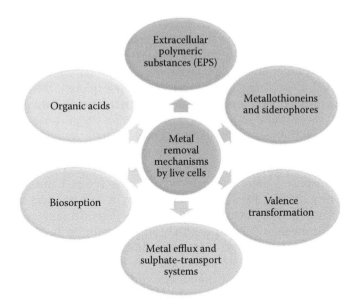

FIGURE 2.8
Various metal removal mechanisms expressed by live cell biosorbents.

al. 2010). As such, metal toxicity is reduced. The role of siderophores was also evident in Dimkpa et al. (2009), who observed that increased Cd and Cr uptake by cowpea was attributed to the addition of siderophore-containing culture filtrate of *Streptomyces tendae* F4 to metal-contaminated soils.

Heavy metal mobilisation was also enhanced by the secretion of organic acids, such as acetate, citrate, gluconate, 2-ketogluconate, malate, oxalate and succinate. Organic acids have the capacity to mobilise heavy metals, which dissolve metals for the subsequent uptake by plants. This was reported by Saravanan et al. (2007); 5-ketogluconic acid produced by the endophytic diazotroph *Gluconacetobacter diazotrophicus* effectively dissolves ZnO, $ZnCO_3$ or $Zn_3(PO_4)_2$ into bioavailable forms for plant uptake. Bacteria have also shown the capability to produce biosurfactants to reduce metal toxicity by forming biosurfactant-complexed metals. These complexed metals are less toxic to microbes, thus enhancing tolerance toward various metals (Maier and Soberon-Chavez 2000).

For metals accumulated in the cytoplasm, live cells produce intracellular metal resistance mechanisms to regulate toxicity. This includes metal efflux systems, production of methallothioneins and the regulation of the sulphate-transport systems. Metal efflux systems are typically reliable mechanisms, which work by pumping out metals from cells. This efflux system is regulated by genes specific to bacteria and the metals. For example, plasmid R773 encoding for arsenic regulator and ArsC arsenate reductase confers resistance to arsenic. For Cd, regulation by P-type ATPase, CadA or ZntA relieves Cd from within the cytoplasm of a variety of microbes, such as *Alcaligenes eutrophus, Bacillus subtilis, Escherichia coli, Listeria, Pseudomonas putida, Staphylococcus aureus*, cyanobacteria, fungi and algae (Roane et al. 2009). In addition to metal efflux, the role of metallothionein, a cysteine-rich protein molecule with high affinity for Cd, Zn, Cu, Hg and Ag, is extensively explored. Metallothioneins are induced by the presence of metal, and this mechanism produced by plants, algae, yeast, bacteria and some fungi is primarily for the purpose of detoxification. The metallothionein-metal deposits can be visually identified as electron-dense areas within the cell matrix (Roane et al. 2009).

In addition to metallothionein, metal detoxification can occur via the sulphate-transport system as observed in some fungi. Chromium (Cr) is reportedly accumulated within the cytoplasm of *Termitomyces clypeatus* by this mechanism (Das and Guha 2009), and *Penicillium* accumulates Cu in the spores via this mechanism (Kapoor and Viraraghavan 1995). Although metal accumulation in live cells aids in detoxification, the repercussion of such an active process resulted in the loss of cellular K, possibly a consequence of bacterial metal sequestration (Chaudhary and Sar 2009). The ionic exchange of other cations, such as Ca, as a result of Cd absorption, has also been documented (Manasi et al. 2014).

Other useful mechanisms displayed by live cells for metal removal include valence transformation and volatilisation mechanisms (Wu et al. 2010). In the valence transformation, specialised redox enzymes are excreted to convert toxic metals to lesser toxic forms. This mechanism was successfully adopted by *Bacillus* sp. SF-1 for the reduction of high concentrations of methylated Se into elemental Se (Kashiwa et al. 2001). This was also demonstrated by the mercury-resistant bacteria, in which organomercurial lyase (MerB) is produced to convert methyl-mercury to Hg, a form that is a hundredfold less toxic than methyl-mercury. In the volatilisation mechanism, metal ions are turned into volatile states to which microbes are less susceptible. Nevertheless, volatilisation is regulated by only a small pool of microbes and has only been reported for limited metals, such as Hg and metalloid Se (Wu et al. 2010).

In addition to the secretion of extracellular polymers and metal regulatory systems, functional groups on the surface of the microbial cells were discovered to influence metal uptake by live cells as well. This is contrary to the general perception that only dead cells rely on the functional groups for cation binding. Choudhary and Sar (2009) examined the role of functional groups in metal sorption and discovered carboxyl and phosphoryl groups to be primarily involved in the binding of Cu, Ni, Co and Cd. Gradual uptake of metals then proceeds intracellularly, often limited in live cells to allow accumulation up to only 6%–8% of cell dry weight. The binding and biosorption of metals on the surface of live cells are supported by several analyses using zeta potential, transmission electron microscopy (TEM), scanning electron microscopy (SEM) coupled with energy dispersive x-ray (EDX) and Fourier transform infrared spectroscopy (FTIR). Huang et al. (2013) revealed that Cd bioaccumulation by *B. cereus* RC-1 depended to a certain extent on the extracellular biosorption. For live cells, however, the biosorption process involved lesser functional groups than in dead cells. The EDX microanalysis employed to estimate the elemental content of the bacterial biomass further concurred with the occurrence of biosorption of metals on the surface of live cells with the spectra for metals (Ni, Co, Cu and Cd) detected. The biosorption mechanism thus is able to keep harmful metal ions out of the cell cytoplasm as well as regulating the bioaccumulation process in live cells (Wu et al. 2010).

Clearly, this section has highlighted that live microbial cells do have some benefits, although they are more susceptible to metal toxicity and adverse operating conditions (temperature, pH, nutrient supplementation) (Figure 2.9). Susceptibility to these conditions results in poor cell viability,

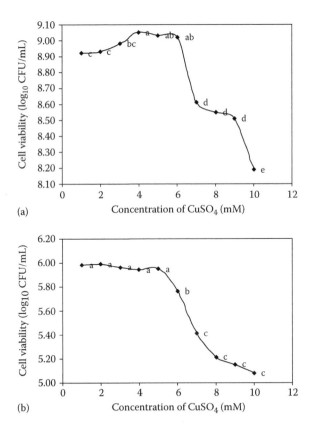

(a)

(b)

FIGURE 2.9

Susceptibility of live cells of (a) *Stenotrophomonas maltophilia* and (b) *Trichoderma asperellum* toward increasing concentrations of Cu (mM). Means with the same letters are not significantly different ($HSD_{(0.05)}$). (Compiled and modified from studies by Ting ASY and Choong CC, Screening for microbial candidates for Cu(II) bioremediation, *Research Report*, Universiti Tunku Abdul Rahman, 145 pp., 2008. [a] From Ting ASY and Choong CC, *World Journal of Microbiology and Biotechnology*, 25, 1431–1437, 2009a. [b] From Ting ASY and Choong CC, *Advances in Environmental Biology*, 3(2); 204–209, 2009b.)

implicating metal removal efficacy. Nevertheless, live microbial cells offer a more complete removal process because live cells can sorb, transport, complex and transform various metals (or metalloids, radionuclides) for removal (via biosorption, bioprecipitation). Live cells, with ongoing metabolic processes, allow the continuous uptake and removal of metals upon physical adsorption. The live cells are therefore most effectively used as a mixed consortium for treatment of samples in sewage treatment plants, in biofilm reactors, phytoremediation and bioremediation of soil and water (Gadd 2009). Live cells also perform better under aerobic conditions as some enzymes, such as metallothionein, are expressed under aerobic conditions. Nevertheless, the use of live microbial cells must be regulated and monitored, particularly with the recent discovery that metal binding activities generate mechanistic complications over time. Concerns arise from the release of EPS and metabolites in addition to the exhaustion of resources by live cells for respiration and nutrient uptake, which alters the microenvironment and may implicate cells. Such prolonged change to the microenvironment subsequently affects the biosorption efficacy of the cells and may also alter the speciation of target metals (Gadd 2009).

2.6 Innovations of Dead and Live Microbial Cells for Metal Removal

Innovations to biosorbents are usually for the purpose of scaling up the biosorption process (under optimum conditions) and improving their efficacy, regeneration and reuse (Wang and Chen 2009). Improvements are necessary because the continuous use of free cell forms is ineffective as free cells clog water systems, have poor durability and complicate solid–liquid separations for cell reuse and metal recovery. Free cell forms are also not feasible for repeated long-term usage as the cells have low density and poor mechanical strength (Gadd 2009). Therefore, in many large-scale industrial applications, immobilised or pelletised forms of microbial cells are highly desirable. This can be achieved via immobilisation with a polymer matrix or by immobilising cells into packed columns or fluidised bed reactors (Gadd 2009; Wang and Chen 2009). With regards to innovations to enhance biosorption processes, surface modifications can be implemented using heat or chemical pretreatments, surface grafting and layer-by-layer fattening. In recent years, unique biosorbent materials used to immobilise microbial cells, such as papaya wood, have also been explored. Innovations to live cells have also been attempted with manipulations to growth requirements and genetic engineering as the two main approaches adopted to enhance biosorption efficacy. Innovations to both live and dead cells are summarised in Figure 2.10.

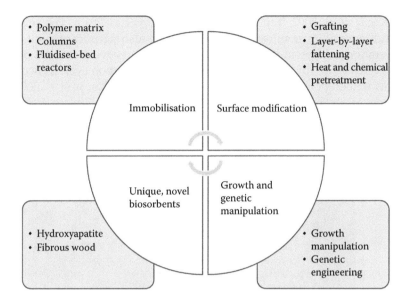

FIGURE 2.10
Innovations to live cells include immobilisation and via growth and genetic manipulations. Dead cells are modified via immobilisation (particularly on unique biosorbents) and surface modifications.

Immobilisation is the most extensively studied approach to improve efficacy and up-scaling the biosorption process. Immobilisation can be achieved via cell immobilisation onto inert carrier materials (activated carbon, sand, glass beads), within a polymer matrix (alginate, polyacrylamide, polymer matrices) and cross-linking to increase functional groups for adsorption. The immobilisation process imparts mechanical strength and resistance to chemical and microbial biodegradation, which is lacking in free cell forms. In immobilised forms, the biosorbent has the desired size, durable mechanical strength, rigidity and porosity for practical processes. Immobilised cells are also easy to use in a manner similar to the application of ion-exchange resins and activated carbons. For microbial cells, immobilisation is often conducted using polymer matrices, which are composed of carbohydrate and noncarbohydrate polymers. Typical examples of carbohydrate polymers are alginate, carboxymethylcellulose, chitin and chitosan, and noncarbohydrate polymers include silica, polyethyleneimine, glutaraldehyde, polypropylene, polyacrylamide, polyvinyl alcohol and polysulfone (Vijayaraghavan and Yun 2008; Park et al. 2010). Polymers selected for immobilisation must be economically feasible, promote biosorption efficacy and be able to withstand successive sorption–desorption cycles and repeated usage.

Alginate is one of the most commonly used immobilising agents that immobilise cells via entrapment. Microbial cells entrapped within alginate experience enhanced adsorptive capacity toward metal ions (Yan and Viraraghavan

2001; Kacar et al. 2002; Tan and Ting 2012; Ting et al. 2013). The regeneration and reuse of immobilised biosorbents were also significantly enhanced. Regeneration using acid (10 mM HCl) yields >90% recovery and has demonstrated reusability for up to three biosorption–desorption cycles, encountering only a negligible decrease in biosorption capacity (Kacar et al. 2002; Tan and Ting 2012). Figure 2.11 illustrates the alginate-immobilised *Trichoderma asperellum* and their regeneration efficiency after three sorption–desorption cycles consecutively. Another commonly used carbohydrate-based polymer

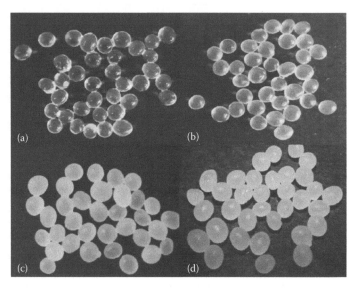

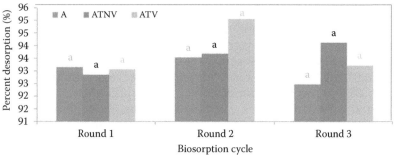

FIGURE 2.11

(See color insert.) Appearance of (a) plain alginate beads and (b) alginate immobilised fungal cells before and after Cu biosorption (c and d). Immobilisation allows recovery of Cu and regeneration of alginate beads for three consecutive cycles. A: Plain alginate beads, ATNV: alginate immobilised dead cells, ATV: alginate immobilised live cells. Means with the same letters within treatments (A, ATNV, ATV) are not significantly different ($HSD_{(0.05)}$). (Compiled and modified from studies by Tan WS and Ting ASY, Development of Cu(II) biosorbents via formulation of *Trichoderma* SP2F1 in alginate-immobilized systems. *Research Report*, Monash University Malaysia, 124 pp., 2011; Tan WS and Ting ASY, *Bioresource Technology*, 123, 290–295, 2012.)

is carboxymethyl cellulose (CMC). Immobilisation with CMC produced good biosorption, desorption and reusability traits as demonstrated by CMC-immobilised *Trametes versicolor* (Bayramoglu et al. 2003). Immobilisation with polymer matrices, such as alginate and CMC, bridges the gap between the biosorption efficacies displayed by live and dead cells as no significant differences in efficacies were found (Yalcinkaya et al. 2002; Bayramoglu et al. 2003; Tan and Ting 2012). In fact, most studies would reveal that the mere entrapment of cells, irrespective of whether dead or live cells are used, is observed to enhance the biosorption efficacy compared to plain polymers. This was observed in *T. versicolor* with immobilised live and heat-inactivated fungal mycelia recording 124 and 153 mg/g Cd, respectively, whereas the amount of Cd adsorbed on the plain CMC beads was only 43 mg/g (Yalcinkaya et al. 2002). Similarly, this was demonstrated by alginate-immobilised dead cells of *T. asperellum* with 134.22 mg/g Cu removed, compared to immobilised live cells and plain alginate beads (control) with 105.96 mg and 94.04 mg/g Cu adsorbed, respectively (Tan and Ting 2012). Immobilisation with the noncarbohydrate polymer matrix polyacrylamide gel also improved the biosorption–desorption of noble metals, such as gold (Au). The biosorption for Au by immobilised *P. maltophilia* cells was significantly enhanced with excellent durability and ability to withstand several biosorption–desorption cycles with 0.1 M thiourea solutions (Tsuruta 2004).

Thus, it appears that immobilisation, irrespective of the polymer matrix used, is the key innovative improvement to up-scale the use of microbial biosorbents. This, in turn, benefits many industries that rely on microbiological activities but are limited by the use of free cell forms. Immobilisation of microbial cells is also a more environmentally friendly strategy to remove heavy metal ions from aqueous solutions compared to conventional physicochemical methods. In retrospect, immobilisation is limited by mass transfer rate and the possible additional process cost from regeneration and removal using complex processes for coexisting ions (Wang and Chen 2006; Vijayaraghavan and Yun 2008). Nevertheless, these can be addressed easily, and if managed accordingly, the benefits of using immobilised microbial biosorbents outweigh the limitations.

The other approach to innovation is the application of chemical or physical treatments on the microbial cells to improve biosorption efficacy. Treatments are usually imposed on live cells to increase metal biosorption capacity through cell-wall modification. Dead cells are rarely treated simply because the treatment process kills cells, generating dead cells for use. Physical treatment methods are mostly based on heat and include heating (boiling/steam), freezing–thawing, autoclaving, and lyophilisation (freeze-drying) (Huang et al. 1988). Heat degrades cells and destroys the cell membranes, offering a larger surface area for metal binding. In some cases, the intracellular components are also exposed, providing more surface binding sites (Lopez-Errasquin and Vazquez 2003). Chemical treatment methods warrant live cells to come into contact with chemical reagents, such as acids (HCl, H_2SO_4,

HNO$_3$), alkalis (NaOH, KOH, detergents) and organic solvents (methanol, ethanol, formaldehyde, acetone toluene) (Park et al. 2010). The chemicals react with the surface characteristics/groups either by removing or masking the groups or by exposing more metal-binding sites, resulting in enhanced biosorption efficacy.

Microbial species are affected differently by the physical and chemical pretreatments applied, with significant improvements (or no changes) in removal efficacy typically observed. As the biosorption process involves mainly cell surface sequestration, modification to the cell walls via pretreatments can greatly alter the metal-binding capacity. For example, *A. niger* pretreated with detergent, sodium hydroxide (NaOH), formaldehyde and dimethyl sulphoxide (DMSO) demonstrated higher metal uptake capacity for Pb, Cd and Cu when compared with untreated cells (Kapoor and Viraraghavan 1995). On the contrary, pretreating *A. oryzae* with potassium hydroxide, formaldehyde and ethanol did not result in significant changes in the biosorption of Cd (Huang et al. 1988). Yeast (*S. cerevisiae*) cells also demonstrated different properties for metal accumulation upon treatment with chemical and physical agents (Lu and Wilkins 1996). This clearly supported the fact that different species are influenced differently by the agents of pretreatments, resulting in varying affinity and sorption efficacy toward different metals. Among the agents, alkali-based treatments are more likely to result in increased metal uptake capacity, whereas acid-based treatments have no influence on metal biosorption (Kapoor and Viraraghavan 1995; Wang 2002; Tan and Ting 2012) (Figure 2.12). Some of the chemical agents are less suitable as they cause detrimental effects in specific microbes. The use of methanol, formaldehyde and glutaraldehyde results in the esterification of carboxyl and methylation of amino groups present in the cell wall, decreasing biosorption capacity. This was observed in *S. cerevisiae* (Wang 2002).

Microbial biosorbents can also be manipulated to enhance biosorption by surface modification via grafting of long polymer chains onto the cell surface. Deng and Ting (2005) modified *P. chrysogenum* by performing graft polymerisation of acrylic acid (AAc) via thermal polymerisation of AAc onto the surface of cells pretreated with ozone, which led to increased sorption capacity for Cu and Cd. In their study, a large number of carboxyl groups were generated on the surface when carboxylic groups were converted to carboxylate ions using NaOH. This grafting exercise is pioneered by the ozone treatment, which generates peroxide and hydroperoxide species on the surface of the biomass. Under thermal induction, the functional groups degrade and form copolymerisation with AAc, resulting in long poly-AAc and increased binding sites. In addition, a novel approach in performing layer-by-layer fattening of functional groups on the surface of microbial cells has also been attempted. Luo et al. (2014) grafted layers of poly(allylamine hydrochloride) (PAA) onto the endophytic *Pseudomonas* sp. Lk9. This was performed by exposing the bacterial cells to PAA and glutaraldehyde (GA) with which cross-linking by GA results in PAA monolayer–modified cells.

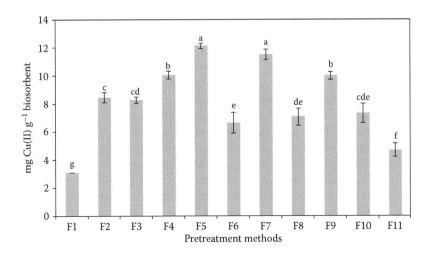

FIGURE 2.12

Cu removal efficacy by *T. asperellum* cells pretreated with acid and alkali reagents. F1: dried viable biomass, F2: autoclaved, F3: boiled in distilled water, F4: incubated in NaOH at room temperature, F5: boiled in NaOH, F6: incubated in detergent at room temperature, F7: boiled in detergent, F8: incubated in DMSO at room temperature, F9: boiled in DMSO, F10: incubated in acetic acid at room temperature, F11: boiled in acetic acid. Means with the same letters are not significantly different ($HSD_{(0.05)}$). Bars indicate standard error of means. (Compiled and modified from studies by Tan WS and Ting ASY, Development of Cu(II) biosorbents via formulation of *Trichoderma* SP2F1 in alginate-immobilized systems. *Research Report*, Monash University Malaysia, 124 pp., 2011; Tan WS and Ting ASY, *Bioresource Technology*, 123, 290–295, 2012.)

This is followed by activation by 4-bromobutyryl chloride (BC) to graft PAA to result in bilayer-modified cells. The success of the stepwise grafting was confirmed by FTIR, x-ray photoelectron spectroscopy (XPS) and elemental analysis. With layer fattening of functional groups, the metal-binding capacity was significantly enhanced with high uptake capacities for Cd and Cu and improved ability to withstand a wider pH range of 3–6. The layer-by-layer fattening approach also promotes stability of the improved biosorbents with stable and high sorption capacity retained even after five successive cycles. When tested with industrial effluents, the layered biosorbents effectively reduced metal concentrations to lower than 0.001 mg/L (Luo et al. 2014).

A possible alternative to surface grafting is the attempt to graft polymeric chains onto the surface of the biosorbents. Although this has not been attempted for microbial biosorbents, this method is attractive as it involves the possibility of grafting polymers to microbial surfaces or to polymer-based carrier/immobilisation materials, such as cellulose. Early discoveries on the benefit of chemically modifying and grafting cellulose on nonmicrobial-based biosorbents have proven to enhance adsorption capacity. This was achieved through direct grafting or the polymerisation of a monomer onto the surface, which introduces functional groups (carboxyl groups) onto the surface (O'Connell et al. 2008). This graft copolymerisation allows the attachment of side chain polymers to the main chain polymer structure covalently, resulting in increased mass and sorption sites. The use of grafting agents, such as acrylic acid, acrylamide and acrylonitrile, on a variety of nonmicrobial biosorbents, that is, sawdusts, banana stalks or cellulose beads, was able to increase adsorption of Cu, Ni, Cd and Pb. The potential of introducing this technology to microbial biosorbents remains to be discovered.

In addition to surface modifications, explorations on the use of unique biosorbent materials have also been attempted with varying degrees of success. Several unorthodox biosorbents have been used to immobilise microbial cells and their capacity for metal sorption evaluated. One such unique biosorbent is the naturally occurring hydroxyapatite $[Ca_{10}(PO_4)_6(OH)_2]$ (HAp) from fish bones. Piccirillo et al. (2013) attempted this by immobilising *Pseudomonas fluorescens* (S3X), *Microbacterium oxydans* (EC29) and *Cupriavidus* sp. (1C2) on the hydroxyapatite to determine their biosorption efficacy on Zn and Cd and in mixed solutions. It was discovered that immobilisation to HAp led to higher adsorption capacity compared to when HAp was used on its own with almost a fourfold increase. Another unique unorthodox biosorbent was developed but for fungal cells. This involved the entrapment of *Phanerochaete chrysosporium* in the structural fibrous network of papaya wood (SFNPW), which was able to remove Zn rapidly and efficiently compared to free cells with a maximum removal capacity of 66.17 mg/g dry cells at equilibrium against the 46.62 mg/g dry cells by free cell forms. SFNPW-immobilised fungal biosorbent was also stable, able to endure five adsorption–desorption cycles with HCl up to 99% recovery of the sorbed metal ions (Iqbal and Saeed 2006).

Improvements to live cells have also been extensively studied, and several beneficial approaches have yielded success. One of the approaches is by growth manipulation. In this approach, modifications or variations are introduced either during the growth of a microorganism or in the pregrown biomass, although it was evident that the former generates biosorbents with more enhanced biosorption activity than the latter (Stoll and Duncan 1996). These variations lead to changes in the composition of the cellular structure (cell components, surface phenol type), which, in turn, affect its biosorption potential. Cultural conditions are established as one of the main drivers of growth changes. Composition of cells and their subsequent biosorption activity are reportedly influenced by glucose, cysteine, glucose, ammonium sulphate, phosphate, ammonium chloride and limitations in elements such as C, N, P, S, Mg and K compositions in the cellular structure (Wang and Chen 2006). Glucose is a primary factor when supplementation of 20 mmol/L glucose to the growth stages of *S. cerevisiae* increased the metal removal efficiency of the cells by 30%–40% for the metals Cd, Cr, Cu, Pb and Zn (Mapolelo and Torto 2004). Manipulation (via glucose supplementation) is most effectively implemented during the active growth stages as introduction of glucose to the effluents (with pregrown biomass) did not enhance uptake of metals (Cu, Cr, Cd, Ni and Zn) (Stoll and Duncan 1996). Other than glucose, addition of L-cysteine into growing cultures is also observed to boost the biosorption capacity for

metals, such as silver (Ag), through the enhancement of protein and sulphydryl group content in the viable yeast cells (Singleton and Simmons 1996).

Contrary to supplementing nutrients as a means of modifying growing
cells and their subsequent biosorption efficacy, starving cells from certain
nutrients is also a strategy adopted to induce metal-binding capacity to the
various metals. Dostalek et al. (2004) performed an interesting analysis of
this matter using *S. cerevisiae*. Cells were grown in K-, Mg-, C-, N-, P- and
S-limited medium. It was observed that yeast cells in various limiting nutrients showed affinity to uptake various metals. K-limiting medium produced
cells that bind most abundantly with Cd and Cu, and affinity to Ag was demonstrated by cells cultured in P-limited medium. S-limited nutrients bind the
least amounts of Cu and Ag. This approach indicates that manipulation to
culture medium could modify the biosorption activity of the cells.

Another prominent improvement strategy implemented on live cells is via
the transgenic or genetic engineering approach in which microbes are tailored
or engineered to show high specificity toward a specific element or group of
elements. Genetic engineering is most useful in cases in which rapid removal
of metals is warranted. This is because, in nature, these microbes develop evolutionary capabilities to adapt to a wide range of chemicals, albeit at a relatively
slow rate, particularly when the acquisition of multiple catalytic activities is necessary. Therefore, genetic engineering accelerates these events. Early attempts
were confined to rhizosphere and symbiotic bacterial communities as they were
localised around roots or tissues, respectively, thereby restricting the transfer
of foreign genes and potential harm of genetic pollution (Wu et al. 2010). In
addition to test microbes, several molecules or genes have also been identified
as desirable for genetic manipulations. One of the early molecules of interest
is metallothioneins (MTs). In nature, MTs bind and sequester metals intracellularly in live cells. To improve metal binding capacity, expression of MTs on
the cell surface was attempted. Sousa et al. (1998) inserted MTs into site 153
of the LamB sequence, which resulted in hybrid proteins with MTs displayed
on the cell surface, consequently improving Cd binding to 15- to 20-fold. Chen
and Wilson (1997) discovered that coexpressing MTs together with their relevant metal (Hg) transport proteins (MerT and MerP) resulted in a significant
increase in the bioaccumulation of Hg. Bae et al. (2003) constructed genetically
engineered MerR in *E. coli*, which exhibited high affinity and selectivity toward
Hg. This was achieved using an ice nucleation protein anchor to the surface,
enabling sixfold higher sorption capacity for Hg compared to the wild type.

Alternatively, Kuroda et al. (2002) engineered histidine hexapeptide to the
cell surface of *S. cerevisiae*, which aided in chelating Cu significantly. The
design of these metal-binding peptides (cysteine-based peptides) leading
to better metal affinity and selectivity has been observed for Cd and Hg
(Klemba et al. 1994; Dieckmann et al. 1997). In addition to MTs and their
corresponding transport proteins, other genes influencing metal uptake or
the detoxification or tolerance to metals have also been identified as potential candidates for genetic engineering. This includes the Arr4p gene from

S. cerevisiae, which has a tolerance to metals, such as As, Co, Cr, Cu and VO (Shen et al. 2003), and genes that regulate cellular thiols, such as glutathione (GSH), phytochelatins [cadystins (γ-Glu-Cys)nGly], labile sulphide (Perego and Howell 1997; Gharieb and Gadd 2004) and the tripeptide glutathione (GSH) (Gharieb and Gadd 2004), which are responsible for detoxifying metal ions.

2.7 Conclusions

Microbial cells as biosorbents have many advantages attributed to their biosorptive nature, amenability to improvement (and reuse) and cost-effectiveness. Biosourcing for these beneficial isolates continuously exhausts diverse communities from both polluted and nonpolluted environments. Therefore, the emerging use of endophytes as biosorbents and the foray into recycling industrial spent biomass are important to meet development and advances in harnessing microbes for biosorbent development. Both types of cells (live and dead cells) make good biosorbents that command versatility, flexibility and wide application capable of reducing metals from ppm to ppb of the drinking water standard. The selection of the type of cells (biomass) to be used depends on the purpose and availability and the metal-removal strategy. Live cells have several physiological limitations absent in dead cells. Nevertheless, both live and dead cells are amenable to continuous innovations, which contribute to improvements in achieving better metal removal efficacy. Improvements based on physicochemical alterations to the surface structures are considered to be of primary focus as the more sophisticated methods involving genetic engineering have met with poor response due to environmental health and safety concerns. In addition, intensive research into modifying biosorbents to enhance metal uptake must also consider the implications of the cost incurred as well as the issues of the wastes (loaded biosorbents) generated, which are toxic and detrimental to the environment. All these can be mitigated by proper planning and the increasing understanding of the fundamental mechanistic principles of microbes as biosorbents, which perhaps in the future would lead to the emergence of novel solutions for improved metal-removal processes.

2.8 Perspectives

Research in developing microbes as biosorbents for metals is driven by the discovery of new microbial communities (endophytes, sulphate-reducing bacteria, yeasts); exploits of their innate ability to tolerate, accumulate and sequester metals via extracellular and intracellular mechanisms; and their amenability to modifications (surface, genetic alterations) to generate durable biomass for repeated use. Every microbial biosorbent is 'unique' in the sense that it responds differently to various metals, expressing a range of metal affinity and sorption capacity. This varying response to metals is influenced by functional groups, pretreatments, modifications to cell surface and other conditions (temperature, pH, initial metal concentration). These fundamental mechanistic principles of microbes as biosorbents can either be interpreted as an option for manipulation of solute conditions to optimise metal sorption or it could be a limiting factor that results in inconsistent sorption efficacy when applied in various environmental conditions. Thus, microbial biosorbents, although they were established many years ago, are still an interesting biosorption alternative that garners a steady interest among researchers today.

Acknowledgements

The author thanks the Malaysian Ministry of Education for the Fundamental research Grant Scheme (FRGS/2/2013/STWN01/MUSM/02/2) awarded to conduct studies on metal biosorption. The author is also grateful to Monash University Malaysia for the facilities and financial assistance that enabled the pursuit of understanding metal biosorption in various microbes.

References

Agarwal GS, Bhuptawat HK and Chaudhari S. 2006. Biosorption of aqueous chromium(VI) by *Tamarindus indica* seeds, *Bioresource Technology* 97: 949–956.

Ahluwalia SS and Goyal D. 2007. Microbial and plant derived biomass for removal of heavy metals from wastewater, *Bioresource Technology* 98: 2243–2257.

Aksu Z, Acikel U, Kabasakal E and Tezer S. 2002. Equilibrium modelling of individual and simultaneous biosorption of chromium(VI) and nickel(II) onto dried activated sludge, *Water Research* 36: 3063–3073.

Aksu Z and Balibek E. 2007. Chromium(VI) biosorption by dried *Rhizopus arrhizus*: Effect of salt (NaCl) concentration on equilibrium and kinetic parameters, *Journal of Hazardous Materials* 145: 210–220.

Aksu Z and Donmez G. 2006. Binary biosorption of cadmium(II) and nickel(II) onto dried *Chlorella vulgaris*: Co-ion effect on mono-component isotherm parameters, *Process Biochemistry* 41: 860–868.

Avery SV, Codd GA and Gadd GM. 1993. Biosorption of tributyltin and other organotin compounds by cyanobacteria and microalgae, *Applied Microbiology and Biotechnology* 39: 812–817.

Babu AG, Kim JD and Oh BT. 2013. Enhancement of heavy metal phytoremediation by *Alnus firma* with endophytic *Bacillus thuringiensis* GDB-1, *Journal of Hazardous Materials* 250–251: 477–483.

Bae W, Wu CH, Kostal J, Mulchandani A and Chen W. 2003. Enhanced mercury biosorption by bacterial cells with surface-displayed MerR, *Applied and Environmental Microbiology* 69(6): 3176–3180.

Bayramoglu G and Arica MY. 2008. Removal of heavy mercury(II), cadmium(II) and zinc(II) metal ions by live and heat inactivated *Lentinus edodes* pellets, *Chemical Engineering Journal* 143: 133–140.

Bayramoglu G, Bekta S and Arica MY. 2003. Biosorption of heavy metal ions on immobilized white-rot fungus *Trametes versicolor*, *Journal of Hazardous Materials* B101: 285–300.

Baysal Z, Ercan C, Bulut Y, Alkan H and Dogru M. 2009. Equilibrium and thermodynamic studies on biosorption of Pb(II) onto *Candida albicans* biomass, *Journal of Hazardous Materials* 161: 62–67.

Chatterjee SK, Bhattacharjee I and Chandra G. 2010. Biosorption of heavy metals from industrial waste water by *Geobacillus thermodenitrificans*, *Journal of Hazardous Materials* 175: 117–125.

Chen S and Wilson DB. 1997. Construction and characterization of *Escherichia coli* genetically engineered for bioremediation of Hg($^{2+}$)-contaminated environments, *Applied Environmental Microbiology* 63(6): 2442–2445.

Choudhary S and Sar P. 2009. Characterization of a metal resistant *Pseudomonas* sp. isolated from uranium mine for its potential in heavy metal (Ni^{2+}, Co^{2+}, Cu^{2+}, and Cd^{2+}) sequestration, *Bioresource Technology* 100: 2482–2492.

Comte S, Guibaud G and Baudu M. 2008. Biosorption properties of extracellular polymeric substances (EPS) towards Cd, Cu and Pb for different pH values, *Journal of Hazardous Materials* 151: 185–193.

Dal Bosco SM, Jimenez RS, Vignado C, Fontana J, Geraldo B, Figueiredo FCA, Mandelli D and Carvalho WA. 2006. Removal of Mn(II) and Cd(II) from wastewaters by natural and modified clays, *Adsorption* 12: 133–146.

Das SK and Guha AK. 2009. Biosorption of hexavalent chromium by *Termitomyces clypeatus* biomass: Kinetics and transmission electron microscopic study, *Journal of Hazardous Materials* 167: 685–691.

Deng Z, Cao LX, Huang HW, Jiang XY, Wang WF, Shi Y and Zhang RD. 2011. Characterization of Cd- and Pb-resistant fungal endophyte *Mucor* sp. CBRF59 isolated from rapes (*Brassica chinensis*) in a metal-contaminated soil, *Journal of Hazardous Materials* 185: 717–724.

Deng S and Ting YP. 2005. Fungal biomass with grafted poly(acrylic acid) for enhancement of Cu(II) and Cd(II) biosorption, *Langmuir* 21(13): 5940–5948.

Dieckmann GR, McRorie DK, Tierney DL, Utschig LM, Singer CP, O'Halloran TV, Penner-Hahn JE, DeGrado WF and Pecoraro VL. 1997. De novo design of mercury-binding two- and three-helical bundles, *Journal of American Chemical Society* 119: 6195–6196.

Dimkpa CO, Merten D, Svatoš A, Büchel G and Kothe E. 2009. Siderophores mediate reduced and increased uptake of cadmium by *Streptomyces tendae* F4 and sunflower (*Helianthus annuus*), respectively, *Journal of Applied Microbiology* 107(5): 1687–1696.

Dostalek P, Patzak M and Matejka P. 2004. Influence of specific growth limitation on biosorption of heavy metals by *Saccharomyces cerevisia*, *International Biodeterioration and Biodegradation* 54(2–3): 203–207.

Dursun AY. 2006. A comparative study on determination of the equilibrium, kinetic and thermodynamic parameters of biosorption of copper(II) and lead(II) ions onto pre-treated *Aspergillus niger, Biochemical Engineering Journal* 28: 187–195.

Febrianto J, Kosasiha AN, Sunarso J, Ju YH, Indraswati N and Ismadji S. 2009. Equilibrium and kinetic studies in adsorption of heavy metals using biosorbent: A summary of recent studies, *Journal of Hazardous Materials* 162: 616–645.

Gabr RM, Hassan SHA and Shoreit AAM. 2008. Biosorption of lead and nickel by living and non-living cells of *Pseudomonas aeruginosa ASU* 6a, *International Biodeterioration and Biodegradation* 62: 195–203.

Gadd GM. 2009. Biosorption: Critical review of scientific rationale, environmental importance and significance for pollution treatment, *Journal of Chemical Technology and Biotechnology* 84:13–28.

Gadd GM and Fomina M. 2011. Uranium and fungi, *Geomicrobiology Journal* 28: 471–482.

Gharieb MM and Gadd GM. 2004. Role of glutathione in detoxification of metal(loid)s by *Saccharomyces cerevisiae, Biometals* 17(2): 183–188.

Gray EJ and Smith DL. 2005. Intracellular and extracellular PGPR: Commonalities and distinctions in the plant–bacterium signalling processes, *Soil Biology and Biochemistry* 37(3): 395–412.

Guo HJ, Luo SL, Chen L, Xiao X, Xi Q, Wei WZ, Zeng GM, Liu CB, Wan Y, Chen JL and He YJ. 2010. Bioremediation of heavy metals by growing hyperaccumulator endophytic bacterium *Bacillus* sp. L14, *Bioresource Technology* 101: 8599–8605.

Halttunen T, Salminen S and Tahvonen R. 2007. Rapid removal of lead and cadmium from water by specific lactic acid bacteria, *International Journal of Food Microbiology* 114: 30–35.

Ho Y. 2006. Second-order kinetic model for the sorption of cadmium onto tree fern: A comparison of linear and non-linear methods, *Water Research* 40: 119–125.

Huang CP, Westman D, Huang C and Morehart AL. 1988. The removal of cadmium(H) from dilute aqueous solutions by fungal adsorbent, *Water Science Technology* 20: 369–379.

Huang F, Dang Z, Guo CL, Lu GN, Gu RR, Liu HJ and Zhang H. 2013. Biosorption of Cd(II) by live and dead cells of *Bacillus cereus* RC-1 isolated from cadmium-contaminated soil, *Colloids and Surfaces B: Biointerfaces* 107: 11–18.

Iqbal M and Saeed A. 2006. Entrapment of fungal hyphae in structural fibrous network of papaya wood to produce a unique biosorbent for the removal of heavy metals, *Enzyme and Microbial Technology* 39: 996–1001.

Kacar Y, Arpa CI, Tan S, Denizli A, Gen O and Arica MY. 2002. Biosorption of Hg(II) and Cd(II) from aqueous solutions: Comparison of biosorptive capacity of alginate and immobilized live and heat inactivated *Phanerochaete chrysosporium, Process Biochemistry* 37: 601–610.

Kapoor A and Viraraghavan T. 1995. Fungal biosorption – An alternative treatment option for heavy metal bearing wastewaters: A review, *Bioresource Technology* 53: 195–206.

Kashiwa M, Ike M, Mihara H, Esaki N and Fujita M. 2001. Removal of soluble selenium by a selenate-reducing bacterium *Bacillus* sp. SF-1, *Journal of Fermentation and Bioengineering* 83: 517–522.

Kieu HTQ, Muller E and Horn H. 2011. Heavy metal removal in anaerobic semi-continuous stirred tank reactors by a consortium of sulfate-reducing bacteria, *Water Research* 45: 3863–3870.

Kiran B and Kaushik A. 2008. Chromium binding capacity of *Lyngbya putealis* exopolysaccharides, *Biochemical Engineering Journal* 38: 47–54.

Klemba M, Gardner KH, Marino S, Clarke ND and Regan L. 1994. Novel metal-binding proteins by design, *Nature Structural Biology* 2: 368–373.

Kuroda K, Ueda M, Shibasaki S and Tanaka A. 2002. Cell surface-engineered yeast with ability to bind, and self-aggregate in response to copper ion, *Applied Environmental Microbiology* 59 (2–3): 259–264.

Li HY, Li DW, He CM, Zhou ZP, Mei T and Xu HM. 2012. Diversity and heavy metal tolerance of endophytic fungi from six dominant plant species in a Pb–Zn mine wasteland in China, *Fungal Ecology* 5: 309–315.

Lin Y, Wang X, Wang B, Mohamad O and Wei G. 2012. Bioaccumulation characterization of zinc and cadmium by *Streptomyces zinciresistens*, a novel actinomycete, *Ecotoxicology and Environmental Safety* 77: 7–17.

Lopez-Errasquin E and Vazquez C. 2003. Tolerance and uptake of heavy metals by *Trichoderma atroviride* isolated from sludge, *Chemosphere* 50: 137–143.

Lu YM and Wilkins E. 1996. Heavy metal removal by caustic-treated yeast immobilized in alginate, *Journal of Hazardous Materials* 49(2–3): 165–179.

Luo SL, Chen L, Chen JL, Xiao X, Xu TY, Wan Y, Rao C, Liu CB, Liu YT, Lai C and Zeng GM. 2011. Analysis and characterization of cultivable heavy metal-resistant bacterial endophytes isolated from Cd-hyperaccumulator *Solanumnigrum* L. and their potential use for phytoremediation, *Chemosphere* 85: 1130–1138.

Luo SL, Li XJ, Chen L, Chen JL, Wan Y and Liu CB. 2014. Layer-by-layer strategy for adsorption capacity fattening of endophytic bacterial biomass for highly effective removal of heavy metals, *Chemical Engineering Journal* 239: 312–321.

Ma Y, Prasad MNV, Rajkumar M and Freitas H. 2011. Plant growth promoting rhizobacteria and endophytes accelerate phytoremediation of metalliferous soils, *Biotechnology Advances* 29: 248–258.

Maier R and Soberon-Chavez G. 2000. *Pseudomonas aeruginosa* rhamnolipids: Biosynthesis and potential applications, *Applied Microbiology and Biotechnology* 54: 625–633.

Manasi RV, Kumar ASK and Rajesh N. 2014. Biosorption of cadmium using a novel bacterium isolated from an electronic industry effluent, *Chemical Engineering Journal* 235: 176–185.

Mapolelo M and Torto N. 2004. Trace enrichment of metal ions in aquatic environments by *Saccharomyces cerevisiae*. *Talanta*. 64(1): 39–47.

Mukhopadhyay M, Noronha SB and Suraiskhumar GK. 2007. Kinetic modeling for the biosorption of copper by pretreated *Aspergillus niger* biomass, *Biosource Technology* 98: 1781–1787.

O'Connell DW, Birkinshaw C and O'Dwyer TF. 2008. Heavy metal adsorbents prepared from the modification of cellulose: A review, *Bioresource Technology* 99: 6709–6724.

Oves M, Khan MS and Zaidi A. 2013. Biosorption of heavy metals by *Bacillus thuringiensis* strain OSM29 originating from industrial effluent contaminated north Indian soil, *Saudi Journal of Biological Sciences* 20: 121–129.

Pardo R, Herguedas M, Barrado E and Vega M. 2003. Biosorption of cadmium, copper, lead and zinc by inactive biomass of *Pseudomonas putida*, *Analytical and Bioanalytical Chemistry* 376: 26–32.

Park D, Yun Y, Jo JH and Park JM. 2006. Biosorption process for treatment of electroplating wastewater containing Cr(VI): Laboratory-scale feasibility test, *Industrial and Engineering Chemistry Research* 45: 5059–5065.

Park D, Yun Y and Park JM. 2010. The past, present, and future trends of biosorption, *Biotechnology and Bioprocess Engineering* 15: 86–102.

Perego P and Howell SB. 1997. Molecular mechanisms controlling sensitivity to toxic metal ions in yeast, *Toxicology and Applied Pharmacology* 147(2): 312–318.

Piccirillo C, Pereira SIA, Marques APGC, Pullar RC, Tobaldi DM, Pintado ME and Castro PML. 2013. Bacteria immobilisation on hydroxyapatite surface for heavy metals removal, *Journal of Environmental Management* 121: 87–95.

Puranik PR and Paknikar KM. 1999. Biosorption of lead, cadmium, and zinc by *Citrobacter* strain MCM B-181: Characterization studies, *Biotechnology Progress* 15(2): 228–237.

Puyen ZM, Villagrasa E, Maldonado J, Diestra E, Esteve I and Solé A. 2012. Biosorption of lead and copper by heavy-metal tolerant *Micrococcus luteus* DE2008, *Bioresource Technology* 126: 233–237.

Rajkumar M, Noriharu A and Freitas H. 2009. Endophytic bacteria and their potential to enhance heavy metal phytoextraction, *Chemosphere* 77: 153–160.

Rajkumar MN, Prasad MNV and Freitas H. 2010. Potential of siderophore-producing bacteria for improving heavy metal phytoextraction, *Trends in Biotechnology* 28: 142–149.

Roane TM, Rensing C, Pepper IL and Maier RM. 2009. Microorganisms and metal pollutants. In Maeir RM and Pepper I (eds.), *Environmental Microbiology*. Academic Press, Inc. pp. 421–441.

Sahoo DK, Kar RN and Das RE. 1992. Bioaccumulation of heavy metal ions by *Bacillus circulans*, *Bioresource Technology* 41: 177–179.

Saravanan S, Madhaiyan M and Thangaraju M. 2007. Solubilization of zinc compounds by the diazotrophic, plant growth promoting bacterium *Gluconacetobacter diazotrophicus*, *Chemosphere* 66(9): 1794–1798.

Satapathy D and Natarajan GS. 2006. Potassium bromate modification of the granular activated carbon and its effect on nickel adsorption, *Adsorption* 12: 147–154.

Schiewer S and Volesky B. 2000. Biosorption processes for heavy metal removal. In: Lovley DR (ed.), *Environmental Microbe-Metal Interactions*. ASM press, Washington, DC, pp. 329–362.

Schütze E, Klos M, Merten D, Nietzsch S, Senftleben D, Roth M and Kothe E. 2014. Growth of streptomycetes in soil and their impact on bioremediation, *Journal of Hazardous Materials* 267: 128–135.

Shen J, Hsu CM, Kang BK, Rosen BP and Bhattacharjee H. 2003. The *Saccharomyces cerevisiae* Arr4p is involved in metal and heat tolerance, *Biometals* 16(3): 369–378.

Shen M, Liu L, Li DW, Zhou WN, Zhou ZP, Zhang CF, Luo YY, Wang HB and Li HY. 2013. The effect of endophytic *Peyronellaea* from heavy metal-contaminated and uncontaminated sites on maize growth, heavy metal absorption and accumulation, *Fungal Ecology* 6: 539–545.

Sim CSF. 2013. Metal tolerance and biosorption potential of endophytic fungi from *Phragmites*. Project Report, Monash University Malaysia, 47 pp.

Singleton I and Simmons P. 1996. Factors affecting silver biosorption by an industrial strain of *Saccharomyces cerevisiae*, *Journal of Chemical Technology and Biotechnology* 65(1): 21–28.

Sousa C, Kotrba P, Ruml T, Cebolla A and De Lorenzo V. 1998. Metallo adsorption by *Escherichia coli* cells displaying yeast and mammalian metallothioneins anchored to the outer membrane protein LamB, *Journal of Bacteriology* 180(9): 2280–2284.

Stoll A and Duncan JR. 1996. Enhanced heavy metal removal from waste water by viable, glucose pretreated *Saccharomyces cerevisiae* cells, *Biotechnology Letters* 18(10): 1209–1212.

Tan WS and Ting ASY. 2011. Development of Cu(II) biosorbents via formulation of *Trichoderma* SP2F1 in alginate-immobilized systems. *Research Report*, Monash University Malaysia, 124 pp.

Tan WS and Ting ASY. 2012. Efficacy and reusability of alginate-immobilized live and heat-inactivated *Trichoderma asperellum* cells for Cu (II) removal from aqueous solution, *Bioresource Technology* 123: 290–295.

Ting ASY, Abdul Rahman NH, Mahamad IMIH and Tan WS. 2013. Investigating metal removal potential by effective microorganisms (EM) in alginate-immobilized and free-cell forms, *Bioresource Technology* 147: 636–639.

Ting ASY and Choong CC. 2008. Screening for microbial candidates for Cu(II) bioremediation, *Research Report*, Universiti Tunku Abdul Rahman, 145 pp.

Ting ASY and Choong CC. 2009a. Bioaccumulation and biosorption efficacy of *Trichoderma* isolate SP2F1 in removing copper (Cu(II)) from aqueous solutions, *World Journal of Microbiology and Biotechnology* 25: 1431–1437.

Ting ASY and Choong CC. 2009b. Utilization of non-viable cells compared to viable cells of *Stenotrophomonas maltophilia* for copper (Cu(II)) removal from aqueous solutions, *Advances in Environmental Biology* 3(2): 204–209.

Ting ASY, Lim SJ and Tan WS. 2011. Diversity and metal tolerance of filamentous fungi from analytical wastewater from laboratory. *Proceedings of the International Congress of the Malaysian Society for Microbiology*. 8–11 December 2011, Penang, Malaysia, pp. 110–113.

Tsuruta T. 2004. Biosorption and recycling of gold using various microorganisms, *Journal of General and Applied Microbiology* 50(4): 221–228.

Vijayaraghavan K, Padmesh TVN, Palanivelu K and Velan M. 2006. Biosorption of nickel(II) ions onto *Sargassum wightii*: Application of two-parameter and three-parameter isotherm models, *Journal of Hazardous Materials* B133: 304–308.

Vijayaraghavan K and Yun YS. 2008. Bacterial biosorbents and biosorption, *Biotechnology Advances* 26: 266–291.

Viraraghavan T and Srinivasan A. 2011. Fungal biosorption and biosorbents. In: Kotrba P, Mackova M and Macek T. (eds.), *Microbial Biosorption of Metals*. Springer: The Netherlands, pp. 143–158.

Wang JL. 2002. Biosorption of copper(II) by chemically modified biomass of *Saccharomyces cerevisiae, Process Biochemistry* 37: 847–850.

Wang JL and Chen C. 2006. Biosorption of heavy metals by *Saccharomyces cerevisiae*: A review, *Biotechnology Advances* 24: 427–451.

Wang J and Chen C. 2009. Biosorbents for heavy metals removal and their future, *Biotechnology Advances* 27: 195–226.

Wu G, Kang HB, Zhang XY, Shao HB, Chu LY and Ru CJ. 2010. A critical review on the bio-removal of hazardous heavy metals from contaminated soils: Issues, progress, eco-environmental concerns and opportunities, *Journal of Hazardous Materials* 174: 1–8.

Yalcinkaya Y, Soysal L, Denizli A, Arica MY, Bektas S and Gen O. 2002. Biosorption of cadmium from aquatic systems by carboxy methylcellulose and immobilized *Trametes versicolor, Hydrometallurgy* 63: 31–40.

Yan G and Viraraghavan T. 2001. Heavy metal removal in a biosorption column by immobilized *M. rouxii* biomass, *Bioresource Technology* 78: 243–249.

Zafar S, Aqil F and Ahmad I. 2007. Metal tolerance and biosorption potential of filamentous fungi isolated from metal contaminated agricultural soil, *Bioresource Technology* 98: 2557–2561.

3

Microbial Degradation of Aromatic Compounds and Pesticides: Challenges and Solutions

Randhir Singh, Rohini Karandikar and Prashant S. Phale

CONTENTS

3.1 Introduction

The Industrial and Green Revolutions brought about worldwide developments in the industrial and agricultural sectors. Since then, aromatic compounds have become an integral part of our environment due to heavy use of various chemicals, fuels and pesticides. Most of these compounds are pollutants and are generated via natural or anthropogenic sources. The natural sources include forest fire, oil spills and seepage, volcanic eruptions and plant exudates, while the anthropogenic sources are pyrolysis of organic

compounds, incomplete combustion of fuel or garbage and industrial effluents (Gilbson 1968; Cerniglia 1993).

The simplest aromatic compound is benzene. However, in complex compounds, the aromatic rings can be fused in linear or angular arrangements
to give more hydrophobic, less water-soluble and recalcitrant compounds
called polycyclic aromatic hydrocarbons (PAHs). The persistence of aromatic compounds in air, water and soil is dependent on their molecular
weight. The high molecular weight compounds are adsorbed easily onto
particulate matters, while the low molecular weight compounds predominantly remain in the gaseous state. The weak solubility of these
compounds in water leads to their accumulation in soil and sediments
(Skupinska et al. 2004). Upon exposure, these compounds easily enter the
body via the oral, nasal, dermal or ocular route and impart neurological,
reproductive and developmental disorders (Zmirou et al. 2000; Skupinska
et al. 2004). Some of these aromatics, such as anthracene, chrysene, benzo(a)
pyrene and benzo(a)anthracene, are toxic, mutagenic and carcinogenic in
nature (Gilbson 1968; Redmond 1970; Blummer and Youngblood 1975). In
the human body, these compounds are metabolised by cytochrome P450-
mediated mixed-function oxidases, yielding epoxides and phenols (Conney
1982; Goldman et al. 2001; Stegeman et al. 2001). These epoxides and phenols further react with DNA bases to form DNA adducts (Garner 1998) and
cause mutations in the oncogenes or tumor suppressor genes leading to
cancer (Goldman et al. 2001).

Due to their toxic, mutagenic and carcinogenic effects, the aromatic
compounds are of environmental concern. Several abiotic (for example,
chemical oxidation, volatilisation, incineration, photo-catalytic oxidation
and immobilised enzymes) and biotic (use of microorganisms) degradation methods have been used for their removal (Bertilsson and Widenfalk
2002; Watts et al. 2002; Wen et al. 2002; Eriksson et al. 2003; Rivas 2006;
Urgun-Demirtas et al. 2006; Acevedo et al. 2010). Compared to biotic methods, abiotic methods are costly and inefficient in complete removal. The
intermediate products produced during abiotic degradation are often more
toxic and recalcitrant than the parent compounds (Elespuru et al. 1974; Shea
and Berry 1983; Wilson et al. 1985; Obulakondaiah et al. 1993). However,
the use of microorganisms appears to be a self-sustaining and inexpensive
cleanup technology (biodegradation) due to their ability to degrade these
compounds into nontoxic products. Detailed description of biodegradation
of various major pollutants, such as mono-, di- or polycyclic aromatic, halogenated and nitrogen-containing pesticides, is given in this chapter (Figure
3.1 and Table 3.1).

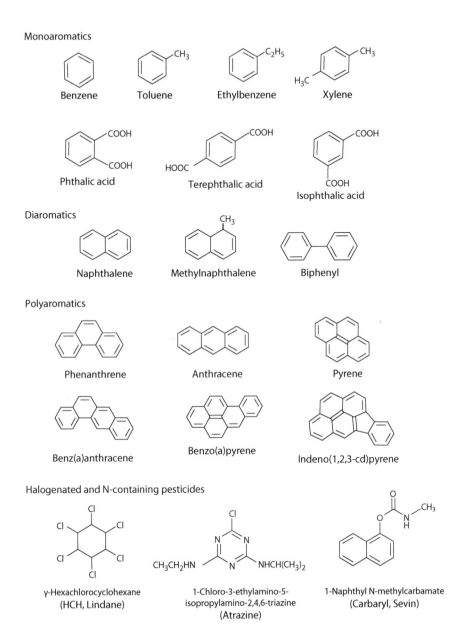

FIGURE 3.1
Structures of various hazardous chemical compounds.

TABLE 3.1

Aromatic Compounds and Pesticides: LC$_{50}$ Values, Hazardous Effects and Microbial Degraders

Compound	LC$_{50}$ in Animal Models via Routes			Hazardous Effects	Degrading Bacteria[v]	References
	Oral	Dermal	Inhalation			
Benzene	930 mg/kg (rat), 4700 mg/kg (mouse)	9400 mg/kg (rabbit), 10,000 ppm 7 h (Rat)	na*	Myeloma, leukaemia, delayed bone formation, reduced fetal body weight, bone marrow damage, etc.	*Polaromonas* sp., *P. monteilii, Acidobacterium* sp., *P. putida* F1, *P. mendocina* KR, *Ralstonia pickettii* PKO1, *Burkholderia cepacia* G4, *Sphingomonas* sp. D3K1, *Arthrobacter* sp., *Rhodococcus erythropolis*	Alagappan and Cowan (2003); Fahy et al. (2008); Xie et al. (2011); Dueholm et al. (2014)
Toluene	636 mg/kg (rat)	8390 mg/kg (rabbit)	26,700 ppm/1 h (rat)	Headache, dizziness, decreased mental ability, irritation to skin and eyes, damage to liver, brain, kidney and CNS	*P. putida, P. alcaligenes, P. mendocina* F1, *P. pickettii* PKO1, *B. cepacia* G4, *Thauera* sp. DNT, *Pseudoxanthomonas spadix* BD, *P. monteilii*	Wackett et al. (1988); Yen et al. (1991); Shaw and Harayama (1992); Shields et al. (1995); Parales et al. (2000); Shinoda et al. (2004); Choi et al. (2013); Dueholm et al. (2014)
Ethylbenzene	3500 mg/kg (rat)	17,800 µl/kg (rabbit)	17.8 mg/4 h (rat)	Eye and throat irritation, vertigo, inner ear, renal damage, cancer	*P. putida* 39/D, *Pseudomonas* NCIB 98164, *P. monteilii, Pseudoxanthomonas spadix*	Gibson et al. (1973); Lee and Gibson (1996); Choi et al. (2013); Dueholm et al. (2014)

Xylene	4988 mg/kg (mouse)	14,000 µl/kg (rabbit)	5267 ppm/6 h (mouse)	Reduced fetal body weight, dizziness, incoordination, tremors, muscular spasms, amnesia, seizures and lung congestion, etc.	*P. cepacia* MB2, *Pseudomonas* Pxy, *Rhodococcus* sp. YU6, *Bacillus subtilis*, *P. aeruginosa, P. monteilii*	Davey and Gibson (1974); Higson and Focht (1992); Jang et al. (2005); Dueholm et al. (2014)
Phthalate isomers	6178–8600 mg/kg (mouse), 9168–31,000 mg/kg (rat), 1000 mg/kg (rabbit)	na	na	Reproductive infertility, birth defects, problems in developmental and nervous system, eye, skin and respiratory tract	*P. cepacia, M. vanbaalenii* PYR-1, *A. keyseri* 12B, *B. cepacia* DBO1, *Rhodococcus* sp. DK17, *Micrococcus* sp. 12B, *P. aeruginosa* strain PP4, *Pseudomonas* sp. strain PPD, *Acinetobacter lwoffii* ISP4, *C. testosterone*, *R. jostii* RHA1	Eaton and Ribbons (1982); Batie et al. (1987); Wang et al. (1995); Choi et al. (2005); Vamsee-Krishna et al. (2006); Kim et al. (2007)
Naphthalene	90 mg/kg (rat)	20 g/kg (rabbit), 2500 mg/kg (rat)	340 mg/m³/1 h (rat)	Hemolytic anaemia, damage to liver and retina, cataract, nasal inflammation, lethargy, laryngeal carcinoma and bronchial adenoma	*P. putida* CSV86, *Bacillus thermoleovorans, Ralstonia* sp, *Nocardia* sp. *Pseudomonas* sp. NCIB, *Arthrobacter* sp. W1, *Herbaspirillum* sp., *Burkholderia* sp., *Polaromonas naphthalenivorans* CJ2	Barnsley (1975); Patel and Barnsley (1980); Mahajan et al. (1994); Annweiler et al. (2000); Zhou et al. (2001); Jeon et al. (2006); Chowdhury et al. (2014); Jauregui et al. (2014)

(Continued)

TABLE 3.1 (CONTINUED)

Aromatic Compounds and Pesticides: LC_{50} Values, Hazardous Effects and Microbial Degraders

Compound	LC$_{50}$ in Animal Models via Routes			Hazardous Effects	Degrading Bacteria[v]	References
	Oral	Dermal	Inhalation			
Phenanthrene	700 mg/kg (mice), 1.8 g/kg (rats)	na	na	Liver congestion, allergic responses, respiratory tract irritation and skin photosensitisation	*Bacillus* sp., *Pseudomonas* sp. strain PPD, *Alcaligenes* sp. strain PPH, *Aeromonas* sp., *Nocardioides* sp. KP7, *Streptomyces flavovirens*, *Cyanobacterium* sp., *Synechococcus* sp. PR-6	Kiyohara et al. (1976); Barnsley (1983a); Iwabuchi and Harayama (1998); Doddamani and Ninnekar (2000); Deveryshetty and Phale (2009, 2010)
Anthracene	4900 mg/kg (mouse)			Carcinogenic, birth defects, reduced body weight, harmful effects on skin, body fluids and immune system	*Pseudomonas* sp., *Rhodococcus* sp., *Mycobacterium* sp. PYR-1 and LB501T, *S. paucimobilis, P. rhodesiae, Sphingomonas* sp.	Cerniglia (1993); Dean-Ross et al. (2001); Moody et al. (2001); van Herwijnen et al. (2003)
γ-HCH	88–190 mg/kg (rats), 59–562 mg/kg (mice), 200 mg/kg (rabbits)	500–900 mg/ kg (rats), 300 mg/kg (mice), 300 mg/kg (rabbits)	na	Blood disorders, changes in the level of sex hormones, kidney and liver damage, possibly carcinogenic	*Sphingobium paucimobilis* UT26A, *S. indicum* B90A, *S. francense, Anabeana* sp., *Nostoc ellipsosporum, Pontibacter indicus* sp., *Devosia lucknowensis, S. chinhatense* strain IP26T, *Pandoraea* sp. strain SD6-2	Sahu et al. (1990); Kuritz and Wolk (1995); Ceremonie et al. (2006); Nagata et al. (2010); Dua et al. (2013); Niharika et al. (2013); Pushiri et al. (2013); Singh et al. (2014)

Compound	LD_{50} (oral)	LD_{50} (dermal)	LC_{50} (inhalation)	Health effects	Bacterial strains	References
Atrazine	500–2000 mg/kg (rats)	4000 mg/kg (rabbits)	5.1 mg/L/4 h (rats)	Endocrine disruptor, intrauterine growth retardation, delayed puberty, affects the function of central nervous, immune and cardiovascular system, carcinogenic	*Pseudomonas* sp. ADP, *Ralstonia* sp. M91-3, *Clavibacter* sp., *Agrobacterium* sp. J14, *Alcaligenes* sp. SG1, *Arthrobacter aurescens* TC1, *Rhodococcus* sp. NI86/21, *Streptomyces* sp. PS1/5, *Nocardia* sp., *Arthrobacter* sp. AK_YN10, *Pseudomonmonas* sp. AK_AAN5, and AK_CAN1	Nagy et al. (1995); Boundy-Mills et al. (1997); de Souza et al. (1998); Sadowsky et al. (1998); Shapir et al. (2005); Fazlurrahman et al. (2009); Sagarkar et al. (2014)
Carbaryl	303–312 mg/kg (rats), 710 mg/kg (rabbits)	>2000 mg/kg (rabbits)	>3.4 mg/L (rats)	Cholinesterase inhibition, broncoconstriction, skin and eye irritation, muscle weakness, headache, memory loss and anorexia	*Nocardia* sp., *Xanthomonas* sp., *Achromobacter* sp., *Pseudomonas* sp., *Rhodococcus* sp. NCIB 12038, *Blastobacter* sp., *Arthrobacter* sp., *Micrococcus* sp., *Pseudomonas* sp. C4, C5, C6, *Burkholderia* sp. C3	Sud et al. (1972); Larkin and Day (1986); Chapalamadugu and Chaudhry (1991); Hayatsu and Nagata (1993); Doddamani and Ninnekar (2001); Hayatsu et al. (2001); Swetha and Phale (2005); Seo et al. (2013)

Note: #na, LC_{50} value of compound not reported in any animal models. LC_{50} values of mentioned compounds are referred from material safety data sheet; ¥Some selected representative bacterial strains are described with references.

3.2 Biodegradation of Pollutants

Microorganisms can degrade a variety of aromatic compounds via aerobic or anaerobic pathways (Bondarenko and Gan 2004; DiDonato et al. 2010). In aerobic degradation, the initial steps are catalysed by a special group of enzymes called 'oxygenase', which incorporate molecular oxygen into the aromatic ring yielding hydroxylated compounds with higher oxidation states. These compounds are more water-soluble than the parent compounds, thus making them more susceptible for further degradation (Malmstrom 1982). Depending on the number of oxygen atom(s) incorporated into the aromatic substrate, the oxygenases are classified into monooxygenases or dioxygenases. Monooxygenases incorporate one atom of the molecular oxygen (O_2) into the substrate, yielding monohydroxylated products. The dioxygenases incorporate both oxygen atoms into the substrate and perform either ring-hydroxylation reactions, yielding dihydroxylated products or ring-cleavage of aromatic diols (Gibson and Parales 2000). Some dioxygenases, such as naphthalene dioxygenase, carry out both dioxygenation as well as mono-oxygenation reactions. Such enzymes are of high industrial importance as they participate in the transformation of a wide variety of chemicals of pharmaceutical, agricultural and environmental significance (Gibson et al. 1995). On the other hand, under anaerobic condition, microorganisms use inorganic compounds, such as nitrate, sulphate or metal ions as electron acceptors to generate cell mass, nitrogen gas (N_2), hydrogen sulphide (H_2S), methane (CH_4) and reduced forms of metals as products (Evans and Fuchs 1988; DiDonato et al. 2010). In this chapter, the aerobic mode of aromatic pollutants and pesticide degradation pathways and enzymes from several bacterial strains is discussed.

3.2.1 Monaromatics

3.2.1.1 BTEX Compounds

BTEX (benzene, toluene, ethylbenzene and xylene) compounds are environmental pollutants that are generally used as parent materials for solvents, plasticisers and pesticides (Jindrova et al. 2002). BTEX compounds are abundantly found in the petroleum products as well as in industrial solvent wastes (Fries et al. 1994). The hazardous effects of BTEX compounds are carcinogenicity, fetotoxicity and neurotoxicity (Table 3.1; IARC 1989, 2000; Kandyala et al. 2010; Rezazadeh et al. 2012; Rajan and Malathi 2014). BTEX compounds are often found as a mixture in polluted sites; hence diverse groups of bacteria appear to be involved in the degradation of these compounds (Olsen et al. 1995; Jindrova et al. 2002). Certain organisms can degrade more than one of the BTEX compounds (Worsey and Williams 1975; Furukawa et al. 1983). The biodegradation pathways of BTEX compounds are as follows.

Benzene

Benzene degradation has been studied in both Gram-negative as well as Gram-positive bacteria (Jindrova et al. 2002). The degradation of benzene follows two different pathways, the first step being common, which is catalysed by benzene dioxygenase to give *cis*-benzene dihydrodiol, which further yields catechol by the action of *cis*-dihydrobenzenediol dehydrogenase. Catechol is metabolized either by catechol 1,2-dioxygenase (intradiol or *ortho*-cleavage) or catechol 2,3-dioxygenase (extradiol or *meta*-cleavage) (Figure 3.2; Smith 1990).

Toulene

Toulene degradation is initiated with the oxidation of methyl group, catalysed by side chain monooxygenase (Figure 3.2, route I; Shaw and Harayama 1992) or by hydroxylation at the *ortho*-, *meta*- or *para*- position by

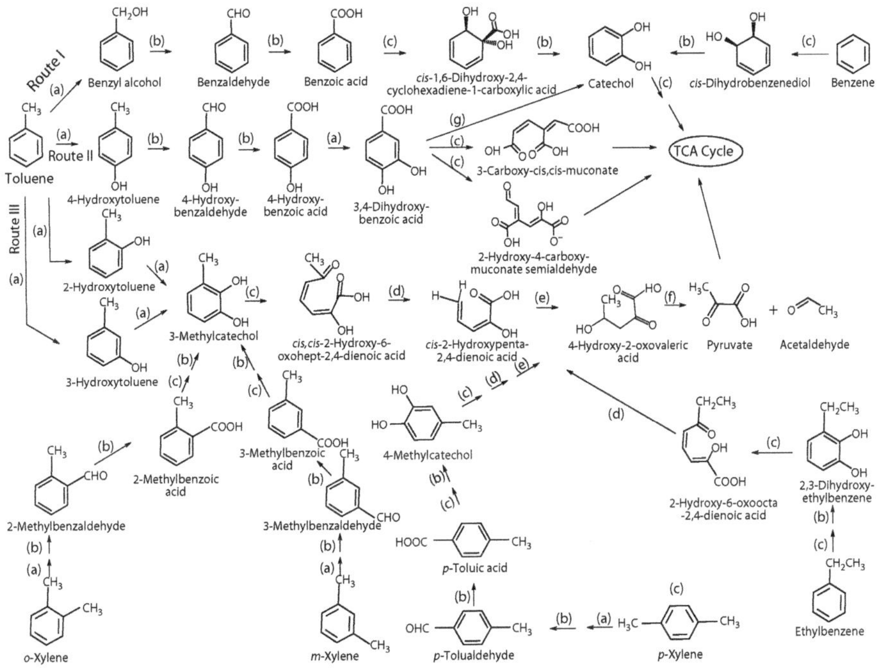

FIGURE 3.2
Pathways for the degradation of BTEX (benzene, toluene, ethylbenzene and xylenes). Enzymes (*in general*) involved in the degradation pathway are (a) monooxygenase, (b) dehydrogenase, (c) dioxygenase, (d) hydratase, (e) hydrolase, (f) aldolase and (g) decarboxylase. Figure is based on reported degradation pathways. (From Gibson DT et al., *Biochemistry* 12, 1520–1528, 1973; Davey JF, and Gibson DT, *J Bacteriol* 119, 923–929, 1974; Wackett LP et al., *Biochemistry* 27, 1360–1367, 1988; Yen KM et al., *J Bacteriol* 173, 5315–5327, 1991; Higson FK, and Focht DD, *Appl Environ Microbiol* 58, 194–200, 1992; Shaw JP, and Harayama S, *Eur J Biochem* 209, 51–61, 1992; Olsen RH et al., *J Bacteriol* 176, 3749–3756, 1994; Shields MS et al., *Appl Environ Microbiol* 61, 1352–1356, 1995; Jindrova E et al., *Folia Microbiol (Praha)* 47, 83–93, 2002; Jang JY et al., *J Microbiol* 43, 325–330, 2005.)

toluene 2-, 3- or 4-monooxygenase, respectively (Figure 3.2, route II; Yen et al. 1991; Olsen et al. 1994; Shields et al. 1995). In another pathway, toluene dioxygenase incorporates two oxygen atoms into toluene, and the degradation finally proceeds via formation of 3-methyl catechol, which gets further metabolised by catechol 2,3-dioxygenase (Figure 3.2, route III; Wackett et al. 1988).

Xylene

In bacterial strains, the xylene isomer (*ortho, meta* and *para*) degradation pathway is initiated by monooxygenase or dioxygenases (Jang et al. 2005). *Pseudomonas cepacia* MB2 degrades *o*-xylene, and *Pseudomonas* Pxy degrades *m*- and *p*-xylene through xylene monooxygenase, with which the –CH$_3$ group is oxidised, and the pathway proceeds through formation of *o-, m-* or *p*-methylbenzyl catechol (Davey and Gibson 1974; Higson and Focht 1992). *Rhodococcus* sp. YU6 metabolized *o*- and *p*-xylene by direct aromatic ring-oxidation and formation of dimethylcatechols, which was further degraded through the *meta*-cleavage pathway (Figure 3.2, route III; Jang et al. 2005).

Ethylbenzene

Ethylbenzene degradation occurs with the help of ethylbenzene dioxygenase to form ethylbenzene dihydrodiol, which, in susbsequent steps, forms propanoic acid or acetaldehyde and pyruvic acid (Figure 3.2; Gibson et al. 1973). An alternative mode of ethylbenzene degradation is through naphthalene dioxygenase; for example, naphthalene dioxygenase from *Pseudomonas* NCIB 98164 displays a relaxed specificity for ethylbenzene and metabolises via styrene or acetophenone (Lee and Gibson 1996).

3.2.1.2 *Phthalate Esters and Isomers*

Phthalate isomers and their esters are used in various industries due to their low cost of manufacturing, ease of fabrication and the flexibility they impart to the finished material. The first industrial application of phthalates was in the manufacture of blood bags containing di-2-ethyl hexylphthalate (DEHP, Gesler 1973). Since then, phthalate and its esters have been extensively used in the manufacture of PVC, food packing, cosmetics and milk bottles (Autian 1973). Phthalate isomers and their esters have low water-solubility but high octane/water coefficients. The hydrophobicity of these esters increases with an increase in alkyl chain length (Autian 1973; Chang et al. 2004). These compounds are loosely bound and hence can easily leach out from the finished product (Jaeger and Rubin 1973). The *para*- and *meta*-isomers of phthalate, terephthalate and isophthalate are also used in industries but not as widely as phthalate. Terephthalate is used in the manufacture of polyethylene terephthalate (PET) bottles (Autian 1973). Upon exposure, phthalates cause the most profound effect on eyes, skin and mucous membranes of the upper respiratory tract (Autian 1973). Repeated exposure may also lead to allergic dermatitis and asthma. Exposure of phthalate esters to rats leads to growth retardation and degeneration of reproductive organs (Gesler 1973). Dimethyl phthalate (DMP) used as an insect repellent and diethyl phthalate (DEP), which is used primarily as a solvent, are shown to have mutagenic effects (Kozumbo et al. 1982).

The pathway of phthalate ester degradation is shown in Figure 3.3. The pathway begins with an esterase, which hydrolyses the ester bond to yield phthalate (Eaton and Ribbons 1982). In Gram-negative bacteria, phthalate is degraded to a dihydrodiol intermediate by a ring-hydroxylating phthalate 4,5-dioxygenase, for example, *Burkholderia cepacia* DBO1 (Batie et al. 1987). However, in Gram-positive bacteria, the same reaction is catalysed by phthalate 3,4-dioxygenase, for example, *Rhodococcus* sp. DK17 (Choi et al. 2005). The phthalate degradation pathway was elucidated from *Pseudomonas aeruginosa* strain PP4 and *Pseudomonas* sp. strain PPD, both of which degrade all three phthalate isomers (Figure 3.3; Vamsee-Krishna et al. 2006). After ring-hydroxylation, the next step is catalysed by a dehydrogenase, which converts phthalate-*cis*-dihydrodiol to a dihydroxyphthalate intermediate, which undergoes decarboxylation to yield 3,4-dihydroxybenzoate or protocatechuate. The latter step is catalysed by dihydroxyphthalate decarboxylase. Isophthalate, terephthalate and their esters are also metabolised in a similar way to form protocatechuate as a metabolic intermediate. However, the degradation pathway for isophthalate and terephthalate lacks the decarboxylase. In this case, respective dihydrodiols are directly converted to protocatechuate. Degradation pathways for all three phthalate isomers converge at protocatechuate. Protocatechuate undergoes a ring-cleavage by protocatechuate dioxygenase to give β-carboxy-*cis,cis* muconic acid, which is further oxidised rapidly to β-ketoadipate that enters into the TCA cycle (Figure 3.3).

3.2.2 Diaromatics

3.2.2.1 Naphthalene

Naphthalene is a bicyclic aromatic hydrocarbon commonly found in the environment. Generally, naphthalene is used in the manufacturing of carbamate insecticides surface-active agents, resins, dye intermediates, synthetic tanning agents, moth repellents, etc. The main source of naphthalene in the environment is the incomplete combustion of industrial, domestic and natural products as well as diesel, jet fuel and cigarette smoke. Various microorganisms can utilise naphthalene as the sole source of carbon and energy (Patel and Gibson 1974; Barnsley 1975, 1976a,b, 1983a; Patel and Barnsley 1980; Mahajan et al. 1994). Degradation of naphthalene in bacteria is initiated by the double hydroxylation of one of the rings to form a dihydrodiol intermediate (Figure 3.4). The first step in naphthalene degradation [formation

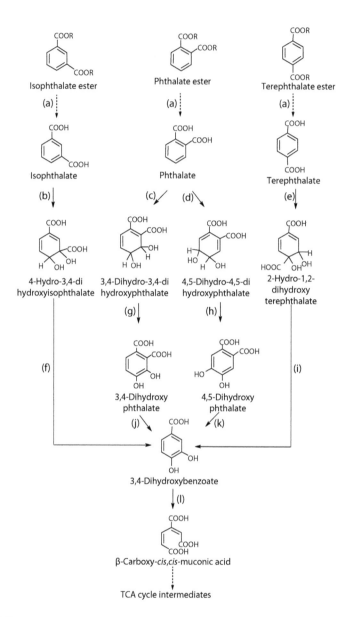

FIGURE 3.3
Pathways for the degradation of phthalate esters. Enzymes involved in the pathway are (a) esterase, (b) isophthalate 3,4-dioxygenase, (c) phthalate 3,4-dioxygenase, (d) phthalate 4,5-dioxygenase, (e) terephthalate 1,2-dioxygenase, (f) isophthalate 3,4-dihydrodiol dehydrogenase, (g) phthalate 3,4-dihydrodiol dehydrogenase, (h) phthalate 4,5-dihydrodiol dehydrogenase, (i) terephthalate 1,2-dihydrodiol dehydrogenase, (j) 3,4-dihydroxyphthalate decarboxylase, (k) 4,5-dihydroxyphthalate decarboxylase and (l) protocatechuate dioxygenase. Figure is based on reported degradation pathways. (From Vamsee-Krishna C et al., *Appl Microbiol Biotechnol* 72, 1263–1269, 2006.)

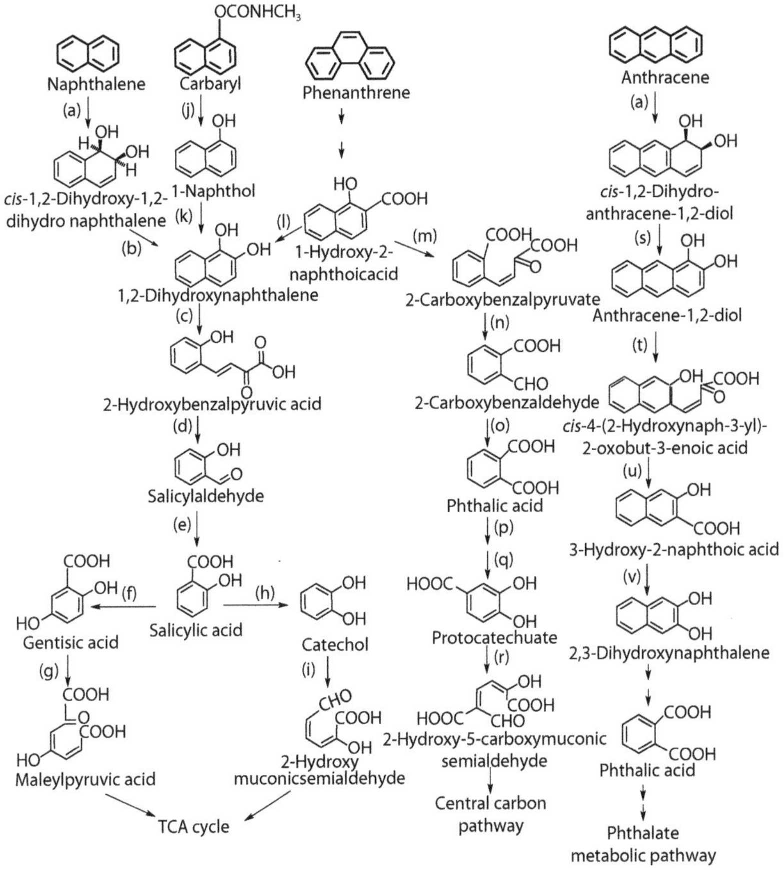

Naphthalene Carbaryl Phenanthrene Anthracene

OCONHCH₃ → $OCONHCH_3$

cis-1,2-Dihydroxy-1,2-dihydro naphthalene

1-Naphthol

1,2-Dihydroxynaphthalene

1-Hydroxy-2-naphthoicacid

2-Carboxybenzalpyruvate

cis-1,2-Dihydroanthracene-1,2-diol

Anthracene-1,2-diol

2-Hydroxybenzalpyruvic acid

2-Carboxybenzaldehyde

cis-4-(2-Hydroxynaph-3-yl)-2-oxobut-3-enoic acid

Salicylaldehyde

Phthalic acid

3-Hydroxy-2-naphthoic acid

Gentisic acid   Salicylic acid   Catechol   Protocatechuate   2,3-Dihydroxynaphthalene

Maleylpyruvic acid   2-Hydroxy muconicsemialdehyde   2-Hydroxy-5-carboxymuconic semialdehyde   Phthalic acid

TCA cycle   Central carbon pathway   Phthalate metabolic pathway

FIGURE 3.4

Degradation pathways for naphthalene, carbaryl, phenanthrene and anthracene. Enzymes involved in the degradation pathway are (a) naphthalene dioxygenase, (b) *cis*-dihydrodiol naphthalene dehydrogenase, (c) 1,2-dihydroxynaphthalene dioxygenase, (d) *trans-o*-hydroxybenzylidenepyruvate hydratase-aldolase, (e) salicylaldehyde dehydrogenase, (f) salicylate 5-hydroxylase, (g) gentisate 1,2-dioxygenase, (h) salicylate 1-hydroxylase, (i) catechol 1,2-dioxygenase, (j) carbaryl hydrolase, (k) 1-naphthol-2-hydroxylase, (l) 1-hydroxy-2-naphthoic acid hydroxylase, (m) 1-hydroxy-2-naphthoic acid dioxygenase, (n) 2-carboxybenzalpyruvate hydratase-aldolase (o) 2-carboxybenzaldehyde dehydrogenase, (p) phthalate 4,5-dioxygenase, (q) decarboxylase, (r) protocatechuate 4,5-dioxygenase, (s) *cis*-1,2-dihydronaphthalene dehydrogenase, (t) anthracene-1,2-diol 1,2-dioxygenase, (u) *cis*-4-(2-hydroxynaph-3-yl)-2-oxobut-3-enoate hydratasealdolase and (v) 3-hydroxy-2-naphthoate hydroxylase. Figure is based on reported degradation pathways. (From Evans WC et al., *Biochem J* 95, 819–831, 1965; Jerina DM et al., *J Am Chem Soc* 98, 5988–5996, 1976; Chapalamadugu S, and Chaudhry GR, *Appl Environ Microbiol* 57, 744–750, 1991; Eaton RW, and Chapman PJ, *J Bacteriol* 174, 7542–7554, 1992; Eaton RW, *J Bacteriol* 176, 7757–7762, 1994; Kiyohara H et al., *J Bacteriol* 176, 2439–2443, 1994; Annweiler E et al., *Appl Environ Microbiol* 66, 518–523, 2000; Dean-Ross D et al., *FEMS Microbiol Lett* 204, 205–211, 2001; Moody JD et al., *Appl Environ Microbiol* 67, 1476–1483, 2001; Swetha VP, and Phale PS, *Appl Environ Microbiol* 71, 5951–5956, 2005; Deveryshetty J, and Phale PS, *Microbiology* 155, 3083–3091, 2009; Deveryshetty J, and Phale PS, *FEMS Microbiol Lett* 311, 93–101, 2010.)

of (+)-*cis*-1(*R*),2(*S*)-dihydroxy-1,2-dihydronaphthalene] is catalysed by multi-component naphthalene dioxygenase (NDO), which is composed of oxygenase, ferrodoxin and reductase components. The oxygenase of NDO consists of α- and β-subunits, in which the α-subunit contains a Rieske [2Fe-2S] center and mononuclear iron at the active site. Electrons from NAD(P)H are transferred to NDO via reductase and ferredoxin component (Kauppi et al. 1998; Karlsson et al. 2003). The second step in the bacterial oxidation of naphthalene is the conversion of *cis*-1,2-dihydroxy-1,2-dihydronaphthalene to 1,2-dihydroxynaphthalene. This reaction is catalysed by NAD-dependent, naphthalene dihydrodiol dehydrogenase (Patel and Gibson 1974). 1,2-Dihydroxynaphthalene is cleaved by a 1,2-dihydroxynaphthalene dioxygenase to yield *cis*-2'-hydroxybenzalpyruvic acid (Patel and Barnsley 1980), which is further metabolised to salicylate via salicylaldehyde by the action of aldolase. In most cases, the salicylate is oxidised to catechol, which is the substrate for ring-fission by the *meta*- or *ortho*-cleavage pathway, giving rise to either 2-hydroxymuconic semialdehyde or *cis,cis*-muconic acid, respectively (Barnsley 1976b). Salicylate metabolism also occurs via gentisic acid (2,5-dihydroxy benzoate). Gentisate is further metabolised by gentisate 1,2-dioxygenase to maleylpyruvate (Figure 3.4; Monticello et al. 1985; Yen and Serdar 1988).

3.2.3 Polyaromatics

3.2.3.1 Phenanthrene

Phenanthrene is commonly found in soil, estuarine water, sediments and other terrestrial as well as aquatic sites. It is found to be toxic to marine diatoms, gastropods, crustaceans and fish and also acts as a human skin photosensitiser, mild allergen, inhibitor of gap junction and inducer of sister chromatid exchanges (Mastrangelo et al. 1996). Several bacterial species are reported to utilise phenanthrene as the sole source of carbon and energy. Phenanthrene metabolism is initiated by double hydroxylation of ring by phenanthrene dioxygenase to yield *cis*-3,4-dihydroxy-3,4-dihydrodihydroxy phenanthrene (Figure 3.4). The dihydrodiol is acted upon by a dehydrogenase to form 3,4-dihydroxyphenanthrene, which is subsequently metabolised to yield 1-hydroxy-2-naphthoic acid (1-H2NA, Barnsley 1983b; Doddamani and Ninnekar 2000; Deveryshetty and Phale 2009). This pathway is referred to as the common upper pathway of the phenanthrene degradation. Two major routes are employed in the degradation of 1-H2NA. In the 'phthalate' route, 1-H2NA is ring-cleaved by a dioxygenase to yield 2-carboxybenzalpyruvate (Kiyohara et al. 1976; Barnsley 1983b; Iwabuchi and Harayama 1998; Deveryshetty and Phale 2009, 2010) and further metabolised to *o*-pthalate and protocatechuic acid, which is cleaved to form pyruvate (Figure 3.4). In the 'naphthalene' route, 1-H2NA is metabolised to 1,2-dihydroxynapthalene (Evans et al. 1965; Deveryshetty and Phale 2010), which is metabolised via salicylaldehyde and salicylic acid. Salicylic acid is further metabolised to either gentisate or catechol, which undergoes ring-cleavage to generate TCA cycle intermediates (Figure 3.4).

3.2.3.2 Anthracene

Anthracene is a persistent and toxic polycyclic aromatic hydrocarbon used in dyes, wood preservatives and insecticides. Anthracene is degraded by several Gram-negative and Gram-positive bacteria and fungi (Cerniglia 1993). The anthracene degradation pathway is initiated by hydroxylation of its aromatic ring to yield cis-1,2-dihydroanthracene-1,2-diol (Evans et al. 1965; Jerina et al. 1976). The intermediate is converted to anthracene-1,2-diol, which is cleaved at the *meta* position to yield 4-(2-hydroxynaph-3-yl)-2-oxobut-3-enoate (Evans et al. 1965). In Gram-negative bacteria, this compound is spontaneously rearranged to produce 6,7-benzocoumarin or converted to 3-hydroxy-2-naphthoate, from which degradation proceeds through 2,3-dihydroxynaphthalene to phthalate (Annweiler et al. 2000; Dean-Ross et al. 2001; Moody et al. 2001). Other strains, such as *M. vanbaalenii* PYR-1, can produce 1-methoxy-2-hydroxy-anthracene from anthracene-1,2-diol, representing a third pathway branch from this intermediate. In addition to initiating anthracene degradation by 1,2-hydroxylation, *M. vanbaalenii* PYR-1 can catalyse 9,10-hydroxylation of anthracene to produce anthracene-9,10-dihydrodiol, which is converted to 9,10-dihydroxyanthracene and gets spontaneously oxidised to 9,10-anthraquinone (Figure 3.4; Moody et al. 2001).

3.2.4 Halogenated and Nitrogen-Containing Pesticides

3.2.4.1 Lindane

Hexachlorocyclohexane is one of the most widely used organochlorine pesticides. HCH causes damage to central nervous, reproductive and endocrine systems (Willett et al. 1998). HCH is synthesised by photochemical chlorination of benzene, which gives rise to five stable isomers α (60%), β (5%–12%), γ (10%–12%), δ (6%–10%) and ε (4%). Out of these, γ-HCH (lindane) has insecticidal activity (Lal et al. 2010). Because of its high toxicity and persistence in the soil, the use of γ-HCH has been banned in many countries (Willett et al. 1998). γ-HCH is degraded rapidly under anaerobic and aerobic conditions (Ohisa et al. 1980). The major studies are focused on aerobic γ-HCH degradation from *Sphingomonas paucimobilis* UT26A, *Sphingomonas indicum* B90A and *Sphingobium francense* (Figure 3.5; Sahu et al. 1990; Ceremonie et al. 2006; Nagata et al. 2010). γ-HCH degradation is initiated with two subsequent dehydrochlorination reactions to produce pentachlorocyclohexene (PCCH). Subsequently, 2,5-dichloro-2,5-cyclohexadiene-1,4-diol (2,5-DDOL) is generated by two rounds of hydrolytic dechlorinations. 2,5-DDOL is then converted by a dehydrogenation reaction to 2,5-dichlorohydroquinone (2,5-DCHQ). This is known as the upstream degradation pathway (Nagata et

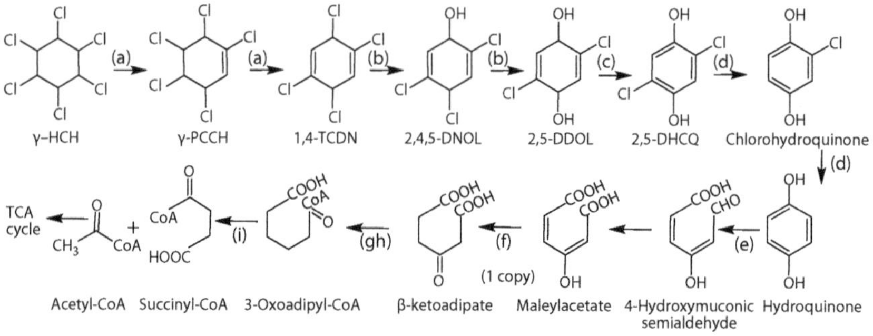

FIGURE 3.5
Pathways for the degradation of γ-HCH. Enzymes involved in the pathway are (a) linA; γ-HCH dehydrochlorinase, (b) linB; 1,3,4,6-tetrachloro-1,4-cyclohexadiene hydrolase, (c) linX/C; 2,5-dichloro-2,5-cyclohexadiene-1,4-diol dehydrogenase, (d) linD; 2,5-dichlorohydroquinone reductive dechlorinase, (e) linE; chlorohydroquinone/hydroquinone 1,2-dioxygenase. (f) linF; maleylacetate reductase, (g, h) linG, H; 3-oxoadipate *CoA*-transferase α-subunit and β-subunit, (i) linJ; acetyl-*CoA*-acetyltransferase. Figure is based on reported degradation pathways. (From Nagata Y et al., *J Bacteriol* 176, 3117–3125, 1994; Endo R et al., *J Bacteriol* 187, 847–853, 2005; Nagata Y et al., *Appl Microbiol Biotechnol* 76, 741–752, 2007.)

al. 1994). In the subsequent downstream pathway, 2,5-DCHQ metabolism begins with two initial dehydrochlorinations to form chlorohydroquinone (CHQ) and hydroquinone (HQ). The HQ is then ring-cleaved to hydroxy-muconic semialdehyde (HMSA), which is further transformed to maleylac-etate (MA). MA is converted to β-ketoadipate (Endo et al. 2005) and then to succinyl-CoA and acetyl-CoA, which are both metabolised by TCA cycle intermediates (Figure 3.5; Nagata et al. 2007).

3.2.4.2 Atrazine

Atrazine [2-chloro-4-(2-propylamino)-6-ethylamino-*S*-triazine] is a member of the triazine ring-containing herbicides. It is one of the widely used her-bicides that have carcinogenic, endocrine-disrupting and teratogenic activi-ties (Fazlurrahman et al. 2009). In bacterial strains, atrazine is metabolised to ammonia and carbon dioxide via different routes, which finally merge into the cyanuric acid metabolism (Figure 3.6; Nagy et al. 1995; Boundy-Mills et al. 1997; de Souza et al. 1998; Sadowsky et al. 1998; Shapir et al. 2002; Fazlurrahman et al. 2009). In *Rhodococcus* sp. strain N186/21, the oxidative removal of ethyl or isopropyl group of atrazine is catalysed by atrazine *N*-dealkylase or atrazine monooxygenase to form deethylatrazine and deiso-propylatrazine, respectively (Figure 3.6, routes I and II; Nagy et al. 1995). Deethylatrazine undergoes a second oxidative removal of the isopropyl group by a monooxygenase to yield deisopropyldeethylatrazine. The hydro-lytic removal of *s*-triazine ring substituents from deisopropylatrazine and deisopropyldeethylatrazine yields cyanuric acid (Figure 3.6, routes I and II).

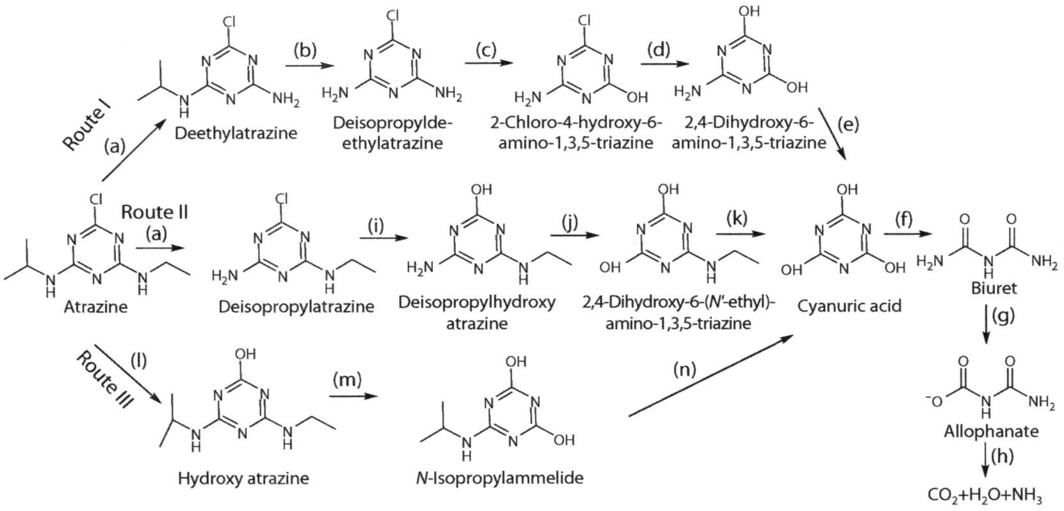

FIGURE 3.6

Pathways for the degradation of atrazine. Enzymes involved in the pathway are (a) atrazine monooxygenase (atrazine N-dealkylase), (b) deethylatrazine monooxygenase, (c) N-ethylammeline chlorohydrolase, (d) hydroxyl dechloro atrazine ethyl amino hydrolase, (e) N-isopropyl ammelide isopropyl amino hydrolase, (f) cyanuric acid amido hydrolase, (g) biuret hydrolase, (h) allophanate hydrolase, (i) N-ethylammeline chlorohydrolase, (j) deisopropyl hydroxyl atrazine amino hydrolase, (k) 2,4-dihydroxy-6-(N'-ethyl) amino-1,3,5-triazine ethylaminohydrolase, (l) atrazine chlorohydrolase, (m) hydroxyl dechloro atrazine ethyl amino hydrolase, (n) N-isopropyl ammelide isopropyl amino hydrolase. Figure is based on reported degradation pathways. (From Nagy I et al., *Appl Environ Microbiol* 61, 2056–2060, 1995; Boundy-Mills KL et al., *Appl Environ Microbiol* 63, 916–923, 1997; de Souza ML et al., *Appl Environ Microbiol* 64, 2323–2326, 1998; Sadowsky MJ et al., *J Bacteriol* 180, 152–158, 1998; Shapir N et al., *J Bacteriol* 184, 5376–5384, 2002; Fazlurrahman, BM et al., *Lett Appl Microbiol* 49, 721–729, 2009.)

In *Ralstonia* sp. and *Pseudomonas* sp., the initial hydrolytic dechlorination of atrazine to hydroxyatrazine is catalysed by atrazine chlorohydrolase. Further, two enzymes from amidohydrolase family catalyse the sequential removal of ethylamine and isopropylamine to yield cyanuric acid (Figure 3.6, route III; Boundy-Mills et al. 1997; de Souza et al. 1998; Sadowsky et al. 1998). The cyanuric acid is converted to biuret by cyanuric acid amidohydrolase. The enzyme biuret hydrolase converts biuret to allophanate. Allophanate is converted to ammonia and carbon dioxide by allophanate hydrolase (Figure 3.6; Shapir et al. 2002, 2005).

3.2.4.3 Carbaryl

Carbaryl (1-naphthyl *N*-methylcarbamate) is used as a broad-spectrum insecticide. The persistence of carbaryl in agricultural soils is due to repeated applications (Felsot et al. 1981; Tal et al. 1989). Carbaryl is toxic in nature due to its ester bond between 1-naphthol and *N*-methylcarbamate (Figure 3.2; Swetha and Phale 2005). The inhibition of cholinesterase is the principle mechanism of carbaryl action. The early symptoms of carbaryl poisoning are bronchial secretions, excessive sweating, salivation, abdominal cramps, headache, anxiety, convulsion and coma. Carbaryl can be degraded by physical or biological means. Physical degradation methods include pH-dependent hydrolysis or ultraviolet light (315–280 nm) to yield toxic 1-naphthol. Biologically, carbaryl is metabolised via aerobic degradation pathway, which is shown in Figure 3.4. The pathway is initiated by the enzyme carbaryl hydrolase (CH) to yield 1-naphthol, which is further metabolised to 1,2-dihydroxynaphthalene by 1-naphthol-2-hydroxylase (Figure 3.4; Swetha and Phale 2005; Sah and Phale 2011). In few bacteria, 1-naphthol is hydroxylated to 4-hydroxy-1-tetralone (Walker et al. 1975), 3,4-dihydro-dihydroxy-1(2H)-naphthalenone (Bollag et al. 1975) or 1,4-naphthoquinone (Rajgopal et al. 1984); however, the metabolic fate of these intermediates is unclear. 1,2-Dihydroxynaphthalene is further metabolized via 2-hydroxybenzalpyruvic acid, salicylaldehyde, salicylate and gentisate to TCA cycle intermediates (Swetha and Phale 2005; Seo et al. 2013). The consortium of *Pseudomonas* sp. strains 50552 and 50581 degrades, carbaryl to salicylate, which is further metabolised via catechol to TCA cycle intermediates (Figure 3.4; Chapalamadugu and Chaudhry 1991).

3.3 Applications and Future Directions

The wealth of information available regarding biodegradation of toxic pollutants has stimulated a lot of interest in remediation of polluted sites using an enriched inoculum of the microorganisms. This process has been in practice for several years for the remediation of wastewater, agricultural lands and forests (Vogel 1996). This section throws light on the application of the microorganisms in degradation of some of the aromatics discussed in this chapter.

Although BTEX compounds are often degraded by consortia, there have been reports in which single cultures can remediate these compounds in situ. *Janibacter* strain SB2 isolated from sea water by the enrichment culture technique could degrade BTEX compounds in a slurry system (Jin et al. 2013). Biodegradation of phthalates has been reported from a wide range of systems, such as freshwater, wastewater, sludge and landfills (Liang et al. 2008). In natural systems, degradation of phthalates occurs mostly in syn-trophy; for example, two strains – *Klebsiella oxytoca* Sc and *Methylobacterium mesophilicum* Sr – were shown to completely degrade phthalates in a coop-erative manner (Gu et al. 2005). Degradation of four phthalates was demon-strated in sludge by a mixture of different bacterial strains – *Enterococcus* sp. strain OM1, *Bacillus benzoevorans* strain S4, *Sphingomonas* sp. strain O18 and *Corynebacterium* sp. strain DK4 (Chang et al. 2004).

Because the metabolic pathway and enzymes involved in atrazine deg-radation are known, several atrazine-degrading microrganisms are also used in the field study (Govantes et al. 2009). Atrazine degradation at the field level has been studied using *Pseudomonas* sp. ADP that contains a large catabolic plasmid pADP-1 harboring the genes for atrazine degradation. In this study, recombinant *E. coli* cells expressing atrazine chlorohydrolase were used for bioremediation and applied to a site with a spill of 1000 lb. of atrazine (Wackett et al. 2002). In another study, *Arthrobacter* sp. AK_YN10, *Pseudomonas* sp. AK_AAN5 and AK_CAN1 were used for bioremediation of soil mesocosms with or without previous exposure to atrazine. A signifi-cant decrease in atrazine concentration was observed within a month for soil samples with or without atrazine exposure (Sagarkar et al. 2014).

Bioaugmentation of phenanthrene is challenging because it becomes unavailable to microorganisms as it ages in soil. However, it is observed that repeated inoculations with bacteria can result in increased mineralisation (Schwartz and Scow 2001). *Sphingomonas paucimobilis* 200006FA has been shown to degrade phenanthrene present in a soil microcosm (Coppotelli et al. 2008). Phenanthrene degradation in mangroves has been studied using *Sphingobium yanoikuyae* and *Mycobacterium parafortuitum*; however, these strains failed to remove phenanthrene in situ (Tam and Wong 2008).

Most of the bacterial strains degrade various aromatic hydrocarbons under laboratory conditions; however, they often fail in field trials due to several fac-tors, such as suboptimal growth conditions, competition with well-adapted indigenous microorganisms and suppression of certain catabolic genes due to environmental factors (Govantes et al. 2009). However, several techniques are used to enhance the degrading ability in natural environments. One such technique involves supplementation of soil with cyclodextrins and inor-ganic nutrients. Cyclodextrins increase the solubility of pollutants in soil and enhance pollutant desorption (Allan et al. 2007). However, surfactants or surface active agents are also used as solubilisers in the remediation of contaminated soil (Lee et al. 2002). The uses of biosurfactants in bioremediation are promising due to their biodegradability, low toxicity and effectiveness in enhancing biodegradation and solubilisation of low-solubility compounds. Biosurfactant producing *P. aeruginosa* strain 64 was used in contaminated soil to enhance the bioremediation. However, the added strain could not degrade PAHs but produced rhamnolipids, which stimulated PAH degradation by other microorganisms (Mulligan 2005). Another approach could be the application of novel organisms, such as *Pseudomonas putida* CSV86, which utilises aromatic compounds in preference to simple carbon sources, such as glucose and organic acids (Basu et al. 2006; Basu and Phale 2008).

Bioremediation of contaminated sites using immobilised enzymes is an alternative tool for the removal of pollutants from the environment. The extracellular enzymes secreted from various microorganisms catalyse the transformation of toxic compounds to less toxic forms. In this approach, the enzymes are immobilised onto a solid carrier, for example, plant polyphenol oxidases used in the decolourisation and removal of textile and non-textile dyes (Khan and Husain 2007). Organophosphate degrading enzyme A (OpdA) was covalently immobilised on highly porous nonwoven polyester fabrics for organophosphate pesticide degradation (Gao et al. 2014). Enzymes degrading a wide range of aromatics can be immobilised and used to treat heavily polluted sites using similar approaches.

3.4 Conclusions

Several microorganisms present in the environment utilise mono-, di- and poly-aromatics and substituted aromatic hydrocarbons and pesticides as sole carbon sources. The wealth of knowledge of their metabolic pathways and enzymes can pave the way for developing remediation of heavily polluted terrestrial and water bodies. Although many studies have successfully demonstrated the in situ degradation, several alternative strategies need to be explored to enhance the complete degradation of these toxic pollutants by microorganisms.

References

Acevedo F, Pizzul L, Castillo MD et al. 'Degradation of polycyclic aromatic hydrocarbons by free and nanoclay-immobilized manganese peroxidase from *Anthracophyllum discolor'*. *Chemosphere* 80 (2010): 271–278.

Alagappan G, and Cowan R. 'Substrate inhibition kinetics for toluene and benzene degrading pure cultures and a method for collection and analysis of respirometric data for strongly inhibited cultures'. *Biotechnol Bioeng* 83 (2003): 798–809.

Allan IJ, Semple KT, Hare R, and Reid BJ. 'Cyclodextrin enhanced biodegradation of polycyclic aromatic hydrocarbons and phenols in contaminated soil slurries'. *Environ Sci Technol* 41 (2007): 5498–5504.

Annweiler E, Richnow HH, Antranikian G et al. 'Naphthalene degradation and incorporation of naphthalene-derived carbon into biomass by the thermophile *Bacillus thermoleovorans*'. *Appl Environ Microbiol* 66 (2000): 518–523.

Autian J. 'Toxicity and health threats of phthalate esters: Review of the literature'. *Environ Health Perspect* 4 (1973): 3–26.

Barnsley EA. 'The induction of the enzymes of naphthalene metabolism in pseudomonads by salicylate and 2-aminobenzoate'. *J Gen Microbiol* 88 (1975): 193–196.

Barnsley EA. 'Naphthalene metabolism by pseudomonads: The oxidation of 1,2-dihydroxynaphthalene to 2-hydroxychromene-2-carboxylic acid and the formation of 2'-hydroxybenzalpyruvate'. *Biochem Biophys Res Commun* 72 (1976a): 1116–1121.

Barnsley EA. 'Role and regulation of the *ortho*- and meta-pathways of catechol metabolism in pseudomonads metabolizing naphthalene and salicylate'. *J Bacteriol* 125 (1976b): 404–408.

Barnsley EA. 'Bacterial oxidation of naphthalene and phenanthrene'. *J Bacteriol* 153 (1983a): 1069–1071.

Barnsley EA. 'Phthalate pathway of phenanthrene metabolism: Formation of 2'-carboxybenzalpyruvate'. *J Bacteriol* 154 (1983b): 113–117.

Basu A, Apte SK, and Phale PS. 'Preferential utilization of aromatic compounds over glucose by *Pseudomonas putida* CSV86'. *Appl Environ Microbiol* 72 (2006): 2226–2230.

Basu A, and Phale PS. 'Conjugative transfer of preferential utilization of aromatic compounds from *Pseudomonas putida* CSV86'. *Biodegradation* 19 (2008): 83–92.

Batie CJ, LaHaie E, and Ballou DP. 'Purification and characterization of phthalate oxygenase and phthalate oxygenase reductase from *Pseudomonas cepacia*'. *J Biol Chem* 262 (1987): 1510–1518.

Bertilsson S, and Widenfalk A. 'Photochemical degradation of PAHs in freshwaters and their impact on bacterial growth influence of water chemistry'. *Hydrobiologia* 469 (2002): 23–32.

Blummer M, and Youngblood WW. 'Polycyclic aromatic hydrocarbons in cells and resin sediments'. *Science* 188 (1975): 53–55.

Bollag JM, Czaplicki EJ, and Minard RD. 'Bacterial metabolism of 1-naphthol'. *J Agric Food Chem* 23 (1975): 85–90.

Bondarenko S, and Gan J. 'Degradation and sorption of selected organophosphate and carbamate insecticides in urban stream sediments'. *Environ Toxicol Chem* 23 (2004): 1809–1814.

Boundy-Mills KL, de Souza ML, Mandelbaum RT, Wackett LP, and Sadowsky MJ. 'The *atzB* gene of *Pseudomonas* sp. strain ADP encodes the second enzyme of a novel atrazine degradation pathway'. *Appl Environ Microbiol* 63 (1997): 916–923.

Ceremonie H, Boubakri H, Mavingui P, Simonet P, and Vogel TM. 'Plasmid-encoded γ-hexachlorocyclohexane degradation genes and insertion sequences in *Sphingobium francense* (ex-*Sphingomonas paucimobilis* Sp+)'. *FEMS Microbiol Lett* 257 (2006): 243–252.

Cerniglia CE. 'Biodegradation of polycyclic aromatic hydrocarbons'. *Current Opinion in Biotechnology* 4 (1993): 331–338.

Chang BV, Yang CM, Cheng CH, and Yuan SY. 'Biodegradation of phthalate esters by two bacterial strains'. *Chemosphere* 55 (2004): 533–538.

Chapalamadugu S, and Chaudhry GR. 'Hydrolysis of carbaryl by a *Pseudomonas* sp. and construction of a microbial consortium that completely metabolizes carbaryl'. *Appl Environ Microbiol* 57 (1991): 744–750.

Choi KY, Kim D, Sul WJ et al. 'Molecular and biochemical analysis of phthalate and terephthalate degradation by *Rhodococcus* sp. strain DK17'. *FEMS Microbiol Lett* 252 (2005): 207–213.

Choi EJ, Jin HM, Lee SH, Math RK, Madsen EL, and Jeon CO. 'Comparative genomic analysis and benzene, toluene, ethylbenzene, and *o*-, *m*-, and *p*-xylene (BTEX) degradation pathways of *Pseudoxanthomonas spadix* BD-a59'. *Appl Environ Microbiol* 79 (2013): 663–671.

Chowdhury PP, Sarkar J, Basu S, and Dutta TK. 'Metabolism of 2-hydroxy-1-naphthoic acid and naphthalene *via* gentisic acid by distinctly different sets of enzymes in *Burkholderia* sp. strain BC1'. *Microbiology* 160 (2014): 892–902.

Conney AH. 'Induction of microsomal enzymes by foreign chemicals and carcinogenesis by polycyclic aromatic hydrocarbons: G. H. A. Clowes Memorial Lecture'. *Cancer Res* 42 (1982): 4875–4917.

Coppotelli BM, Ibarrolaza A, Del Panno MT, and Morelli IS. 'Effects of the inoculant strain *Sphingomonas paucimobilis* 20006FA on soil bacterial community and biodegradation in phenanthrene-contaminated soil'. *Microb Ecol* 55 (2008): 173–183.

Davey JF, and Gibson DT. 'Bacterial metabolism of *para*- and *meta*-xylene: Oxidation of a methyl substituent'. *J Bacteriol* 119 (1974): 923–929.

de Souza ML, Wackett LP, and Sadowsky MJ. 'The *atzABC* genes encoding atrazine catabolism are located on a self-transmissible plasmid in *Pseudomonas* sp. strain ADP'. *Appl Environ Microbiol* 64 (1998): 2323–2326.

Dean-Ross D, Moody JD, Freeman JP, Doerge DR, and Cerniglia CE. 'Metabolism of anthracene by a *Rhodococcus* species'. *FEMS Microbiol Lett* 204 (2001): 205–211.

Deveryshetty J, and Phale PS. 'Biodegradation of phenanthrene by *Pseudomonas* sp. strain PPD: Purification and characterization of 1-hydroxy-2-naphthoic acid dioxygenase'. *Microbiology* 155 (2009): 3083–3091.

Deveryshetty J, and Phale PS. 'Biodegradation of phenanthrene by *Alcaligenes* sp. strain PPH: Partial purification and characterization of 1-hydroxy-2-naphthoic acid hydroxylase'. *FEMS Microbiol Lett* 311 (2010): 93–101.

DiDonato RJ, Jr., Young ND, Butler JE et al. 'Genome sequence of the *Deltaproteobacteria* strain NaphS2 and analysis of differential gene expression during anaerobic growth on naphthalene'. *PLoS One* 5 (2010): e14072.

Doddamani HP, and Ninnekar HZ. 'Biodegradation of phenanthrene by a *Bacillus* species'. *Curr Microbiol* 41 (2000): 11–14.

Doddamani HP, and Ninnekar HZ. 'Biodegradation of carbaryl by a *Micrococcus* species'. *Curr Microbiol* 43 (2001): 69–73.

Dua A, Malhotra J, Saxena A, Khan F, and Lal R. '*Devosia lucknowensis* sp. nov., a bacterium isolated from hexachlorocyclohexane (HCH) contaminated pond soil'. *J Microbiol* 51 (2013): 689–694.

Dueholm MS, Albertsen M, D'Imperio S et al. 'Complete genome sequences of *Pseudomonas monteilii* SB3078 and SB3101, two benzene-, toluene-, and ethylbenzene-degrading bacteria used for bioaugmentation'. *Genome Announc* 2 (2014): e00524-14.

Eaton RW, and Ribbons DW. 'Metabolism of dibutylphthalate and phthalate by *Micrococcus* sp. strain 12B'. *J Bacteriol* 151 (1982): 48–57.

Eaton RW, and Chapman PJ. 'Bacterial metabolism of naphthalene: Construction and use of recombinant bacteria to study ring cleavage of 1,2-dihydroxynaphthalene and subsequent reactions'. *J Bacteriol* 174 (1992): 7542–7554.

Eaton RW. 'Organization and evolution of naphthalene catabolic pathways: Sequence of the DNA encoding 2-hydroxychromene-2-carboxylate isomerase and trans-o-hydroxybenzylidenepyruvate hydratase-aldolase from the NAH7 plasmid'. *J Bacteriol* 176 (1994): 7757–7762.

Elespuru R, Lijinsky W, and Setlow JK. 'Nitrosocarbaryl as a potent mutagen of environmental significance'. *Nature* 247 (1974): 386–387.

Endo R, Kamakura M, Miyauchi K et al. 'Identification and characterization of genes involved in the downstream degradation pathway of γ-hexachlorocyclohexane in *Sphingomonas paucimobilis* UT26'. *J Bacteriol* 187 (2005): 847–853.

Eriksson M, Sodersten E, Yu Z, Dalhammar G, and Mohn WW. 'Degradation of polycyclic aromatic hydrocarbons at low temperature under aerobic and nitrate-reducing conditions in enrichment cultures from northern soils'. *Appl Environ Microbiol* 69 (2003): 275–284.

Evans WC, Fernley HN, and Griffiths E. 'Oxidative metabolism of phenanthrene and anthracene by soil pseudomonads. The ring-fission mechanism'. *Biochem J* 95 (1965): 819–831.

Evans WC, and Fuchs G. 'Anaerobic degradation of aromatic compounds'. *Annu Rev Microbiol* 42 (1988): 289–317.

Fahy A, Ball AS, Lethbridge G, McGenity TJ, and Timmis KN. 'High benzene concentrations can favour Gram-positive bacteria in groundwaters from a contaminated aquifer'. *FEMS Microbiol Ecol* 65 (2008): 526–533.

Fazlurrahman BM, Pandey J, Suri CR, and Jain RK. 'Isolation and characterization of an atrazine-degrading *Rhodococcus* sp. strain MB-P1 from contaminated soil'. *Lett Appl Microbiol* 49 (2009): 721–729.

Felsot A, Maddox JV, and Bruce W. 'Enhanced microbial degradation of carbofuran in soils with histories of Furadan use'. *Bull Environ Contam Toxicol* 26 (1981): 781–788.

Fries MR, Zhou J, Chee-Sanford J, and Tiedje JM. 'Isolation, characterization, and distribution of denitrifying toluene degraders from a variety of habitats'. *Appl Environ Microbiol* 60 (1994): 2802–2810.

Furukawa K, Simon JR, and Chakrabarty AM. 'Common induction and regulation of biphenyl, xylene/toluene, and salicylate catabolism in *Pseudomonas paucimobilis*'. *J Bacteriol* 154 (1983): 1356–1362.

Gao Y, Truong YB, Cacioli P, Butler P, and Kyratzis IL. 'Bioremediation of pesticide contaminated water using an organophosphate degrading enzyme immobilized on nonwoven polyester textiles'. *Enzyme Microb Technol* 54 (2014): 38–44.

Garner RC. 'The role of DNA adducts in chemical carcinogenesis'. *Mutat Res* 402 (1998): 67–75.

Gesler RM. 'Toxicology of di-2-ethylhexyl phthalate and other phthalic acid ester plasticizers'. *Environ Health Perspect* 3 (1973): 73–79.

Gibson DT, Gschwendt B, Yeh WK, and Kobal VM. 'Initial reactions in the oxidation of ethylbenzene by *Pseudomonas putida*'. *Biochemistry* 12 (1973): 1520–1528.

Gibson DT, and Parales RE. 'Aromatic hydrocarbon dioxygenases in environmental biotechnology'. *Curr Opin Biotechnol* 11 (2000): 236–243.

Gibson DT, Resnick SM, Lee K et al. 'Desaturation, dioxygenation, and monooxygenation reactions catalyzed by naphthalene dioxygenase from *Pseudomonas* sp. strain 9816-4'. *J Bacteriol* 177 (1995): 2615–2621.

Gilbson DT. 'Microbial degradation of aromatic compounds'. *Science* 161 (1968): 1093–1097.

Goldman R, Enewold L, Pellizzari E et al. 'Smoking increases carcinogenic polycyclic aromatic hydrocarbons in human lung tissue'. *Cancer Res* 61 (2001): 6367–6371.

Govantes F, Porrua O, Garcia-Gonzalez V, and Santero E. 'Atrazine biodegradation in the lab and in the field: Enzymatic activities and gene regulation'. *Microb Biotechnol* 2 (2009): 178–185.

Gu JD, Li J, and Wang Y. 'Biochemical pathway and degradation of phthalate ester isomers by bacteria'. *Water Sci Technol* 52 (2005): 241–248.

Hayatsu M, Mizutani A, Hashimoto M, Sato K, and Hayano K. 'Purification and characterization of carbaryl hydrolase from *Arthrobacter* sp. RC100'. *FEMS Microbiol Lett* 201 (2001): 99–103.

Hayatsu M, and Nagata T. 'Purification and characterization of carbaryl hydrolase from *Blastobacter* sp. strain M501'. *Appl Environ Microbiol* 59 (1993): 2121–2125.

Higson FK, and Focht DD. 'Degradation of 2-methylbenzoic acid by *Pseudomonas cepacia* MB2'. *Appl Environ Microbiol* 58 (1992): 194–200.

IARC. 'Toluene', (71) 1989. IARC Monographs on the Evaluation of Carcinogenic Risks to Humans.

IARC. 'Some industrial chemicals. Ethylbenzene'. (77). 22-8-2000. IARC Monographs on the Evaluation of Carcinogenic Risks to Humans.

Iwabuchi T, and Harayama S. 'Biochemical and molecular characterization of 1-hydroxy-2-naphthoate dioxygenase from *Nocardioides* sp. KP7'. *J Biol Chem* 273 (1998): 8332–8336.

Jaeger RJ, and Rubin RJ. 'Extraction, localization, and metabolism of di-2-ethylhexyl phthalate from PVC plastic medical devices'. *Environ Health Perspect* 3 (1973): 95–102.

Jang JY, Kim D, Bae HW et al. 'Isolation and characterization of a *Rhodococcus* sp. strain able to grow on *ortho-* and *para*-xylene'. *J Microbiol* 43 (2005): 325–330.

Jauregui R, Rodelas B, Geffers R, Boon N, Pieper DH, and Vilchez-Vargas R. 'Draft genome sequence of the naphthalene degrader *Herbaspirillum* sp. strain RV1423'. *Genome Announc* 2 (2014).

Jerina DM, Selander H, Yagi H et al. 'Dihydrodiols from anthracene and phenanthrene'. *J Am Chem Soc* 98 (1976): 5988–5996.

Jeon CO, Park M, Ro HS, Park W, and Madsen EL. 'The naphthalene catabolic (*nag*) genes of *Polaromonas naphthalenivorans* CJ2: Evolutionary implications for two gene clusters and novel regulatory control'. *Appl Environ Microbiol* 72 (2006): 1086–1095.

Jin HM, Choi EJ, and Jeon CO. 'Isolation of a BTEX-degrading bacterium, *Janibacter* sp. SB2, from a sea-tidal flat and optimization of biodegradation conditions'. *Bioresour Technol* 145 (2013): 57–64.

Jindrova E, Chocova M, Demnerova K, and Brenner V. 'Bacterial aerobic degradation of benzene, toluene, ethylbenzene and xylene'. *Folia Microbiol (Praha)* 47 (2002): 83–93.

Kandyala R, Raghavendra SP, and Rajasekharan ST. 'Xylene: An overview of its health hazards and preventive measures'. *J Oral Maxillofac Pathol* 14 (2010): 1–5.

Kauppi B, Lee K, Carredano E et al. 'Structure of an aromatic-ring-hydroxylating dioxygenase-naphthalene 1,2-dioxygenase'. *Structure* 6 (1998): 571–586.

Karlsson A, Parales JV, Parales RE, Gibson DT, Eklund H, and Ramaswamy S. 'Crystal structure of naphthalene dioxygenase: Side-on binding of dioxygen to iron'. *Science* 299 (2003): 1039–1042.

Khan AA, and Husain Q. 'Potential of plant polyphenol oxidases in the decolorization and removal of textile and non-textile dyes'. *J Environ Sci (China)* 19 (2007): 396–402.

Kim SJ, Kweon O, Jones RC, Freeman JP, Edmondson RD, and Cerniglia CE. 'Complete and integrated pyrene degradation pathway in *Mycobacterium vanbaalenii* PYR-1 based on systems biology'. *J Bacteriol* 189 (2007): 464–472.

Kiyohara H, Torigoe S, Kaida N, Asaki T, Iida T, Hayashi H, and Takizawa N. 'Cloning and characterization of a chromosomal gene cluster, pah, that encodes the upper pathway for phenanthrene and naphthalene utilization by *Pseudomonas putida* OUS82'. *J Bacteriol* 176 (1994): 2439–2443.

Kiyohara H, Nagao K, and Nomi R. 'Degradation of phenanthrene through o-phthalate by an *Aeromonas* sp'. *Agric Biol Chem* 40 (1976): 1075–1082.

Kozumbo WJ, Kroll R, and Rubin RJ. 'Assessment of the mutagenicity of phthalate esters'. *Environ Health Perspect* 45 (1982): 103–109.

Kuritz T, and Wolk CP. 'Use of filamentous cyanobacteria for biodegradation of organic pollutants'. *Appl Environ Microbiol* 61 (1995): 234–238.

Lal R, Pandey G, Sharma P et al. 'Biochemistry of microbial degradation of hexachlorocyclohexane and prospects for bioremediation'. *Microbiol Mol Biol Rev* 74 (2010): 58–80.

Larkin MJ, and Day MJ. 'The metabolism of carbaryl by three bacterial isolates, *Pseudomonas* spp. (NCIB 12042 & 12043) and *Rhodococcus* sp. (NCIB 12038) from garden soil'. *J Appl Bacteriol* 60 (1986): 233–242.

Lee DH, Cody RD, Kim DJ, and Choi S. 'Effect of soil texture on surfactant-based remediation of hydrophobic organic-contaminated soil'. *Environ Int* 27 (2002): 681–688.

Lee K, and Gibson DT. 'Toluene and ethylbenzene oxidation by purified naphthalene dioxygenase from *Pseudomonas* sp. strain NCIB 9816-4'. *Appl Environ Microbiol* 62 (1996): 3101–3106.

Liang DW, Zhang T, Fang HH, and He J. 'Phthalates biodegradation in the environment'. *Appl Microbiol Biotechnol* 80 (2008): 183–198.

Mahajan MC, Phale PS, and Vaidyanathan CS. 'Evidence for the involvement of multiple pathways in the biodegradation of 1- and 2-methylnaphthalene by *Pseudomonas putida* CSV86'. *Arch Microbiol* 161 (1994): 425–433.

Malmstrom BG. 'Enzymology of oxygen'. *Annu Rev Biochem* 51 (1982): 21–59.

Mastrangelo G, Fadda E, and Marzia V. 'Polycyclic aromatic hydrocarbons and cancer in man'. *Environ Health Perspect* 104 (1996): 1166–1170.

Monticello DJ, Bakker D, Schell M, and Finnerty WR. 'Plasmid-borne Tn5 insertion mutation resulting in accumulation of gentisate from salicylate'. *Appl Environ Microbiol* 49 (1985): 761–764.

Moody JD, Freeman JP, Doerge DR, and Cerniglia CE. 'Degradation of phenanthrene and anthracene by cell suspensions of *Mycobacterium* sp. strain PYR-1'. *Appl Environ Microbiol* 67 (2001): 1476–1483.

Mulligan CN. (2005). Environmental applications for biosurfactants. *Environ. Pollut.* 133, 183–198.

Nagata Y, Endo R, Ito M, Ohtsubo Y, and Tsuda M. 'Aerobic degradation of lindane (γ-hexachlorocyclohexane) in bacteria and its biochemical and molecular basis'. *Appl Microbiol Biotechnol* 76 (2007): 741–752.

Nagata Y, Ohtomo R, Miyauchi K, Fukuda M, Yano K, and Takagi M. 'Cloning and sequencing of a 2,5-dichloro-2,5-cyclohexadiene-1,4-diol dehydrogenase gene involved in the degradation of γ-hexachlorocyclohexane in *Pseudomonas paucimobilis*'. *J Bacteriol* 176 (1994): 3117–3125.

Nagata Y, Ohtsubo Y, Endo R et al. 'Complete genome sequence of the representative γ-hexachlorocyclohexane-degrading bacterium *Sphingobium japonicum* UT26'. *J Bacteriol* 192 (2010): 5852–5853.

Nagy I, Compernolle F, Ghys K, Vanderleyden J, and De MR. 'A single cytochrome P-450 system is involved in degradation of the herbicides EPTC (*S*-ethyl dipropylthiocarbamate) and atrazine by *Rhodococcus* sp. strain NI86/21'. *Appl Environ Microbiol* 61 (1995): 2056–2060.

Niharika N, Sangwan N, Ahmad S, Singh P, Khurana JP, and Lal R. 'Draft genome sequence of *Sphingobium chinhatense* strain IP26T, isolated from a hexachlorocyclohexane dumpsite'. *Genome Announc* 1 (2013): e00680-13.

Obulakondaiah M, Sreenivasulu C, and Venkateswarlu K. 'Nontarget effects of carbaryl and its hydrolysis product, 1-naphthol, towards *Anabaena torulosa*'. *Biochem Mol Biol Int* 29 (1993): 703–710.

Ohisa N, Yamaguchi M, and Kurihara N. 'Lindane degradation by cell-free extracts of *Clostridium rectum*'. *Arch Microbiol* 125 (1980): 221–225.

Olsen RH, Kukor JJ, and Kaphammer B. 'A novel toluene-3-monooxygenase pathway cloned from *Pseudomonas pickettii* PKO1'. *J Bacteriol* 176 (1994): 3749–3756.

Olsen RH, Mikesell MD, Kukor JJ, and Byrne AM. 'Physiological attributes of microbial BTEX degradation in oxygen-limited environments'. *Environ Health Perspect* 103 Suppl 5 (1995): 49–51.

Parales RE, Ditty JL, and Harwood CS. 'Toluene-degrading bacteria are chemotactic towards the environmental pollutants benzene, toluene, and trichloroethylene'. *Appl Environ Microbiol* 66 (2000): 4098–4104.

Patel TR, and Barnsley EA. 'Naphthalene metabolism by pseudomonads: Purification and properties of 1,2-dihydroxynaphthalene oxygenase'. *J Bacteriol* 143 (1980): 668–673.

Patel TR, and Gibson DT. 'Purification and propeties of (plus)-*cis*-naphthalene dihydrodiol dehydrogenase of *Pseudomonas putida*'. *J Bacteriol* 119 (1974): 879–888.

Pushiri H, Pearce SL, Oakeshott JG, Russell RJ, and Pandey G. 'Draft genome sequence of *Pandoraea* sp strain SD6-2, isolated from lindane-contaminated Australian soil'. *Genome Announc* 1 (2013): e00415-13.

Rajan T, and Malathi N. 'Health hazards of xylene: A literature review'. *J Clin Diagn Res* 8 (2014): 271–274.

Rajgopal BS, Rao VR, Nagendraappa G, and Sethunathan N. 'Metabolism of carbaryl and carbofuran by soil enrichment and bacterial cultures'. *Can J Microbiol* 30 (1984): 1458–1466.

Redmond DE, Jr. 'Tobacco and cancer: The first clinical report, 1761'. *N Engl J Med* 282 (1970): 18–23.

Rezazadeh AM, Naghavi KZ, Zayeri F, Salehpour S, and Seyedi MD. 'Occupational exposure of petroleum depot workers to BTEX compounds'. *Int J Occup Environ Med* 3 (2012): 39–44.

Rivas FJ. 'Polycyclic aromatic hydrocarbons sorbed on soils: A short review of chemical oxidation based treatments'. *J Hazard Mater* 138 (2006): 234–251.

Sadowsky MJ, Tong Z, de SM, and Wackett LP. '*AtzC* is a new member of the amidohydrolase protein superfamily and is homologous to other atrazine-metabolizing enzymes'. *J Bacteriol* 180 (1998): 152–158.

Sagarkar S, Nousiainen A, Shaligram S et al. 'Soil mesocosm studies on atrazine bioremediation'. *J Environ Manage* 139 (2014): 208–216.

Sah S, and Phale PS. '1-naphthol 2-hydroxylase from *Pseudomonas* sp. strain C6: Purification, characterization and chemical modification studies'. *Biodegradation* 22 (2011): 517–526.

Sahu SK, Patnaik KK, Sharmila M, and Sethunathan N. 'Degradation of α-, β-, and γ-hexachlorocyclohexane by a soil bacterium under aerobic conditions'. *Appl Environ Microbiol* 56 (1990): 3620–3622.

Schwartz E, and Scow KM. 'Repeated inoculation as a strategy for the remediation of low concentrations of phenanthrene in soil'. *Biodegradation* 12 (2001): 201–207.

Seo JS, Keum YS, and Li QX. 'Metabolomic and proteomic insights into carbaryl catabolism by *Burkholderia* sp. C3 and degradation of ten *N*-methylcarbamates'. *Biodegradation* (2013): Epub ahead of print 10.1007/s10532-013-9629-2-doi.

Shapir N, Osborne JP, Johnson G, Sadowsky MJ, and Wackett LP. 'Purification, substrate range, and metal center of *atz*C: The *N*-isopropylammelide aminohydrolase involved in bacterial atrazine metabolism'. *J Bacteriol* 184 (2002): 5376–5384.

Shapir N, Sadowsky MJ, and Wackett LP. 'Purification and characterization of allophanate hydrolase (*atzF*) from *Pseudomonas* sp. strain ADP'. *J Bacteriol* 187 (2005): 3731–3738.

Shaw JP, and Harayama S. 'Purification and characterisation of the NADH: Acceptor reductase component of xylene monooxygenase encoded by the TOL plasmid pWW0 of *Pseudomonas putida* mt-2'. *Eur J Biochem* 209 (1992): 51–61.

Shea TB, and Berry ES. 'Toxicity and intracellular localization of carbaryl and 1-naphthol in cell cultures derived from goldfish'. *Bull Environ Contam Toxicol* 30 (1983): 99–104.

Shields MS, Reagin MJ, Gerger RR, Campbell R, and Somerville C. 'TOM, a new aromatic degradative plasmid from *Burkholderia (Pseudomonas) cepacia* G4'. *Appl Environ Microbiol* 61 (1995): 1352–1356.

Shinoda Y, Sakai Y, Uenishi H et al. 'Aerobic and anaerobic toluene degradation by a newly isolated denitrifying bacterium, *Thauera* sp. strain DNT-1'. *Appl Environ Microbiol* 70 (2004): 1385–1392.

Singh AK, Garg N, Lata P et al. '*Pontibacter indicus* sp. nov., isolated from hexachlorocyclohexane-contaminated soil'. *Int J Syst Evol Microbiol* 64 (2014): 254–259.

Skupinska K, Misiewicz I, and Kasprzycka-Guttman T. 'Polycyclic aromatic hydrocarbons: Physicochemical properties, environmental appearance and impact on living organisms'. *Acta Pol Pharm* 61 (2004): 233–240.

Smith MR. 'The biodegradation of aromatic hydrocarbons by bacteria'. *Biodegradation* 1 (1990): 191–206.

Stegeman JJ, Schlezinger JJ, Craddock JE, and Tillitt DE. 'Cytochrome P450 1A expression in midwater fishes: Potential effects of chemical contaminants in remote oceanic zones'. *Environ Sci Technol* 35 (2001): 54–62.

Sud RK, Sud AK, and Gupta KG. 'Degradation of sevin (1-naphthyl-*N*-methyl carbamate) by *Achromobacter* sp'. *Arch Mikrobiol* 87 (1972): 353–358.

Swetha VP, and Phale PS. 'Metabolism of carbaryl via 1,2-dihydroxynaphthalene by soil isolates *Pseudomonas* sp. strains C4, C5, and C6'. *Appl Environ Microbiol* 71 (2005): 5951–5956.

Tal A, Rubbin B, and Katan J. 'Accelerated degradation of thiocarbamate herbicides in Israeli soils following repeated use of vernolate'. *Pestic Sci* 25 (1989): 343–353.

Tam NF, and Wong YS. 'Effectiveness of bacterial inoculum and mangrove plants on remediation of sediment contaminated with polycyclic aromatic hydrocarbons'. *Mar Pollut Bull* 57 (2008): 716–726.

Urgun-Demirtas M, Stark B, and Pagilla K. 'Use of genetically engineered microorganisms (GEMs) for the bioremediation of contaminants'. *Crit Rev Biotechnol* 26 (2006): 145–164.

Vamsee-Krishna C, Mohan Y, and Phale PS. 'Biodegradation of phthalate isomers by *Pseudomonas aeruginosa* PP4, *Pseudomonas* sp.strain PPD and *Acinetobacter lwoffii* ISP4'. *Appl Microbiol Biotechnol* 72 (2006): 1263–1269.

van Herwijnen R, Springael D, Slot P, Govers HA, and Parsons JR. 'Degradation of anthracene by *Mycobacterium* sp. strain LB501T proceeds *via* a novel pathway, through *o*-phthalic acid'. *Appl Environ Microbiol* 69 (2003): 186–190.

Vogel TM. 'Bioaugmentation as a soil bioremediation approach'. *Curr Opin Biotechnol* 7 (1996): 311–316.

Wackett LP, Kwart LD, and Gibson DT. 'Benzylic monooxygenation catalyzed by toluene dioxygenase from *Pseudomonas putida*'. *Biochemistry* 27 (1988): 1360–1367.

Wackett LP, Sadowsky MJ, Martinez B, and Shapir N. 'Biodegradation of atrazine and related *s*-triazine compounds: From enzymes to field studies'. *Appl Microbiol Biotechnol* 58 (2002): 39–45.

Wang YZ, Zhou Y, and Zylstra GJ. 'Molecular analysis of isophthalate and terephthalate degradation by *Comamonas testosteroni* YZW-D'. *Environ Health Perspect* 103 Suppl 5 (1995): 9–12.

Walker N, Janes NF, Spokes JR, and van Berkum P. 'Degradation of 1-naphthol by a soil pseudomonad'. *J Appl Bacteriol* 39 (1975): 281–286.

Watts RJ, Stanton PC, Howsawkeng J, and Teel AL. 'Mineralization of a sorbed polycyclic aromatic hydrocarbon in two soils using catalyzed hydrogen peroxide'. *Water Res* 36 (2002): 4283–4292.

Wen S, Zhao J, Sheng G, Fu J, and Peng P. 'Photocatalytic reactions of phenanthrene at TiO_2/water interfaces'. *Chemosphere* 46 (2002): 871–877.

Willett KL, Ulrich EM, and Hites RA. 'Differential toxicity and environmental fates of hexachlorocyclohexane isomers'. *Environ Sci Technol* 32 (1998): 2197–2207.

Wilson GD, d'Arcy DM, and Cohen GM. 'Selective toxicity of 1-naphthol to human colorectal tumour tissue'. *Br J Cancer* 51 (1985): 853–863.

Worsey MJ, and Williams PA. 'Metabolism of toluene and xylenes by *Pseudomonas putida* mt-2: Evidence for a new function of the TOL plasmid'. *J Bacteriol* 124 (1975): 7–13.

Xie S, Sun W, Luo C, and Cupples AM. 'Novel aerobic benzene degrading microorganisms identified in three soils by stable isotope probing'. *Biodegradation* 22 (2011): 71–81.

Yen KM, Karl MR, Blatt LM et al. 'Cloning and characterization of a *Pseudomonas mendocina* KR1 gene cluster encoding toluene-4-monooxygenase'. *J Bacteriol* 173 (1991): 5315–5327.

Yen KM, and Serdar CM. 'Genetics of naphthalene catabolism in pseudomonads'. *Crit Rev Microbiol* 15 (1988): 247–268.

Zhou NY, Fuenmayor SL, and Williams PA. '*nag* genes of *Ralstonia* (formerly *Pseudomonas*) sp. strain U2 encoding enzymes for gentisate catabolism'. *J Bacteriol* 183 (2001): 700–708.

Zmirou D, Masclet P, Boudet C, Dor F, and Dechenaux J. 'Personal exposure to atmospheric polycyclic aromatic hydrocarbons in a general adult population and lung cancer risk assessment'. *J Occup Environ Med* 42 (2000): 121–126.

4

Laccases and Their Role in Bioremediation of Industrial Effluents

Vijaya Gupta, Neena Capalash and Prince Sharma

CONTENTS

4.1 Laccase: An Enzyme with Many Functions

Laccase is a multi-copper containing a lignolytic enzyme that catalyses the oxidation of substrates using oxygen from the environment and produces water as a by-product. The biggest advantage of this enzyme is that it requires only oxygen unlike other lignolytic enzymes that require hazardous and expensive hydrogen peroxide and thus do not meet industrial requirements. Laccase is present in all forms of life, such as plants, fungi and bacteria. Laccase was first identified in 1883 by Yoshida in the plant *Rhus vernicifera* and was designated as a *p*-diphenol oxidase until 1962. It was identified in 1993 that laccase was involved in the lignification process in plants (O'Malley et al. 1993). Plant laccases have not been well studied due to difficulty in their purification from plant extracts. Fungal laccases are well studied and applied to the industries. Bertrand and Labored recognised laccase as a fungal enzyme in 1896 (Thurston 1994). In the 1980s, laccase was isolated from the fungus *Pycnoporous chrysosporium* (Tien and Kirk 1983). The first bacterial laccase was found in *Azospirillum lipoferum* in 1994 (Diamantidis et al. 2000). An extensive search of the bacterial genome database has revealed wide occurrence of the laccase gene in bacteria (Alexandre and Zhulin 2000; Claus 2004).

4.1.1 *In Vivo* Functions of Laccases

Laccases are widely distributed taxonomically (Hoegger et al. 2006). Expression of laccase varies with the source and environmental conditions (Courty et al. 2006) participating in a variety of *in vivo* functions.

4.1.1.1 *Plants*

Lignification is the primary function in the cell wall of plants. Laccase catalyses the oxidation of monolignols to phenoxy radicals, which couple to form the lignin polymer. Many physiological roles of laccases in plants are probably yet to be identified; for example, in the case of *Arabidopsis thaliana*, the enzyme is expressed in various tissues that are not extensively lignified. Some of the known roles of laccases are iron metabolism and oxidative polymerisation of flavonoids (Pourcel et al. 2005) to produce seed pigment. Some examples of plant sources from which laccases have been isolated include sycamore maple (*Acer pseudoplatanus*), loblolly pine (*Pinus taeda*) (O'Malley et al. 1993), *Rhus vernicifera* (Benfield et al. 1964) and *Populus euramericana* (Ranocha et al. 1999). Besides lignification, plant laccases play an important role in wound healing and in the mechanism of defence against external conditions (Dwivedi et al. 2011).

4.1.1.2 *Fungi*

Basidiomycetes, duteromycetes and ascomycetes are known for producing laccases. On the contrary, lower classes of fungi are not known as laccase producers. Fungi mostly produce laccase extracellularly, and laccase activities have been determined in soil and litter due to their involvement in lignin degradation and transformation of lignin and humic acids. Fruiting body formation, fungal morphogenesis and sporulation are some other *in vivo* functions performed by laccases in fungi. Laccase activity in *Daldinia concentrica* and *Lentinus edodes* is associated with pigment formation in structures that are more rigid than a simple mycelial aggregate. This pigment formation strengthens cell-to-cell adhesion in association with oxidative polymerisation of cell wall components that may allow formation of fruiting bodies (Gary and Mark 1981; Leatham and Stahmann 1981). Laccase from plant pathogenic fungus detoxifies the toxic compounds produced by plants to protect itself and cause infection. *Cryptococcus neoformans* laccase converts host catecholamine into melanine, induces damage to the host and works as a virulence factor (Riva et al. 2006).

Laccases are commonly found in white-rotting fungi with other enzymes, such as different peroxidases. High laccase activity was measured in soil and litter associated with the degradation of lignocellulosic biomass. Laccases with these enzymes delignify and transform lignin. Some strains of the white-rot fungus *Phanerochaete chrysosporium* degrade lignin but do not synthesise laccase. Therefore, lignin can be degraded without laccase. Still, we cannot deny the role of laccase in lignin degradation because there are strains known to produce laccase only as a lignolytic enzyme (Eggert et al. 1996a). Ander and Eriksson (1976) eliminated the lignin degradability of *Sporotrichum pulverulentum* in laccase-minus mutants and recovered the ability with laccase-plus revertants. This function of laccases was further validated by a series of studies showing that laccase can take part in many of the reactions required for ligninolysis (Lundquist and Kristersson 1985; Bourbonnais and Paice 1990, 1992; Kersten et al. 1990).

4.1.1.3 *Bacteria*

In silico analysis of genomic databases revealed the occurrence of the laccase gene in bacteria (Ausec et al. 2011). The first bacterial laccase was isolated from *Azospirillum lipoferum*. The CotA protein from the *Bacillus subtilis* spore has been recognised as laccase (Hullo et al. 2001). In the case of *Bacillus*, CotA protects the spores against UV radiation or hydrogen peroxide. This role was confirmed by CotA mutants defunct in the ability to synthesise brownish spore pigment. Laccase enforces the polymerisation of residues, such as tyrosine, to dityrosine in microorganisms that function for the assembly of the heat- and UV-resistant *Bacillus* spores. On the other hand, laccases cause detoxification of phenolic compounds in nodules and roots and defend nitrogenase from oxygen in *Azotobacter* alternatively. Singh et al. (2008) showed that xenobiotics of environmental interest and natural products increase the laccase activity, which indicated that laccase in γ-*Proteobacterium* protects the cells from respiratory stress. A laccase such as phenol oxidase from *Streptomyces griseus* was closely associated with cellular differentiation of this organism. This phenol oxidase was considered as a laccase type oxidase attributable to sequence similarity and substrate selectivity. This laccase does not oxidise tyrosine but oxidises dihydroxyphenylalanine (DOPA) to generate melanin pigment (Endo et al. 2002).

4.2 Structural Properties

One monomer of laccase contains four copper atoms to perform its catalytic activity. These four copper atoms are classified into three copper centres on the basis of electronic paramagnetic resonance (EPR) signals (Figure 4.1). Type I and Type II copper ions display EPR signals. The type III copper centre contains two copper ions that are anti-ferromagnetically coupled through a bridging ligand (Piontek et al. 2002). This copper centre does not give any EPR signal. Type I copper ion coordinates with two histidines and one cysteine as conserved ligands and a fourth variable ligand. Earlier it was thought that variation in this fourth ligand influences the oxidation potential of the enzyme (Durao et al. 2006). Redox potential was divided into three ranges of values, which are high (730–780 mV), mid (470–710 mV) and low (340–490 mV), depending on the type of ligand present. However, mutation of this ligand does not affect the redox potential of laccase. When leucine was replaced with phenylalanine at the T1 centre, it did not influence the redox potential of laccases derived from *Myceliophthora thermophile* and *Rhizoctonia solani* (Xu 1997). Another possibility was that the Cu–N bond length may influence the redox potential due to the unavailability of a free electron pair of nitrogen due to an increase in distance. In case of high potential laccase from *Trametes versicolor*, ligand His458 on a short α-helix is in close proximity to the Cu ion (Piontek et al. 2002), whereas this ligand is far from the Cu ion in medium potential laccase from *Coprinopsis cinerea* due to shifting of the whole helix by the formation of the hydrogen bond between Glu460 and Ser113 (Ducros et al. 1998).

The T1 copper provides a blue colour to the protein due to its intense electronic absorption at around 600 nm. Type II copper is colourless as a result of weak absorption in the visible spectrum (Messerschmidst 1990). Type II copper is coordinated by two histidines and a water molecule. Type III copper showed electronic absorption at 330 nm. Each of the Type III copper atoms is coordinated with three histidines. Type II and Type III form a tri-nuclear centre at which the reduction of molecular oxygen takes place.

Nevertheless, it is possible to find non-blue laccases in nature (Palmieri et al. 1997); the 'white' laccases, as they are called, have been structurally

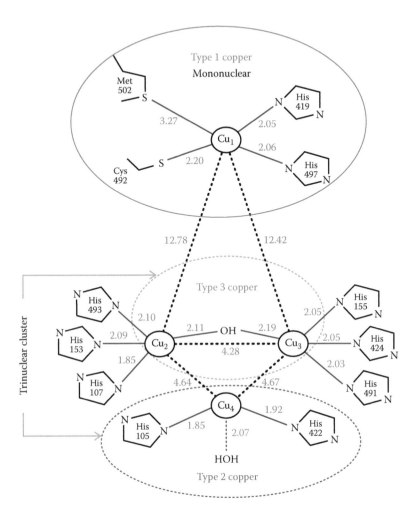

FIGURE 4.1
(See color insert.) Structure of laccase copper centres and ligands with interatomic distances. (Reprinted from *J Mol Cat B Enz*, 68, 2, Dwivedi, U. N. et al., Singh, P., Pandey, V. P., and Kumar, A., 117–128, Copyright 2011, with permission from Elsevier.)

characterised and atypically show the presence of one copper, one iron and two zinc atoms per molecule.

4.2.1 Three-Dimesional Structure of Laccases

Dwivedi et al. (2011) analysed the three-dimensional structures of laccases derived from fungi, bacteria and plants (Figure 4.2), which revealed that all were composed of three sequentially arranged cuprodoxin-like domains. These domains are primarily structured by β-barrels consisting of β-sheets and β-strands. The reaction mechanism of the laccase enzyme to oxidize

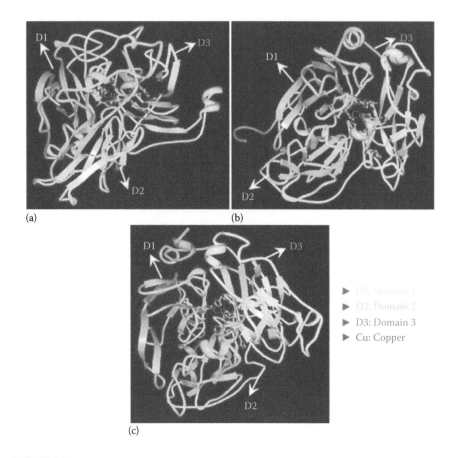

FIGURE 4.2
(See color insert.) Three-dimensional structure of (a) bacterial laccase (*Bacillus subtilis*), (b) fungal laccase (*Trametes versicolor*) and (c) plant laccase (*Populus trichocarpa*). (Reprinted from *J Mol Cat B Enz*, 68, 2, Dwivedi, U. N. et al., Singh, P., Pandey, V. P., and Kumar, A., 117–128, Copyright 2011, with permission from Elsevier.)

copper and reduce oxygen molecules is associated with its structural conservation of N- and C-terminal regions, corresponding to domains 1 and 3. Domain 3 also helps in the formation of a putative substrate binding site. Analysis of this site in all laccases revealed that bacterial laccases have larger binding cavities in comparison to those of plants and fungi.

4.3 Mechanism of Catalysis

We have studied earlier that laccase contains four copper atoms, which form three copper centres. One copper atom is named Cu T1, where the substrate is reduced, and the remaining three copper atoms form a tri-nuclear copper centre named T2/T3, where oxygen binds and is reduced to water. Laccase enzyme abstracts electrons from the substrate and carries out the reduction of molecular oxygen to water. It is assumed that four electrons queue up in the enzyme by oxidising the substrate due to the requirement of four electrons for complete reduction of oxygen to water (Equation 4.1) (Bourbonnais and Paice 1990).

$$4SH + O_2 \rightarrow 4S^{\cdot} + 2H_2O \qquad (4.1)$$

Substrate oxidation by laccase generates free radicals by one electron reaction. These reactive radicals are unstable intermediates and can lead to two kinds of reactions. One is a laccase catalysed reaction, i.e. formation of quinone from phenol, and the other is a nonenzymatic reaction, which includes dimer and trimer formations. These two laccase and nonlaccase mediated reactions perform cleavage as well as synthesis. Laccase serves this function by oxidising phenolic and nonphenolic substances; forming reactive radicals, which can separate small substituents or parts of the large molecule from the parent compound; and performing coupling reactions with other radicals or molecules that are not oxidisable by laccase. Generalised mechanism of laccase on phenol is given in Figure 4.3.

4.3.1 Nonenzymatic Reactions

Radical formation can further catalyse nonenzymatic reactions such as polymerisation and disproportionation. Dimers, trimers and polymers are synthesised after enzymatic oxidation of phenolic compounds and anilines. This cross-coupling of monomers occurs due to formation of C–C, C–O and C–N covalent bonds from the radicals formed after oxidation. This mechanism is already adopted by various reactions in the environment. For example,

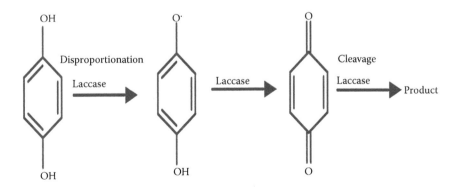

FIGURE 4.3
General mechanism for oxidation of substrate with laccase.

natural and xenobiotic phenols bound with natural phenol humus constituents in soil. The oxidation of these compounds is followed by partial decarboxylations, demethylations and dehalogenations (Sjoblad and Bollag 1977). Laccases gain importance in the bioremediation process due to their potential to detoxify the xenobiotic compounds.

Laccases carry out the lignification process in plants, oxidative coupling of catechol for cuticle sclerotisation in insects and assembly of heat- and UV-resistant spores in bacteria by utilising its capability of polymerisation. Medically (Agematu et al. 1993) and chemically important compounds and dyes (Setti et al. 1999) have been synthesised using this property of laccases.

4.3.1.1 Degradation of Lignin

As laccase generates reactive radicals, these radicals can lead to cleavage of covalent bonds and release monomers. This capacity of laccases is the basis for their potential to degrade natural polymers. For example, laccase-associated degradation of lignin generates phenoxy radicals from the phenolic hydroxyl group of lignin (Bourbonnais and Paice 1990). This radical further carried out the degradation of β-1 and β-O-4 dimers by C_α–C_β cleavage, C_α oxidation and alkyl-aryl cleavage (Figure 4.4).

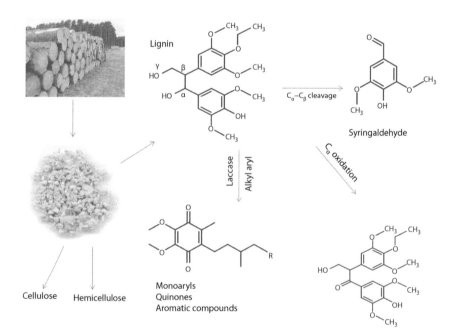

FIGURE 4.4
Schematic representation of lignin degradation by laccase.

4.3.2 Laccase-Mediator System (LMS)

Laccase has become one of the most important applied enzymes due to its simple requirement of substrate and oxygen for catalysis. Laccases oxidise a range of substrates, which are further expanded with the use of small natural low molecular weight molecules with high redox potential known as mediators. A laccase-mediated oxidation reaction depends on the redox potential of laccase and its substrate. Laccase should have a higher redox potential than the substrate to be oxidised. It is not always possible that every laccase enzyme has higher redox potential. The oxidation of non-natural substrates and those with higher redox potential is possible with the use of mediators. These mediators work as electron shuttles and help the laccase enzyme to oxidise substrate with high potential. The enzyme generates a strong oxidising intermediate called the comediator, which disperses from the enzymatic reaction centre and further oxidises the substrate (Figure 4.5), which has not been oxidised by laccase earlier due to its large size.

Small natural low molecular weight compounds or derivatives of natural compounds present in the environment and considered as natural mediators are vanillin, acetovanillone, sinapic acid, ferulic acid and p-coumaric acid. Development of new and efficient synthetic mediators, namely, 2,2'-azino-bis-(3-ethylbenzothiazoline-6-sulphonic acid) (ABTS) and 1-hydroxybenzotriazole (HBT), raised the laccase catalysis toward recalcitrant compounds (Bourbonnais and Paice 1990; Eggert et al. 1996b; Camarero et al. 2005).

4.3.3 Selection of Substrate by Laccase

The substrate specificity and affinity of laccases depend on factors such as pH, compound substitution and type of laccase. Laccase activity reduces as pH increases for substrates whose oxidation does not involve proton exchange (such as ferrocyanide). Whereas for substrates whose oxidation involves proton exchange (such as phenol), the pH activity profile of laccase can exhibit an optimal pH whose value depends on the type of laccase rather than substrate

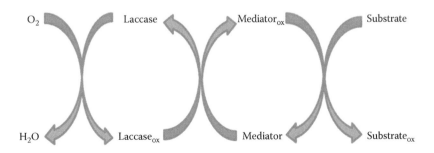

FIGURE 4.5
Laccase Mediator System.

(Rosenberg et al. 1976; Hoffmann and Esser 1977; Bourbonnais and Paice 1992; Fukushima and Kirk 1995; Xu 1996, 1997). For example, optimum pH of fungal laccases is in the range of 3 to 7 and plant laccases work at elevated pH that is above 9 (Bollag and Leonowicz 1984; Xu 1996).

Dec and Bollag (1994) investigated that the oxidation potential of laccase depends on the substituent present and the position of substituent. The oxidation potential of laccase decreased with increasing molecular weight of substrate and varied with substitution position (para > meta > ortho) (Dec and Bollag 1994). Oxidation of phenolic substrate becomes difficult when the phenoxy radical is substituted with groups such as 4-NO_2 and 4-$COCH_3$ to surrender an electron to the T1 copper. Bulky o-substituents (2,6-di-t-butyl-phenols) cause steric hindrance and interfere with interaction among the substrate and substrate pocket in laccase (Xu 1996).

Oxidation potential of laccase also depends on the redox potential difference between laccase and substrate. Xu (1996) observed that first electron transfer occurs with the 'outer sphere' mechanism. In this mechanism, the rate of reaction is mainly regulated by redox potential difference (Marcus and Sutin 1985). Toluene, fluorobenzene and anisole have been shown to be less active unlike their phenol analogs. Aryl F and aryl CH_3 with high oxidation potential (IV) (Kersten et al. 1990) have not been oxidised with laccase (Xu 1996).

However, the LMS can help overcome this steric hindrance and high redox potential of substrates. Information regarding stability, redox potential of laccase and substrates and substitution of compounds can help identify and optimise their potential for the industry accordingly.

4.4 Industrial Applications of Laccase

4.4.1 Food Industry

Laccases in the food industry have been applied for removal of phenolic compounds responsible for browning, haze formation and turbidity in clear fruit juice, beer and wine (Rodríguez and Toca 2006). Laccase from *Trametes hirsuta* has been used to improve the resistance of dough and decreased the dough extensibility in both flour and gluten dough due to its ability to carry out polymerisation (Selinheimo et al. 2006).

4.4.2 Pulp and Paper Industry

Lignin removal is a central issue in paper manufacturing, which has been traditionally handled by chemical pulping and bleaching. But an increase in pollution has led to new approaches, which can delignify pulp, and enzymes make this possible. Xylanases, peroxidases and laccases are the enzymes used for pulp bio-bleaching. Among these enzymes, laccases

seem to be promising because they oxidise a wide variety of lignin derivatives and phenolic compounds requiring molecular oxygen and a mediator
(Eugenio et al. 2010). There are several reports on delignification of pulp
using fungal laccases from *Trametes versicolor* (Archibald et al. 1997), *Coriolus
versicolor* (Balakshin et al. 2001), *Pleurotus eryngii* (Camarero et al. 2005) and
Pycnoporus cinnabarinus (Georis et al. 2003) at acidic pH. Bacterial laccase
from *Streptomyces cyaneus* (Arias et al. 2003) and *Rheinheimera indica* (Virk
et al. 2013) have been used for bleaching of kraft pulp at a laboratory level.
Bleaching of agro-based pulp (wheat straw pulp) to reduce chlorine consumption was also done with laccase from ϒ-proteobacterium (Singh et
al. 2008). Laccase treatment of pulp was also done to improve or add novel
properties (Chandra and Ragauskas 2001; Buchert et al. 2005; Schroeder et al.
2008). Laccase-mediated grafting of phenolic compounds to improve hydrophobicity of paper was also done (Chandra and Ragauskas 2002; Suurnakki
et al. 2006; Elegir et al. 2008).

The detachment of ink particles as well as contaminants with additional
chemical modifications induced by the laccase enzyme could enhance
brightness and reduce residual ink concentration (Leduc et al. 2011). There
are reports on the use of laccase-mediator systems alone and in combination
with cellulases and hemicellulases (Ibarra et al. 2011; Xu et al. 2011). Recently,
deinking of old newsprint with laccase from *Rheinheimera indica*, which does
not require a mediator, was reported (Virk et al. 2013).

4.4.3 Synthetic Chemistry

Chemically (polyaniline as conducting material, urushiol as polymeric film)
and medically important (benzofuran and napthoquinones) polymers have
been synthesised using laccases (Kobayashi et al. 2001; Karamyshev et al.
2003; Witayakran et al. 2007). Cosmetic pigments and textile and hair dyes
have also been synthesised (Jeon et al. 2010). Laccases are more advantageous for dye formulation because they eliminate the requirement of hydrogen peroxide as an oxidising agent (Lang and Cotteret 1999). Moreover,
aminonaphthoquinone compounds having anticancer activity have been
synthesised recently. Laccase catalysed the nuclear monoamination of a
1,4-naphthohydroquinone with primary aromatic amines (Wellington et al.
2012).

4.4.4 Biosensors

Laccases catalyse the electron transfer reactions without additional cofactors. Biosensors have gained importance due to their rapid detection and
estimation of molecules present in clinical and environmental samples.
Biosensors have been developed with laccase to detect various phenolic
compounds such as adrenaline (Franzoi et al. 2010), morphine and codeine
(Bauer et al. 1999), catecholamines (Quan and Shin 2004) and plant flavonoids

(Jarosz-Wilkołazka et al. 2004). Rosmarinic acid, inhibiting the HIV-1 virus and blood clots, was detected in plant extracts using a laccase-based biosensor. High sensitivity, low detection limit, low cost, renewability, simplicity and fast construction of the biosensor are some characteristics of these laccase-based biosensors (Franzoi et al. 2009).

4.5 Bioremediation of Industrial Effluents with Laccase (Table 4.1)

4.5.1 Pulp and Paper Mill Effluent

Pulp and paper mills use a large amount of chemicals to manufacture paper from its raw material. Chemicals such as NaOH, Na_2S and Na_2SO_3 are used to dissolve lignin followed by bleaching with chlorinated compounds. Recycling of paper also requires NaOH, silicates and surfactants and other chemicals. Moreover, the physical properties of paper, such as strength and hydrophobicity, are enhanced with the use of chemicals. As a result, the pulp and paper industry effluent contains a large amount of hazardous compounds. Different stages of paper manufacturing led to formation of different compounds. An aromatic biopolymer present in a plant for its protection from insects and microbial attack and lignins are released during the pulping stage.

The bio-bleaching stage discharges many chlorinated substitutes and derivatives of lignin to the effluent. These are bioaccumulative compounds and can cause hormonal disturbances and reproductive deformities. Lignin is a heterogenous, complex polymer, composed of monolignols commonly known as aromatic alcohols. The chemicals used for bleaching react with lignin and its derivatives and form highly toxic and recalcitrant compounds. Trichlorophenol, trichloro guaiacol, dichloroguaiacol and pentachlorophenol are major contaminants formed in the effluent of pulp and paper mills (Pokhrel and Viraraghavan 2004). Endocrine-disrupting compounds have been found in pulp and paper industry effluent. These compounds are dimethyl phthalate (DMP), diethyl phthalate (DEP), dibutyl phthalate (DBP) and benzyl butylphthalate (BBP) (Lattore et al. 2005; Carabias-Martínez et al. 2006). These compounds increase BOD and COD and disturb natural processes. Colour in water bodies inhibits the sunlight reaching toward aquatic plants, which disturbs the whole food chain of the ecosystem. Anaerobic treatments, such as activated sludge and biological lagoons, are ineffective for effluent treatment due to the persistence of lignins and their derivatives.

Laccase is commonly found in many lignin-degrading white-rot fungi, and this has led to speculation that laccase plays a role in wood and pulp delignification. It is believed that laccase alone catalyses the oxidation of

TABLE 4.1

Laccase-Mediated Transformation of Industrial Pollutants

Pollutants	Sources of Laccase	References
Dye (Azo, Anthraquinone, Triphenylmethane dyes)	*Trametes hirsuta, Trametes villosa, Pleurotus florida, Lentines edodes, Cerrena unicolor, T. hirsuta, Lentinus polychrous, Ganoderma lucidu, Polyporus* sp., *Cryptococcus albidus, Bacillus, Galerina*	Abadulla et al. (2000); Zille et al. (2005); Moorthi et al. (2007); D'Souza et al. (2009); Murugesan et al. (2009); Shanmugam et al. (2009); Suwannawong et al. (2010); Hadibarata et al. (2011); Mendoza et al. (2011); Thakur et al. (2014)
Textile mill effluent	*Aspergillus oryzae*	Rathnan et al. (2013); Govindwar et al. (2014)
Soda paper mill effluent, paper mill effluent	*T. versicolor, Trichoderma harzianum*	Font et al. (2003); Sadhasivam et al. (2010)
Olive oil mill effluent	*Pycnoporous coccineus, T. versicolor*	Tsioulpas et al. (2002); Aggelis et al. (2003); Jaouani et al. (2003); Chiacchierini et al. (2004); Minussi et al. (2007); Aytar et al. (2011)
Aromatic amines	*Myceliophthora thermophile*	Francíscon et al. (2010)
Phenol, anilines, chlorophenol, benzenethiols	*Rhizoctonia proticola, Recombinant laccase, T. versicolor*	Bollag et al. (1983, 1988); Xu et al. (1996); Menale et al. (2012)
2,4-Dichlorophenol, Bisphenol A	*Trametes* sp., *P. pulmonarius, Coriolus versicolor*	Nakamura et al. (2003); Rodriguez et al. (2004); Jhang et al. (2009)
Nonyphenol	*UHH-1-6-18-4 Clavariopsis*	Junghanns et al. (2005)
Benzo[a]pyrene	*P. eryngii, Fusarium santarosense, T. versicolor*	Rodriguez et al. (2004); Li et al. (2010)
Polyaromatic hydrocarbons	*Ganoderma lucidum chaain-001, T. versicolor*	Han et al. (2004); Canas et al. (2007)
Isoxaflutole	*Phanerochaete chrysosporium*	Mougin et al. (2000)
Glyphosate and isoproturon	*T. versicolor*	Farragher (2013)
Lindane and endosulfan	*Bacterial cotA*	Ulčnik et al. (2013)

phenolic lignin moieties via a one-electron abstraction. Laccase is considered to be the highly efficient enzyme for detoxification of effluent because lignin and lignin derivatives and phenolic compounds are present as natural substrates for laccase. Limited use of laccase was known for biodegradation of lignin due to steric hindrance caused by its bulky size. A laccase mediator system (LMS) overcomes this problem by increasing the oxidation potential of laccase. The role of mediators in lignin biodegradation has been well covered by Cresitini (2003). *Pycnosporus cinnabarinus* does not secrete either LiP or MnP and does secrete only laccase required for lignin breakdown by producing a metabolite that works as a mediator for enhancing the redox potential for the degradation of nonphenolic lignin and synthetic lignin by laccase (Eggert et al. 1996b).

T. versicolor laccase decreased the colour intensity of effluent by 70% of a paper mill, from brown to a clear light yellow and reduced COD. Laccase production was correlated with toxicity reduction (Font et al. 2003). Mediator usage further enhanced the decolourisation rates of effluents. *T. versicolor* laccase displayed 70% removal efficiency against chlorophenols in a bed reactor at pH 5. This degradation was independent of substitution position as revealed by the order (trichlorophenol > pentachlorophenol > dichlorophenol) in which chlorophenols were degraded. Davis and Burns (1992) polymerized phenols present in the wastewater of a paper mill using *Coriolus versicolor* laccase. Laccases generally get deactivated at the time of reaction resulting in an increase in the cost and operational instability of the process. Entrapment of laccase can increase the treatment efficiency. Therefore, Davis and Burns (1990) covalently immobilised laccase to activated carbon and eliminated colour from the pulp mill bleach effluent. Laccase-mediated biobleaching (Virk et al. 2012) and deinking (Virk et al. 2013) minimised the use of chlorine and alkali and brought down the load of pollutants indirectly in the effluent of a paper mill. The mechanism of laccase on these compounds needs to be investigated further.

4.5.2 Olive Oil Mill Effluent

Olive oil mills release large amounts of dark liquid effluents during the extraction of olive oil. This olive oil wastewater forms 50% of total biomass of olives used. Solid residue (30%) and olive oil (20%) account for the remaining 50% of the total biomass. Low pH and high organic load (lipids, pectin, polysaccharides and polyphenols) are characteristics of effluent generated by olive oil mills (Jaouani et al. 2003). Our major concern is the removal of these wastes from the environment, preventing natural resources from getting contaminated and decreasing the cost of the treatment. Detoxification using conventional physicochemical processes (simple evaporation, reverse osmosis and ultrafiltration) is costly and less effective. Olive oil mill wastewater (OMW) has a high concentration of phenolic compounds, a dark color and acidic pH (Jaouani et al. 2003). This wastewater can disturb the aquatic ecosystem of water bodies, and this food chain can cause bioaccumulation of these toxic and recalcitrant compounds to higher trophic levels. Phenolic compounds present in OMW are similar to derivatives of lignin, which makes their degradation difficult. Generally, the fate of these compounds is governed by three phenomena, namely, polymerisation, degradation and leaching. Their absorption in soil is reported, and this movement exposes the water tables to these hazardous compounds, especially in soil conditions that favour leaching (Spandre and Dellomonaco 1996). Abiotic factors promote polymerisation and oxidation of these compounds to less toxic compounds. The degradation of these compounds in the OMW (decolourisation) using *Pleurotus ostreatus, Trametes versicolor, S. Phanerochaete* spp. and other fungi has been reported (Aggelis et al. 2003; Aytar et al. 2011). In fungi, lignolytic enzymes such as manganese-dependent peroxidase and the phenol oxidase (laccase) are involved in the OMW decolourisation process. Laccases can detoxify and transform phenolic compounds in OMW (Chiacchierini et al. 2004). Laccase, the sole ligninolytic enzyme, was detected during OMW treatment using *Pleurotus ostreatus*. Tsioulpas et al. (2002) observed that *Pleurotus* strains were able to grow in OMW without any pretreatment and without additional nutrients. The decrease in dark colour of OMW was measured with the growth of the *Pleurotus* strain, and 69%–76% of the initial phenolic compounds were found to be removed.

4.5.3 Textile and Printing Mill Effluent

Dyes are the major contaminants in industrial effluent having nondegradable structures as they are utilised in various industries such as printing, textiles and pharmaceuticals. Dyes are made up of auxochromes that provide solubility in water and chromophore groups that impart colour and help in classification of dyes. Adsorption on solid support, sedimentation, filtration, electrochemical methods, sonochemical techniques and photocatalysis are some physicochemical methods used to degrade dyes. Formation of hazardous compounds, high energy requirement and high cost are some of the issues related to these physicochemical methods. However, biological agents for dye decolourisation have an advantage of less or no ill effects on the ecosystem and low energy requirements (Michniewicz et al. 2008).

Biological treatments are sometimes ineffective for degradation due to the complex structure and synthetic origin of these dyes, for example, azo dyes that resist degradation due to the presence of an N=N bond (Figure 4.6) and/or suphonate groups (SO_3H) (Steffan et al. 2005). Decolourisation of azo dyes by anaerobic treatment leads to formation of toxic aromatic amines (Martins et al. 2001). Further, these anaerobic treatments were combined with other treatments, such as aerobic approaches (Lourenço et al. 2006), electrochemical techniques (Carvalho et al. 2007) and cultivating a mixture of bacterial strains (Saratale et al. 2011) to degrade the generated aromatic amines. Azo dyes are major contributors of industrial application due to a number of chromophore groups and diverse structures. Separation of dyes from wastewater using fungi is possible due to adsorption capacity of fungal mycelium. Requirement of growth supplements, maintenance of fungal monoculture and elimination of generated biomass are some disadvantages of fungal systems. An alternative of fungal culture is the extracellular enzymes, such as manganese peroxidase and laccase. *T. trogii*, producing enzymes manganese peroxidase and laccase, decolourised dyes Ponceau 2R, malachite green and anthraquinone blue. It was seen that laccase enzyme decolourised Ponceau 2R, and most of the malachite green was decolourised by manganese peroxidase (Levin et al. 2005). Laccases offer the possibility of decolourising/

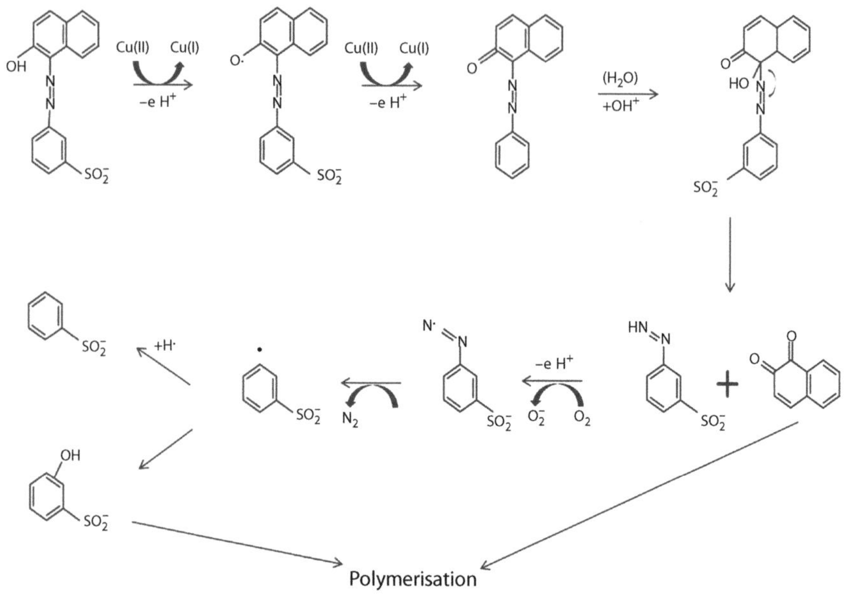

FIGURE 4.6
Mechanism of Azo dye (3-(2-hydroxy-1-nephthylazobenzenesulphonic acid)) degradation by laccase. (From Zille, A. et al., *J Mol Cat B Enz*, 33, 23–28, 2005.)

degrading dyes belonging to different classes, and this is attributed to their broad spectrum of substrates (Wesenberg et al. 2003). Laccase caused decolourisation of dyes by polymerising (Zille et al. 2005) or cleaving (Chivukula and Renganathan 1995).

The decolourisation or degradative potential of laccases depends on various factors such as pH, temperature, reaction time and redox potential of dye and laccase (Chivukula and Renganathan 1995; Zille et al. 2005). For example, the mechanism of decolourisation of azo dyes and triphenylmethane dyes is different. Laccase provides ecofriendly degradation of azo dyes into phenolic compounds unlike other peroxidases, which leads to the formation of aromatic amines. Azo dyes are subjected to oxidation by laccase, form phenoxy radicals and take forward the reaction by cleavage of azo linkage, which lead to the formation of quinone and sulphophenyldiazene-containing structures. This latter unstable structure produces a phenyldiazene radical in the presence of oxygen. The radicals formed will stabilise or react with quinone (Chivukula and Renganathan 1995; Zille et al. 2005) (Figure 4.6). Laccase activity is also affected by substituent groups of dyes. Chloro-, nitro-, 2-methyl-, 2,3-dimethyl- and 2,3-dimethoxy-substituted azo dyes have been found to be poorly decolourised in comparison to methoxy-substituted azo dyes.

In case of triphenylmethane dye, laccase promotes N-demethylation of the structure in the course of decolourisation (Figure 4.7) (Murugesan et al. 2009). This is followed by oxidation of the dye structure, which leads to its

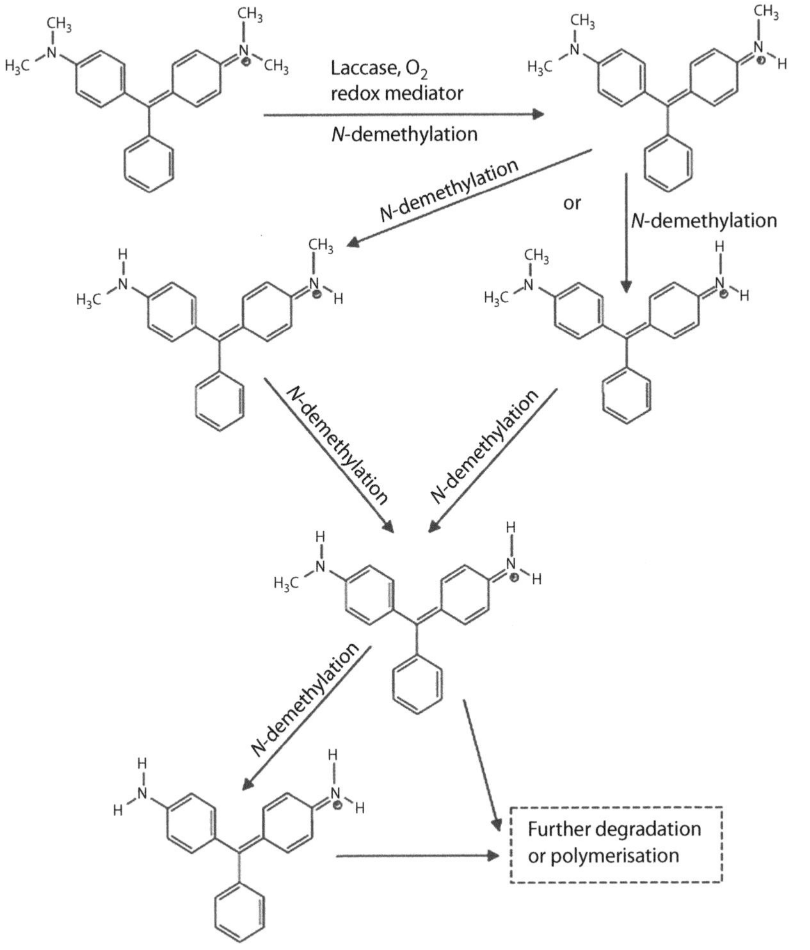

FIGURE 4.7
Degradation pathway of triphenylmethane dye (Malachite Green) by laccase. (With kind permission from Springer Science+Business Media: *J Hazard Mat*, 168, 1, 2009, Murugesan, K., Kim, Y. M., Jeon, J. R., and Chang, Y. S., 523–529.)

cleavage (Casas et al. 2009). However, Chhabra et al. (2008) showed that carbinol oxidation encouraged the breakdown of the dye structure after demethylation when the laccase is accompanied by a mediator.

Trametes versicolor laccase is the extensively studied basidiomycete. Mendoza et al. (2011) used syringaldehyde as a mediator with *T. versicolor* laccase for decolourisation of dyes Red FN-2BL, Red BWS, Remazol Blue RR and Blue 4BL, and 100%, 85%, 80% and 78% were the decolourisation yields obtained, respectively. More than 90% decolourisation of reactive yellow 15 using commercial laccase (109.8 U/L) of *T. versicolor* was identified (Tavares et al. 2009). Immobilised *T. versicolor* decolourised Reactive blue 19, an anthraquinone dye, with 97.5% removal yield (Champagne and Ramsay 2007). Laccase from *Lentines edodes* decolourised reactive yellow (74.1%), reactive blue (77.5%) and Ramazol brilliant blue R (75%) as well as detoxified (90%) the textile and paper effluent containing the dyes. Fungal laccases are advantageous due to their high redox potential, but slow growth and difficulty in controlling the degree of glycosylation prohibits application of fungal laccases. *B. subtilis* produces CotA laccase enzyme associated with spore coat surface formation. Spore laccases are quite attractive due to their stability and self-immobilisation, which avoids an additional immobilisation step. One milligram of cotA spores with laccase activity (3.4×10^2 U for ABTS as a substrate) decolourised 44.6 μg indigo carmine in 2 h. Laccase releases azo linkage upon oxidation of azo dye, which inhibits the formation of a toxic compound (Chivukula and Renganathan 1995), but no direct relationship has been observed between colour removal and reduction in toxicity of effluent (Abadulla et al. 2000). Olukanni et al. (2013) decolourised the malachite green using laccase from *Bacillus thuringiensis* RUN1 and found a direct relationship between decolourisation and laccase production. The mechanism of enzymatic degradation of dyes and toxicity of products needs to be explored (Wesenberg et al. 2003). Recently, Kumar et al. (2014) constructed a bioremediation tool (magnetic cross-linked laccase aggregates) applicable for commercial application, which can be helpful for implementation of this laccase-based technology to the industrial level.

4.5.4 Bioremediation of Polluted Soil

Soil pollution results because of the occurrence of xenobiotics, toxic altered natural compounds and agricultural and industrial wastes. Petroleum hydrocarbon, polynuclear aromatic hydrocarbons (such as naphthalene and benzo-[a]pyrene), pesticides and phenolic compounds are common soil pollutants.

4.5.4.1 *Bioremediation of Polycyclic Aromatic Hydrocarbons*

Polycyclic aromatic hydrocarbons (PAHs) are formed during the incomplete burning of oil, coal and other organic compounds. PAHs are made up of fused benzene rings with different stereochemical structures and are toxic compounds with potentially carcinogenic and mutagenic properties. Low water solubility, subsequent low degradation rates and formation of more toxic compounds are the barriers in bioremediation of PAH. Therefore, alternative ecofriendly methods having high PAH-degrading capabilities are essential. Phenanthrene is a polycyclic aromatic hydrocarbon usually present as a pollutant. Lignolytic enzymes from white-rot fungus have broad substrate specificity and potential for degradation of xenobiotics. *Phanerochaete chrysosporium* and *Trametes versicolor* have a higher removal ability of PAHs. Laccase from *T. versicolor* oxidised PAH, which was further enhanced with the addition of redox mediator (ABTS). Benzo-[a]pyrene is considered a dominant pollutant among PAHs. Products generated during the degradation of benzo-[a]pyrene by *T. versicolor* laccase displayed damage to human DNA. But these compounds are prone to further microbial degradation due to their increased solubility than the parent compound. Degradation potential against benzo-[a]pyrene has been further improved with the use of a mediator. Canas et al. (2007) showed enhanced laccase activity of *Pycnoporus cinnabarinus* against PAHs in the presence of synthetic as well as natural mediators. They reported 100% degradation of anthracene and 15% of both benzo-[a]pyrene and pyrene in 24 h without a mediator. On the other hand, with mediators (ABTS and HBT), degradation efficiency was 95% for anthracene and benzo-[a]pyrene and 50% for pyrene, respectively. Laccase from *G. lucidum* appeared to be highly effective against benzo-[a]pyrene (71.71%), acenaphthene (80.49%) and acenaphthylene (85.85%) and less effective for benzo[a]anthracene (9.14%) in the absence of a mediator (Punnapayak et al. 2009). Casas et al. (2009) revealed different mechanisms followed by different mediators, namely, ABTS, p-coumaric acid and HBT. They found varied products with the oxidation of anthracene and benzo-[a]pyrene depending on the laccase mediator system used. The laccase-ABTS system follows an electron transfer route. In contrast, HBT (nitroxyl) radicals oxidise the aromatic substrate by a hydrogen atom transfer (HAT) route (Cantarella et al. 2003). Acetosyringone and syringaldehyde did not promote PAH oxidation by laccase, which may be due to the formation of stable phenoxyl radicals of these mediators. This is due to the presence of electron-donating groups, in *ortho* positions to the phenol group of both the mediators (Marzullo et al. 1995). Sinapic, ferulic and p-coumaric acid (PCA) were considered as worse laccase substrates for *P. cinnabarinus* laccase with catalytic efficiencies 3.9, 1.4 and 0.01 s^{-1} M^{-1}, respectively. On the other hand, *P. cinnabarinus* laccase along with these mediators promoted PAH oxidation. PCA phenoxy radicals cleaved C–H in the PAH and formed RO–H in the mediator (Canas et al. 2007).

4.5.4.2 Biodegradation of Endocrine Disrupting Compounds in the Soil

Endocrine-disrupting compounds are hormone-like compounds, used in factories, farmland, golf courses and households. These compounds gained attention due to their potential to mimic natural hormones in living beings and disrupt the signal of natural hormones. These chemicals are complex in structure and contain aromatic rings due to which their degradation is difficult. Attempts were made to degrade these compounds by laccases having potential to degrade aromatic compounds. Seventy percent of bisphenol A and 60% of nonyphenol removal was observed using laccase from *T. versicolor* (Tsutsumi et al. 2001). Another study showed the use of laccase for biodegradation of bisphenol A, 2,4-dichlorophenol and diethyl phthalate that resulted in complete disappearance of bisphenol A and 2,4-dichlorophenol but not of diethyl phthalate (Nakamura et al. 2003). Laccase in the reversed micellar (RM) system with high enzymatic activity and better catalytic efficiency than laccase in aqueous media was applied for bisphenol A degradation (Chhaya and Gupte 2013). Junghanns et al. (2005) assessed the degradation of xenoestrogen nonylphenol (NP) using laccases from the mitosporic fungal strain UHH 1-6-18-4 and *Clavariopsis aquatic* and proposed mechanisms for NP degradation. Laccases promote hydroxylation of individual t-NP isomers at their branched nonyl chains followed by side chain degradation of certain isomers. Action of laccase can also lead to oxidative coupling of primary radical products to compounds with higher molecular masses.

4.5.4.3 Bioremediation of Herbicides/Pesticides in Soil

Isoproturon (selective) and glycophosate (nonselective) are herbicides used to kill weeds. Castillo et al. (2001) observed the role of ligninases for degradation of isoproturon using the white-rot fungus *Phanerochaete chrysosporium*. It was found that laccase was also responsible for degradation. This degradation ability of laccase was further investigated by Pizzul et al. (2009) on glycophosate. Optimisation of factors such as pH, immobilisation and screening of natural mediators for efficient degradation of isoproturon and glycophosate was done by Farragher (2013). Laccase-producing white-rot fungus was also shown to degrade atrazine. Vasil-chenko et al. (2002) investigated the degradation of atrazine with laccase-positive and laccase-negative *Mycelia sterialia* to determine the role of laccase. Degradation of atrazine was 80% with laccase-positive and 60%–70% with laccase-negative strains. Diketonitrile, a herbicide activated in soil and plants, is a derivative of isoxaflutole. The conversion of this diketone to benzoic acid using *Phanerochaete chrysosporium* and *Trametes versicolor* was done (Mougin 2000). Correlation between the metabolism of herbicide and production of laccase enzyme was observed. Recently Ulčnik et al. (2013) showed that degradation of two insecticides (lindane and endosulfan) was better with bacterial laccase (Bacillus CotA protein) than fungal laccase. They also considered that the degradation mechanism for these insecticides could be their absorption on mycelial biomasses in case of fungi. Partial degradation of herbicide forms aniline and phenolic compounds (Smith 1985). Bollag et al. (1983) examined the cross-coupling of anilines with phenolic humus constituents. Laccase catalysed the formation of polymers in the presence of phenolic acids with aniline or their enzymatic products, whereas laccase from *Rhizoctonia praticola* was not able to transform aniline alone. Phenolic compounds were also released from various other industries, such as wood, dye, resin and plastic as well as from petroleum refineries. Xu (1996) did the comparative analysis for oxidation of phenol, aniline and benzenethiols using fungal laccases and observed their K_m and K_{cat} values depend on the structure of the substrate. Laccases from *Rhizoctonia praticola* have been found to reverse the toxic effect of cresol and 2,6-xylenol phenols, whereas no reversal of the effect of p-chlorophenol and 2,4,5-trichlorophenol was observed (Bollag et al. 1988). Annadurai et al. (2008) optimised the medium components and growth conditions to maximise phenol degradation using laccase from *Pseudomonas putida*.

4.6 Future Direction

This chapter summarises the applications of laccase for bioremediation of industrial pollutants. However, lack of stability, production of laccase in large amounts and requirement of expensive mediators are some issues that need to be tackled for successful implementation of laccase-based remediation at the pilot or industrial scale. Molecular techniques, such as mutagenesis and directed evolution, can help overcome these problems. Also, there is a need to search for novel laccases with better bioremediation potential and capability to work with natural mediators or without mediators.

References

Abadulla, E., Tzanov, T., Costa, S., Robra, K. H., Cavaco-Paulo, A., and Gubitz, G. M. 2000. Decolorization and detoxification of textile dyes with a laccase from *Trametes hirsuta*. *Appl Environ Microbiol* 66:3357–3362.

Agematu, H., Tsuchida, T., Kominato, K. et al. 1993. Enzymatic dimerization of penicillin-X. *J Antibiot* 46:141–148.

Aggelis, G., Iconomou, D., Christou et al. 2003. Phenolic removal in a model olive oil mill wastewater using *Pleurotus ostreatus* in bioreactor cultures and biological evaluation of the process. *Water Res* 37(16):3897–3904.

Alexandre, G., and Zhulin, I. B. 2000. Laccases are widespread in bacteria. *Trends Biotechnol* 18:41–42.

Ander, P., and Eriksson, K. E. 1976. The importance of phenol oxidase activity in lignin degradation by the white-rot fungus *Sporotrichum pulverulentum*. *Arch Microbiol* 109:1–8.

Annadurai, G., Ling, L. Y., and Lee, J. 2008. Statistical optimization of medium components and growth conditions by response surface methodology to enhance phenol degradation by *Pseudomonas putida*. *J Hazard Mater* 151:171–178. doi:10.1016/j.jhazmat.2007.05.061.

Arias, M. E., Arenas, M., Rodriguez, J., Soliveri, J., and Ball, A. S. 2003. Kraft pulp biobleaching and mediated oxidation of a non-phenolic substrate by laccase from *Streptomyces cyaneus* CECT 3335. *Appl Environ Microbiol* 69:1953–1958.

Archibald, F. S., Bourbonnais, R., Jurasek, L., Paice, M. G., and Reid, I. D. 1997. Kraft pulp bleaching and delignification by *Trametes versicolor*. *J. Biotechnol* 53:215–236.

Ausec, L., Zakrzewski, M., Goesmann, A., Schlüter, A., and Mandic-Mulec, I. 2011. Bioinformatic analysis reveals high diversity of bacterial genes for laccase-like enzymes. *PLOS ONE* 6(10):1–10, e25724. doi:10.1371/journal.pone.0025724.

Aytar, P., Gedikli, S., Çelikdemir, M. et al. 2011. Dephenolization of olive mill waste-water by pellets of some white rot fungi. *J Biol Chem* 39 (4):379–390.

Balakshin, M., Chen, C. L., Gratzl, J. S., Kirkman, A. G., and Jakob, H. 2001. Biobleaching of pulp with dioxygen in laccase-mediator system effect of variables on the reaction kinetics. *J Mol Catal B: Enzym* 16:205–215.

Bauer, C. G., Kuhn, A., Gajovic, N., Skorobogatko, O., Holt, P. J., Bruce, N. C. et al. 1999. New enzyme sensors for morphine and codeine based on morphine dehy-drogenase and laccase. *Fresenius' J Anal Chem* 364:179–183.

Benfield, G., Bocks, S. M., Bromley, K., and Brown, B. R. 1964. Studies in fungal and plant laccases. *Phytochemistry* 3:79–88.

Bollag, J., and Leonowicz, A. 1984. Comparative studies of extracellular fungal laccases. *Appl Environ Microbiol* 29:849–854.

Bollag, J., Mlnard, R. D., and Llu, S. 1983. Cross-linkage between anilines and phenolic humus constituents. *Environ Sci Technol* 17:72–80.

Bollag, J. M., Shuttleworth, K. L., and Anderson, D. H. 1988. Laccase-mediated detoxification of phenolic compounds. *Appl Environ Microbiol* 54:3086–3091.

Bourbonnais, R., and Paice, M. 1990. Oxidation of non-phenolic substrates: An expanded role for laccase in lignin biodegradation. *FEBS Lett.* 267:99–102.

Bourbonnais, R., and Paice, M. G. 1992. Demethylation and delignification of kraft pulp by *Trametes versicolor* laccase in the presence of 2,2 -azinobis-(3-ethylbenz-thiazoline-6-sulphonate). *Appl Microbiol Biotechnol* 36:823–827.

Buchert, J., Gronqvist, S., Mikkonen, H. et al. 2005. Process for producing a fibrous product. European Patent WO2005061790.

Camarero, S., Ibarra, D., Martínez, M. J., Martínez, Á. T., Camarero, S., and Ibarra, D. 2005. Lignin-derived compounds as efficient laccase mediators for decolorization of different types of recalcitrant dyes. *Appl Environ Microbiol* 71:1775–1784. doi:10.1128/AEM.71.4.1775.

Carabias-Martínez, R., Rodríguez-Gonzalo, E., and Revilla-Ruiz, P. 2006. Determination of endocrine-disrupting compounds in cereals by pressurized liquid extraction and liquid chromatography-mass spectrometry. *J Chromatogr A* 1137:207–215.

Canas, A. I., Alcade, M., Plou, F., Martínez, M. J., Martínez, A. T., and Camarero, S. 2007. Transformation of polycyclic aromatic hydrocarbons by laccase is strongly enhanced by phenolic compounds present in soil. *Environ Sci Technol* 41:2964–2971.

Cantarella, G., Galli, C., and Gentili, P. 2003. Free radical versus electron-transfer routes of oxidation of hydrocarbons by laccase-mediator systems. *J Mol Catal B: Enzym* 22:135.

Carvalho, C., Fernandes, A., Lopes, A., Pinheiro, H., and Goncalves, I. 2007. Electro-chemical degradation applied to the metabolites of Acid Orange 7 anaerobic bio-treatment. *Chemosphere* 67(7):1316–1324. doi:10.1016/j.chemosphere.2006.10.062.

Casas, N., Parella, T., Vicent, T., and Caminal, G. 2009. Chemosphere metabolites from the biodegradation of triphenylmethane dyes by *Trametes versicolor* or laccase. *Chemosphere* 75(10):1344–1349. doi:10.1016/j.chemosphere.2009.02.029.

Castillo, M. P., Andersson, A., Ander, P., Stenstron, J., and Torstensson, L. 2001. Establishment of the white rot fungus *Phanerochaete chrysosporium* on unsterile straw of solid subtrate fermentation systems intended for degradation of pesti-cides. *World J Microbiol Biotechno* 17:627–633.

Champagne, P. P., and Ramsay, J. A. 2007. Reactive blue 19 decolouration by laccase immobilized on silica beads. *Appl Microbiol Biotechnol* 77(4):819–823.

Chandra, R. P., and Ragauskas, A. J. 2001. Laccase: The renegade of fiber modification. *Tappi Pulping Confer* 1041–1051.

Chandra, R. P., and Ragauskas, A. J. 2002. Evaluating laccase-facilitated coupling of phenolic acids to high-yield kraft pulps. *Enzyme Microb Technol* 30:855–861.

Chhabra, M., Mishra, S., and Sreekrishnan, T. R. 2008. Mediator-assisted decolorization and detoxification of textile dyes/dye mixture by *Cyathus bulleri* laccase. *Appl Biochem Biotechnol* 151:587–598.

Chhaya, U., and Gupte, A. 2013. Possible role of laccase from *Fusarium incarnatum* UC-14 in bioremediation of Bisphenol A using reverse micelles system. *J Hazard Mater* 254–255:149–156. doi:10.1016/j.jhazmat.2013.03.054.

Chiacchierini, E., Restuccia, D., and Vinci, G. 2004. Bioremediation of food effluents: Recent applications of free and immobilized polyphenoloxidases. *Food Sci Technol Int* 10:373–382.

Chivukula, M., and Renganathan, V. 1995. Phenolic azo dye oxidation by laccase from *Pyricularia oryzae*. *Appl Environ Microbiol* 61:4374–4377.

Claus, H. 2004. Laccases: Structure, reactions, distribution. *Micron* 35:93–96. doi: 10.1016/j.micron.2003.10.029.

Courty, P. E., Pouysegur, R., Buée, M., and Garbaye, J. 2006. Laccase and phosphatase activities of the dominant ectomycorrhizal types in a lowland oak forest. *Soil Biol Biochem* 38(6):1219–1222. doi:10.1016/j.soilbio.2005.10.005.

Cresitini, C., Jurasek, L., and Argyropoulos, D. S. 2003. On the mechanism of the laccase-mediator system in the oxidation of lignin. *Chem Eur J* 9:5371–5378.

Davis, S., and Burns, R. G. 1990. Decolorization of phenolic effluents by soluble and immobilized phenol oxidases. *Appl Microbiol Biotechnol* 32:721–726

Davis, S., and Burns, R. G. 1992. Covalent immobilization of laccase on activated carbon for phenolic effluent treatment. *Appl Microbiol Biotechnol* 37:474–479.

Dec, J., and Bollag, J. M. 1994. Dehalogenation of chlorinated phenols during oxidative coupling. *Environ Sci Technol* 28:484–490.

Diamantidis, G., Effosse, A., Potier, P., and Bally, R. 2000. Purification and characterization of the first bacterial laccase in the rhizospheric bacterium *Azospirillum lipoferum*. *Soil Biol Biochem* 32:919–927.

D'Souza-Ticlo, Sharma, D., and Raghukumar, C. A. 2009. Thermostable metaltolerant laccase with bioremediation potential from a marine derived fungus. *Mar Biotechnol* 6:725–737.

Ducros, V., Brzozowski, A. M., Wilson K. S. et al. 1998. Crystal structure of the type-2 Cu depleted laccase from *Coprinus cinereus* at 2.2 resolution. *Nature Structural Biol* 5:310–316.

Durao, P., Bento, I., Fernandes, A. T., Melo, E. P., Lindley, P. F., and Martins, L. O. 2006. Perturbations of the T1 copper site in the CotA laccase from *Bacillus subtilis*: Structural, biochemical, enzymatic and stability studies. *J Biol Inorg Chem* 11(4):514–526.

Dwivedi, U. N., Singh, P., Pandey, V. P., and Kumar, A. 2011. Enzymatic structure – Function relationship among bacterial, fungal and plant laccases. *J Mol Cat B Enz* 68(2):117–128. doi:10.1016/j.molcatb.2010.11.002.

Eggert, C., Temp, U., Eriksson, K. E., and Temp, U. 1996a. The ligninolytic system of the white rot fungus *Pycnoporus cinnabarinus*: Purification and characterization of the laccase. *Appl Environ Microbiol* 62(4):1151–1158.

Eggert, C., Temp, U., Dean, J. F. D., and Eriksson, K. L. 1996b. A fungal metabolite mediates degradation of non-phenolic lignin structures and synthetic lignin by laccase. *FEBS Lett* 391:144–148.

Elegir, G., Kindl, A., Sadocco, P., and Orlandi, M. 2008. Development of antimicrobial cellulose packaging through laccase-mediated grafting of phenolic compounds. *Enzyme Microb Technol* 43:84–92.

Endo, K., Hosono, K., Beppu, T., and Ueda, K. 2002. A novel extracytoplasmic phenol oxidase of *Streptomyces*: Its possible involvement in the onset of morphogenesis. *Microbiology* 148:1767–1776.

Eugenio, M. E., Santos, S. M., Carbajo, J. M., Martín, J. A., Martín-sampedro, R., González, A. E., and Villar, J. C. 2010. Kraft pulp biobleaching using an extracellular enzymatic fluid produced by *Pycnoporus sanguineus*. *Bioresour Technol* 101(6):1866–1870. doi:10.1016/j.biortech.2009.09.084.

Farragher, N. 2013. Degradation of pesticides by the ligninolytic enzyme laccase screening for natural mediators. PhD Diss, Swedish University of Agricultural Sciences.

Font, X., Caminal, G., Gabarrell, X., Romero, S., and Vicent, M. T. 2003. Black liquor detoxification by laccase of *Trametes versicolor* pellets. *J Chem Technol Biotechnol* 554:548–554. doi:10.1002/jctb.834.

Franciscon, E., Piubeli, F., Fantinatti-garboggini, F. et al. 2010. Enzyme and Microbial Technology Polymerization study of the aromatic amines generated by the biodegradation of azo dyes using the laccase enzyme. *Enz Microbial Technol* 46(5):360–365. doi:10.1016/j.enzmictec.2009.12.014.

Franzoi, A. C., Dupontb, J., Spinelli, A., and Vieiraa, I. C. 2009. Biosensor based on laccase and an ionic liquid for determination of rosmarinic acid in plant extracts. *Talanta* 77:1322–1327.

Franzoi, A. C., Vieira, I. C., and Dupont, J. 2010. Biosensors of laccase based on hydrophobic ionic liquids derived from imidazolium cation. *J Braz Chem Soc* 21(8):1451–1458.

Fukushima, Y., and Kirk, T. K. 1995. Laccase component of the *Ceriporiopsis subvermispora* lignin-degrading system. *Appl Environ Microbiol* 61:872–876.

Gary, F. L., and Mark, A. S. 1981. Studies on the laccase of *Lentinus edodes*: Specificity, localization and association with the development of fruiting bodies. *J Gen Microbiol* 125:147–157.

Georis, J., Lomascolo, A., Camarero, S. et al. 2003. *Pycnoporus cinnabarinus* laccases: An interesting tool for food or nonfood applications. *Commun Agric Appl Biol Sci* 68:263–266.

Govindwar, S. P., Kurade, M. B., Tamboli, D. P., Kabra, A. N., Kim, P. J., and Waghmode, T. R. 2014. Decolorization and degradation of xenobiotic azo dye Reactive Yellow-84A and textile effluent by *Galactomyces geotrichum*. *Chemosphere* 109:234–238. doi:10.1016/j.chemosphere.2014.02.009.

Hadibarata, T., Rahim, A., and Yusoff, M. 2011. Decolorization and metabolism of anthraquinonone-type dye by laccase of white-rot fungi *Polyporus* sp. *Water Air Soil Pollut* 223:933–941. doi:10.1007/s11270-011-0914-6.

Han, M., Choi, H., and Song, H. 2004. Degradation of phenanthrene by *Trametes versicolor* and its laccase. *J Microbiol* 42(2):94–98.

Hoegger, P. J., Kilaru, S., James, T. Y., Thacker, J. R., and Ku, U. 2006. Phylogenetic comparison and classification of laccase and related multicopper oxidase protein sequences. *FEBS J* 273:2308–2326. doi:10.1111/j.1742-4658.2006.05247.x.

Hoffmann, P., and Esser, K. 1977. The phenol oxidases of the ascomycete *Podospora anserina*. *Arch Microbiol* 112:111–114.

Hullo, M. F., Moszer, I., Danchin, A., and Martin-Verstraete, I. 2001. CotA of *Bacillus substilis* is a copper-dependent laccase. *J Bact* 183:5426–5430.

Ibarra, D., Monte, M. C., Blanco, A., Martinez, A. T., and Martinez, M. J. 2011. Enzymatic deinking of secondary fibers: Cellulases/hemicellulases versus laccase-mediator system. *J Ind Microbiol Biotechnol* 39:1–9.

Jaouani, A., Sayadi, S., Vanthournhout, M., and Penninckx, M. J. 2003. Potent fungi for decolorization of olive oil mill wastewaters. *Enzyme Microb Technol* 33:802–809.

Jarosz-Wilkołazka, A., Ruzgas, T., and Gorton, L. 2004. Use of laccase-modified electrode for amperometric detection of plant flavonoids. *Enzyme Microb Technol* 35:238–234.

Jeon, J. R., Kim, E. J., Murugesan, K. et al. 2010. Laccase-catalysed polymeric dye synthesis from plant-derived phenols for potential application in hair dyeing: Enzymatic colourations driven by homo- or hetero-polymer synthesis. *Microbial Biotechnol* 3:324–335.

Junghanns, C., Moeder, M., Krauss, G., and Martin, C. 2005. Degradation of the xenoestrogen nonylphenol by aquatic fungi and their laccases. *Microbiology* 151:45–57. doi:10.1099/mic.0.27431-0.

Karamyshev, A. V., Shleev, S. V., Koroleva, O. V., Yaropolov, A. I., and Sakharov, I. Y. 2003. Laccase-catalyzed synthesis of conducting polyaniline. *Enzyme Microb Technol* 33:556–564.

Kersten, P. J., Kalyanaraman, B., Hammel, K. E., Reinhammar, B., and Kirk, T. K. 1990. Comparison of lignin peroxidase, horseradish peroxidase and laccase in the oxidation of methoxybenzenes. *Biochem J* 268(2):475–480.

Kobayashi, S., Uyama, H., and Kimura, S. 2001. Enzymatic polymerization. *Chem Rev* 101:3793–3818.

Kumar, V. V., Sivanesan, S., and Cabana, H. 2014. Magnetic cross-linked laccase aggregates – Bioremediation tool for decolorization of distinct classes of recalcitrant dyes. *Sci of the Total Environ* 487:830–839.

Lang, G., and Cotteret, J. 1999. Hair dye composition containing a laccase. (L'Oreal, Fr.). *Int Pat Appl* WO9936036.

Lattore, A., Rigol, A., and Lacorte, S. 2005. Organic compounds in paper mill wastewaters. *Handb Environ Chem* 5:25–51.

Leatham, G. F., and Stahmann, M. A. 1981. Studies on the laccase EC-1.10.3.2 of *Lentinus edodes*: Specificity, localization, and association with the developments of fruiting bodies. *J Gentile Crobiol* 125:147–158.

Leduc, C., Marie, L., Roch, L., and Claude, D. 2011. Use of enzymes of enzymes in deinked pulp bleaching. *Cell Chem Technol* 45:657–663.

Levin, L., Forchiassin, F., and Viale, A. 2005. Ligninolytic enzyme production and dye decolorization by *Trametes trogii*: Application of the Plackett-Burman experimental design to evaluate nutritional requirements. *Process Biochem* 40:1381–1387.

Li, X., Lin, X., Yin, R., Wu, Y., Chu, H., Zeng, J., and Yang, T. 2010. Optimization of laccase-mediated benzo[a]pyrene oxidation and the bioremedial application in aged polycyclic aromatic hydrocarbons-contaminated soil. *J Health Sci* 56:534–540.

Lourenço, N. D., Novais, J. M., and Pinherio, H. M. 2006. Kinetic studies of reactive azo dye decolorization in anaerobic/aerobic sequencing batch reactors. *Biotechnol Lett* 28:733–739.

Lundquist, K., and Kristersson, P. 1985. Exhaustive laccase-catalysed oxidation of a lignin model compound (vanillyl glycol) produces methanol and polymeric quinoid products. *Biochem J* 229(1):277–279.

Marcus, R. A., and Sutin, N. 1985. Electron transfers in chemistry and biology. *Biochim Biophys Acta* 811:265–322.

Martins, M. A. M., Ferreira, I. C., Santos, I. M., Queiroz, M. J., and Lima N. 2001. Biodegradation of bioaccessible textile azo dyes by *Phanerochaete chrysosporium*. *J. Biotechnol* 89:91–98.

Marzullo, L., Cannio, R., Giardina, P., Santini, M. T., and Sannia, G. 1995. Veratryl alcohol oxidase from *Pleurotus ostreatus* participates in lignin biodegradation and prevents polymerization of laccase-oxidized substrates. *J Biol Chem* 270:3823–3827.

Menale, C., Nicolucci, C., Catapane, M., Rossi, S., Bencivenga, U., Mita, D. G., and Diano, N. 2012. Enzymatic optimization of operational conditions for biodegradation of chlorophenols by laccase-polyacrilonitrile beads system. *J Mol Cat B Enz* 78:38–44. doi:10.1016/j.molcatb.2012.01.021.

Mendoza, L., Jonstrup, M., Hatti-kaul, R., and Mattiasson, B. 2011. Enzyme and microbial technology azo dye decolorization by a laccase/mediator system in a membrane reactor: Enzyme and mediator reusability. *Enzyme Microbial Technol* 49(5):478–484. doi:10.1016/j.enzmictec.2011.08.006.

Messerschmidt, A. 1990. The blue oxidases, ascorbate oxidase, laccase and ceruloplasmin. Modelling and structural relationships. *Eur J Biochem* 187:341–352.

Michniewicz, A., Ledakowicz, S., Ullrich, R., and Hofrichter, M. 2008. Kinetics of the enzymatic decolorization of textile dyes by laccases from *Cerrena unicolor*. *Dyes Pigments* 77:295–302.

Minussi, R. C., Miranda, M. A., Silva, J. A., Ferreira, C. V., Aoyama, H., Marangoni, S., Rotilio, D., Pastore, G.M., and Durán N. 2007. Purification, characterization and application of laccase from *Trametes versicolor* for colour and phenolic removal of olive mill wastewater in the presence of 1-hydroxybenzotriazole. *Afr J Biotechnol* 6:1248–1254.

Moorthi, S., Periyar Selvam, S., Sasikalaveni, A., Murugesan, K., and Kalaichelvan, P. T. 2007. Decolorization of textile dyes and their effluents using white rot fungi. *Afr J Biotechnol* 6:424–429.

Mougin, C. 2000. Cleavage of the diketonitrile derivative of the herbicide isoxaflutole by extracellular fungal oxidases. *J Agric Food Chem* 48(10):4529–4534.

Mougin, C., Boyer, F. D., Caminade, E., and Rama, R. 2000. Cleavage of the diketonitrile derivative of the herbicide isoxaflutole by extracellular fungal oxidases. *J Agric Food Chem* 48:4529–4534.

Murugesan, K., Kim, Y. M., Jeon, J. R., and Chang, Y. S. 2009. Effect of metal ions on reactive dye decolorization by laccase from *Ganoderma lucidum*. *J Hazard Mat* 168(1):523–529. doi:10.1016/j.jhazmat.2009.02.075.

Nakamura, K., Kawabata, T., Yura, K., and Go, N. 2003. Novel types of two-domain multi-copper oxidases: Possible missing links in the evolution. *FEBS Lett* 553:239–244.

Olukanni, O. D., Adenopo, A., Awotula, A. O., and Osuntoki, A. A. 2013. Biodegradation of malachite green by extracellular laccase producing *Bacillus thuringiensis* RUN1. *J Basic Appl Sci* 9:543–549.

O'Malley, D. M., Whetten, R., Bao, W. L., Chen, C. L., and Sederoff, R. R. 1993. The role of laccase in lignification. *Plant J* 4:751–757.

Palmieri, G., Giardina, P., Bianco, C., Scaloni, A., Capasso, A., and Sannia, G. 1997. A novel white laccase from *Pleurotus ostreatus*. *J Biol Chem* 272:31301–31307.

Piontek, K., Antorini, M., and Choinowski, T. 2002. Crystal structure of a laccase from the fungus *Trametes versicolor* at 1.90 Å resolution containing a full complement of coppers. *J Biol Chem* 277(40):37663–37669.

Pizzul, L., Castillo, M. D. P., and Stenström, J. 2009. Degradation of glyphosate and other pesticides by ligninolytic enzymes. *Biodegradation* 20(6):751–759. doi:10.1007/s10532-009-9263-1.

Pokhrel, D., and Viraraghavan, T. 2004. Treatment of pulp and paper mill effluent – A review. *Sci Total Environ* 333:37–58.

Pourcel, L., Routaboul, J. M., Kerhoas, L., Caboche, M., Lepiniec, L., and Debeaujon I. 2005. TRANSPARENT TESTA10 encodes a laccase-like enzyme involved in oxidative polymerization of flavonoids in arabidopsis seed coat. *Plant Cell* 17:2966–2980.

Punnapayak, H., Prasongsuk, S., Messner, K., Danmek, K., and Lotrakul, P. 2009. Polycyclic aromatic hydrocarbons (PAHs) degradation by laccase from a tropical white rot fungus *Ganoderma lucidum*. *African J Biotechnol* 8(21):5897–5900.

Quan, D., and Shin, W. 2004. Amperometric detection of catechol and catecholamines by immobilized laccase from DeniLite. *Electroanalysis* 16(19):1576–1582. doi:10.1002/elan.200302988.

Ranocha, P., McDougall, G., Hawkins, S., Sterjiades, R., Borderies, G., Stewart, D., Cabanes-Macheteau, M., Boudet, A. M., and Goffner, D. 1999. Biochemical characterization, molecular cloning and expression of laccases – A divergent gene family in poplar. *Eur J Biochem* 259:485–495.

Rathnan, R. K., Anto, S. M., Rajan, L., Sreedevi, E. S., and Balasaravanan, A. M. 2013. Comparative studies of decolorization of toxic dyes with laccase enzymes producing mono and mixed cultures of fungi. *J Nehru Arts Sci Coll* 1:21–24.

Riva, S., Molecolare, R., and Bianco, V. M. 2006. Laccases: Blue enzymes for green chemistry. *Trends Biotechnol* 24(5). doi:10.1016/j.tibtech.2006.03.006.

Rodríguez, E., Nuero, O., Guillén, F., Martínez, A. T., and Martínez, M. J. 2004. Degradation of phenolic and non-phenolic aromatic pollutants by four *Pleurotus* species: The role of laccase and versatile peroxidase. *Soil Biol Biochem* 36:909–916.

Rodríguez Couto, S., and Toca Herrera, J. L. 2006. Industrial and biotechnological applications of laccases: A review. *Biotechnol Adv* 24:500–513.

Rosenberg, R. C., Wherland, S., Holwerda, R. A., and Gray, H. B. 1976. Ionic strength and pH effects on the rates of reduction of blue copper proteins by $Fe(EDTA)^{2-}$ – Comparison of the reactivities of *Pseudomonas aeruginosa*. *J Am Chem Soc* 98(20):6364–6369.

Sadhasivam, S., Savitha, S., Swaminathan, K., and Lin, F.H. 2010. Biosorption of RBBR by *Trichoderma harzianum* WL1 in stirred tank and fluidized bed reactor models. *J Taiwan Inst Chem Eng* 41:326–332.

Saratale, R. G., Saratale, G. D., Chang, J. S., and Govindvar, S. P. 2011. Bacterial decolorization and degradation of azo dyes: A review. *J Taiwan Inst Chem* 42:138–157. doi:10.1016/j.jtice.2010.06.006.

Schroeder, M., Pöllinger-Zierler, B., Aichernig, N., Siegmund, B., and Guebitz, G. M. 2008. Enzymatic removal of off-flavors from apple juice. *J Agric Food Chem* 56:2485–2489.

Selinheimo, E., Kruus, K., Buchert, J., Hopia, A., and Autio, K. 2006. Effects of laccase, xylanase and their combination on the rheological properties of wheat doughs. *J Cereal Sci* 43:152–159.

Setti, L., Giuliani, S., Spinozzi, G., and Pifferi, P. G. 1999. Laccase catalyzed oxidative coupling of 3-methyl 2-benzothiazolinone hydrazine and methoxyphenols. *Enzyme Microb Technol* 25:285–289.

Singh, G., Ahuja, N., Batish, M., Capalash, N., and Sharma, P. 2008. Biobleaching of wheat straw-rich soda pulp with alkalophilic laccase from c-proteobacterium JB: Optimization of process parameters using response surface methodology. *Bioresour Technol* 99:7472–7479. doi:10.1016/j.biortech.2008.02.023.

Sjoblad, R. D., and Bollag, J. M. 1977. Oxidative coupling of aromatic pesticide intermediates by a fungal phenol oxidase. *Appl Environ Microbiol* 33(4):906–910.

Shanmugam, S., Rajasekaran, P., and Thanikal, J. V. 2009. Synthetic dye decolourization, textile dye and paper industrial effluent treatment using white rot fungi *Lentines edodes*. *Desal Water Treat* 4:143–147.

Smith, A. E. 1985. Identification of 4-chloro-2-methylphenol as a soil degradation product of ring-labelled [^{14}C] Mecoprop. *Bull Environ Contam Toxicol* 34:646–660.

Spandre, R., and Dellomonaco, G. 1996. Polyphenols pollution by olive mill wastewaters, Tuscany, Italy. *J Environ Hydrol* 4:1–13.

Steffan, S., Bardi, L., Marzona, M. 2005. Azo dye biodegradation by microbial cultures immobilized in alginate beads. *Environ Int* 31:201–205.

Suurnakki, A. B., Buchert, J., Gronqvist, S., Mikkonen, H., Peltonen, S., and Viikari, L. 2006. Bringing new properties to lignin rich fiber materials. In: *Symp on Appl Mater Proc* 244:61–70.

Suwannawong, P., Khammuang, S., and Sarnthima, R. 2010. Decolorization of rhodamine B and congo red by partial purified laccase from *Lentinus polychrous*. *J Biochem Technol* 3:182–186.

Tavares, A. P. M., Cristóvão, R. O., Loureiro, J. M., Boaventura, R. A. R., and Macedo, E. A. 2009. Application of statistical experimental methodology to optimize reactive dye decolourization by commercial laccase. *J Hazard Mater* 162:1255–1260. doi:10.1016/j.jhazmat.2008.06.014.

Thakur, J. K., and Sangeeta, T. 2014. Degradation of sulphonated azo dye red HE7B by *Bacillus* sp. and elucidation of degradative pathways. *Curr Microbiol.* 69(2):183–191. doi:10.1007/s00284-014-0571-2.

Thurston, C. F. 1994. The structure and function of fungal laccases. *Microbiology* 140:19–26.

Tien, M., and Kirk T. K. 1983. Lignin-degrading enzyme from the hymenomycete *Phanerochaete chrysosporium*. *Science* 326:520–523.

Tsioulpas, A., Dimou, D., Iconomou, D., and Aggelis, G. 2002. Phenolic removal in olive mill waste water by strains of *Pleurotus ostreatus* spp. in respect to their phenol oxidase (laccase) activity. *Bioresource Technol* 84:251–257.

Tsutsumi, Y., Haneda, T., and Nishida, T. 2001. Removal of estrogenic activities of bisphenol A and nonylphenol by oxidative enzymes from lignin-degrading basidiomycetes. *Chemosphere* 42(3):271–276.

Ulčnik, A., Kralj, C. I., and Pohleven, F. 2013. Degradation of lindane and endosulfan by fungi, fungal and bacterial laccases. *World J Microbiol and Biotechnol* 29(12):2239–2247. doi:10.1007/s11274-013-1389-y.

Vasil-chenko, L. G., Khromonygina, V. V., Koroleva, O. V., Landesman, E. O., Gaponenko, V. V., Kovaleva, T. A., Kozlov, I. P., and Rabinovich, M. L. 2002. Consumption of the triazine herbicide atrazine by laccase and laccase-free variants of the soil fungus *Mycelia sterilia* INBI-2-26. *Prikl Biokhim Mikrobiol* 38(5):534–539.

Virk, A. P., Capalash, N., and Sharma, P. 2012. An alkalophilic laccase from *Rheinheimera* sp. isolate: Production and biobleaching of kraft pulp. *Biotechnol Prog* 28:1426–1431.

Virk, A. P., Puri, M., Gupta, V., Capalash, N., and Sharma, P. 2013. Combined enzymatic and physical deinking methodology for efficient eco-friendly recycling of old newsprint. *PLOS ONE* 8(8):1–9.

Wellington, K. W., and Kolesnikova, N. I. 2012. A laccase-catalysed one-pot synthesis of aminonaphthoquinones and their anticancer activity. *Bioorg Med Chem* 20:4472–4481. doi:10.1016/j.bmc.2012.05.028.

Wesenberg, D., Kyriakides, I., and Agathos, S. N. 2003. White-rot fungi and their enzymes for the treatment of industrial dye effluents. *Biotechnol Adv* 22:161–187.

Witayakran, S., Zettili, A., and Ragauskas, A. J. 2007. Laccase-generated quinones in naphthoquinone synthesis via Diels-Alder reaction. *Tetrahedron Lett* 48:2983–2987.

Xu, F. 1996. Oxidation of phenols, anilines, and benzenethiols by fungal laccases: Correlation between activity and redox potentials as well as halide inhibition. *Biochemistry* 35(23):7608–7614.

Xu, F. 1997. Effects of redox potential and hydroxide inhibition on the pH activity profile of fungal laccases. *J Biol Chem* 272:924–928.

Xu, Q. H., Wang, Y. P., Qin, M. H., Fu, Y. J., Li, Z. Q., Zhang, F. S., and Li, J. H. 2011. Fiber surface characterization of old newsprint pulp deinked by combining hemicellulase with laccase-mediator system. *Bioresour Technol* 102:6536–6540.

Yoshida, H. 1883. Chemistry of lacquer (urushi). *J Chem Soc* 43:472–486.

Zille, A., Munteanu, F., Georg, M. G., and Cavaco-paulo, A. 2005. Laccase kinetics of degradation and coupling reactions. *J Mol Cat B Enz* 33:23–28. doi:10.1016/j.molcatb.2005.01.005.

5

Biosurfactants and Bioemulsifiers for Treatment of Industrial Wastes

Zulfiqar Ahmad, David Crowley,
Muhammad Arshad and Muhammad Imran

CONTENTS

5.1 Introduction

5.1.1 Overview and State of the Art

Surfactants are surface-active molecules that mediate the solubility of hydrophobic chemicals in aqueous media by forming micelles and emulsions that physically arrange to suspend hydrocarbons, solvents and metals in water. As such, these compounds have fundamental roles in chemical synthesis and industrial processes in which they are used to physically separate and concentrate chemicals that are targeted for disposal, recycling or further processing. The most common uses of surfactants are as detergents to desorb oil and other hydrophobic chemicals from surfaces and suspend the contaminants in water during equipment washing. The second main use is as emulsifiers and demulsifiers to create or break emulsions that gather and concentrate hydrophobic chemicals and metals previously dispersed in water. Lastly, surfactants can be used to alter the bioavailability of chemicals to degrader microorganisms that are used for bioremediation of organic pollutants either in a bioreactor or for treatment of contaminated soils and sludges that are generated during industrial processes. Given all of these applications, studies on the properties of surfactants and their optimisation for specific applications comprise a large body of knowledge with hundreds of research articles published each year in dedicated journals on surfactant chemistry, which include both synthetically produced surfactants and biologically produced biosurfactants.

Although both synthetic surfactants and biosurfactants are used for the same general purposes, biosurfactants are distinguished from synthetic surfactants in that they are naturally produced by certain bacteria, yeasts and plants, whereas synthetic surfactants are chemically synthesised from different feedstocks, mainly petroleum, but also from materials of plant and animal origin. The relative importance of surfactants and biosurfactants today is indicated by the size of the markets for these materials and their market growth rate. In 1995, annual worldwide production of surfactants was estimated at 3 million tonnes with a value of US$4 billion, of which 54% was used for production of household detergents (Sarney and Vulfson 1995). As of 2011, surfactant production increased to 15 million tonnes, worth US$25 billion (Transparency Market Research 2012). Currently, biosurfactants comprise 476,000 tonnes of this market but have a high value worth US$1.7 billion. Analysts suggest that the biosurfactant market will continue to grow with a 3.5% annual growth rate to US$2.2 billion in 2018. Part of the shift from synthetic surfactants to biosurfactants is due to environmental problems with synthetic surfactants that are slow to degrade and the increasing cost of petrochemical feedstocks. There also has been considerable research demonstrating the use of low-cost feedstocks, such as agricultural wastes, for production of biosurfactants. As highly pure chemicals are not necessary for treating soils or industrial waste streams, downstream processing and purification can be streamlined to reduce their costs. This is particularly true when surfactants can be produced in situ, either by inoculating the soil with surfactant-producing microorganisms or by additions of agricultural waste materials that stimulate biosurfactant production by inoculated or indigenous microorganisms. Lastly, the range of chemical diversity and functional properties of biosurfactants open many possible applications for waste treatment. In this chapter, we specifically examine the properties and uses of biosurfactants for treatment of industrial wastes with the view that biosurfactants and bioemulsifiers will continue to become more affordable and increasingly used as production costs decrease in the future.

5.1.2 Chemistry, Diversity and Production of Biosurfactants

Although many different molecules produced by plants, bacteria and fungi have surfactant properties and can thus be classified as biosurfactants based on their origin, from a microbiological perspective, biosurfactants are more specifically defined as molecules that are produced by certain species of bacteria, fungi and plants. With some microorganisms, biosurfactant production is inducible, occurring under conditions with which they are supplied with hydrophobic substances that are used as substrates for microbial growth. In this case, biosurfactants are excreted primarily to modify the hydrophobicity of their cell envelope or to gain access to hydrophobic chemicals that are adsorbed to surfaces or suspended in water. A less restrictive definition would include various types of fatty acids, phospholipids, glycolipids, lipopolysaccharides and cyclodextrins that have surfactant properties but that primarily serve other purposes, for example, as components of cell membranes or for production of the extracellular polysaccharide that coats the cell envelope. Other functions of biosurfactants are related to their properties as wetting agents that enable gliding motility in certain bacteria that swarm on the surface of water films or as antibiotics that disrupt the membranes of competing organisms. Lastly, biosurfactants also include molecules that are fortuitously produced during growth, for example, as metabolic products that temporarily accumulate in the growth medium during growth on substrates having a high carbon-to-nitrogen ratio. Whereas biosurfactants can be obtained from almost any plant and microbial biomass, microorganisms with inducible biosurfactant secretion comprise a relatively small portion of the total culturable bacteria in soils and produce particularly interesting molecules that are of interest to biotechnology. In this review, we focus in particular on the strains of bacteria and yeasts that specifically produce biosurfactants, including the various screening and assay methods that are used for their detection, methods for production and purification and applications using purified or partially purified surfactants or in situ production.

5.1.3 Types of Biosurfactants

5.1.3.1 Microbially Derived Biosurfactants

Microbially derived surfactants are categorised based on their chemical composition and origin and are further distinguished based on their molecular weights. Low molecular weight surfactants are generally more efficient in lowering the surface and interfacial tension of water with hydrophobic substances, whereas high molecular weight biosurfactants are effective as emulsifiers and emulsion stabiliser agents (Banat 1995; Karanth et al. 1999; Rosenberg and Ron 1999; Youssef et al. 2005). Low molecular weight (LMW) biosurfactants are typically composed of a hydrophobic lipid moiety that is linked to a hydrophilic substance, such as a sugar or peptide. LMW biosurfactants include glycolipids, sophorolipids, trehalose lipids, lipopeptides and surfactins. Other low molecular weight biosurfactants include substances with antibiotic properties, such as gramicidin, polymyxins, streptofactin and corynomycolic acids, or molecules, such as serrawettin and viscosin, that are produced to improve the gliding motility of bacteria. Of the many different types of biosurfactants, glycolipids and especially the rhamnolipids are the most widely investigated for their multifunctional properties and abilities to form metal complexes, modify cell surface hydrophobicity and serve as either emulsifiers or demulsifiers for industrial processes. Rhamnolipid-producing *Pseudomonas* sp. also have been used as soil inoculants to facilitate bioremediation via in situ production of biosurfactants in soils and sludges that have become contaminated as a result of chemical spills or industrial processes (Figure 5.1).

The second general type of biosurfactants are the high molecular weight biosurfactants, which include glycoproteins and high molecular weight sugar polymers, such as emulsan, alasan, liposan, emulsifier lipoproteins, lipopolysaccharides and mixtures of these biopolymers. These biosurfactants

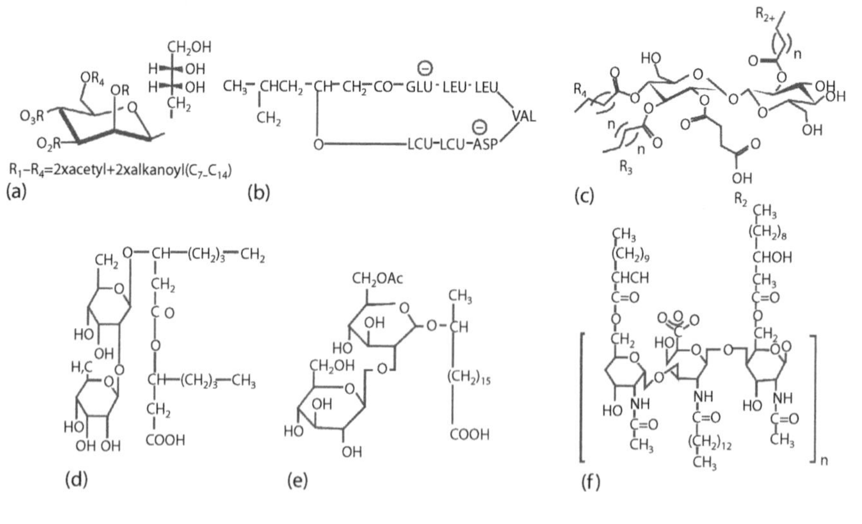

FIGURE 5.1
Chemical structures of some common biosurfactants (a) mannosylerythritollipid, (b) surfactin, (c) trehalose lipids, (d) sophorolipids, (e) rhamnolipids and (f) emulsan.

are well established for their abilities to form stable oil-in-water emulsions (Rosenberg and Ron 1999; Calvo et al. 2009).

As with synthetic surfactants, biosurfactants are further classified as anionic, cationic, neutral or amphoteric, depending on the electrical charge that is associated with the polar head portion of the molecule (Mulligan et al. 2001b). Anionic surfactants have been the most widely used surfactants, although the long-term trend is toward neutral biosurfactants (Transparency Market International 2012). The primary molecular feature that determines whether a chemical has surfactant properties is the existence of both hydrophobic and hydrophilic moieties on the same molecule that can simultaneously interact with both water molecules and hydrophobic compounds. On a thermodynamic basis, surfactants serve to reduce the ordering of water molecules that normally takes place at interfacial surfaces, around hydrophobic molecules and on molecules containing hydrophobic patches in their structures, thereby increasing the free energy (entropy) of the chemical system. Water has a polar structure in which the oxygen atom established an electro-negative region and the two hydrogen molecules that are oriented toward the other end as electropositive. The resulting molecular attractions and cohesion of water molecules result in a tensile strength that requires force in order to break through the water film at the air–water interface. The ability of a surfactant to reduce the surface tension of water is physically measured in units of dynes per centimetre, using an instrument that measures the force per unit length. A common instrument for measuring surface tension is the DuNuoy tensiometre, which measures the force required to pull a 1-cm metal ring through the surface of water under standardised conditions. Pure water has a surface tension of 72 dynes at 25°C. Addition of surfactant to pure water lowers the surface tension to values typically in the range of 27–50 dynes/cm.

Other important parametres used to characterise a surfactant are the critical micelle concentration (CMC), the emulsification index (E24%) and the hydrophilic–lipophilic balance (HLB) values. The CMC is the concentration at which the surfactant molecules undergo self-organisation into spherical or tubular structures in which the hydrophobic portion of the molecules orients inward toward the centre of the sphere to form micelles. During micelle formation, hydrophobic contaminants migrate into the micelle cores where they have the lowest free energy state. Substances that are sequestered into micelles will have altered bioavailability to degrader organisms, depending on the abilities of individual species to interact with the micelle, in some cases, enhancing biodegradation and, in other cases, resulting in an increase in solubility but a reduction in bioavailability. The second parametre, the emulsification index (E24%), is a simple test in which a biosurfactant is mixed with a standard hydrocarbon, such as kerosene, to produce an emulsion. The emulsion is allowed to stand for 24 h, after which the percentage of stable emulsion in relation to total volume is measured. Lastly, the general properties of surfactants with respect to their potential applications can also be predicted by examining the relative sizes of the hydrophilic and lipophilic portions of the molecule (HLB) (Griffin 1954; also see review by Müller et al. 2012). Low HLB values are associated with water in oil emulsifiers and wetting agents, whereas compounds with higher HLB values have properties that make them useful as detergents.

In the case of rhamnolipid-producing bacteria, biosurfactants are produced as mixtures of structurally similar compounds that vary in the number of fatty acid chains and the length of the fatty acids, both of which affect their interactions with various substrates and their bioavailability of different strains of bacteria (Zhang and Miller 1995).

Rhamnolipids are produced by a wide variety of bacteria and have four main structural types that include mono- and di-rhamnolipids and the methyl esters of mono- and di-rhamnolipids. All four types can occur in the same mixture with different compositions depending on the growth substrate and producing bacterium. The effect of structural variations on bioavailability and surface tension reduction has recently been examined systematically for lipopeptides that vary in the length of the hydrocarbon chain and degree of saturation for different lipids (Youssef et al. 2007). In the cited research, the degree of reduction in interfacial activity with toluene was improved for mixtures of biosurfactants having heterogeneous mixtures of different fatty acids. Mixtures of rhamnolipid and lipopeptide were shown to be more effective for lowering interfacial tensions. Also mixtures of surfactants with specific fatty acids behaved in different ways: interfacial tensions between toluene and water decreased as the proportion of a C14 lipid increased in the mixture, whereas lipopeptides mixed with a C12 branched lipid were more effective for solubilising hexane and decane. These results highlight the potential ability to create specific formulations of surfactants for different purposes and the ongoing effort to predict the properties of different surfactant mixtures based on their composition and the characteristics of individual components of the mixture.

5.1.3.2 Plant-Derived Biosurfactants

Many surface-active molecules are derived from plant materials. The most common plant-derived biosurfactants are saponins, lecithin, soy protein, cyclodextrins and a category referred to as humic-like substances (Table 5.1). Different plant-derived biosurfactants along with their sources are summarised in Table 5.2. Among the plant-derived biosurfactants, lecithin is the most widely used and is predominantly manufactured from soybean oil seed, which is an abundant, low-cost feedstock. There is also interest in the use of biosurfactant-producing plants for phytoremediation of soil contaminants or for addition of plant wastes containing biosurfactants for biostimulation of pollutant degraders.

TABLE 5.1

Microbially Derived Biosurfactants

Class	Producing Microorganism	Application/Use	References
Low Molecular Weight Surfactants			
Rhamnolipids	*P. aeruginosa* *Pseudomonas* spp.	Enhanced degradation and dispersion of different classes of hydrocarbons Emulsification of hydrocarbons Removal of metals from soil	Rendell et al. (1990) Sim et al. (1997) Lang and Wullbrandt (1999) Arino et al. (1996)
Trehalose lipids	*R. erythropolis* *A. paraffineus* *Corynebacterium* spp. *Mycobacterium* spp.	Enhanced bioavailability of hydrocarbons	Ristau and Wagner (1983) Kim et al. (1990) Lang and Philip (1998)
Sophorolipids	*Rothobacter* sp. *Mycobacterium* sp.	Soil decontamination	Inoue et al. (1986) Lemal et al. (1994)
Glycolipids	*A. borkumensis* *Rhodococcus* sp. *Candida apicola* *Serratia marcesecens* *P. aeruginosa*	Bioremediation of oil contaminated soil Removal of heavy metals	Mulligan et al. (2001a) McCray et al. (2001) Noordman et al. (2000) Golyshin et al. (1999) Sandrin et al. (2000)
High Molecular Weight Surfactants			
RAG-1 Emulsan BD4 Emulsan	*A. calcoaceticus* RAG-1 *A. calcoaceticicus* BD413 *P. fluorescens*	Stabilise emulsions Cleaning and recovery of hydrocarbon residues	Rosenberg et al. (1979) Kaplan and Rosenberg (1982) Persson et al. (1988) Chamanrokh et al. (2008)
Alasan	*A. radiorestence* KA53	Hydrocarbon compounds PAH degradation	Navon-Venezia et al. (1995) Ben Ayed et al. (2013) Kobayashi et al. (2012)
Manann lipid protein	*C. tropicalis*	Emulsification of various hydrocarbon	Kaeppeli et al. (1984)
Liposan	*C. lipoytica*	Efficient stabilisation oil-in-water emulsions with a variety of commercial vegetable oils	Cirigliano and Carman (1984, 1985)
Protein complex	*M. thermo-autotrophium*	Well-bore cleanup and mobility agent in saline or thermophilic oil reserves	De Acevedo and McInemey (1996)
Lipopeptide Viscosin Surfactin	*Arthrobacter* spp. *Bacillus polymxya* *Streptomyces tendae* *B. licheniformis* *B. subtilis*	Bioemulsification Surface motility Biofilm formation and colonisation	Horowitz and Griffin (1991) Lin et al. (1994) Yakimov et al. (1995) Neu and Poralla (1990) Wei and Chu (1998)

(Continued)

TABLE 5.1 (CONTINUED)

Microbially Derived Biosurfactants

Class	Producing Microorganism	Application/Use	References
Streptofactin Corynomycolic acids	*N. erythropolis* *C. lepus*	Biodegradation of tetrahydrofuran cellulose production	MacDonald et al. (1981) Cooper et al. (1981)
Phospholipids Fatty acids	*Acinetobacter* spp. *T. thiooxidans*	Metal ion sequestration	Kaeppeli and Finnerty (1980) Beebe and Umreit (1971) Hong et al. (1998)
PM factor	*P. marginalis*	Emulsifier Polycyclic aromatic hydrocarbon (PAH)	Burd and Ward (1996b) Burd et al. (1996a)

Source: Parthasarathi R, and Sivakumaar PK, *Global Journal of Environmental Research,* 3, 99–101, 2009.

TABLE 5.2

Plant-Derived Biosurfactants

Class	Source	Application/Use	References
Saponin	Soapberry	Removal of Ni, Cr and Mn from contaminated soils	Maity et al. (2013)
	Quillaja	Chromium recovery	Kiliç et al. (2011)
	Sigma Chemical Co.	Enhanced desorption of PCB and trace metal elements (Pb and Cu) from contaminated soils	Cao et al. (2013)
	Soapberry	Removal of Cu, Pb and Zn from contaminated industrial soils	Maity et al. (2013)
	Chinese soapberry	For enhancing washing of phenanthrene contaminated soil	Zhou et al. (2013)
	Quillaja	Desorption of Cu(II) and Ni(II) from kaolinite	Chen et al. (2008)
	Fruit pericarps of Ritha	Enhanced solubility and desorption of hexachlorobenzene	Kommalapati et al. (1997)
	Tea seeds	Enhanced adsorption capacity for Cd (II) by *P. simplicissimum*	Liu et al. (2011)
	Tea	Remove cadmium, lead and copper from aqueous solution	Yuan et al. (2008)
	Quillaja bark	Removal of heavy metals (Cd and Zn)	Hong et al. (2002)
Humic acid-like substance	Sigma Chemical Co.	Removal of phenanthrene and Cd from contaminated soils	Song et al. (2008)
	Soapnut plant	Removal of As(V) from Fe rich soil	Mukhopadhyay et al. (2013)
Cyclodextrins	Maize plant	Tetrachloroethene solubilisation/ degradation	Adani et al. (2010)

5.2 Production Media

5.2.1 Carbon Sources

A number of carbon sources have been used as growth substrates for bio-surfactant production, including diesel, crude oil, glucose, sucrose and glycerol (Desai and Banat 1997; Ilori et al. 2005). Production of biosurfactants is usually induced by growth of microorganisms on hydrocarbons that are not miscible with water. This includes plant and animal oils and petroleum hydrocarbons. Biosurfactants also can be produced by some bacteria using water-soluble compounds, such as glucose, glycerol or ethanol (Guerra-Santo et al. 1984; Cooper and Goldenberg 1987; Palejwala and Desai 1989). In an effort to lower the cost of production, much research has been conducted on the use of various agricultural products and waste materials as growth substrates for biosurfactant-producing bacteria that use these types of substrates for growth. Examples of the surfactant properties of biosurfactant mixtures that were produced by a strain of *Pseudomonas* sp. when supplied with different carbon sources are shown in Table 5.3 (adapted from Parthasarathi and Sivakumaar 2009). General results from these and other studies show that both the quality and quantity of biosurfactants are affected by the types of carbon sources that are used in the production process (Rahman and Gakpe 2008). Olive oil also is a very good substrate that is commonly used in many laboratory studies, but it is very expensive. In the study highlighted in Table 5.3, the highest production of rhamnolipids that was produced by the strain of *Pseudomonas* used for this research was achieved using cashew apple juice (6 g L^{-1}), whereas growing the same bacterium on fructose resulted in a surfactant mixture with the lowest critical micelle concentration but also the lowest yield as compared to other substrates.

TABLE 5.3

Production of Rhamnolipid on Different Carbon Sources

Carbon Sources	Surface Tension (mN/m)	CMC^{-1}	Surfactant Produced (g L^{-1})
Glucose	28.5	16.2	4.8
Fructose	31.7	1.7	2.4
Glycerol	29.0	14.2	3.5
Mannitol	28.4	14.9	3.9
Olive oil	28.5	20.0	5.0
Cashew apple juice	40.0	24.0	6.0

Source: Data from Parthasarathi R, and Sivakumaar PK, *Global Journal of Environmental Research*, 3, 99–101, 2009.

Although different carbon sources affect the types of biosurfactants that are produced in biosurfactant mixtures, they have no effect on the chain lengths of the fatty acids moieties part of individual types of biosurfactant molecules (Syldatk et al. 1985). Yields are also highly variable and depend on many factors, especially the carbon-to-nitrogen ratio of the culture medium. Duvnjak and Kosaric (1985) observed that a large amount of biosurfactant can be bound to the producing cells when grown in a medium containing glucose as a carbon source but that addition of hexadecane facilitated the release of surfactant from the cells. Biosurfactant production also can be manipulated by switching substrates, using one carbon source to produce a large biomass, followed by addition of a water-immiscible compound to trigger biosurfactant production (Robert et al. 1989). Yields of 5 to 10 g L^{-1} are typical for many production systems but can be much higher. Lee and Kim (1993) showed that, in a batch culture of the marine yeast *Torulopsis bombicola*, 37% of the carbon input used was channelled to produce 80 g L^{-1} of sophorolipid biosurfactants.

5.2.2 C/N Ratios

One of the main factors affecting surfactant production is the carbon-to-nitrogen (C/N) ratio of the growth substrate. It is speculated that depletion of nitrogen triggers biosurfactant production, enabling partial transformation of the substrate into a temporary carbon storage pool that can be further metabolised once nitrogen becomes available again. In this regard, biosurfactant production can also be induced by depletion of phosphorus and other nutrient elements (Amezcua-Vega et al. 2004). Nitrogen limitation not only causes the overproduction of biosurfactant but also changes the composition of the biosurfactant mixture that is produced. Results of one such study (Khopade et al. 2012) showed the effects of different C/N ratios on the emulsification and surface tension properties of biosurfactant mixtures produced by the marine actinomycetes *Nocardiopsis* sp. strain B4 and found that a C/N ratio of 20:1 gave the best production and surfactant with the highest emulsification activity. In this case, the bacterium was grown using olive oil and phenylalanine as carbon and nitrogen sources, respectively. Use of ammonium salts to supply nitrogen resulted in loss of pH control. Thus, both the C/N ratio and carbon and nitrogen source need to be optimised individually for maximum production (Figure 5.2).

Ghribi and Ellouze-Chaabouni (2011) studied different carbon-to-nitrogen ratios on biosurfactant production using *B. subtilis*, and they found a significant increase up to 900 mg/L by using a C/N ratio of 7:1. Cultures produced at greater C/N ratios had decreased biosurfactant production.

5.2.3 Growth Stage

Biosurfactant production is generally growth-linked with maximum production often occurring during late log phase and stationary phase in cell

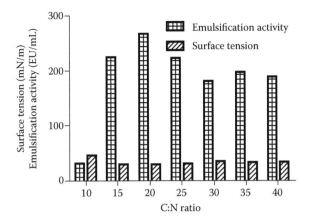

FIGURE 5.2
Effect of C/N ratio on biosurfactant production by *Nocardiopsis st.* B4 during growth on olive oil and phenylalanine as carbon and nitrogen sources.

cultures (Desai and Desai 1993). In many bacteria, the production of biosurfactants is cell density–dependent and is regulated by quorum sensing in which small signal molecules accumulate in the growth medium to a threshold level that induces the production of biosurfactants and other secondary metabolites (Brint and Ohman 1995; Ron and Rosenberg 2001). Nonetheless, there are no fixed rules for cell cultures as many strain-specific factors affect both the amount and timing of maximum biosurfactant accumulation in the medium, including the carbon source, C/N ratio, pH and metal ions.

To optimise biosurfactant production using a previously uncharacterised microorganism, it is necessary to carry out a series of studies to determine the factors that are associated with peak production values by measuring decreases in surface tension and emulsification properties of the culture medium over time. An example of the dynamics in biosurfactant production is shown in Figure 5.3, which is taken from a study aimed at optimisation of biosurfactant production by *Azotobacter vinelandi* (Qomarudin et al. 2008). In this research, experiments were conducted to optimise the glucose and alkane concentrations in the growth medium, along with nitrogen concentrations, during which the presence of biosurfactants was monitored over time. In this case, the production of EPS is growth-linked, corresponding with increases in optical density. Maximum levels of EPS and fatty acid biosurfactants in the medium occurred at 16 h during the early exponential phase, after which the fatty acid–based biosurfactant levels rapidly decreased and then held steady. The EPS component of the biosurfactant mixture remained relatively level between 16 and 40 h and then decreased as the culture entered the stationary phase. In similar types of studies on other organisms, peak production is observed at other times with many studies showing maximum accumulation during the stationary phase. In the case of

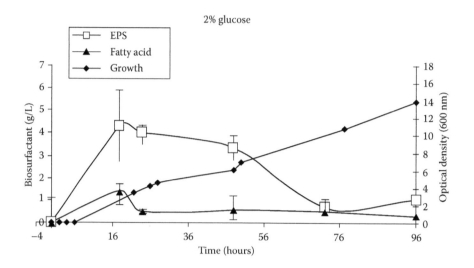

FIGURE 5.3
Time course study of biosurfactant production with 2% glucose as a carbon source. (From Qomarudin H et al., *International Journal of Civil & Environmental Engineering*, 10, 1–6, 2008.)

pseudomonads, forced entry into the stationary phase by nitrogen depletion results in maximum accumulation (Velraeds et al. 1996).

5.2.4 Genetics and Biosynthesis of Biosurfactant Production

The genetics of biosurfactant production has been studied for a few microorganisms using naturally occurring mutants or by transposition-induced gene knockouts (Georgiou et al. 1990; Desai and Desai 1993; Reiser et al. 1993; Desai et al. 1994) and, most recently, using cDNA expression analysis (see review: Mohammad et al. 2011). The process of selection and screening of such mutants has been difficult as many genes are involved in biosurfactant production, and the main criterion for evaluating the contribution of individual genes is reflected downstream only by changes in biosurfactant production, which is already affected by various interacting nutritional and environmental factors. Furthermore, mutations to genes affecting biosurfactant production can be lethal, especially those that affect production of high molecular weight biosurfactants that constitute essential components of the cell envelope.

The best-studied organisms in which the genetics of biosurfactant production have been investigated are the pseudomonads that produce rhamnolipids. Biosurfactant production by *Pseudomonas* sp. is regulated in relation to cell density via quorum sensing and in *P. aeruginosa* is expressed during pathogenesis. It is presumed that similar regulatory mechanisms function in nonpathogenic pseudomonads that are of interest for biotechnology. With further understanding of the molecular signals that regulate biosurfactant production, it may be possible to obtain better control of surfactant production to produce large quantities or to trigger biosurfactant production by indigenous bacteria in contaminated soils and sediments. The genetics of rhamnolipids biosynthesis in *P. aeruginosa* are under control of the *rhl*ABR gene cluster, which is responsible for the synthesis of RhlR regulatory protein and rhamnosyl transferase (Ochsner et al. 1994, 1995). The rhamnosyl transferase complex is located in the cytoplasmic membrane and consists of a dual protein complex having a 32.5-kDa RhlA protein that has two putative membrane-spanning domains and an exterior domain that functions as a signal receptor. The second component of the complex is a 47-kDa RhlB protein that is located in the periplasm. Two other genes involved in regulation of rhamnolipid production are *rhl*R, which encodes a 28-kDa transcriptional activator, and *rhl*I, which encodes the autoinducer synthetase enzyme that produces the quorum sensor signal molecule.

A recent study showed that genes encoding lipopolysaccharide biosurfactant production in the bacterium *Bacillus subtilis* SK320 could be cloned into *E. coli* to produce recombinant strains that produced the surfactant when grown on olive oil (Sekhon et al. 2011). Production of the biosurfactants was also dependent on esterase activity, which was closely correlated with the amount of extracellular biosurfactant produced in culture. This suggests that it may be possible to generate highly efficient microbial factories for biosurfactant production using genes obtained from unculturable bacteria and inserting them into the genome of culturable bacteria that can tolerate different ranges of environmental conditions.

5.3 Uses of Biosurfactants for Waste Treatment

5.3.1 Bioremediation of Hydrophobic Substances

Remediation of soil can be done either by excavation or in situ remediation, that is, soil washing (Mulligan et al. 2001a). Biosurfactants represent a promising tool for the treatment of nonaqueous-phase and soil-bound organic pollutants (Robinson et al. 1996). Rhamnolipids are the most widely studied biosurfactants for enhancing bioavailability and biodegradation of organic contaminants. These include polycyclic aromatic hydrocarbons (PAHs), which are priority pollutants that are generated by incomplete combustion. Low molecular weight PAHs are easily degraded but become increasingly insoluble as the number of aromatic rings increases. PAHs with four and five rings are largely insoluble but still are of concern as mutagens and carcinogens. Low molecular weight PAHs can be degraded by many different bacteria, but their concentrations are often too low to support a high cell density of degrader bacteria and thus are considered recalcitrant. When mobilised by a biosurfactant, PAHs with three and four rings can be degraded, either by the surfactant-producing bacterium or by other PAH-degrading bacteria that have the requisite enzymes but that do not produce surfactants themselves. In this sense, biosurfactant-producing bacteria can be considered as keystone microorganisms in that they can control the bioavailability of the pollutant as a carbon substrate for microbial growth.

There are two mechanisms by which biosurfactants increase hydrocarbon bioavailability. When biosurfactants are present in the medium at concentrations below their CMC, the biosurfactant intercalates between the hydrophobic substance and the soil matrix to displace the substance from the soil. The surfactant also binds to the cell walls of microorganisms and alters the hydrophobicity and attachment of bacteria. At concentrations above the CMC, the hydrophobic substance partitions into the micellar cores (Deshpande et al. 1999; Wang and Mulligan 2004a; Mulligan 2005). Concentrations above the CMC are generally used for soil washing, whereas sub-CMC concentrations are often more effective for stimulating biodegradation by promoting desorption, but do not result in physical protection of the contaminant within the micelles that form above the CMC (Herman et al. 1997).

The relative bioavailability of the target compound depends upon the type of surfactant and its charge, the surfactant concentrations (Barkay et al. 1999), hydrophobicity and interactions of surfactants with the soil (Makkar and Rockne 2003). Anionic surfactants have the lowest adsorption and are repelled from the soil cation exchange complex and thus are the most commonly used. At concentrations above the CMC, surfactants can inhibit the biodegradation process (Mulligan et al. 2001b; Makkar and Rockne 2003; Mulligan and Yong 2004). Various reasons for this phenomenon include toxicity to microbial cells, enzyme inhibition, decreased bioavailability within the micelle core and the production/accumulation of toxic compounds that inhibit the biodegradation process. The efficiency and cost of surfactant-enhanced bioremediation can be increased by lowering the concentration of surfactants to optimal levels as determined by feasibility studies in the laboratory (Cameotra and Randhir 2010).

In the field, mixtures of surfactants can also be useful for stimulating bioremediation (Nguyen et al. 2008). In a study with rhamnolipids used in combination with three alkyl propoxylated sulfate synthetic surfactants, the addition of rhamnolipids decreased the interfacial tension values by one to two orders of magnitude over that achieved with individual surfactants. Das and Mukherjee (2007) investigated the effect of three biosurfactant-producing bacteria, *Bacillus subtilis* DM-04, *Pseudomonas aeruginosa* st M and *Pseudomonas aeruginosa* st. NM, on the biodegradation of crude oil–contaminated soils. They found that the bioaugmentation of studied soils with the selected consortium gave a reduction in TPH levels from 84 to 21 g kg^{-1} soil over an uninoculated control (83 g kg^{-1} of soil). Wang et al. (2008) conducted a study on biodegradation of diesel-contaminated wastewater using two different biosurfactants, surfactin (from *Bacillus subtilis*) and rhamnolipids (*Pseudomonas aeruginosa*). They found that addition of 40 mg/L of surfactin or rhamnolipids to diesel/water mixtures significantly enhanced the biodegradation of diesel (94%) and increased biomass growth over the unamended control treatment. In studies testing the effects of biosurfactant on degradation of different PAHs varying in molecular weight and the number of aromatic rings, Martinez-Checa et al. (2007) isolated a bioemulsifier-producing strain of *Halomonas eurihalina* and tested it for a bioremediation process for 96 h in a liquid culture medium supplemented with naphthalene, phenanthrene, fluoranthene and pyrene. Efficiency of the isolated strain was confirmed by measuring the residual concentration of fluoranthene (56.6%) and pyrene (44.5%), which was higher than naphthalene (13.6%) and phenanthrene (15.6%). Thus, although the surfactant improved bioavailability of all of the PAHs, the degradation rates followed the same expected pattern in which low molecular weight PAHs are more easily degraded than higher molecular weight PAHs.

5.3.2 Biostimulation Removal of Heavy Metals from Contaminated Environments

Introduction of heavy metals into soil and water via industrial waste streams is very hazardous for the environment and human health (Pacwa-Płociniczak et al. 2011). There are many remediation technologies available, including excavation, landfill and use of plants and microorganisms. Heavy metals are not biodegradable, but they can be transformed from toxic to less toxic forms by redox reactions that determine the types of minerals that form from the reaction of metals with counter ions. One of the major problems in remediation of heavy metals is their toxicity to microorganisms. Some biosurfactants and bioemulsifiers form complexes with metals that enable soil washing (Figure 5.4).

The use of biosurfactants for heavy metal removal from soils is still a very active area of research. Juwakar et al. (2008) investigated the effect of rhamnolipid biosurfactants on the removal of multi-metal contaminated soil in packed columns to which 0.1% di-rhamnolipids solution was applied. They found that di-rhamnolipids selectively removed the heavy metals from soil in the order Cd = Cr > Pb = Cu > Ni. A similar study by Das et al. (2009) used biosurfactants isolated from marine yeasts and studied their efficacy for removal of heavy metals from solution. Wang and Mulligan (2004b) investigated a rhamnolipid foam technology for metal removal from sandy soils. They suggested that application of a rhamnolipid applied as a foam could significantly increase metal removal over that obtained using a solution form of the surfactant. They found that the foam rhamnolipid increased the efficiency of extraction of Cd and Ni by 73% and 68%, respectively, as compared to 62% and 51% removal, respectively, when rhamnolipid was used in a solution form.

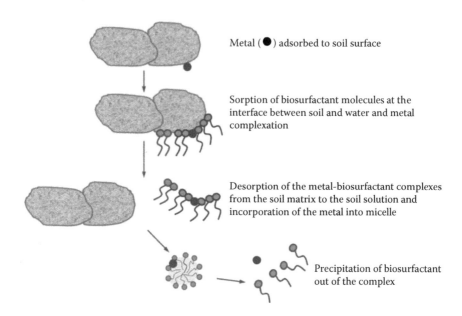

Metal (●) adsorbed to soil surface

Sorption of biosurfactant molecules at the interface between soil and water and metal complexation

Desorption of the metal-biosurfactant complexes from the soil matrix to the soil solution and incorporation of the metal into micelle

Precipitation of biosurfactant out of the complex

FIGURE 5.4
Mechanism of biosurfactants for removal of heavy metals from contaminated soil. (Adapted from Pacwa-Płociniczak M et al., *International Journal of Molecular Sciences*, 12, 1, 633–654, 2011; Mulligan CN, *Environmental Pollution*, 133, 183–198, 2005.)

Biosurfactants play a significant role in arsenic mobilisation by reducing the surface and interfacial tension between arsenic and mine tailings as they are helpful in developing micelles that sequester the arsenic and increase the wettability of the mine tailings (Wang and Mulligan 2009). Gnanamani et al. (2010) studied the potential application of biosurfactants produced by *Bacillus* sp. MTCC 5514 for bioremediation of chromium-contaminated soils. They observed that the chromium was first reduced from the Cr (VI) to less soluble Cr (III) by the synthesis of an extracellular chromium reductase, which was then followed by entrapment of Cr (III) by the biosurfactant. The combined action of these processes was that the bacterial cell provided increased tolerance to high concentrations of chromium.

Chen et al. (2011) investigated the effect of rhamnolipids (biosurfactants) and chemically synthesised surfactants (SDS, Tween) on separation of mercury ions from artificially contaminated water by a foam fractionation process. They found that the highest mercury recovery by surfactants was achieved by using biosurfactants as compared to the synthetic surfactants, SDS and Tween 80. Zeftawy et al. (2011) conducted a study to investigate the role of rhamnolipids to remove heavy metals from wastewater by micelle-enhanced ultrafiltration. After optimisation of different factors by response surface methodology, they found that initial biosurfactant concentration played a key role in developing a system suitable for removal of heavy metals. They concluded that a rhamnolipid-based ultrafiltration technique is efficient for removal of cadmium (Cd), lead (Pb), copper (Cu), zinc (Zn) and nickel (Ni) from contaminated industrial wastewater. Gao et al. (2012) conducted a similar type of study of removal of different heavy metals (Pb, Cr and Ni) from the sludge of industry water by the application of biosurfactants (saponins and sorpholipids). They found that the saponins having a carboxyl group showed better metal removal efficiency (90%–100%) in polluted sludge than sorpholipids, using a precipitation method with an alkaline solution to remove the metals.

Zouboulis et al. (2003) conducted a study on the use of biosurfactants for flotation processes for removal of heavy metals. They used a two-stage separation process called sorptive flotation. Under an optimised condition, better flotabilities were obtained with biosurfactants as compared to synthetic surfactants under similar conditions. Kiliç et al. (2011) conducted a comparative study for the chromium recovery from tannery sludge by using saponins, a plant-derived biosurfactant, over a pH range from 2 to 3. They found that treatment of tannery sludge with saponins extracted 24% of the Cr. They suggested that the extraction efficiency of saponins is strongly dependent on the organic matter content of the sample, which affects chromium mobility due to its high adsorption capacity.

5.4 Biostimulation and In Situ Production

Remediation can be carried out by ex situ excavation and land farming (Straube et al. 1999) or by in situ methods using soil inoculants or biostimulants (Mulligan et al. 2001a). Most research on in situ biosurfactant production has been for use of biosurfactants in microbially enhanced oil recovery (MEOR) (Al-Bahry et al. 2013). In situ production of biosurfactants can be accomplished by biostimulation of indigenous bacteria using hydrocarbon-rich substrates, such as soybean oil, or by addition of superior biosurfactant-producing bacteria (bioaugmentation) and addition of appropriate growth substrates to support growth and biosurfactant production. In a study comparing the rates of degradation of petroleum hydrocarbons in soil amended with molasses and inoculated with the surfactant-producing and oil-degrading bacterium *P. cepacia*, addition of the bacterium and surfactant together resulted in the fastest degradation, followed by the treatment receiving the biosurfactant, and the slowest degradation occurred in soil treated only with the degrader bacterium (Silva et al. 2014). Nonetheless, although degradation rates were significantly increased over the first 30 days, the difference in degradation after 1 month may not warrant the added cost of using purified biosurfactant. In the cited experiment, all of the treatments achieved the same degradation endpoint, with 95% hydrocarbon degradation at 1 month in all treatments. Similarly Kang et al. (2010) investigated the biodegradation potential of sorpholipids on aliphatic and aromatic hydrocarbon and Iranian light crude oil under lab conditions. They observed that surfactants enhanced the total petroleum hydrocarbon degradation from 85% in nontreated soils to 97% in treated soils and suggested that sorpholipids have potential for facilitating the bioremediation of hydrocarbon-contaminated sites having limited water solubility and bioavailability to microorganisms.

The use of surfactants and biosurfactants for bioremediation of soils has recently been reviewed (Megharaj et al. 2011). Earlier studies on the potential applications of biosurfactants that were conducted in the 1990s are reviewed by Banat et al. (2000). An inadequate supply of carbon and other nutrient sources to sustain the growth of microorganisms may affect the bioremediation process (Odokuma and Dickson 2003; Ward and Singh 2004). For this reason, various inducing substrates or low-cost nutrient amendments are often added to stimulate the degradation process; for example, peanut or soybean oil cake and yeast extract are common soil biostimulants. Abiotic factors can also affect the bioremediation process because oxygen is the most limiting factor among all. Oxygen can be manipulated through physical processes (landforming, composting) or addition of chemicals, such as hydrogen peroxide and manganese peroxide, techniques to stimulate the microbial communities (Ward and Singh 2004). One of the challenges of supplementing mineral nutrient fertilisers to stimulate biodegradation of hydrocarbons is to introduce the fertilisers in a form that will partition into the hydrocarbons. To this end, Churchill et al. (1995) found that addition of rhamnolipids along with oleophilic fertiliser inipol EAP-22 was effective for enhancing the degradation of hexane, benzene, toluene and naphthalene both in liquid and soil cultures.

Another approach to in situ use of surfactants is to cultivate plant species that produce biosurfactants or phytoremediation. Plants vary in their ability to produce surfactants and to stimulate hydrocarbon- and PAH-degrading bacteria. Among natural biosurfactants that are produced by plants are root mucilages, which are similar to bacterial extracellular polysaccharides. Mucilage facilitates soil wetting and also can mobilise PAHs. Another chemical that serves as a powerful biostimulant for PAH degradation is linoleic acid, which is a component of plant cell membranes. Plants with high concentrations of linoleic acid in their root tissues, such as tuberous plants, radish, carrot and celery root, can be added directly to soils, or linoleic acid can be added directly to soils to enhance PAH degradation (Yi and Crowley 2007).

Co-contaminated matrices represent a complex problem in the bioremediation processes because the biodegradation of organic pollutants could be suppressed by the simultaneous presence of high levels of metals, thus imposing a double stress on the microbial populations. Remediation of persistent polycyclic aromatic hydrocarbons along with heavy metal contamination from soils is generally a slow and expensive process. In such conditions, biosurfactant-assisted phytoremediation provides a good tool to deal with an existing problem. To remediate heavy metal–contaminated sites, phytoremediation could be enhanced by inoculation of biosurfactant-producing microorganisms along with a heavy metal–resistant one. In this dual strategy, the toxicity of metals can be reduced enhancing remediation of co-contaminated sites (Pacwa-Płociniczak et al. 2011). In one study, Sheng et al. (2008) reported the use of *Bacillus* sp. J119 strain as a plant growth–promoting bacterium that enabled improved plant tolerance of Cd uptake by rape, maize, sudangrass and tomato. They found that the tested strain has the potential to colonise the roots of all the tested plants but that it was most effective with tomato plants. Further work on biosurfactant-assisted phytoremedation is required to determine the role of biosurfactants in plant microbial interactions.

5.5 Conclusions

Biosurfactants are composed of a wide range of chemical substances that have amphiphilic properties that enable them to solubilise and sequester hydrophobic substances by formation of micelles and emulsions. They also function to increase the bioavailability of organic pollutants to degrader microorganisms, and some biosurfactants can be used to extract heavy metals from soils. Given these many functions, they play a pivotal role as detergents and emulsifiers in industrial processes and for treatment of industrial wastes. Biosurfactants offer several advantages in comparison to synthetic chelates, including the ability to produce these chemicals using low-cost substrates, their inherent biodegradability and their diverse chemistry. Disadvantages are that production of biosurfactants requires optimisation of the growth conditions and substrate preferences for individual strains of bacteria and fungi that produced these compounds. Although many microorganisms produce biosurfactants, novel biosurfactants must be evaluated empirically for each proposed purpose. In situ production of biosurfactants is possible either by soil inoculation, by biostimulation with specific substrates or by providing a suitable facilitating environment, such as the plant rhizosphere in which compounds such as polysaccharides in root mucilage and fatty acids in plant root tissues can also function as surfactants.

Much of the research on biosurfactants carried out over the past decade has sought to increase the cost competitiveness of biosurfactants with lower cost synthetic surfactants. It has been shown that commercial production can be accomplished using renewable low-cost substrates, such as agricultural waste materials. Many industrial waste materials and contaminated soils are co-contaminated with metals and organic pollutants, so use of heavy metal–resistant microorganisms/plants could prove useful for bioremediation.

Other research directions include studies of the mechanisms involved in biosurfactant-assisted bioremediation, including their adsorption behaviour in soils and micellar and submicellar solubilisation of organic hydrocarbons and metals. It will also be relevant to study the role of biosurfactant-producing microorganisms and biostimulants on microbial community structures on which such organisms play a keystone role in shaping the composition of the pollutant degrader community. As this technology advances, biosurfactants will become increasingly important for industrial processes and waste treatment.

References

Adani F, Tambone F, Davoli E, and Scaglia B. 2010. Surfactant properties and tetrachloroethene (PCE) solubilisation ability of humic acid-like substances extracted from maize plant and from organic wastes: A comparative study. *Chemosphere* 78(8): 1017–1022.

Al-Bahry SN, Al-Wahaibi YM, Elshafie AE, Al-Bemani AS, Joshi SJ, Al-Makhmari HS, and Al-Sulaimani HS. 2013. Biosurfactant production by *Bacillus subtilis B20* using date molasses and its possible application in enhanced oil recovery. *International Biodeterioration and Biodegradation* 81: 141–146.

Amezcua-Vega C, Ferrera-Cerrato R, Esparza-Garcia F, Rios-Leal E, and Rodriguez-Vazquez. 2004. Effect of combined nutrients on biosurfactant produced by *Pseudomonas putida*. *Journal of Environmental Science and Health: A. Toxic and Hazardous Substances and Environmental Engineering* 39(11–12): 2983–2991.

Arino S, Marchal R, and Vandecasteele JP. 1996. Identification and production of a rhamnolipidic biosurfactant by a *Pseudomonas species*. *Applied Microbiology and Biotechnology* 45: 162–168.

Banat IM. 1995. Biosurfactants production and possible uses in microbial enhanced oil recovery and oil pollution remediation: A review. *Bioresource Technology* 51(1): 1–12.

Banat IM, Makkar RS, and Cameotra SS. 2000. Potential commercial applications of microbial surfactants. *Applied Microbiology and Biotechnology* 53: 495–508.

Barkay T, Navon-Venezia S, Ron EZ, and Rosenberg E. 1999. Enhacement of solubilization and biodegradation of polyaromatic hydrocarbons by the bioemulsifier alasan. *Applied and Environmental Microbiology* 65: 2697–2702.

Beebe JL, and Umbreit WW. 1971. Extracellular lipid of *Thiobacillus thiooxidans*. *Journal of Bacteriology* 108: 612–515.

Ben Ayed H, Jridi M, Maalej H, Nasri M, and Hmidet N. 2013. Characterization and stability of biosurfactant produced by *Bacillus mojavensis A21* and its application in enhancing solubility of hydrocarbon. *Journal of Chemical Technology and Biotechnology* 89(7): 1007–1014.

Brint JM, and Ohman DE. 1995. Synthesis of multiple exoproducts in *Pseudomonas aeruginosa* is under the control of RhlR-RhlI, another set of regulators in strain PAO1 with homology to the autoinducer-responsive LuxR-LuxI family. *Journal of Bacteriology* 177: 7155–7163.

Burd G, and Ward OP. 1996a. Bacterial degradation of polycyclic aromatic hydro-carbons on agar plates: The role of biosurfactants. *Biotechnology Techniques* 10: 371–374.

Burd G, and Ward OP. 1996b. Physicochemical properties of PM-factor, a surface-active agent produced by *Pseudomonas marginalis*. *Canadian Journal of Microbiology* 42: 243–252.

Calvo C, Manzanera M, Silva-Castro GA, Uad I, and González-López J. 2009. Application of bioemulsifiers in soil oil bioremediation processes. Future pros-pects. *Science of the Total Environment* 407: 3634–3640.

Cameotra SS, and Randhir SM. 2010. Biosurfactant-enhanced bioremediation of hydrophobic pollutants. *Pure Applied Chemistry* 82(1): 97–116.

Cao M, Hu Y, Sun Q, Wang L, Chen J, and Lu X. 2013. Enhanced desorption of PCB and trace metal elements (Pb and Cu) from contaminated soils by saponin and EDDS mixed solution. *Environmental Pollution* 174: 93–99.

Chamanrokh P, Mazaheri Assadi M, Noohi A, and Yahyai S. 2008. Emulsan analy-sis produced by locally isolated bacteria and *Acinetobacter calcoaceticus*. *Iranian Journal of Environmental Health Science and Engineering* 5(2): 101–108.

Chen HR, Chen CC, Reddy AS, Chen CY, Li WR, Tseng MJ, and Atla SB. 2011. Removal of mercury by foam fractionation using surfactin, a biosurfactant. *International Journal of Molecular Sciences* 12(11): 8245–8258.

Chen WJ, Hsiao LC, and Chen KKY. 2008. Metal desorption from copper (II)/nickel (II)-spiked kaolin as a soil component using plant-derived saponin biosurfac-tant. *Process Biochemistry* 43(5): 488–498.

Churchill SA, Griffin RA, Jones LP, and Churchill PF. 1995. Biodegradation rate enhancement of hydrocarbons by an oleophilic fertilizer and a rhamnolipid bio-surfactant. *Journal of Environmental Quality* 24(1): 19–28.

Cirigliano MC, and Carman GM. 1984. Purification and characterization of liposan, a bioemulsifier from *Candida lipolytica*. *Applied and Environmental Microbiology* 50(4): 846–850.

Cirigliano MC, and Carman GM. 1985. Purification and characterization of liposan, a bioemulsifier from *Candida lipolytica*. *Applied and Environmental Microbiology* 50(4): 846–850.

Cooper DG, and Goldenberg BG. 1987. Surface active agents from two *Bacillus species*. *Applied and Environmental Microbiology* 53: 224–229.

Cooper DG, MacDonald CR, Duff SJB, and Kosaric N. 1981. Enhanced production of surfactin of *B. subtilis* by continuous product removal and metal cation addi-tions. *Applied and Environmental Microbiology* 42: 408–412.

Das K, and Mukherjee AK. 2007. Crude petroleum-oil biodegradation efficiency of *Bacillus subtilis* and *Pseudomonas aeruginosa* strains isolated from a petroleum-oil contaminated soil from North-East India. *Bioresource Technology* 98: 1339–1345.

Das P, Mukherjee S, and Sen R. 2009. Biosurfactant of marine origin exhibiting heavy metal remediation properties. *Bioresource Technology* 100: 4887–4890.

De Acevedo GT, and McInerney MJ. 1996. Emulsifying activity in thermophilic and extremely thermophilic microorganism. *Journal of Industrial Microbiology* 16: 17–22.

Desai AJ, Patel RM, and Desai JD. 1994. Advances in production of biosurfactants and their commercial applications. *Pakistan Journal of Scientific and Industrial Research* 53: 619–629.

Desai J, and Desai A. 1993. Production of biosurfactants. In: *Biosurfactants: Production, Properties and Applications,* Kosaric N (ed.). Marcel Dekker, New York, pp. 65–97.

Desai JD, and Banat IM. 1997. Microbial production of surfactants and their commercial potential. *Microbiology and Molecular Biology Reviews* 61: 47–64.

Deshpande S, Shiau BJ, Wade D, Sabatini DA, and Harwell H. 1999. Surfactant selection for enhancing ex situ soil washing. *Water Resource* 33(2): 351–360.

Duvnjak Z, and Kosaric N. 1985. Production and release of surfactant by *Corynebacterium lepus* in hydrocarbon and glucose media. *Biotechnology Letters* 7: 793–796.

Gao L, Kano N, Sato Y, Li C, Zhang S, and Imaizumi H. 2012. Behavior and distribution of heavy metals including rare earth elements, thorium, and uranium in sludge from industry water treatment plant and recovery method of metals by biosurfactants application. *Bioinorganic Chemistry and Applications* 10: 1155.

Georgiou G, Lin SC, and Sharma MM. 1990. Surface active compounds from microorganisms. *Biology/Technology* 10: 60–65.

Ghribi D, and Ellouze-Chaabouni S. 2011. Enhancement of *Bacillus subtilis* lipopeptide biosurfactants production through optimization of medium composition and adequate control of aeration. *Biotechnology Research International* 653654. doi: 10.4061/2011/653654.

Gnanamani A, Kavitha V, Radhakrishnan N, Rajakumar GS, Sekaran G, and Mandal AB. 2010. Microbial products (biosurfactant and extracellular chromate reductase) of marine microorganism are the potential agents reduce the oxidative stress induced by toxic heavy metals. *Colloidal and Surfaces* 79: 334–339.

Golyshin PM, Fredrickson HL, Giuliauo L, Rothmel R, Timmis KN, and Yakimov MM. 1999. Effect of novel biosurfactants on biodegradation of polychlorinated biphenyls by pure and mixed bacterial cultures. *Microbiologica* 22: 257–267.

Griffin WC. 1954. Calculation of the HLB values of non-ionic surfactants. *Journal of the Society of Cosmetic Chemists* 5(4): 249–256.

Guerra-Santos LH, Kappeli O, and Fiechter A. 1984. *Pseudomonas aeruginosa* biosurfactant production in continuous culture with glucose as carbon source. *Applied and Environmental Microbiology* 48: 301–305.

Herman DC, Zhang YM, and Miller RM. 1997. Rhamnolipid (bio-surfactant) effects on cell aggregation and biodegradation of residual hexadecane under saturated flow conditions. *Applied and Environmental Microbiology* 63: 3622–3627.

Hong KJ, Choi YK, Tokunaga S, Ishigami Y, and Kajiuchi T. 1998. Removal of cadmium and lead from soil using asecin as a biosurfactant. *Journal of Surfactants and Detergents* 2: 247–250.

Hong KJ, Tokunaga S, and Kajiuchi T. 2002. Evaluation of remediation process with plant-derived biosurfactant for recovery of heavy metals from contaminated soils. *Chemosphere* 49(4): 379–387.

Horowitz S, and Griffin WM. 1991. Structural analysis of *Bacillus licheniformis 86* surfactant. *Journal of Industrial Microbiology* 7: 45–52.

Ilori MO, Amobi CJ, and Odocha AC. 2005. Factors affecting biosurfactant production by oil degrading *Aeromonas spp.* isolated from a tropical environment. *Chemosphere* 61(7): 985–992.

Inoue S, Kimura Y, and Utsunomiya T. 1986. Neuer mikroorganismus: *Torulopsis bombicola* KSM-36. *DE 3525411A1.*

Juwarkar AA, Dubey KV, Nair A, and Singh SK. 2008. Bioremediation of multi-metal contaminated soil using biosurfactant – A novel approach. *Indian Journal of Microbiology* 48: 142–146.

Kaeppeli O, and Finnerty WR. 1980. Characteristics of hexadecane partition by the growth medium of *Acinetobacter sp. Biotechnology and Bioengineering* 22: 495–501.

Kaeppeli O, Walther P, Mueller M, and Fiechter A. 1984. Structure of cell surface of the yeast *Candida tropicalis* and its relation to hydrocarbon transport. *Archives of Microbiology* 138: 279–282.

Kang SW, Kim YB, Shin JD, and Kim EK. 2010. Enhanced biodegradation of hydrocarbons in soil by microbial biosurfactant, sophorolipid. *Applied Biochemistry and Biotechnology* 160: 780–790.

Kaplan N, and Rosenberg E. 1982. Exopolysaccharide distribution and bioemulsifier production in *Acinetobacter calcoaceticus BD4* and *BD413. Applied Environmental Microbiology* 44: 1335–1341.

Karanth NGK, Deo PG, and Veenanadig NK. 1999. Microbial production of biosurfactants and their importance. *Current Science* 77(1): 116–126.

Khopade A, Biao R, Liu X, Mahadik K, Zhang L, and Kokare C. 2012. Production and stability studies of the biosurfactant isolated from marine *Nocardiopsis sp. B4. Desalination* 285: 198–204.

Kiliç E, Font J, Puig R, Çolak S, and Çelik D. 2011. Chromium recovery from tannery sludge with saponin and oxidative remediation. *Journal of Hazardous Materials* 185(1): 456–462.

Kim JS, Powalla M, Lang S, Wagner F, Lünsdorf H, and Wray V. 1990. Microbial glycolipid production under nitrogen limitation and resting cell conditions. *Journal of Biotechnology* 13(4): 257–266.

Kobayashi T, Kaminaga H, Navarro RR, and Iimura, Y. 2012. Application of aqueous saponin on the remediation of polycyclic aromatic hydrocarbons-contaminated soil. *Journal of Environmental Science and Health Part A* 47(8): 1138–1145.

Kommalapati RR, Valsaraj KT, Constant WD, and Roy D. 1997. Aqueous solubility enhancement and desorption of hexachlorobenzene from soil using a plant-based surfactant. *Water Research* 31(9): 2161–2170.

Lang S, and Philip JC. 1998. Surface-active lipids in *rhodococci. Antonie van leeuwenhoek* 74(1–3): 59–70.

Lang S, and Wullbrandt D. 1999. Rhamnose lipids – biosynthesis, microbial production and application potential. *Applied Microbiology and Biotechnology* 51(1): 22–32.

Lee LH, and Kim JH. 1993. Distribution of substrate carbon in sophorose lipid production by *Torulopsis bombicola. Biotechnology Letters* 15: 263–266.

Lemal J, Marchal R, Davila AM, and Sulzer C. 1994. Fermentative production of sophorolipid composition. *FR 2692593.*

Lin SC, Minton MA, Sharma MM, and Georgiou G. 1994. Structural and immunological characterization of a biosurfactant produced by *Bacillus licheniformis JF-2. Applied Environmental Microbiology* 60: 31–38.

Liu ZF, Zeng GM, Zhong H, Yuan XZ, Jiang LL, Fu HY, and Zhang JC. 2011. Effect of saponins on cell surface properties of *Penicillium simplicissimum*: Performance on adsorption of cadmium (II). *Colloids and Surfaces B: Biointerfaces* 86(2): 364–369.

MacDonald CR, Cooper DG, and Zajic JE. 1981. Surface-active lipids from *Nocardia erythropolis* grown on hydrocarbons. *Applied Environmental Microbiology* 41: 117–123.

Maity JP, Huang YM, Fan CW, Chen CC, Li CY, Hsu CM, and Jean JS. 2013. Evaluation of remediation process with soapberry derived saponin for removal of heavy metals from contaminated soils in Hai-P. Taiwan. *Journal of Environmental Sciences* 25(6): 1180–1185.

Makkar RS, and Rockne KJ. 2003. Comparison of synthetic surfactants and biosur-
    factants in enhancing biodegradation of polycyclic aromatic hydrocarbons.
    Environmental Toxicology and Chemistry 22: 2280–2292.
Martínez-Checa F, Toledo FL, El Mabrouki K, Quesada E, and Calvo C. 2007.
    Characteristics of bioemulsifier V2-7 synthesized in culture media added of
    hydrocarbons: Chemical composition, emulsifying activity and rheological
    properties. *Bioresource Technology* 98(16): 3130–3135.
McCray JE, Bai G, Maier RM, and Brusseau ML. 2001. Biosurfactants enhanced solu-
    bilization of NAPL mixtures. *Journal of Contaminant Hydrology* 48: 45–68.
Megharaj M, Ramakrishnan B, Venkateswarlu K, Sethunathan N, and Naidu R.
    2011. Bioremediation approaches for organic pollutants: A critical perspective.
    Environment International 37(8): 1362–1375.
Mohammad A, Mawgoud A, Hausmann R, Leipine F, Muller MM, and Deziel E. 2011.
    Rhamnolipids: Detection, analysis, biosynthesis, genetic regulation and bioen-
    gineering of production. In: *Biosurfactants: From Genes to Applications*, Soberon-
    Chavez G (ed.). Springer, New York, pp. 13–56.
Mukhopadhyay S, Hashim MA, Sahu JN, Yusoff I, and Gupta BS. 2013. Comparison
    of a plant based natural surfactant with SDS for washing of As (V) from Fe rich
    soil. *Journal of Environmental Sciences* 25(11): 2247–2256.
Müller MM, Kugler JH, Henkel M, Gerlitzki M, Hormann B, Pohnlein M, Syldatk C,
    and Hausmann R. 2012. Rhamnolipids – Next generation surfactants. *Journal of
    Biotechnology* 162: 366–380.
Mulligan CN. 2005. Environmental application for biosurfactants. *Environmental
    Pollution* 133: 183–198.
Mulligan CN, Raymond NY, and Gibbs BE. 2001a. Heavy metal removal from sedi-
    ments by biosurfactants. *Journal of Hazardous Materials* 85: 112–125.
Mulligan CN, Yong RN, and Gibbs BF. 2001b. Surfactant-enhanced remediation of
    contaminated soil: A review. *Engineering Geology* 60(1): 371–380.
Mulligan CN, and Yong RN. 2004. Natural attenuation of contaminated soils.
    Environment International 30: 587–60.
Navon-Venezia S, Zosim Z, Gottlieb A, Legmann R, Carmeli S, Ron EZ, and Rosenberg
    E. 1995. Alasan, a new bioemulsifier from *Acinetobacter radioresistens*. *Applied
    Environmental Microbiology* 61: 3240–3244.
Neu TR, and Poralla K. 1990. Emulsifying agent from bacteria isolated during screening
    for cells with hydrophobic surfaces. *Applied Environmental Microbiology* 32: 521–525.
Nguyen TT, Youssef NH, McInerney MJ, and Sabatini DA. 2008. Rhamnolipid biosur-
    factant mixtures for environmental remediation. *Water Research* 42(6): 1735–1743.
Noordman W II, Bruining JP, Wietzes P, and Janssen DK. 2000. Facilitated transport of
    a PAH mixture by a rhamnolipid biosurfactant in porous silica matrices. *Journal
    of Contaminant Hydrology* 44: 119–140.
Ochsner UA, Koch AK, Fiechter A, and Reiser J. 1994. Isolation and characterization of
    a regulatory gene affecting rhamnolipid biosurfactant synthesis in *Pseudomonas
    aeruginosa*. *Journal of Bacteriology* 176: 2044–2054.
Ochsner UA, Reiser J, Fiechter A, and Witholt B. 1995. Production of *Pseudomonas aeru-
    ginosa* rhamnolipid biosurfactants in heterogeneous host. *Applied Environmental
    Microbiology* 61: 3503–3506.
Odokuma LO, and Dickson AA. 2003. Bioremediation of a crude oil polluted tropical
    rain forest soil. *Global Journal of Environmental Sciences* 2: 29–40.

Pacwa-Płociniczak M, Płaza GA, Piotrowska-Seget Z, and Cameotra SS. 2011. Environmental applications of biosurfactants: Recent advances. *International Journal of Molecular Sciences* 12(1): 633–654.

Palejwala S, and Desai JD. 1989. Production of extracellular emulsifier by a Gram negative bacterium. *Biotechnology Letters* 11: 115–118.

Parthasarathi R, and Sivakumaar PK. 2009. Effect of different carbon sources on the production of biosurfactant by *Pseudomonas fluorescens* isolated from mangrove forests (Pichavaram), Tamil Nadu, India. *Global Journal of Environmental Research* 3: 99–101.

Persson A, Oesterberg E, and Dostalek M. 1988. Biosurfactant production by *Pseudomonas fluorescens 378*: Growth and product characteristics. *Applied Environmental Microbiology* 29: 1–4.

Qomarudin H, Kardena E, Funamizu N, and Wisjnuprapto. 2008. Application of biosurfactant produced by *Azotobacter Vinelandii AV01* for enhanced oil recovery and biodegradation of oil sludge. *International Journal of Civil and Environmental Engineering* 10: 1–6.

Rahman PKSM, and Gakpe E. 2008. Production, characterization and application of Biosurfactants – Review. *Biotechnology* 7: 360–370.

Reiser J, Koch AK, Ochsner UA, and Fiechter A. 1993. Genetics of surface active compounds. In: *Biosurfactants: Production, Properties, Applications*, Kosaric N (ed.). Marcel Dekker, Inc., New York, pp. 231–249.

Rendell NB, Taylor GW, Somerville M, Todd H, Wilson R, and Cole PJ. 1990. Characterisation of *Pseudomonas rhamnolipids*. *Biochimica et Biophysica Acta (BBA) – Lipids and Lipid Metabolism* 1045(2): 189–193.

Ristau E, and Wagner F. 1983. Formation of novel anionic trehalose tetraesters from *Rhodococcus erythropolis* under growth limiting conditions. *Biotechnology Letters* 5(2): 95–100.

Robert M, Mercade ME, Bosch M, Parra JL, Espuny MJ, Manresa MA, and Guinea J. 1989. Effect of the carbon source on biosurfactant production by *Pseudomonas aeruginosa 44T*. *Biotechnology Letters* 11: 871–874.

Robinson KG, Ghosh MM, and Shi Z. 1996. Mineralization enhancement of non-aqueous phase and soilbound PCB using biosurfactant. *Water Science and Technology* 34(7): 303–309.

Ron EZ, and Rosenberg E. 2001. Natural roles of biosurfactants. *Environmental Microbiology* 3: 229–236.

Rosenberg E, Perry A, Gibson DT, and Gutnick D. 1979. Emulsifier of *Arthrobacter RAG-1*: Specificity of hydrocarbon substrate. *Applied Environmental Microbiology* 37: 409–413.

Rosenberg E, and Ron EZ. 1999. High-and low-molecular-mass microbial surfactants. *Applied Microbiology and Biotechnology* 52(2): 154–162.

Sandrin TR, Cbech AM, and Maier RM. 2000. A rhamnolipid biosurfactatnt reduces cadmium toxicity during naphthalene biodegradation. *Applied Environmental Microbiology* 66: 4585–4588.

Sarney DB, and Vulfson EN. 1995. Application of enzymes to the synthesis of surfactants. *Trends in Biotechnology* 13(5): 164–172.

Sekhon, KS, Khanna S, and Cameotra SS. 2011. Enhanced biosurfactant production through cloning of three genes and role of esterase in biosurfactant release. *Microbial Cell Factories* 10: 49–59.

Sheng X, He L, Wang Q, Ye H, and Jiang C. 2008. Effects of inoculation of biosurfactant-producing *Bacillus sp. J119* on plant growth and cadmium uptake in a cadmium-amended soil. *Journal of Hazardous Materials* 155(1): 17–22.

Silva EJ, Rocha e Silva NMP, Rufino RD, Luna JM, Silva RO, and Sarubbo LA. 2014. Characterization of a biosurfactant produced by *Pseudomonas cepacia CCT6659* in the presence of industrial wastes and its application in the biodegradation of hydrophobic compounds in soil. *Colloids and Surfaces B: Biointerfaces* 117: 36–41.

Sim L, Ward OP, and Li ZY. 1997. Production and characterisation of a biosurfactant isolated from *Pseudomonas aeruginosa* UW-1. *Journal of Industrial Microbiology and Biotechnology* 19(4): 232–238.

Song S, Zhu L, and Zhou W. 2008. Simultaneous removal of phenanthrene and cadmium from contaminated soils by saponin, a plant-derived biosurfactant. *Environmental Pollution* 156(3): 1368–1370.

Straube WL, Jones-Meehan J, Pritchard PH, and Jones WR. 1999. Bench-scale optimization of bioaugmentation strategies for treatment of soils contaminated with high molecular weight polyaromatic hydrocarbons. *Resources, Conservation and Recycling* 27(1): 27–37.

Syldatk C, Lang S, and Wagner F. 1985. Chemical and physical characterization of four interfacial-active rhamnolipids from *Pseudomonas sp. DSM 2874* grown on n-alkanes. *Zeitschrift für Naturforschung* 40C: 51–60.

Transparency Market Research. 2012. *Specialty Surfactants Market – Global Scenario, Raw Material and Consumption Trends, Industry Analysis, Size, Share and Forecast, 2011–2017*, 82 pp. Available at http://www.transparencymarketresearch.com /specialty-surfactants-market.html (accessed January 9, 2015).

Velraeds MM, Van der Mei HC, and Reid G. 1996. Physicochemical and biochemical characterization of biosurfactants released by *Lactobacillus* strains. *Colloid Surface and Biointerface* 8: 51–61.

Wang S, and Mulligan CN. 2004a. An evaluation of surfactant foam technology in remediation of contaminated soil. *Chemosphere* 57: 1079–1089.

Wang S, and Mulligan CN. 2004b. Rhamnolipid foam enhanced remediation of cadmium and nickel contaminated soil. *Water Air Soil Pollution* 157: 315–330.

Wang S, and Mulligan CN. 2009. Rhamnolipid biosurfactant-enhanced soil flushing for the removal of arsenic and heavy metals from mine tailings. *Process Biochemistry* 44:296–301.

Wang ZY, Gao DM, Li FM, Zhao J, Xin YZ, Simkins S, and Xing BS. 2008. Petroleum hydrocarbon degradation potential of soil bacteria native to the Yellow River delta. *Pedosphere* 18(6): 707–716.

Ward O, and Singh A. 2004. Soil bioremediation and phytoremediation – An overview. In: *Applied Bioremediation and Phytoremediation*, Singh A and Ward OP (eds.). Springer, Berlin, Germany, pp. 1–12.

Wei YH, and Chu IM. 1998. Enhancement of surfactin production in iron-enriched media by *Bacillus subtilis ATCC 21332*. *Enzyme and Microbial Technology* 22: 724–728.

Yakimov MM, Timmis KN, Wray V, and Fredrickson HL. 1995. Characterization of a new lipopeptide surfactant produced by thermotolerant and halotolerant subsurface *Bacillus licheniformis BAS50*. *Applied Environmental Microbiology* 61: 1706–1713.

Yi H, and Crowley DE. 2007. Biostimulation of PAH degradation with plants containing high concentrations of linoleic acid. *Environmental Science and Technology* 41(12): 4382–4388.

Youssef NH, Duncan KE, and McInerney MJ. 2005. Importance of 3-hydroxy fatty acid composition of lipopeptides for biosurfactant activity. *Applied and Environmental Microbiology* 71(12): 7690–7695.

Youssef NH, Nguyen T, Sabatini DA, and McInerney MJ. 2007. Basis for formulating biosurfactant mixtures to achieve ultra low interfacial tension values against hydrocarbons. *Journal of Industrial Microbiology and Biotechnology* 34(7): 497–507.

Yuan XZ, Meng YT, Zeng GM, Fang YY, and Shi JG. 2008. Evaluation of tea-derived biosurfactant on removing heavy metal ions from dilute wastewater by ion flotation. *Colloids and Surfaces A: Physicochemical and Engineering Aspects* 317(1): 256–261.

Zeftawy E, Monem MA, and Mulligan CN. 2011. Use of rhamnolipid to remove heavy metals from wastewater by micellar-enhanced ultrafiltration (MEUF). *Separation and Purification Technology* 77(1): 120–127.

Zhang Y, and Miller RM. 1995. Effect of rhamnolipid (biosurfactant) structure on solubilization and biodegradation of n-alkanes. *Applied and Environmental Microbiology* 61(6): 2247–2251.

Zhou W, Wang X, Chen C, and Zhu L. 2013. Enhanced soil washing of phenanthrene by a plant-derived natural biosurfactant, Sapindus saponin. *Colloids and Surfaces A Physicochemical and Engineering Aspects* 425: 122–128.

Zouboulis AI, Matis KA, Lazaridis NK, and Golyshin PN. 2003. The use of biosurfactants in flotation: Application for the removal of metal ions. *Minerals Engineering* 16(11): 1231–1236.

6

Biodegradation of Lignocellulosic Waste in the Environment

Monika Mishra and Indu Shekhar Thakur

CONTENTS

6.1 Lignocellulose Structure

Lignocellulosic biomass, composed of forestry, agricultural and agro-industrial wastes, are abundant, renewable and inexpensive energy sources. Such wastes include a variety of materials: sawdust; poplar trees; sugarcane bagasse; waste paper; brewer's spent grains; switch grass; and straws, stems, stalks, leaves, husks, shells and peels from cereals, such as rice, wheat, corn, sorghum and barley. Lignocellulose wastes are accumulated every year in large quantities, causing environmental problems. However, due to their chemical composition based on sugars and other compounds of interest, they could be utilised for the production of a number of value-added products, such as ethanol, food additives, organic acids, enzymes and others. Therefore, besides the environmental problems caused by their accumulation in nature, the nonuse of these materials results in a loss of potentially valuable sources (Mishra and Thakur 2011).

The major constituents of lignocellulose are cellulose, hemicellulose and lignin, polymers that are closely associated with each other, constituting the cellular complex of the vegetal biomass. Basically, cellulose forms a skeleton, which is surrounded by hemicellulose and lignin (Table 6.1).

6.2 Ecology of Lignocellulose Biodegradation

Lignocellulose degradation is essentially a race between cellulose and lignin degradation (Reid 1995). This contest is even more extensive and complex in nature (Rayner and Boddy 1988). Decomposition of complex substrates

TABLE 6.1

Main Components of Lignocellulose Wastes

Lignocellulose Waste	Cellulose (wt%)	Hemicellulose (wt%)	Lignin (wt%)
Barley straw	33.8	21.9	13.8
Corn cobs	33.7	31.9	6.1
Cotton stalks	35.0	16.8	7.0
Oat straw	39.4	27.1	17.5
Rice straw	36.2	19	9.9
Rye straw	37.6	30.5	19.0
Soya stalks	34.5	24.8	19.8
Sugarcane baggase	40	27	10
Sunflower stalks	42.1	29.7	13.4
Wheat straw	32.9	24	8.9

incubated in soil, such as plant residues, usually yield a multislope decomposition curve (Paul and Clark 1989; Van Veen et al. 1984).

Fungi with restricted metabolic capabilities (for example, soft rots like Mucor sp.) develop mutualistic relationships with and flourish alongside fungi-degrading cellulose and lignin. Microorganisms unable to overcome the lignin or physical barrier can obtain energy from the low molecular weight intermediates released from lignocellulose by the true white-rot fungi. Such complex associations have been observed under natural conditions (Blanchette et al. 1978).

6.2.1 Actinomycetes and Bacteria

Lignocellulose biodegradation by prokaryotes is essentially a slow process characterised by the lack of powerful lignocellulose-degrading enzymes, especially lignin peroxidases. Grasses are more vulnerable to actinomycete attack than wood (McCarthy 1987). Together with bacteria, actinomycetes play a significant role in the humification processes associated with soils and composts (Trigo and Ball 1994). The enzymatic ability to cleave alkyl–aryl ether bonds enables bacteria to degrade oligomeric and monomeric aromatic compounds released during fungal lignin degradation (Vicuna et al. 1993; Vicuna 2000; White et al. 1996). Therefore, lignocellulose biodegradation by prokaryotes is of ecological significance, but lignin biodegradation by fungi, especially white-rot fungi, is of commercial importance.

6.2.2 Fungi

Most fungi are proficient cellulose degraders. However, their ability to facilitate rapid lignocellulose degradation attracted attention from scientists and entrepreneurs alike. White-rot fungi comprise powerful lignin-degrading enzymes that enable them in nature to bridge the lignin barrier and, hence, overcome the rate-limiting step in the carbon cycle (Elder and Kelly 1994). Of these, *Phanerochaete chrysosporium* is the best studied. The fungi, by means of the enzymes secreted from their hyphae, attack and penetrate the wood very rapidly, up to 1 mm/h (Eriksson 1990). New information regarding the identities of the cellulose-, hemicellulose- or lignin-degrading enzymes; their unique catalytic capabilities; and the physiological conditions required for optimum secretion or activity, etc., is constantly being added to an already impressive volume of work and varies between fungi and bacterial genera, species and even strains. Anaerobic fungi (*Piromyces* sp., *Neocalli-mastix* sp. and *Orpinomyces* sp.) form part of the rumen microflora. These fungi produce active polymer-degrading enzymes, including cellulases and xylanases (Hodrova et al. 1998). Their cellulases are among the most active reported to date and are able to solubilise both amorphous and crystalline cellulose (Wubah et al. 1993). These fungi can be used in situations in which process settings impose anaerobic conditions. In such a scenario, ruminant manure will serve as inoculum, and this waste product will meet a crucial requirement in biotechnology: cost-effectiveness versus optimum utility.

On the basis of ligninase enzymes produced, fungi are classified into five main groups (Tuor et al. 1995):

1. White-rot fungi expressing Lip, MnP and laccase. This group contains the best-known white-rot fungi *Coriolus versicolor, Phanerochaete chrysosporium* and *Phlebia radiata. P. chrysosporium* is listed within this group because laccase production was reported (Ander et al. 1980; Eriksson et al. 1990). However, this fungus is generally considered not to produce laccase. All of them usually colonise deciduous trees; only *Phlebia radiata* occasionally degrades conifers.

2. White-rot fungi simultaneously produce both types of phenoloxidases, MnP and laccase, but reportedly do not secrete detectable levels of lignin peroxidase. Nevertheless, these fungi are strong lignin degraders. *Dichomitus squalens* and the edible fungus *Lentinulu edodes* belong to this group.

3. White-rot fungi with LiP and one of either phenoloxidases. Lactase is the predominant phenoloxidase produced; only in the case of *Coriolus* pruinosum was MnP production reported. These fungi grow on hardwood. As an exception, only *Phlebia tremellosus* degrades coniferous wood.

4. Four white-rot fungi secrete LiP without phenoloxidases. Again, with one exception, they are hardwood degraders.

5. The last group probably consists of fungi that are incompletely characterized. Notably, *Fames lignosus* and *Trametes cingulata* are white-rot degraders, but neither of the oxidative enzymes was detected.

6.3 Barriers to Lignocellulose Biodegradation

6.3.1 Microorganism Access to Substrate

The physical barrier concept is best described by making use of the rumen as an example. The rumen being an anaerobic environment and having discussed the distinctiveness of cellulose-degrading systems under anaerobic conditions, it is clear that access and attachment of the microorganism to the substrate are vital if efficient cellulose hydrolysis is to be effected. The waxes and cuticle of an intact epidermis prevent microorganism access to the interior of leaves and stems (Wilson and Mertens 1995). The ruminant circumvents the physical barrier effect imposed by the lignocellulose higher order structure by physically reducing the particle size of the plant materials during ingestion. Anaerobic fungi alleviate the physical barrier effect further by physically disrupting lignified tissues, allowing microorganisms greater access to the digestible portions of the plant fibre (Varga and Kolver 1997). According to Wilson and Mertens (1995), degradation of the middle lamella–primary wall region by rumen bacteria is prevented not only by the chemical nature (lignin concentration) but also by physical structure and architecture.

Therefore, in order for man to successfully exploit lignocellulosics for commercial purposes, treatments that increase the accessibility of the catalyst (whether microbial or enzymic) to the substrate have been studied (Fan et al. 1982). These include mechanical, chemical, thermal and biological pretreatments. Again, the principles of cost and profit govern the choice of method to be used. Usually, choices are influenced not only by available funds and time but also by the value of the end product, which may warrant the use of a more expensive procedure. Certainly for bioremediation purposes, practical issues relating to the type of substrate available (for example, grasses versus wood), cost of on-site pretreatment (transport to and installation of equipment on-site), etc., will impact the budget of the project as will the value of the end product, which may warrant the use of a more expensive procedure.

6.3.2 Enzyme Access to Substrate

Aerobic fungi and bacteria secrete cellulose- and hemicellulose-hydrolysing enzymes into the culture medium. Evidence also suggests the same situation applies to most lignin-degrading enzymes. Therefore, unlike anaerobic microorganisms, direct physical contact of the enzyme-delivery agent with the substrate is not essential to facilitate polymer hydrolysis. However, it is at the lower order level structures (tertiary and secondary) of the substrate that enzyme preclusion can occur. Polymer hydrolysis can be complicated by various substrate and enzymatic factors. Crystalline cellulose is highly recalcitrant, yet it is completely hydrolysable through the concerted action of all endo- and exo-acting enzymes. However, endo- and exoglucanases are inhibited by cellobiose, and the presence or absence of cellobiose is the rate-limiting step. Lignin, on the other hand, contains no chains of repeating subunits, thereby making the enzymatic hydrolysis of this polymer extremely difficult.

Despite the above-mentioned complications associated with enzymatic polymer hydrolysis, the primary problem remains the tertiary architecture of the lignocellulose complex. Lignin–carbohydrate complexes (LCCs) are recognized as key structures determining forage digestibility. These intricate associations between cellulose, hemicellulose and lignin prevent polymer-hydrolysing enzymes access to its substrates (Cornu et al. 1994).

According to Tomme et al. (1995) and Grethlein (1985), the accessibility of enzymes to wood and fibres is limited by factors such as adsorption to surface areas, low fibre porosity and low median pore size of fibres. Reid (1995) suggested that physical contact between enzyme and substrate is the rate-limiting step in lignin degradation. Grethlein (1985) indicated that substrate pretreatment (using dilute sulphuric acid in this example) is necessary to increase the number of available sites for cellulase action.

The factors discussed in this section will have a profound impact on the outcome of biotechnological applications. Low-cost and low-technology bioremediation projects are likely to suffer from this dilemma because it represents an engineering challenge as well. Improving access of microorganisms to the substrate by substrate pretreatment alone might not be sufficient. Lignocellulosic substrates with small pore volumes not only negate biological catalysts, complete access to the substrate, but they incur hydraulic problems as well. This dilemma presents itself as a target for future research initiatives. Alternatively, if the substrate quality cannot be improved, then the desired substrate must be defined by means of a thorough lignocellulose screening procedure and procured if available.

6.3.3 Toxicity of Lignin–Carbohydrate Complexes and Lignocellulose Degradation Intermediates

Careful consideration must be made regarding the choice of the lignocellulose material to be incorporated into the biotechnology process. Toxic substances released from plant cells during the process might result in catalyst inhibition or decay followed by process failure. After biodegradation of any compound, it forms an array of secondary metabolites, which may cause toxicity. Pulp and paper mill effluent, having lignocellulosic compounds, has not been classified as a potent carcinogenic, teratogenic or genotoxic compound. However, compounds present in it may bind an AhR (aromatic hydrocarbon receptor) and hence caused the toxicity in terms of CYP activity and apoptosis.

6.4 Decomposition of Lignocellulose Ingredients

6.4.1 Cellulose Biodegradation

Most of the cellulolytic microorganisms belong to eubacteria and fungi, even though some anaerobic protozoa and slime moulds able to degrade cellulose have also been described. Cellulolytic microorganisms can establish synergistic relationships with noncellulolytic species in cellulosic wastes. The interactions between both populations lead to complete degradation of cellulose, releasing carbon dioxide and water under aerobic conditions and carbon dioxide, methane and water under anaerobic conditions (Béguin and Aubert 1994; Leschine 1995).

Microorganisms capable of degrading cellulose produce a battery of enzymes with different specificities working together (Figure 6.1). Cellulases hydrolyse the β-1,4-glycosidic linkages of cellulose. Traditionally, they are divided into two classes referred to as endoglucanases and cellobiohydrolases.

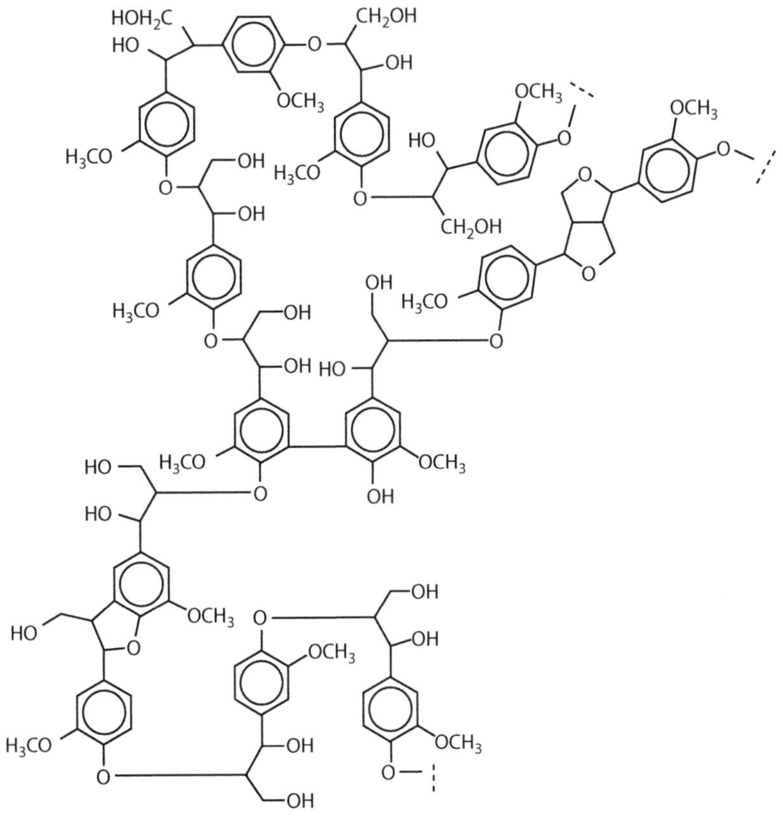

FIGURE 6.1
Lignin from gymnosperms showing the different linkages between the phenyl propane units. Angiosperm lignin is very similar, but phenyl propane units contain two methoxyl groups in ortho position to oxygen.

Endoglucanases (endo-1,4-β-glucanases, EGs) can hydrolyse internal bonds (preferably in cellulose-amorphous regions) releasing new terminal ends. Cellobiohydrolases (exo-1,4-β-glucanases, CBHs) act on the existing or endoglucanase-generated chain ends. Both enzymes can degrade amorphous cellulose, but with some exceptions, CBHs are the only enzymes that efficiently degrade crystalline cellulose. CBHs and EGs release cellobiose molecules. An effective hydrolysis of cellulose also requires β-glucosidases, which break down cellobiose releasing two glucose molecule interactions occurring in such environments (Leschine 1995). To function correctly, endoglucanases, exoglucanases and β-glycosidases must be stable in the exocellular environment and may form a ternary complex with the substrate, the cellulase systems of the mesophilic fungi. *Trichoderma reesei* and *Phanerochaete chrysosporium* are the most thoroughly studied. The EGs of these consortiums have two structural domains: the catalytic domain and the union domain. Their molecular masses range from 25 to 50 kDa, and they have optimum activities at acidic pH. CBHs act synergistically with EGs to solubilise high molecular weight cellulose molecules. They are also glycosylated and present an optima activity at acidic pH. *P. chrysosporium* has several β-glucosidases, whereas only one isoenzyme Q has been described in *T. reesei*. All of them have high molecular masses ranging from 165 to 182 kDa. Several thermophilic fungi can degrade cellulose faster than *T. reesei*. The EGs of thermophilic fungi are thermostable, and their molecular masses range from 30 to 100 kDa. They show optimal activity between 55°C and 80°C at pH 5.0–5.5. Exoglucanases (40–70 kDa) are optimally active at 50°C–75°C. The molecular characteristics of β-glucosidases are variable; their molecular masses range from 45 to 250 kDa, optimal pH from 4.1 to 8.1 and optimal temperature from 35°C to 71°C (Mishra and Thakur 2011). Among aerobic cellulolytic bacteria, species from the genera *Cellulomonas, Pseudomonas* and *Streptomyces* are the best studied (Béguin and Aubert 1994).

6.4.2 Hemicellulose Biodegradation

Hemicelluloses are biodegraded to monomeric sugars and acetic acid. Hemicellulases are frequently classified according to their action on distinct substrates. Xylan is the main carbohydrate found in hemicellulose. Its complete degradation requires the cooperative action of a variety of hydrolytic enzymes. An important distinction should be made between endo-1,4-β-xylanase and xylan 1,4-β-xylosidase. The former generates oligosaccharides from the cleavage of xylan; the latter works on xylan oligosaccharides, producing xylose (Jeffries 1994). In addition, hemicellulose biodegradation needs accessory enzymes, such as xylan esterases, ferulic and p-coumaric esterases, α-l-arabinofuranosidases and α-4-O-methyl glucuronosidases, acting synergistically to efficiently hydrolyse wood xylans and mannans. In the case of O-acetyl-4-O-methylglucuronxylan, one of the most common hemicelluloses, four different enzymes are required for degradation: endo-1,4-β-xylanase (endoxylanase), acetyl esterase, α-glucuronidase and β-xylosidase. The degradation of O-acetyl galactoglucomannan starts with a rupture of the polymer by endomannases. Acetyl glucomannan esterases remove acetyl groups, and α-galactosidases eliminate galactose residues. Finally, β-mannosidase and β-glycosidase break down the endomannase-generated oligomers β-1,4 bonds.

Xylanases, the major component of hemicellulases, have been isolated from many ecological niches in which plant material is present. Due to the important biotechnological exploitations of xylanases, especially in biopulping and bleaching, many publications have appeared in recent years (Kulkarni et al. 1999; Mishra and Thakur 2011). The white-rot fungus *Phanerochaete chrysosporium* has been shown to produce multiple endoxylanases (Kapich et al. 1999). Also, bacterial xylanases have been described in several aerobic species and some ruminal genera (Blanco et al. 1999; Kulkarni et al. 1999; Mishra and Thakur 2011).

Hydrolysis of β-glycosidic linkages is carried out by acid catalytic reactions common to all glycanases. Many microorganisms contain multiple loci encoding overlapping xylanolytic functions. Xylanases, like many other cellulolytic and hemicellulolytic enzymes, are highly modular in structure. They consist of either a single domain or a number of different domains, classified as catalytic and noncatalytic domains. Based on the homology of the conserved amino acids, xylanases can be grouped into two different families: family 10 (F), with relatively high molecular weight, and family 11 (G), with lower molecular weight. The catalytic domains for the two families differ in their molecular masses, net charge and isoelectric points (Kulkarni et al. 1999) and may play a major role in determining specificity and reactivity. Biochemically and structurally, the two families are unrelated.

The release of reducing sugars from purified xylan is highly dependent on the xylanase pI. Isoelectric points for endoxylanases from various microorganisms vary from 3 to 10. Optimum temperature for xylanases from bacterial and fungal origin ranges from 40°C to 60°C. Fungal xylanases are generally less thermostable than bacterial xylanases (Mishra and Thakur 2011).

Researchers have paid special attention to thermostable hemicellulases because of their biotechnological applications. Thermophilic xylanases have been described in actinobacteria (formerly actinomycetes), such as *Thermomonospora* and *Actinomadura* (George et al. 2001). Also, a very thermostable xylanase has been isolated from the hyperthermophilic primitive bacterium *Thermotoga* (Simpson et al. 1991). Xylanases of thermophilic fungi are also receiving considerable attention. As in mesophilic fungi, a multiplicity of xylanases differing in stability, catalytic efficiency and activity on substrates has been observed (Maheshwari et al. 2000). The optimal temperatures vary from 60°C to 80°C, and the pI ranges from 3.7 to 9.0. This diversity of xylanase isoenzymes of different molecular masses might be to allow their diffusion into the plant cell walls. The use of xylanases in bleaching pulps has stimulated the search for enzymes with alkaline pH optima.

Most xylanases from fungi have pH optima between 4.5 and 5.5. Xylanases from actinobacteria are active at pH 6.0–7.0 (Jefferies 1994). However, xylanases active at alkaline pH have been described from *Bacillus* sp. or *Streptomyces viridosporus* (Blanco et al. 1999). Genes encoding several xylanases have been cloned in homologous and heterologous hosts in order to overproduce the enzyme with the goal of altering its properties so that it can be used for commercial applications (Kulkarni et al. 1999).

6.4.3 Lignin Biodegradation

The structural complexity of lignin, its high molecular weight and its insolubility make its degradation very difficult. Extracellular, oxidative and unspecific enzymes that can liberate highly unstable products that further undergo many different oxidative reactions catalyse the initial steps of lignin depolymerisation. This nonspecific oxidation of lignin has been referred to as 'enzymatic combustion' (Kirk and Farrel 1987). White-rot fungi are the microorganisms that most efficiently degrade lignin from wood. Of these, *Phanerochaete chrysosporium* is the most extensive.

For recent reviews on lignin biodegradation by white-rot fungi and advances in the molecular genetic of ligninolytic fungi, see Vicuna (2000). Two major families of enzymes are involved in ligninolysis by white-rot fungi: peroxidases and laccases.

Apparently, these enzymes act using low molecular weight mediators to carry out lignin degradation. Several classifications of fungi have been proposed based on their ligninolytic enzymes. Some of them produce all of the major enzymes; others only two of them or even only one. In addition, reductive enzymes, including cellobiose oxidising enzymes, aryl alcohol oxidases and aryl alcohol dehydrogenases, seem to play major roles in ligninolysis.

Two groups of peroxidases, lignin peroxidases (LiPs) and manganese-dependent peroxidases (MnPs), have been well characterised. LiP has been isolated from several white-rot fungi. The catalytic, oxidative cycle of LiP has been well established and is similar to those of other peroxidases. In most fungi, LiP is present as a series of isoenzymes encoded by different genes. LiP is a glycoprotein with a heme group in its active centre. Its molecular mass ranges from 38 to 43 kDa and its pI from 3.3 to 4.7. So far, it is the most effective peroxidase and can oxidise phenolic and non-phenolic compounds, amines, aromatic ethers and polycyclic aromatics with appropriate ionisation potential. Because LiP is too large to enter the plant cell, its degradation is carried out only in exposed regions of lumen. This kind of degradation is found in simultaneous wood decay. However, microscopic studies of selective lignin biodegradation reveal that white-rot fungi remove the polymer from inside the cell wall. An indirect oxidation by LiP of low molecular weight diffusible compounds capable of penetrating the cell wall and oxidising the polymer has been suggested. However, this theory lacks evidence because low molecular weight intermediates, such as veratryl alcohol cation radical, are too short-lived to act as mediators (Kapich et al. 1999). MnPs are molecularly very similar to LiPs and are also glycosylated proteins, but they have slightly higher molecular masses, ranging from 45 to 60 kDa. MnPs oxidise Mn(II) to Mn(III). They have a conventional peroxidase catalytic cycle but with Mn(II) as a substrate. This Mn(II) must be chelated by organic acid chelators, which stabilise the product Mn(III) (Pérez and Jeffries 1992).

Mn(III) is a strong oxidant that can leave the active centre and oxidise phenolic compounds, but it cannot attack nonphenolic units of lignin. MnP generates phenoxy radicals, which, in turn, undergo a variety of reactions, resulting in depolymerisation (Gold et al. 2000). In addition, MnP oxidises nonphenolic lignin model compounds in the presence of Mn(II) via peroxidation of unsaturated lipids (Mishra and Thakur 2011). A novel versatile peroxidase (VP), which has both manganese peroxidase and lignin peroxidase activities and which is involved in the natural degradation of lignin, has been described (Camarero et al. 1999). VP can oxidise hydroquinone in the absence of exogenous H_2O_2 when Mn(II) is present in the reaction. It has been suggested that chemical oxidation of hydroquinones promoted by Mn(II) could be important during the initial steps of wood biodegradations because ligninolytic enzymes are too large to penetrate into nonmodified wood cell walls (Gómez-Toribio et al. 2001).

Laccases are blue-copper phenoloxidases that catalyse the one-electron oxidation mainly of phenolic compounds and nonphenolics in the presence of mediators (Gianfreda et al. 1999). The phenolic nucleus is oxidised by removal of one electron, generating phenoxy free-radical products, which can lead to polymer cleavage. Wood-rotting fungi are the main producers of laccases, but this oxidase has been isolated from many fungi, including *Aspergillus* and the thermophilic fungi *Myceliophora thermophile* and *Chaemotium thermophilium* (Leonowicz et al. 2001). Recently, bacterial laccase-like proteins have been found (Kaushik and Thakur 2013). These enzymes polymerised a low molecular weight, water-soluble organic matter fraction isolated from compost into high molecular weight products, suggesting the involvement of laccase in humification during composting (Maheshwari et al. 2000). The role of laccases in lignin biodegradation has been discussed recently (Kaushik and Thakur 2013; Leonowicz et al. 2001).

The potential biotechnological applications of white-rot fungi or their ligninolytic enzymes are many. The most promising applications may be biopulping and bleaching of chemical pulps. White-rot fungi can degrade/ mineralise a wide variety of toxic xenobiotics, including polycyclic aromatic hydrocarbons, chlorophenols, nitrotoluenes, dyes and polychlorinated biphenyls. It has been found that Basidiomycetes, such as *Grammathels fuligo* and *Phanerochaete crassa*, have the capability to degrade chlorinated phenols and pentachlorophenol. At appropriate temperature and pH, *G. fuligo* and *P. crassa* showed superior mycelial growth. Reduction in chlorophenols was due to adsorption. Other fields under research are the use of these fungi for biocatalysis in the production of fine chemicals and natural flavors (for example, vanillin) and the biotreatment of several wastewaters, such as bleach plant effluents or other wastewater containing lignin-like polymers, as in the cases of dye industry effluents and olive oil mill wastewaters Martínez et al. (1998) (Figure 6.2).

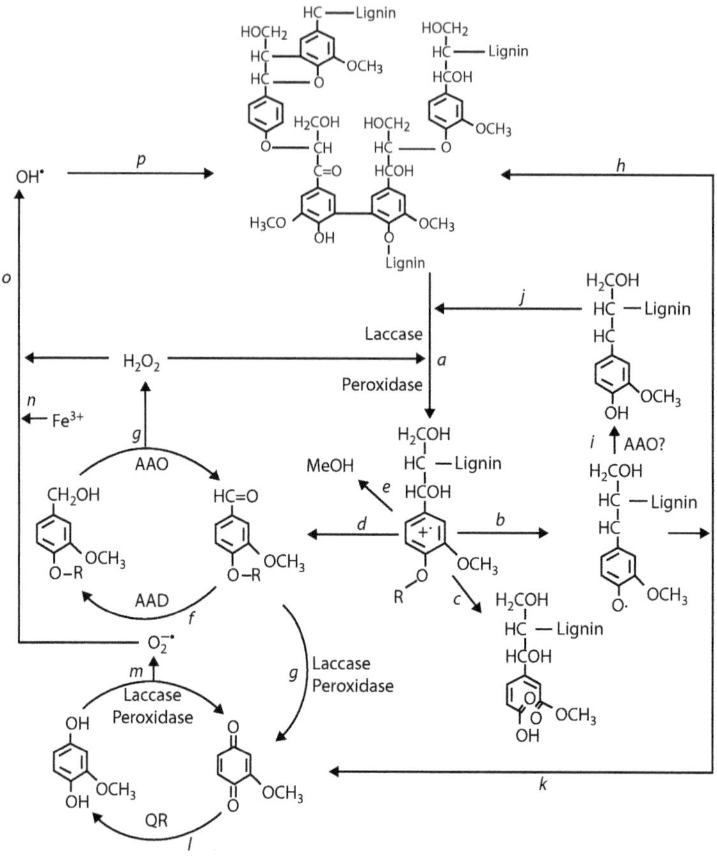

FIGURE 6.2
A scheme for lignin biodegradation including enzymatic reactions and oxygen activation. As shown in Figure 6.4, laccases or ligninolytic peroxidases (LiP, MnP and VP) produced by white-rot fungi oxidise the lignin polymer, thereby generating aromatic radicals (a). These evolve in different nonenzymatic reactions, including C4 ether breakdown (b), aromatic ring cleavage (c), Cα-Cβ breakdown (d) and demethoxylation (e). The aromatic aldehydes released from Cα-Cβ breakdown of lignin or synthesised de novo by fungi (f, g) are the substrate for H_2O_2 generation by AAO in cyclic redox reactions involving also AAD. Phenoxy radicals from C4 ether breakdown (b) can repolymerise on the lignin polymer (h) if they are not first reduced by oxidases to phenolic compounds (i) as reported for AAO. The phenolic compounds formed can be again reoxidised by laccases or peroxidases (j). Phenoxyradicals can also be subjected to Cα-Cβ breakdown (k), yielding p-quinones. Quinones from (g) and/or (k) contribute to oxygen activation in redox-cycling reactions involving QR, laccases and peroxidases (l, m). This results in reduction of the ferric iron present in wood (n), either by superoxide cation radical or directly by the semiquinone radicals, and its reoxidation with concomitant reduction of H_2O_2 to hydroxyl free radical (OH') (o). The latter is a very strong oxidiser that can initiate the attack on lignin (p) in the initial stages of wood decay when the small size of pores in the still-intact cell wall prevents the penetration of ligninolytic enzymes. Then, lignin degradation proceeds by oxidative attack of the enzymes described above. In the final steps, simple products from lignin degradation enter the fungal hyphae and are incorporated into intracellular catabolic routes.

6.5 Enzymatic Processes for Lignin and Hemicelluloses Degradation

In the pulp and paper industry, cellulose is used for paper production, and lignin and hemicellulose end up in the effluent. The bacteria capable of degrading lignin and hemicellulose can be used for the treatment of effluent. The degradation process involves the use of a number of enzymes collectively called ligninase. Ligninase is a generic name for a group of isozymes that catalyse the oxidative depolymerisation of lignin. They include lignin peroxidase, manganese peroxidase and laccase. Another enzyme, xylanase, plays a crucial role in hemicellulose degradation. These enzymes are extracellular, non-substrate-specific and aerobic in nature. This is an essential requirement for lignin degradation as it is a randomly synthesised biopolymer that cannot enter inside the cell, and degradation involves the cleavage of a carbon–carbon or ether bond that links various subunits in the oxidative environment (Breen and Singleton 1999). The mechanism of action of these enzymes is as follows.

6.5.1 Xylanase

Xylan, a major constituent of hemicellulose, is composed of β-1,4-linked xylopyranosyl residues, which can be substituted with arabinosyl and methylglucuronyl side chains. Xylanases (endo-1,4-β-D-xylan xylanohydrolase; EC 3.2.1.8) are a group of enzymes that hydrolyse xylan backbone into small oligomers (Kiddinamoorthy et al. 2008). The xylanolytic enzyme system carrying out the xylan hydrolysis is usually composed of a repertoire of hydrolytic enzymes: β-1,4-endoxylanase, β-xylosidase, α-L-arabinofuranosidase, α-glucuronidase, acetyl xylan esterase and phenolic acid (ferulic and *p*-coumaric acid) esterase (Figure 6.3). The presence of such a multifunctional xylanolytic enzyme system is quite widespread among fungi, actinomycetes and bacteria (Beg et al. 2001).

Due to xylan heterogeneity, the enzymatic hydrolysis of xylan requires different enzymatic activities. Two enzymes, β-1,4-endo-xylanase (EC 3.2.1.8) and β-xylosidase (EC 3.2.1.37), are responsible for hydrolysis of the main chain, the first attacking the internal main-chain xylosidic linkages and the second releasing xylosyl residues by endwise attack of xylooligosaccharides (Subramaniyan and Prema 2002). These two enzymes are the major components of xylanolytic systems produced by biodegradative microorganisms, such as *Trichoderma*, *Aspergillus*, *Schizophyllum*, *Bacillus*, *Clostridium* and *Streptomyces* sp. (Bedard et al. 1987). However, for complete hydrolysis of the molecule, side-chain cleaving enzyme activities are also necessary.

Xylanases have several different industrial applications, including kraft pulp bleaching in the paper industry; biodegradation of lignocellulose in animal feed, foods and textiles; and biopulping in the paper and pulp industry (Madlala et al. 2001).

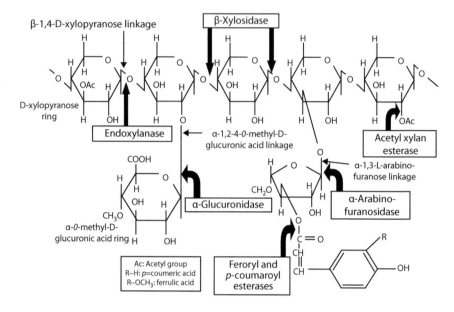

FIGURE 6.3
A hypothetical plant xylan structure showing different substituent groups with sites of attack by microbial xylanases. (From Beg QK et al., *Appl Microbiol Biotechnol* 56: 326–338, 2001.)

6.5.2 Lignin Peroxidase (LiP)

Lignin peroxidase is a heme-containing glycoprotein, which requires hydrogen peroxide as an oxidant. Fungi secrete several isoenzymes into their cultivation medium, although the enzymes may also be cell-wall bound (Lackner et al. 1991). LiP oxidises nonphenolic lignin substructures by abstracting one electron and generating cation radicals, which are then decomposed chemically (Figure 6.4). Reactions of LiP using a variety of lignin model compounds and synthetic lignin have thoroughly been studied, catalytic mechanisms have been elucidated and its capability for C~–C~ bond cleavage, ring opening and other reactions have been demonstrated (Eriksson et al. 1990). LiP is secreted during secondary metabolism as a response to nitrogen limitation. They are strong oxidisers capable of catalysing the oxidation of phenols, aromatic amines, aromatic ethers and polycyclic aromatic hydrocarbons (Breen and Singleton 1999).

6.5.3 Manganese Peroxidase (MnP)

Manganese peroxidase is also a heme-containing glycoprotein, which requires hydrogen peroxide as an oxidant. MnP oxidises Mn(II) to Mn(III), which then oxidises phenol rings to phenoxy radicals, which lead to decomposition of compounds (Figure 6.5). Evidence for the crucial role of MnP in lignin biodegradation is accumulating, for example, in depolymerisation of lignin (Wariishi et al. 1992) and chlorolignin, in demethylation of lignin and

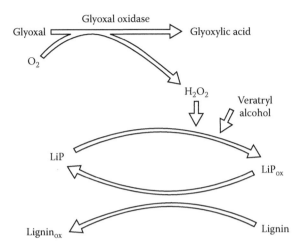

FIGURE 6.4
Mechanism of action for lignin peroxidase, LiP; ox stands for oxidised state of enzyme. (From Breen A and Singleton FL, *Curr Opin Biotechnol* 10: 252–258, 1999.)

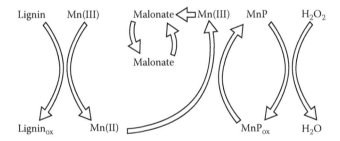

FIGURE 6.5
Mechanism of action for manganese peroxidase, MnP; ox stands for oxidised state of enzyme. (From Breen A and Singleton FL, *Curr Opin Biotechnol* 10: 252–258, 1999.)

delignification and bleaching of pulp (Paice et al. 1993) and in mediating initial steps in the degradation of high molecular mass lignin.

6.5.4 Laccase (Lac)

Laccase (EC No. 1.10.3.2; benzenediol: 0, oxidoreductase) is a true phenoloxidase with broad substrate specificity. It is a copper-containing glycoprotein widely reported in fungi and plants. Most famous are rot fungi, such as *Phanerochaete chrysosporium, Ceriporiopsis subvermispora, Coriolus versicolor* var. antarcticus, *Pycnoporus sanguineus, Trametes elegans, Bjerkandera adusta, Pleurotus eryngii, Phlebia radiata,* etc. (Baldrian 2006). It has also been reported in some plants, such as *Acer pseudoplantanus, Aesculus parviflora, Populus euramericana* etc. In plants, laccase participates in the radical-based mechanisms of lignin polymer formation (Sterjiades et al. 1992), whereas, in fungi, laccases probably have more roles, including morphogenesis, fungal plant pathogen/host interaction, stress defence and lignin degradation. The presence of laccase has been reported in bacteria; however, such reports remain controversial (Diamantidis et al. 2000).

The reactions catalysed by laccases proceed by the monoelectronic oxidation of a suitable substrate molecule (phenols and aromatic or aliphatic amines) to the corresponding reactive radical. The redox process takes place with the assistance of a cluster of four copper atoms that form the catalytic core of the enzyme (Figure 6.6); they also confer the typical blue colour to

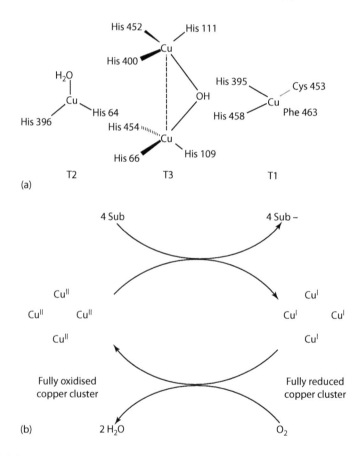

FIGURE 6.6
Catalytic action of laccases. Laccases: active-site structure and catalytic cycle. (a) Model of the catalytic cluster of the laccase from *Trametes versicolor* made of four copper atoms. Type I (T1) copper confers the typical blue colour to the protein and is the site where substrate oxidation takes place. Type 2 (T2) and Type 3 (T3) copper form a trinuclear cluster, where reduction of molecular oxygen and release of water takes place. (b) Schematic representation of a laccase catalytic cycle producing two molecules of water from the reduction of one molecule of molecular oxygen and the concomitant oxidation (at the T1 copper site) of four substrate molecules to the corresponding radicals. (From Riva S, *Trends Biotechnol* 24:221–226, 2006.)

these enzymes because of the intense electronic absorption of the Cu–Cu linkages (Piontek et al. 2002).

Lignin is formed via the oxidative polymerisation of monolignols within the plant cell-wall matrix. Peroxidases, which are abundant in virtually all cell walls, have long been held to be the principal catalysts for this reaction. Recent evidence shows, however, that laccases secreted into the secondary walls of vascular tissues are equally capable of polymerising monolignols in the presence of O_2. The role of laccases in lignification has often been debated. Laccase from *Acer pseudoplantanus* was able to polymerise monolignols in the complete absence of peroxidase (Sterjiades et al. 1992). This shows that laccase was involved in the early stages of lignification, and peroxidases were involved later.

Laccases are able to catalyse electron transfer reactions without additional cofactors; hence, their use has been studied in biosensors to detect various phenolic compounds, oxygen or azides. An enzyme electrode based on the coimmobilisation of an osmium redox polymer and a laccase from *T. versicolor* on glassy carbon electrodes has been applied to ultrasensitive amperometric detection of the catecholamine neurotransmitters dopamine, epinephrine and norepinephrine, attaining nanomolar detection limits (Ferry and Leech 2005). Laccase can also be immobilised on the cathode of biofuel cells that could provide power, for example, for small transmitter systems (Calabrese et al. 2002).

6.6 Molecular Aspects

Due to their potential use as industrial biocatalysts, the catalytic mechanisms of lignin-degrading oxidoreductases (including peroxidases, oxidases and laccases) have been extensively investigated. LiP and MnP were the second and third peroxidases whose crystal structure was solved (Poulos et al. 1993; Sundaramoorthy et al. 1994) just 10 years after their discovery in *P. chrysosporium*. These peroxidases catalyse the oxidation of the recalcitrant nonphenolic lignin units by H_2O_2. This is possible because of the formation of a high redox potential oxo-ferryl intermediate during the reaction of the heme cofactor with H_2O_2. This two-electron reaction allows the activated enzyme to oxidise two substrate units, being reduced to the peroxidase resting state (which reacts again with peroxide). The catalytic cycle, consisting of the resting peroxidase and compounds I (two-electron oxidised form) and II (one-electron oxidised form), is common to other peroxidases. However, two aspects in their molecular structure provide ligninolytic peroxidases their unique catalytic properties: (i) a heme environment, conferring high redox potential to the oxo-ferryl complex, and (ii) the existence of specific binding sites (and mechanisms) for oxidation of their characteristic substrates, including nonphenolic aromatics in the cases of LiP, manganous iron in the case of MnP and both types of compounds in the case of the new VP.

Similar heme environments in the above three peroxidases have been evidenced by 1H-NMR, which allows the signals of both the heme cofactor protons and several amino acid residues forming the heme pocket to be identified (Banci et al. 2003). This is possible due to the paramagnetic effect caused by the cofactor iron, which displaces the signals of neighbour protons outside the region in which most protein protons overlap. In ligninolytic peroxidases, this residue is displaced away from the heme iron, increasing its electron deficiency and increasing the redox potential of the oxo-ferryl complex. In addition, recent studies have contributed to identification of the substrate binding sites in ligninolytic peroxidases.

The aromatic substrate binding site and the manganese binding site were first identified in LiP and MnP (Doyle et al. 1998; Gold et al. 2000) and then confirmed in the crystal structure of VP solved at atomic resolution. These studies revealed that the novel catalytic properties of VP are due to its hybrid molecular architecture as suggested several years before (Camarero et al. 1999; Ruiz-Dueñas et al. 1999). Mn^{2+} oxidation is produced at a binding site near the cofactor, at which this cation is bound by the carboxylates of three acidic residues, which enables direct electron transfer to one of the heme propionates. By contrast, veratryl alcohol and lignin model substrates are oxidised at the surface of the protein by a long-range electron transfer mechanism that initiates at an exposed tryptophan residue. The rationale of the existence of this electron transfer mechanism is related to the fact that many LiP/VP aromatic substrates, including the lignin polymer, cannot penetrate inside the protein to transfer electrons directly to the cofactor. Therefore, these substrates are oxidised at the enzyme surface, and electrons are transferred to the heme by a protein pathway.

The H_2O_2 responsible for oxidative degradation of lignin is generated by extracellular fungal oxidases, which can reduce dioxygen to peroxide in a catalytic reaction. Flavin cofactors are generally involved in this reaction, as in the Pleurotus flavoenzyme AAO, although glyoxal oxidase from *P. chrysosporium* is a copper-containing oxidase (Whittaker et al. 1996). Among flavoenzyme oxidases, fungal glucose oxidase has been crystallised (Wohlfahrt et al. 1999), but this is an intracellular enzyme that is not involved in lignin degradation. However, its crystal structure has been used as a template to predict the molecular structure of AAO (Varela et al. 2000). It has been shown that AAO is a unique oxidase due both to its spectroscopic characteristics (flavin intermediates and reactivity) and to the wide range of aromatic and aliphatic polyunsaturated primary alcohols (and even aldehydes) that it is able to oxidise (Ferreira et al. 2005). The molecular structure of AAO includes two catalytically active histidines near the N5 of the flavin ring, which might help electron transfer to or from the cofactor by acting as bases in the oxidation of aromatic alcohols (which would proceed via a hydride transfer mechanism) and as acids in the reduction of oxygen to H_2O_2.

As noted, laccases were the first ligninolytic enzymes to be investigated and had been known in plants for many years. Nevertheless, the first molecular structure of a complete fungal laccase was published only in 2002. That year, the crystal structures of the laccases from the basidiomycete *T. versicolor* and the ascomycete *Melanocarpus albomyces* were reported (Bertrand et al. 2002; Hakulinen et al. 2002; Piontek et al. 2002). The first structure of a bacterial laccase was published one year later (Enguita et al. 2003).

The active site of laccases includes four copper ions. Type I copper acts as an electron acceptor from substituted phenols or amines (the typical laccase substrates), and type II copper transfers the electrons to the final acceptor, dioxygen, which is reduced to water. The two type III coppers act as intermediates in the electron transfer pathway that also includes one cysteine and two histidine protein residues. The molecular environment of laccase type I copper seems to regulate the redox potential of the enzyme (Piontek et al. 2002). The fact that laccase can use atmospheric oxygen as the final electron acceptor represents a considerable advantage for industrial and environmental applications compared with peroxidases, which require a continuous supply of H_2O_2. Taking into account that the advantage of peroxidases is their higher redox potential, engineering the active site of laccases to obtain high redox potential variants would be of considerable biotechnological interest.

Most enzymes involved in wood lignin degradation (a multienzymatic process that includes, among others, peroxidases, oxidases and laccases acting synergistically) have been identified, and the mechanisms of action of several of them have been established at a considerably precise level. These enzymes, however, cannot penetrate the compact structure of sound wood tissues due to their comparatively large molecular size. Therefore, small chemical oxidisers, including activated oxygen species and enzyme mediators, are probably involved in the initial steps of wood decay.

6.7 Biotechnological Application of Lignocellulose and Its Biodegradation

The primary objective of lignocellulose pretreatment by the various industries is to access the potential of the cellulose and hemicellulose encrusted by lignin within the lignocellulose matrix. The combination of solid-state fermentation (SSF) technology with the ability of white-rot fungi to selectively degrade lignin has made possible industrial-scale implementation of lignocellulose-based biotechnologies. SSF offers the advantages of a robust technology and outperforms conventional fermentation technologies with respect to simplicity, cost-effectiveness and maintenance requirements. These advantages make SSF an attractive technology for environmental problems for which money and expertise are limited. Problems commonly associated with SSF are heat buildup, bacterial contamination, scale-up, biomass growth estimation and control of substrate moisture content (Lonsane et al. 1985).

6.7.1 Lignocellulose-Based Technologies Using Unsterile Substrates

Silage manufacture is a good example of a working technology based on unsterile lignocellulose substrates. Silage is the material produced by the controlled fermentation of moist plant material (McDonald 1981). Water authorities consider silage effluent a serious threat to natural water supplies (Haigh 1994). Arnold et al. (2000) used *Candida utilis* and *Galactomyces geotrichum* to reduce the polluting potential of silage effluent with an initial chemical oxygen demand (COD) of 80,000 mg/l. COD, phosphate and ammonia concentrations were reduced by 74%–95%, 82%–99% and 16%–64%, respectively. A similar effluent is produced during oyster mushroom cultivation (personal observation). These effluents represent sources of carbon and nutrients and might even, in the future, serve as cheap sources of fermentation adjuvants.

6.7.2 Bio-Pulping

Lignin becomes problematic to cellulose-based wood processing because it must be separated from cellulose at enormous energy, chemical and environmental expense. Bio-pulping is therefore a solid-state fermentation process during which wood chips are treated with white-rot fungi to improve the delignification process. Biological pulping has the potential to reduce energy costs and environmental impact relative to traditional pulping operations (Breen and Singleton 1999). The benefits of biopulping were demonstrated by Scott et al. (1998) using 40-ton scale experiments: tensile, tear and burst indexes of the resulting paper were improved (indicative of a higher degree of cellulose conservation during the pulping process); brightness of the pulp was increased (indicating improved lignin removal); and improved energy savings of 30%–38% were realized.

Problems endemic to SSF still plague this concept. Inoculation, aeration and heat removal are key parameters that influence fungal activity. Also, poor colonisation of wood chips by white-rot fungi has been attributed to competition with naturally occurring microorganisms or to inhibition by wood chemical components. Substrate sterilisation is usually a major expense, and secondary contamination by airborne microorganisms must be prevented at additional costs as well. Breen and Singleton (1999) summarised decades of dissatisfaction associated with high capital costs to make SSF viable for the pulp and paper industry: 'Overcoming these challenges will determine, in a large part, if biopulping becomes a reality'.

6.7.3 Animal Feed

Cellulose is the most important source of carbon and energy in a ruminant's diet, although the animal itself does not produce cellulose-hydrolysing enzymes. Rumen microorganisms utilise cellulose and other plant carbohydrates as their source of carbon and energy. Thus, the microorganisms convert the carbohydrates in large amounts of acetic, propionic and butyric acids, which the higher animal can use as its energy and carbon sources. The concept of preferential delignification of lignocellulose materials by white-rot fungi has been applied to increase the nutritional value of forages (Zadrazil and Isikhuemhen 1997). This increased digestibility provides organic carbon that can be fermented to organic acids in an anaerobic environment, such as the rumen. However, upgrading of animal feed by white-rot fungi failed to reach industrial proportions. A possible explanation can be that the animals' instincts prevent them from ingesting mushrooms for they can contain toxicants or they can be toxic to their rumen microflora and, hence, toxic to the animal also.

6.7.4 Potential of Lignocellulose in Space Exploration

Advances in lignocellulose research will enable scientists to contribute to space science/exploration. Space travel will benefit from this research in the near future as the transport of lignocellulose to space can result in substantial cost savings. Lignocelluloses can be a feedstock to provide for all basic needs: fuel, energy, feedstock chemicals, food and water. Recycling of inedible plant material by white-rot fungi (*Pleurotus ostreatus*) has been investigated in a Closed Ecological Life Support System (CELSS) (Sarikaya and Ladish 1997). Lignocellulose can therefore be the 'super fuel' of the future – being a compact natural polymer containing enough potential energy to sustain man and machine in space.

6.7.5 Organic Acids

Several organic acids have already been produced by SSF from lignocellulose wastes. Citric acid is one of these examples. Almost the entire production of this acid has been obtained using crops and crop residues as substrates and *Aspergillus niger* as the production strain. When comparing sugarcane bagasse, coffee husk and cassava bagasse as a solid substrate for citric acid production by *Aspergillus niger,* cassava bagasse led to the highest production results. Lactic acid may be produced by SSF using fungal as well as bacterial cultures. Strains of *Rhizopus* sp. have been common among the fungal cultures and that of *Lactobacillus* sp. among the bacterial cultures.

6.7.6 Single Cell Protein

Some processes have focused on the direct conversion of lignocellulosic wastes to single-cell protein. The microorganisms' strains *Chaetomium cellulolyticum* mutant, *Pleurotus sajor-caju,* strains of *Aspergillus* and *Penicillium* spp., may be used for this purpose. This cocultivation of fungi has the ability to utilize cellulose and hemicellulose after lignin degradation for single-cell protein production. There is no need for any treatment in the raw material before use in this system as together these fungi are capable of separating lignocellulose into its individual components. The cellulose obtained may be also used for paper production or as single-cell protein for animal or human feed (Nigam et al. 2009).

6.7.7 Bioactive Compounds

Several bioactive compounds may be produced by SSF from different lignocellulose wastes. Some examples include (i) the production of gibberellic acid by *Giberella fujikuroi* and *Fusarium moniliforme* from corn cobs, (ii) the production of tetracycline from cellulosic substrates, (iii) the production of oxytetracycline by *Streptomyces rimosus* from corn cobs, (iv) the production of destrucxins A and B (cyclodepsipeptides) by *Metarhizium anisopliae* from rice husk and (v) the production of ellagic acid by *Aspergillus niger* from pomegranate peel and creosote bush leaves.

6.8 Conclusions

This review aims to empower scientists with the vision to apply science to lignocellulose wastes taken for granted. Natural biodegradation processes and the consortia involved (for example, rumen, termite hindgut, etc.) are treasure troves filled with the potential for lateral applications of scientific discoveries. Current working lignocellulose-based technologies produce potential polluting effluents, but when properly managed, these can overcome financial barriers preventing the success of innovative processes. The ultimate beneficiaries of this approach will be the local and national economies of communities – and the people themselves.

References

Ander P, Hatakka A and Eriksson K-E (1980) Vanillic acid metabolism by the white-rot fungus *Sporotrichum pulverulentum*. *Arch Microbiol* 125: 189–202.

Arnold JL, Knapp JS and Johnson CL (2000) The use of yeasts to reduce the polluting potential of silage effluent. *Wat Res* 34: 3699–3708.

Baldrian P (2006) Fungal laccases – occurrence and properties. *FEMS Microbiol Rev* 30: 215–242.

Banci L, Camarero S, Martínez AT, Martínez MJ, Pérez-Boada M, Pierattelli R and Ruiz-Dueñas FJ (2003) NMR study of Mn(II) binding by the new versatile peroxidase from the white-rot fungus Pleurotus eryngii. *J Biol Inorg Chem* 8: 751–760.

Béguin P and Aubert JP (1994) The biological degradation of cellulose. *FEMS Microbiol Rev* 13: 25–58.

Bedard DL, Haberl ML, May RJ and Brennan MJ (1987) Evidence for novel mechanisms of polychlorinated biphenyl metabolism in *Alcaligenes eutrophus* H 850. *Appl Environ Microbiol* 53: 1103–1112.

Beg QK, Kapoor M, Mahajan L and Hoondal GS (2001) Microbial xylanases and their industrial applications: A review. *Appl Microbiol Biotechnol* 56: 326–338.

Bertrand T, Jolivalt C, Briozzo P, Caminade E, Joly N, Madzak C and Mougin C (2002) Crystal structure of a four-copper laccase complexed with an arylamine: Insights into substrate recognition and correlation with kinetics. *Biochemistry* 41: 7325–7333.

Blanchette RA, Shaw CG and Cohen AL (1978) A SEM study of the effects of bacteria and yeasts on wood decay by brown- and white-rot fungi. *Scan Elec Micros* 2: 61–68.

Blanco A, Díaz P, Zueco J, Parascandola P and Pastor JF (1999) A multidomain xylanase from a Bacillus sp. with a region homologous to thermostabilizing domains of thermophilic enzymes. *Microbiology* 145: 2163–2170.

Breen A and Singleton FL (1999) Fungi in lignocellulose breakdown and biopulping. *Curr Opin Biotechnol* 10: 252–258.

Calabrese BS, Pickard M, Vazquez-Duhalt R and Heller A (2002) Electroreduction of O_2 to water at 0.6 V (NHE) at pH 7 on the 'wired' *Pleurotus ostreatus* laccase cathode. *Biosens Bioelectron* 17: 1071–1074.

Camarero S, Sarkar S, Ruiz-Dueñas FJ, Martínez MJ and Martínez AT (1999) Description of a versatile peroxidase involved in the natural degradation of lignin that has both manganese peroxidase and lignin peroxidase. *J Biol Chem* 274: 10324–10330.

Cornu A, Besle JM, Mosoni P and Grent E (1994) Lignin-carbohydrate complexes in forages: Structure and consequences in the ruminal degradation of cell-wall carbohydrates. *Reprod Nutr Dev* 24: 385–398.

Diamantidis G, Effiosse AE, Potier P and Bally R (2000) Purification and characterization of the first bacterial laccase in the rhizospheric bacterium *Azospirillum lipoferum*. *Soil Biol Biochem* 32: 919–927.

Doyle WA, Blodig W, Veitch NC, Piontek K and Smith AT (1998) Two substrate interaction sites in lignin peroxidase revealed by site-directed mutagenesis. *Biochemistry* 37: 15097–15105.

Elder DJ and Kelly DJ (1994) The bacterial degradation of benzoic acid and benzenoid compounds under anaerobic conditions: Unifying trends and new perspectives. *FEMS Microbiol Rev* 13: 441–468.

Enguita FJ, Martins LO, Henriques AO and Carrondo MA (2003) Crystal structure of a bacterial endospore coat component: A laccase with enchanced thermostability properties. *J Biol Chem* 278: 19416–19425.

Eriksson K-EL, Blanchette RA and Ander P (1990) *Microbial and enzymatic degradation of wood and wood components*. Springer-Verlag: Berlin Heidelberg, 407 pp.

Eriksson KEL 1990. Biotechnology in the pulp and paper industry. *Wood Sci Technol* 24: 79–101.

Fan LT, Lee YH and Gharpuray MM (1982) The nature of lignocellulosics and their pretreatments for enzymatic hydrolysis. *Adv Biochem Eng* 23: 157–187.

Ferreira P, Medina M, Guillén F, Martínez MJH, van Berkel WJ and Martínez AT (2005) Spectral and catalytic properties of aryl-alcohol oxidase, a fungal flavoenzyme acting on polyunsaturated alcohols. *Biochem J* [doi:10.1042/BJ20041903].

Ferry Y and Leech D (2005) Amperometric detection of catecholamine neurotransmitters using electrocatalytic substrate recycling at a laccase electrode. *Electroanalysis* 17: 2113–2119.

George SP, Ahmad A and Rao MB (2001) A novel thermostable xylanase from Thermomonospora sp.: Influence of additives on thermostability. *Bioresour Technol* 78: 221.

Gianfreda L, Xu F and Bollag J-M (1999) Laccases: A useful group of oxidoreductive enzymes. *Bioremediation J* 3: 1–25.

Gómez-Toribio V, Martínez A, Martínez MJ and Guillén F (2001) Oxidation of hydroquinones by versatile ligninolytic peroxidase from Pleurotus eryngii. H_2O_2 generation and the influence of Mn^{2+}. *Eur J Biochem* 268: 4787–4793.

Gold MH, Youngs HL and Gelpke MD (2000) Manganese peroxidase. *Met Ions Biol Syst* 37: 559–586.

Grethlein HE (1985) The effect of pore size distribution on the rate of enzymatic hydrolysis of cellulosic substrates. *Bio/Technol* 3: 155–160.

Haigh PM (1994) A review of agronomic factors influencing grass silage effluent production in England and Wales. *J Agric Engng Res* 57: 73–87.

Hakulinen N, Kiiskinen LL, Kruus K, Saloheimo M, Paananen A, Koivula A and Rouvinen J (2002) Crystal structure of a laccase from *Melanocarpus albomyces* with an intact trinuclear copper site. *Nature Struct Biol* 9: 601–605.

Hodrova B, Kopecny J and Kas J (1998) Cellulolytic enzymes of rumen anaerobic fungi Orpinomyces joyonii and Caecomyces communis. *Res Microbiol* 149: 417–427.

Jeffries TW (1994) Biodegradation of lignin and hemicelluloses, In: Ratledge C (ed.) *Biochemistry of microbial degradation.* Kluwer, Dordrecht, pp. 233–277.

Kapich AN, Jensen KA and Hammel KE (1999) Peroxyl radicals are potential agents of lignin biodegradation. *FEBS Lett* 461: 115–119.

Kaushik G and Thakur IS (2013) Purification, characterization and usage of thermotolerant laccase from bacillus sp. for biodegradation of synthetic dyes. *Appl Biochem Micro* 49 (4): 352.

Kiddinamoorthy J, Anceno JA, Haki DG and Rakshit SK (2008) Production, purification and characterization of Bacillus sp. GRE7 xylanase and its application in eucalyptus Kraft pulp biobleaching. *World J Microbiol Biotechnol* 24: 605–612.

Kirk TK and Farrel RL (1987) Enzymatic 'combustion': The microbial degradation of lignin. *Annu Rev Microbiol* 41: 465–505.

Kulkarni N, Shendye A and Rao M (1999) Molecular and biotechnological aspect of xylanases. *FEMS Microbiol Rev* 23: 411–456.

Lackner R, Srebotnik E and Messner K (1991) Oxidative degradation of high molecular weight chlorolignin by manganese peroxidase of Phanerochaete cht3, sosporittm. *Biochem Biophys Res Commun* 178: 1092–1098.

Leonowicz A, Cho NS, Luterek J, Wilkolazka A, Wojtas-Wasilewska M, Matuszewska A, Hofrichter M, Wesenberg D and Rogalski J (2001) Fungal laccase: Properties and activity on lignin. *J Basic Microbiol* 41: 185–182.

Leschine SB (1995) Cellulose degradation in anaerobic environments. *Annu Rev Microbiol* 49: 399–426.

Lonsane BK, Ghildyal NP, Budiatman S and Ramakrishna SV (1985) Engineering aspects of solid state fermentation. *Enzyme Microb Technol* 7: 258–265.

Madlala AM, Bissoon S, Singh S and Christove L (2001) Xylanase induced reduction of chlorine dioxide consumption during elemental chlorine-free bleaching of different pulp types. *Biotechnol Letts* 23: 345–351.

Maheshwari R, Bharadwaj G and Bhat MK (2000) Thermophilic fungi: Their physiology and enzymes. *Microb Mol Biol Rev* 64: 461–488.

Martínez J, Pérez J and de la Rubia T (1998) Olive oil mill waste water degradation by ligninolytic fungi In: Pandalai SG (ed) *Recent developments in microbiology.* Research Signpost, Trivandrum, India.

McCarthy AJ (1987) Lignocellulose-degrading actinomycetes. *FEMS Microbiol. Rev.* 46: 145–163.

McDonald P (1981) *The Biochemistry of Silage.* John Wiley & Sons, Chichester, UK, pp. 11–12, 168–178.

Mishra M and Thakur IS (2011) Purification, characterization and mass spectroscopic analysis of thermo-alkalotolerant b-1, 4 endoxylanase from *Bacillus* sp. and its potential for dye decolorization. *Int Biodeterioration Biodegradation* 65: 301–308.

Nigam PS, Gupta N and Anthwal A (2009) Pre-treatment of agro-industrial residues. In: Nigam PS, Pandey A, eds. *Biotechnology for agro-industrial residues utilization,* 1st edn. Netherlands: Springer, pp. 13–33.

Paice MG, Reid ID, Boubonnais R, Archibald FS and Jurasek L (1993) Manganese peroxidase, produced by *Trametes cersicolor* during pulp bleaching, demethylates and delignifies kraft pulp. *Appl Environ Microbiol* 59: 260–265.

Paul EA and Clark FE (1989) *Soil Microbiology and Biochemistry.* Academic Press, Inc. San Diego.

Pérez J and Jeffries TW (1992) Roles of manganese and organic acid chelators in regulating lignin biodegradation and biosynthesis of peroxidases by *Phanerochaete chrysosporium. Appl Environ Microbiol* 58: 2402–2409.

Piontek K, Antorini M and Choinowski T (2002) Crystal structure of a laccase from the fungus Trametes versicolor at 1.90 Å resolution containing a full complement of coppers. *J Biol Chem* 277: 37663–37669.

Poulos TL, Edwards SL, Wariishi H and Gold MH (1993) Crystallographic refinement of lignin peroxidase at 2 Å. *J Biol Chem* 268: 4429–4440.

Rayner ADM and Boddy L (1988) Fungal communities in the decay of wood. *Adv Microb Ecol* 10: 115–166.

Reid ID (1995) Biodegradation of lignin. *Can J Bot* 73 (Suppl 1): S1011–S1018.

Riva S (2006) Laccases: Blue enzymes for green chemistry. *Trends Biotechnol* 24: 221–226.

Ruiz-Dueñas FJ, Martínez MJ and Martínez AT (1999) Molecular characterization of a novel peroxidase isolated from the ligninolytic fungus Pleurotus eryngii. *Mol Microbiol* 31: 223–235.

Sarikaya A and Ladisch MR (1997) Mechanism and potential applications of bioligninolytic systems in a CELSS. *Appl Biochem Biotechnol* 62: 131–149.

Scott GM, Akhtar M, Lentz MJ, Kirk TK and Swaney R (1998) New technology for papermaking: Commercializing biopulping. *Tappi J* 81: 220–225.

Simpson HD, Haufler UR and Daniel, RM (1991) An extremely thermostable xylanase from the thermophilic eubacterium Thermotoga. *Biochem J* 227: 413–417.

Sterjiades R, Dean JFD and Eriksson KE (1992) Laccase from sycamore maple (Acer pseudoplatanus) polymerizes monolignols. *Plant Physiol* 99: 1162–1168.

Subramaniyan S and Prema P (2002) Biotechnology of microbial Xylanases: Enzymology, molecular biology and application. *Crit Rev Biotechnol* 22: 33–46.

Sundaramoorthy M, Kishi K, Gold MH and Poulos TL (1994) The crystal structure of manganese peroxidase from Phanerochaete chrysosporiumat 2.06 Å resolution. *J Biol Chem* 269: 32759–32767.

Tomme P, Warren RA and Gilkes NR (1995) Cellulose hydrolysis by bacteria and fungi. *Adv Microb Physiol* 37: 1–81.

Trigo C and Ball AS (1994) Is the solubilized product from the degradation of lignocellulose by actinomycetes a precursor of humic substances? *Microbiology* 140: 3145–3152.

Tuor U, Winterhalter K and Fiechter A (1995) Enzymes of white-rot fungi involved in lignin degradation and ecological determinants for wood decay. *J Biotechnol* 41: 1–17.

Van Veen JA, Ladd J and Frissel MJ (1984) Modelling C&N turnover through the microbial biomass in soil. *Plant Soil* 76: 257–274.

Varela E, Martínez MJ and Martínez AT (2000) Aryl-alcohol oxidase protein sequence: A comparison with glucose oxidase and other FAD oxidoreductases. *Biochim Biophys Acta* 1481: 202–208.

Varga GA and Kolver ES (1997) Microbial and animal limitations to fiber digestion and utilization. *J. Nutr.* 127: 819S–823S.

Vicuna R (2000) Ligninolysis. A very peculiar microbial process. *Mol Biotechnol* 14: 173–176.

Vicuna R, Gonzalez B, Seelenfreund D, Ruttimann C and Salas L (1993) Ability of natural bacterial isolates to metabolize high and low molecular weight lignin-derived molecules. *J Biotechnol* 30: 9–13.

Wariishi H, Valli K and Gold MH (1992) Manganese (II) oxidation by manganese peroxidase from the basidiomycete *Phanerochaete chrysosporium*: Kinetic mechanism and role of chelators. *J Biol Chem* 267: 23688–23695.

White GF, Russell NJ and Tidswell EC (1996) Bacterial scission of ether bonds. *Microbiol Rev* 60: 216–232.

Whittaker MM, Kersten PJ, Nakamura N, Sanders-Loehr J, Schweizer ES and Whittaker JW (1996) Glyoxal oxidase from Phanerochaete chrysosporium is a new radical-copper oxidase. *J Biol Chem* 271: 681–687.

Wilson JR and Mertens DR (1995) Cell wall accessibility and cell structure limitations to microbial digestion of forage. *Crop Sci* 35: 251–259.

Wohlfahrt G, Witt S, Hendle J, Schomburg D, Kalisz HM and Hecht H-J (1999) 1.8 and 1.9 Å resolution structures of the Penicillium amagasakiense and Aspergillus niger glucose oxidase as a basis for modelling substrate complexes. *Acta Crystallogr D* 55: 969–977.

Wubah DA, Akin DE and Borneman WS (1993) Biology, fiber-degradation, and enzymology of anaerobic zoosporic fungi. *Crit Rev Microbiol* 19: 99–115.

Zadrazil F and Isikhuemhen O (1997) Solid state fermentation of lignocellulosics into animal feed with white rot fungi. In: Roussos S, Lonsane BK, Raimbault M and Viniegra-Gonzalez G (eds) *Advances in Solid State Fermentation*. Kluwer Academic Publishers, Dordrecht, The Netherlands, pp. 23–38.

7

Microbial Degradation of Hexachlorocyclohexane (HCH) Pesticides

Hao Chen, Bin Gao, Shengsen Wang and June Fang

CONTENTS

7.1 Introduction

Hexachlorocyclohexane (HCH) is a six-chlorine substituted cyclohexane; it collectively represents the eight isomers of 1,2,3,4,5,6-hexachlorocyclo-hexane. Named by Greek letters (α, β, γ, δ, ζ, η and θ), these isomers differ in the spatial orientations and their axial equatorial substitution around the ring (Figure 7.1 and Table 7.1). Among the isomers, α-HCH is constituted of two enantiomers.

HCHs are manufactured by benzene photochlorination under UV light. This way, HCHs form five stable isomers: α (60% to 70%), β (5% to 12%), γ (10% to 12%), δ (6% to 10%) and ε (3% to 4%). The γ-isomer of HCH (CAS No. 58-89-9), also known as lindane, is the isomer with the highest pesticidal

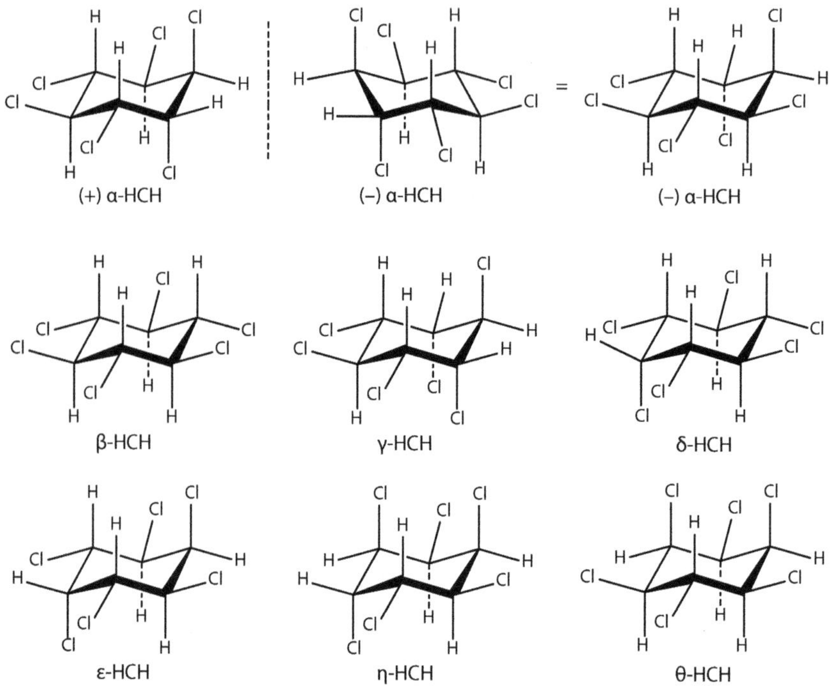

FIGURE 7.1
Structure and configuration of eight isomers of hexachlorocyclohexane. Configuration of Cl (a – axial, c – equatorial).

activity and is used as an insecticide in agriculture and forestry administration. However, mixtures of all isomers have been often used as commercial pesticides. Production of 1 ton of lindane generates 8 to 12 tons of other HCH isomers; these isomers can be more problematic than lindane itself to the environment.

TABLE 7.1

Size of the Hexachlorocyclohexane Isomers

Isomer	Melting Point (°C)	Molecular Diameter in Plane of Ring	Molecule Thickness
β	297	9.5 9.5 9.5	5.4
δ	130	8.5 9.5 9.5	6.3
α	157	8.5 8.5 9.5	7.2
γ	112	8.5 8.5 8.5	7.2
ε	219	7.5 9.5 9.5	7.2
η	90	7.5 9.5 8.5	7.2
θ	124	8.5 9.5 8.5	6.3

Source: Mullins, L. J., *Science*, 122, 3159, 118–119, 1955.

α-, β-, γ- and δ-HCHs are considered to be four major HCH isomers for environmental pollution. In general, these HCH isomers are stable to light, high temperatures and acid. At 5°C and pH 8, the hydrolytic half-lives of α- and γ-HCH were 26 and 42 years, respectively (Ngabe, Bidleman et al. 1993). Compared to other organochlorines (such as DDT), HCH isomers are generally more water soluble and volatile, which explains why HCHs are widely distributed and can be detected in all environmental compartments, including water, sediments, air and animals.

The physical and chemical properties of the four major HCH isomers are quite different from one another as illustrated in Table 7.2. For example, β-HCH has a much lower vapour pressure and a much higher melting point compared to α-HCH. These properties are largely dictated by the axial and equatorial positions of the chlorine atoms on each molecule. As shown in Figure 7.1, all of the chlorines on β-HCH are in the equatorial positions, which seems to confer the greatest physical and metabolic stability to this isomer. This stability of β-HCH is reflected in its environmental and biological persistence. For instance, the amount of HCH in human fat that is β-HCH is nearly 30 times higher than that of γ-HCH (Table 7.2). The γ-isomer has three chlorines in axial positions creating two ways that HCl can be eliminated, generating pentachlorocyclohexene (PCCH) metabolites (Buser and Mueller 1995).

About 600,000 tons of HCHs were used worldwide between the 1940s and the 1990s (Weber, Gaus et al. 2008; Lal, Pandey et al. 2010). Growing concerns about HCHs' persistence and nontarget toxicity caused them to be deregistered in most countries (Waliszewski, Villalobos-Pietrini et al. 2003). As of 1992, γ-HCH was banned in several countries, such as India, Sudan and Columbia; however, due to its great persistence, HCH can still be present in the surrounding environment (Kole, Banerjee et al. 2001; Kannan, Ramu et al. 2005; Singh, Malik et al. 2007). However, the usage of lindane

TABLE 7.2

Physical and Chemical Properties of Four Major HCH Isomers

Property	α-	β-	γ-	δ-
CAS registry no.	319-84-6	319-85-7	58-89-9	319-86-8
Colour	Brownish	White	White	White
Density	1.87 g cm^{-3}	1.89 g cm^{-3}	1.89 g cm^{-3}	No data
Melting point	159°C–160°C	314°C–315°C	112°C	141°C–142°C
Vapour pressure	1.6×10^{-2}	4.2×10^{-5}	5.3×10^{-3}	2.1×10^{-3}
Water solubility	10 mg L^{-1}	5 mg L^{-1}	17 mg L^{-1}	10 mg L^{-1}
Partition coefficients (K_{ow})	3.9	3.9	3.7	4.1
Bioconcentration factor in humans	20	527	19	8.5

Source: Willett, K. L. et al., *Environmental Science & Technology*, 32, 15, 2197–2207, 1998.

still remains in North America and Europe as a seed dressing and as a human medicine.

Today, HCHs face two major problems: (1) high-concentration point source contamination and (2) diffusion, which results in HCHs being widely spread. Intensive high-concentration HCH pollution involves the point source contamination during lindane manufacturing. Sites heavily contaminated with HCHs have been reported from The Netherlands (Langenhoff, Staps et al. 2013), Brazil (Torres, Froes-Asmus et al. 2013), Germany (Ricking and Schwarzbauer 2008), Spain (Fernandez, Arjol et al. 2013), China (Zhang, Luo et al. 2009), the United States (Walker, Vallero et al. 1999) and India (Jit, Dadhwal et al. 2011). Around 4 to 6 million tonnes of HCHs have been dumped worldwide (Weber, Gaus et al. 2008), which has contributed the majority of persistent organic pollutants.

The diffusion of HCHs from high to low concentration results in HCHs being widely spread in the surrounding environment. The HCH waste has been discarded mainly near the production sites (Santos, Silva et al. 2003; Jit, Dadhwal et al. 2011; Fernandez, Arjol et al. 2013; Alamdar, Syed et al. 2014). At many of the sites, HCH residues have percolated into soils and hence contaminated groundwater (Frische, Schwarzbauer et al. 2010; Fernandez, Arjol et al. 2013). HCH residues have now been reported for many countries in a variety of samples: air (Halse, Schlabach et al. 2012; Syed, Malik et al. 2013; Alamdar, Syed et al. 2014), water (McNeish, Bidleman et al. 1999; Tian, Li et al. 2009), soil (Rissato, Galhiane et al. 2006; Jit, Dadhwal et al. 2011; Vijgen, Abhilash et al. 2011), food (Battisti, Caminiti et al. 2013; Song, Ma et al. 2013), fish (Singh and Singh 2008; Gonzalez-Mille, Ilizaliturri-Hernandez et al. 2010) and human blood samples (Porta, Fantini et al. 2013; Wang, Chen et al. 2013). The HCH residuals in remote sites, such as the Arctic, Antarctica and the Pacific Ocean, also have been reported (Galban-Malagon, Cabrerizo et al. 2013; Ge, Woodward et al. 2013; Newton, Bidleman et al. 2014). Due to its persistence, HCHs continue to generate serious residue problems in a variety of circumstances around the world. Despite the extensive data available on HCHs, the fate of HCHs in the environment is still not fully understood.

7.2 Structure Toxicity Relationship of HCHs

Due to the differences in the molecular structure and physical properties of HCH isomers (Figure 7.1), the bioactivity of HCHs differs significantly. The greatest differences in the molecular structure and physical properties in isomers are between β-HCH and γ-HCH. β-HCH has a relatively plane shape with weak physiological activity as an inert or weak depressant, and γ-HCH has a relatively spherical shape with strong insecticidal action (Mullins 1955). β-HCH and γ-HCH act differently within the membrane. A membrane is mainly composed of phospholipid macromolecules arranged in a regular hexagonal packing. The interspaces of these macromolecules can act as transport pathways of the membrane, by which the HCHs can be introduced. Based on plane orientation, the cyclohexane ring γ-HCH (8.5 Å) is smaller than that of β-HCH (9.6 Å). It can be easier for γ-HCH to penetrate the membranes. Thus, bioactivity difference between these two isomers can be expected.

Small molecules filling up the membrane interspaces can cause the depression of excitable tissue or narcosis. This process is thermodynamic; the thermodynamic activities are inversely proportional to the melting points. For four major HCHs, they follow the order δ > γ > α > β (Table 7.2). Based on the melting points, in the membranes, δ-HCH can act as a strong depressant. If size-comparable molecules are introduced into the membrane interspace, the effects of these molecules will be determined by the attractive forces between the introduced molecules and the membrane molecules and the attractive forces between the membrane molecules themselves. If these attractive forces are comparable, the molecule in the interspace will remain without distortion of the membrane structure. If the attractive force between the inserted molecule and the membrane molecules were stronger, it would appear possible that the mean position of the membrane molecules around the interspace could be distorted. The distortion of the position of these molecules may lead to instability of the membrane structure and to ion leaks that are a prelude to excitation. In the case of the HCHs, chlorine atoms may have a strong attraction to the membrane and possibly cause distortion. Thus, the potential toxicity of HCHs is largely dependent on the orientation of the chlorine atoms.

7.3 Toxicity of HCHs

HCHs primarily affect the central nervous system. Hypotheses suggest that convulsions are mediated by the inhibition of γ-aminobutyric acid neurotransmission or stimulation related to neurotransmitter release (Abalis, Eldefrawi et al. 1985). In insects, γ-HCH stimulates the central nervous system and causes rapid, violent convulsions that are generally followed by either death or recovery within 24 h (Agrawal, Sultana et al. 1991). Other physiological systems affected by HCH isomers include renal and liver function, hematology and biochemical homeostasis (Sauviat and Pages, 2002; Spinelli, Ng et al. 2007; Sonne, Maehre et al. 2013).

Rats treated with diets containing 20 mg kg^{-1} of α- or γ-HCH for 15 or 30 days showed increased levels of liver cytochrome P-450 followed by increased production of both thiobarbituric acid reactants by liver homogenates and microsomes and superoxide anion production by liver microsomes. The increased activity of the cytochrome P-450 system can sometimes lead to the formation of free radicals that are hazardous to the liver cell (Barros, Simizu et al. 1991). Feeding rats γ-HCH can also cause severe liver and kidney congestion (Fidan, Cigerci et al. 2008). Another HCH toxicity–to-rats study showed that technical HCHs had significantly lower red blood cell counts, white blood cell counts and hemoglobin concentrations as compared to controls (Joseph, Viswanatha et al. 1992).

The reproductive effects for male rats with γ-HCH intake were observed: decrease in testes weight, decrease in testicular sperm head count, increase in abnormal tail morphology, abnormal head morphology and decrease in sperm motility as compared to control (Simic, Kmetic et al. 2012). The toxic effects of γ-HCH were still evident even after 14 days of metabolism (Sharma and Singh 2010). Studies also indicate that γ-HCH has adverse effects on female reproduction, such as alterations in sexual receptivity, disrupted ovarian cyclicity and reduction in uterine weight (Cooper, Chadwick et al. 1989).

Among four major HCHs, the α-HCH isomer is considered the most tumourgenic (Agrawal, Sultana et al. 1991). Neoplastic nodules and tumours developed after continuous exposure to HCHs (500 mg L^{-1}) for 4 and 6 months, respectively (Bhatt and Nagda 2012). In mice, it was demonstrated that β-HCH can act as a breast cancer promoter, which exerts its tumorigenic activity (Wong and Matsumura 2007). Organochlorines, such as HCHs, can enhance human breast cancer cell proliferation, and the effect was considered to be cumulative (Payne, Scholze et al. 2001). HCH toxicological reports in humans are still largely limited to occupational exposures and accidental poisonings.

7.4 HCH Exposure and Bioaccumulation

Due to their lipophilic nature, HCHs can accumulate in animal bodies. The β-HCH is considered the most metabolically and inactively persistent with a bioconcentration factor of 527 in human fat (Table 7.2). Occupational exposures and accidental poisonings are not the only ways humans are exposed to HCHs (Figure 7.2). Humans can also be exposed by consuming food contaminated by HCHs. Several studies report HCH intake from milk, fruit, vegetables, fish, etc. (Liu, Qiu et al. 2009; Salem, Ahmad et al. 2009; Wang, Chen et al. 2013). Human accumulation of HCHs has been evidenced by the blood levels of HCHs among people living close to an industrial area, which is highly related to the HCH contamination in the surrounding area (Porta, Fantini et al. 2013). Studies have shown that HCHs can accumulate in breast-fed children to concentrations about twice as high as those fed with formula (Grimalt, Carrizo et al. 2010).

FIGURE 7.2
(See color insert.) Hexachlorocyclohexane exposure pathways.

Wildlife exposed to HCHs can also accumulate HCHs in the same was as humans. In fact, several wildlife species have been used as HCH contamination monitors to study temporal trends of HCHs (Ito, Yamashita et al. 2013). For example, in recent years, significant decreasing of HCHs in the eggs of seabirds around the Barents Sea indicated the decreasing use of HCHs (Borga, Gabrielsen et al. 2001). Mussels along the contaminated coast used as sentinel species also have shown a clearly decreasing trend of HCHs (Sturludottir, Gunnlaugsdottir et al. 2013). Global HCH transport is also evident in residues found in marine mammals and Arctic species (Kucklick, Krahn et al. 2006; Sonne, Letcher et al. 2012). For example, total HCHs in polar bears showed statistically significant average yearly declines in East Greenland (Dietz, Riget et al. 2013).

7.5 HCH Degradation and Reaction Pathways

The key reaction to HCH degradation is dehalogenation. During this step, chloride, which is responsible for the toxicity of HCHs, is most commonly replaced by a hydrogen atom or a hydroxyl group. After chloride removal, the intermediates with lower toxicity are more feasible for further degradation. Basic HCH degradation pathways include substitution, nonreductive elimination and reductive elimination. Except for the structure of the HCHs, the factors governing the degradation process include solvent effects, solution pH and the nature of the nucleophilic or departing groups.

Theoretically, the nucleophilic substitution of HCHs includes both SN1 and SN2 mechanisms, yielding alcohols or polyols. For HCHs, the SN1 mechanism contains two steps: an initial loss of chloride to form a carbenium ion followed by rapid uptake of an available nucleophile; the SN2 mechanism includes the backside attack on an electron-deficient carbon centre by a suitable nucleophile (or electron-rich species) coupled with the simultaneous departure of chloride. The most important nucleophiles in the environment are water and hydroxide, especially in the alkaline condition. Besides, sulfide is also a major nucleophile in the subsurface. The substitution of the HCHs doesn't easily occur naturally, so it did not play a major role in the initial transformation of HCHs. However, for intermediates with less chloride, substitution is a major degradation mechanism.

HCH nonreductive elimination involves the loss of chloride and a neighbouring proton to form an alkene. Without the oxidation state change of the reacting carbon centres, this reaction is considered to be nonreductive. Insofar as environmental eliminations are concerned, the predominant basic species is generally hydroxide. For the HCHs, three steps of nonreductive elimination can be achieved due to the relative acidity of the chloride atom on the HCHs (Figure 7.3). The first dehydrohalogenation step produces PCCH, which has the chemical formula $C_6H_5Cl_5$; similar to the parent HCHs, PCCH can exist in one or more isomeric forms. Successively, PCCH can lose chlorine to form tetrachlorodiene. Due to the instability of tetrachlorodiene, successive nonreductive eliminations would be expected to produce relatively stable trichlorobenzenes. Thus, after three chlorine elimination reactions, complete degradation of HCHs can be achieved.

The reductive elimination of HCHs can also occur with a net reduction at the involved carbon centres. The reductive elimination includes dihaloelimination and hydrogenolysis.

In dihaloelimination, schematically depicted in Figure 7.4, an external reducing agent or electron donor supplies a pair of electrons resulting in the loss of halogen from adjacent (that is, vicinal) carbon centres. Thus, dihaloelimination constitutes a reduction–oxidation (redox) reaction in which the

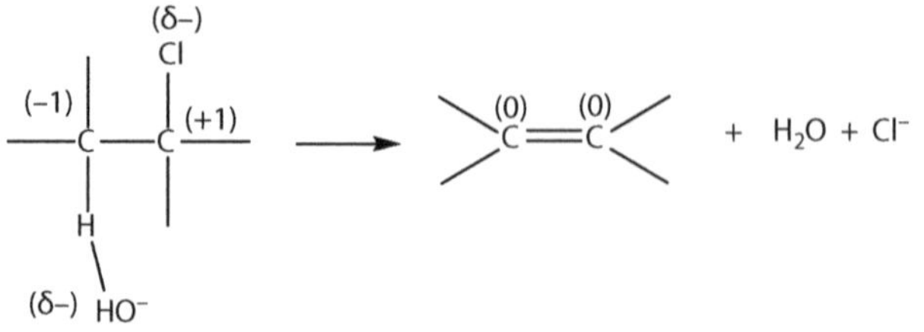

FIGURE 7.3
Schematic of nonreactive elimination reaction.

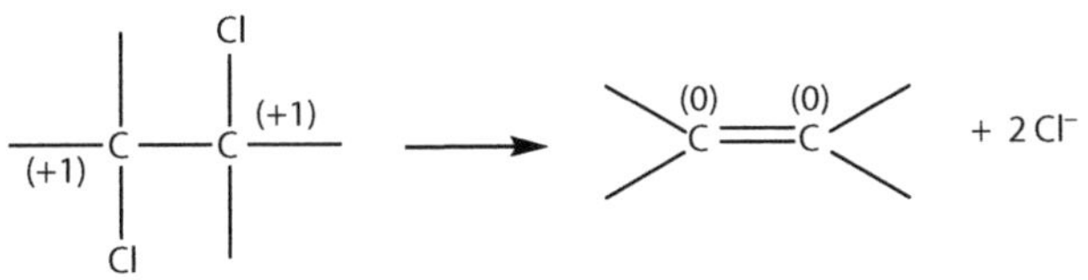

FIGURE 7.4
Schematic of dihaloelimination reaction.

HCHs are the electron acceptor and another species serves as the electron donor. During the course of the reduction, the reacting carbon centres acquire increasing electron density, helping to drive the ejection of the chloride anions. As a result, the oxidation state of these carbon centres decreases to zero (0) from a relatively oxidised state (+1). Unlike the nonreductive elimination, the dihaloelimination pathway is not strongly influenced by pH.

Dihaloelimination is a favoured transformation pathway for polyhalogenated alkyl halides. Similar to nonreductive elimination, the orientation of the halogen substituents influences the observed reactivity of the parent molecule. Specifically, alkyl halides with axially oriented vicinal halogens undergo much more rapid dihaloelimination than would be the case if one or both were equatorial. The diaxial, antiperiplanar orientation of departing halogens features a dihedral angle of 180°, resulting in the most favourable positioning for reaction among the involved substituents.

Because the HCHs have three vicinal chlorine pairs, they can theoretically undergo three sequential dihaloelimination steps to form benzene. The initial dihaloelimination would produce tetrachlorocyclohexene with a molecular formula of $C_6H_6Cl_4$. As was the case with PCCH isomers and the parent HCH isomers, several tetrachlorocyclohexene isomers are known. Subsequent steps would proceed through an unstable highly reactive dichlorocyclohexadiene intermediate to benzene, the stable end product. As compared to nonreactive elimination, an even greater driving force is provided here as all of the reactive C–Cl bonds are reduced and the ring carbons are stabilised by the delocalised resonance of the aromatic system (Orloff 1954).

In hydrogenolysis, the departing halogen is replaced by a proton with an alkane as the principal product (Figure 7.5). Hydrogenolysis readily proceeds under acidic conditions, whereas nonreaction elimination prefers alkaline conditions. Two external electron donors are required to drive the

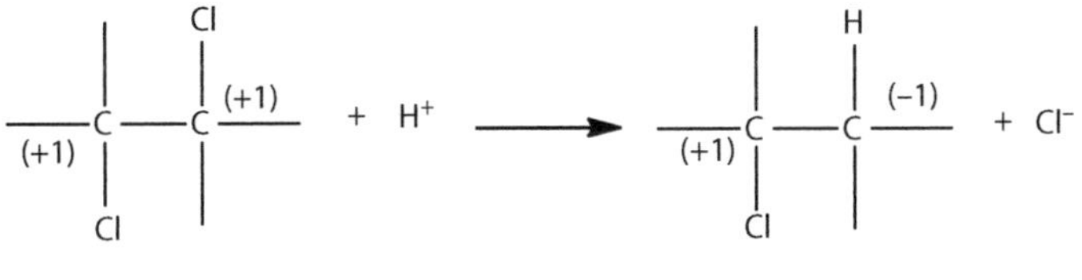

FIGURE 7.5
Schematic of hydrogenolysis reaction.

reaction to reduce its oxidation state. Under acidic conditions, the sequential hydrogenolysis of HCHs would be expected to produce a series of less-chlorinated alicyclic hydrocarbons terminating with cyclohexane itself. In the environment, however, hydrogenolysis is not a significant transformation pathway for HCH degradation.

7.6 Bioremediation of HCH-Contaminated Soils

After entering soil, HCHs can bind to the mineral or soil organic matter through a combination of physical and chemical processes. The strong soil HCH sorption may decrease HCH bioavailability, thereby influencing effectiveness of the bioremediation. Generally, the HCH bioremediation process is largely influenced by the soil's physical and chemical characteristics. For instance, pH, redox potential and soil organic matter all might contribute to differences in HCH removal rates (Phillips, Seech et al. 2005).

The two major approaches taken to soil bioremediation are biostimulation and bioaugmentation. Biostimulation involves the modification of the environment by the addition of various forms of rate-limiting nutrients and electron acceptors, such as phosphorus, nitrogen, oxygen or carbon, to stimulate existing bacteria, which is capable of bioremediation. Bioaugmentation involves the introduction of HCH-degrading microorganisms that are not necessarily native to the site. Both approaches have various levels of success for HCH remediation purposes. Combinations of different strategies might be used to further enhance the effectiveness of HCH bioremediation. Besides, the rhizosphere microbial community combined with phytoremediation has been recently reported as a new efficient HCH degradation technique.

7.6.1 Biostimulation

Phillips et al. (2006) reported using proprietary biostimulation with Daramend derived from natural plant fibres to treat 1100 tonnes of HCH-contaminated soil (~5000 mg kg^{-1}). Based on successful laboratory-scale degradation, two treatments have been adopted: half of the contaminated site was treated by anoxic/oxic cycling, and the other half was only treated under a strictly aerobic condition. Some tillage was also applied to enhance aerobic degradation. After 371 days, total HCH concentrations were reduced in the most contaminated soil, and the site with strict aerobic treatment had better degradation efficiency. The average HCH reductions for anoxic/oxic cycling and aerobic treatment were 40% and 47%, respectively. Elevated chloride ion concentrations have also been observed after treatment-demonstrated HCH removal. This full-scale project demonstrated the potential for solid phase bioremediation treatment of soil containing high HCH concentrations.

Rubinos et al. (2007) used the combination of nutrient biostimulation and periodic tillage treatment for a soil with HCH contamination (>5000 mg kg^{-1}). HCHs are degraded, transformed and immobilised by means of biotic and abiotic reactions. The soil showed a significant decrease in α- and γ-HCH (up to 89% and 82%, respectively) after 11 months of treatment, although the β-HCH remained unaffected.

Most recently, Gupta, Lal et al. (2013) studied the ex situ biostimulation approach for HCH bioremediation of heavily contaminated soil (~84 g kg^{-1}). Molasses and ammonium phosphate were used in different combinations as nutrients to stimulate the indigenous microbial community. Maximum reduction was seen in the pit that received a combination of molasses and ammonium phosphate. Substantial HCH reduction with a change in the microbial community was observed in 12 months. The dominated degradation bacteria observed were Sphingomonads. The study provided the prospects of biostimulation in decontaminating soils heavily contaminated with HCHs.

Ex situ biostimulation practice also has been reported by Dadhwal et al. (2009). Their study was based on a heavily hexachlorocyclohexane-contaminated site; after the identification of indigenous HCH degraders, an ex situ biostimulation experiment was conducted. HCH-contaminated soil with indigenous HCH degraders from a dump site was mixed with pristine garden soil. Aeration, moisture and nutrients were then provided randomly. HCHs were reduced to less than 30% of the original in 24 days and less than 3% in 240 days. The alteration of the microbial community structure was also observed with a genetic markers comparison during the experiment.

7.6.2 Bioaugmentation

The potential benefit of bioaugmentation is maintenance of high levels of biodegradation without the need for initial inoculations or ongoing supplementation. However, bioaugmentation processes involve the release of foreign microorganisms, so survival and activity of the inoculum are not always guaranteed. Few studies addressed the bioaugmentation remediation. Bidlan et al. (2004) demonstrated in the lab that the addition of an HCH-degrading microbial consortium shows significant removal of all four major HCH isomers from spiked soils. No detectable HCHs remained after 5 days starting with the initial 0.25 mg kg^{-1} HCHs. Mertens et al. (2006) then reported that half of γ-HCH was removed (50 mg L^{-1} added every few days) by a slow-release inoculation approach. Their work showed that under aerobic conditions, a slow-release inoculation approach using a catabolic strain encapsulated in open-ended tubes can ensure prolonged biodegradation activity in comparison with traditional inoculation strategies. Raina et al. (2008) proposed the possibility of the removal of HCHs by the inoculation of aerobic bacterium *Sphingobium indicum* B90A both in lab-scale transplanted HCH-contaminated soils and in situ with contaminated agricultural soils.

Their study tested immobilisation, storage and inoculation procedures and determined the survival and HCH-degradation activity of inoculated cells in soil. In an HCH-contaminated agricultural site, up to between 85% and 95% HCHs were degraded by strain B90A applied via corncob; also up to 20% of the inoculated B90A cells survived under field conditions after 8 days.

7.6.3 HCH Removal by Rhizosphere Microorganisms

The rhizosphere is defined as a zone of high intense bacterial activity that is driven in part by plant root exudates (Bowen and Rovira 1999). Plant roots exude an enormous range of potentially valuable small molecular weight compounds into the rhizosphere. Organic acids, sugars, amino acids, lipids, coumarins, flavonoids, proteins, enzymes, aliphatics and aromatics are examples of the primary substances from root exudates. Among them, the organic acids have been considered to play a key role in simulating microbial metabolism and for serving as intermediates for biogeochemical reactions in soils (Hinsinger, Plassard et al. 2006). For rhizosphere contaminant degradation, the rhizospheric bacteria are responsible for the elimination of the contaminants, and the root exudates are responsible for providing nutrients for microorganisms to proliferate (Böltner, Godoy et al. 2008) (Figure 7.6).

For the cleanup of contaminated sites, special microbes can be selected to degrade the target organic pollutants. An enrichment method for the microbes' isolation has been used to choose the microbe with excellent root colonisation and efficient contaminant degradation. Interactions between

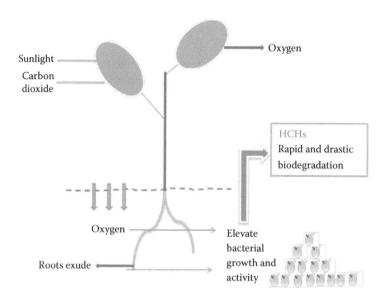

FIGURE 7.6
(See color insert.) HCH removal by rhizosphere microorganisms.

plants and soil microbes are highly dynamic in nature and based on co-evolutionary pressures (Dobbelaere, Vanderleyden et al. 2003; Duffy, Keel et al. 2004; Morgan, Bending et al. 2005). Consequently, rhizosphere microbial communities differ between plant species (Batten, Scow et al. 2006), between genotypes within species (Kowalchuk, Hol et al. 2006) and between different developmental stages of a given plant (Mougel, Offre et al. 2006).

Recent studies show that phytostimulation of HCHs through microorganism degradation can be a successful remediation strategy. Because HCHs can inhibit plant performance, the key processes for this technology are the selection of plant species with high HCH tolerance. Only a few plants can adapt well to the acidic conditions generated during HCH degradation. Benimeli, Fuentes et al. (2008) showed that γ-HCH concentration as high as 400 µg kg^{-1} soils did not affect the germination and vigour index of maize plants seeded. When a microorganism named *Streptomyces* sp. M7 was inoculated to the same conditions, a better maize vigour index was observed. With maize, the γ-HCH removal efficiency of *Streptomyces* sp. M7 also improved.

The leguminous species *Cytisus striatus*, one of the species that grows naturally on HCH-contaminated sites, has been proposed as a candidate for the cleanup of HCH contamination (Kidd, Prieto-Fernandez et al. 2008). Becerra-Castro, Prieto-Fernandez et al. (2013) studied the leguminous shrub *Cytisus striatus* (Hill) Rothm. as a rhizoremediation candidate for HCH remediation. Inoculation with the previously isolated bacteria *Rhodococcus erythropolis* (ET54b) and *Sphingomonas* sp. D4 (ETD4), both bacterial strains, resulted in decreased HCH phytotoxicity and improved plant growth. In the presence of *C. striatus*, HCH dissipation was enhanced. Inoculating *C. striatus* with this combination of bacteria is a promising approach for HCH remediation in the contaminated sites.

Alvarez, Yanez et al. (2012) evaluated γ-HCH removal by *Streptomyces* sp. A5 and *Streptomyces* sp. M7 under the influence of maize root exudates. Results showed that both strains were able to grow with maize root exudates on minimal medium supplement, suggesting maize root exudates could be used as a carbon source to support HCH aerobic degradation. γ-HCH removal by both *Streptomyces* sp. A5 and *Streptomyces* sp. M7 increased when root exudates were present in the culture medium, suggesting that maize root exudates promote the γ-HCH degradation by both strains.

7.7 Metabolic Pathways of HCH Degradation

In general, the HCH isomers can be biodegraded to a series of less chlorinated organic compounds under both aerobic and anaerobic conditions, and in some cases, HCHs can be used as the sole carbon source for bacterial growth (Siddique, Okeke et al. 2002; Rubinos, Villasuso et al. 2007). During the microbial degradation, the chloride atoms of HCHs, which are usually considered to be toxic and xenobiotic, may most commonly be replaced by hydrogen or hydroxyl groups.

Reports on the anaerobic microbial degradation of HCH isomers started appearing in the 1970s, and subsequently, aerobic bacteria capable of degrading HCH isomers were also found. Over the last few years, the focus of biologically mediated HCH transformations by specific bacterial strains has been studied on the genetic aspects and at the molecular level. Genes involved in degradation (*Lin* genes) have been isolated from aerobic bacteria and the gene products have been characterised. There is growing interest in using them to develop bioremediation technologies for decontamination of HCH-contaminated sites (Wu, Xu et al. 1997; Endo, Ohtsubo et al. 2007). The full-scale in situ bioremediation of HCHs had been recently reported by Langenhoff, Staps et al. (2013).

7.7.1 Anaerobic Degradation

In the 1970s, the anaerobic degradation of γ-HCH was observed by Jagnow et al. (1977). All four major HCH isomers can be degraded under anaerobic conditions (Macrae, Raghu et al. 1969; Heritage and Macrae 1977; van Doesburg, van Eekert et al. 2005; Ge, Woodward et al. 2013). γ-HCH was the most readily degradable isomer, followed by α-HCH > δ-HCH $\approx$ β-HCH, which is correlated with the general reactivity (Macrae, Raghu et al. 1967; Johri, Dua et al. 1998; Quintero, Moreira et al. 2005a). The degradation rates of the HCHs can differ significantly due to the nature of the HCH structure and the anaerobic degradation cultivative environment. Buser and Mueller (1995) studied HCH anaerobic degradation in sewage sludge using a pseudo-first-order degradation model and found that the half-lives for $(+)\alpha$, $(-)\alpha$, γ, β and δ isomers were calculated to be 35, 99, 20.4, 178 and 126 h, respectively, with pseudo-first-order rate constants of 1.95×10^{-2}, 6.76×10^{-3}, 3.38×10^{-2}, 3.9×10^{-3} and 5.5×10^{-3} h^{-1}, respectively. After 10 days, more than 90% of α-HCH and more than 99% of γ-HCH had been degraded.

There were two sequential bioprocess trains that have been suggested by recent studies for HCH anaerobic degradation: With chlorobenzenes as one of the anaerobic degradation products, successive dichloroeliminations and then dehydrochlorination have been suggested to be part of the process of α-, β- and δ-HCH degradation (Heritage and Macrae 1977; Middeldorp, Jaspers et al. 1996; Manickam, Misra et al. 2007; Rodriguez-Garrido, Lu-Chau et al. 2010); the other anaerobic pathway could also generate chlorobenzene, and Quintero, Moreira et al. (2005b) proposed the HCH anaerobic degradation followed by the product of PCCH, 1,2- and 1,3-dichlorobenzene, and then chlorobenzene. As a result, most studies of anaerobic HCH degradation reported the accumulation of chlorobenzene and benzene (Buser and Mueller 1995; Middeldorp, Jaspers et al. 1996; Van Eekert, Van Ras et al. 1998). Under aerobic conditions, both chlorobenzene and benzene can be mineralised

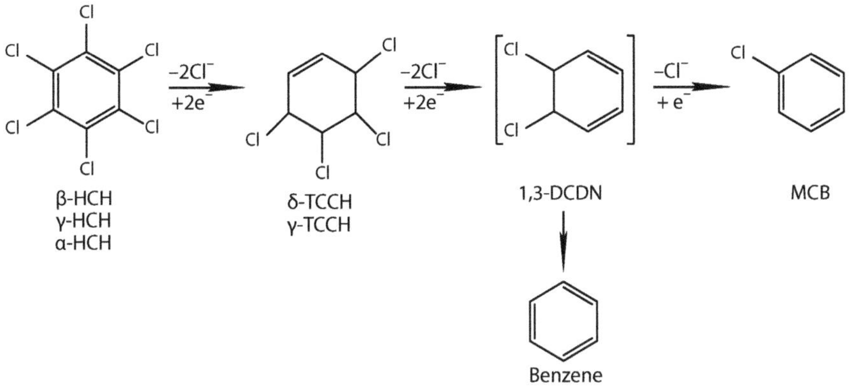

FIGURE 7.7
Proposed anaerobic transformation pathway of β-HCH and formation of chlorobenzene and benzene based on our experiments. (From Middeldorp, P. J. M. et al., *Environmental Science & Technology*, 30, 7, 2345–2349, 1996.)

easily (Fairlee, Burback et al. 1997; Mars, Kasberg et al. 1997; Deeb, Spain et al. 1999). Middeldorp et al. (1996) proposed the HCH degradation pathway in the environment shown in Figure 7.7. In the environment, HCH soil sorption can greatly limit their microbial degradation (Rooney-Varga, Anderson et al. 1999); HCH degradation is much accelerated under liquid conditions (Rooney-Varga, Anderson et al. 1999; Quintero, Moreira et al. 2005a).

7.7.2 Aerobic Degradation

The four major HCHs all can be degraded under aerobic conditions (Bachmann, Debruin et al. 1988; Nagasawa, Kikuchi et al. 1993; Sahu, Patnaik et al. 1995). The majority of HCH-degrading aerobes belong to the family Sphingomonadaceae (Lal, Dogra et al. 2006; Lal, Dadhwal et al. 2008), for example, *Sphingobium japonicum* UT26, *Sphingobium indicum* B90A and *Sphingobium francense* Sp+ (Mohn, Mertens et al. 2006; Nagata, Endo et al. 2007). Genes encoding the HCH degradation enzymes have been cloned, sequenced and characterised (Boltner, Moreno-Morillas et al. 2005; Manickam, Misra et al. 2007). With a nearly identical set of genes in all the degraders, the *Lin* genes have been identified as those responsible for the HCH-degrading sphingomonads tested (Boltner, Moreno-Morillas et al. 2005; Ceremonie, Boubakri et al. 2006; Lal, Dogra et al. 2006; Ito, Prokop et al. 2007; Yamamoto, Otsuka et al. 2009).

The pathway of HCH degradation in *Sphingobium japonicum* UT26 (Nagata, Endo et al. 2007) consists of the following steps (Figure 7.8): LinA, encoding a dehydrochlorinase (Imai, Nagata et al. 1991), and LinB, encoding a haloalkane dehalogenase (Nagata, Nariya et al. 1993), catalyse the dehydrochlorinase and hydrolytic dechlorinase reactions, respectively; LinC, encoding

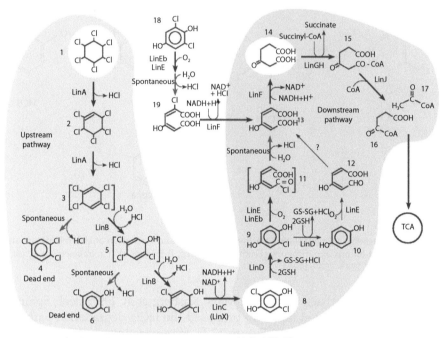

1. γ-hexachlorocyclohexane
2. Pentachlorocyclohexene
3. 1,3,4,6-tetrachloro-1,4-cyclohexadiene
4. 1,2,4-trichlorobenzene
5. 2,4,5-trichloro-2,5-cyclohexadiene-1-o1
6. 2,5-dichlorophenol
7. 2,5-dichloro-25-cyclohexadiene
8. 2,5-dichlorohydroquinone
9. Chlorohydroquinone
10. Hydroquinone

11. Acylchloride
12. γ-hydroxymuconic semialdehyde
13. Maleylacetate
14. β-ketoadipate
15. 3-oxoadipyl-CoA
16. Succinyl-CoA
17. Acetyl-CoA
18. 2,6-dichlorohydroquinone
19. 2-chloromaleylacetate

Compounds: (1) γ-hexachlorocyclohexane (γ-HCH), (2) pentachlorocyclohexene (γ-PCCH), (3) 1,3,4,6-tetrachloro-1,4-cyclohexadiene (1,4-TCDN), (4) 1,2,4-trichlorobenzene (1,2,4-TCB), (5) 2,4,5-trichloro-2,5-cyclohexadiene-1-o1 (2,4,5-DNOL), (6) 2,5-dichlorophenol (2,5-DCP), (7) 2,5-dichloro-25-cyclohexadiene-1,4-diol (2,5-DDOL), (8) 2,5-dichlorohydroquinone (2,5-DCHQ), (9) chlorohydroquinone (CHQ), (10) hydroquinone (HQ), (11) acylchloride, (12) γ-hydroxymuconic semialdehyde, (13) maleylacetate (MA: 2-maleylacate, 4-oxohex-2-enedioate), (14) β-ketoadipate (3-oxoadipate), (15) 3-oxoadipyl-CoA, (16) succinyl-CoA, (17) acetyl-CoA, (18) 2,6-dichlorohydroquinone (2,6-DCHQ), and (19) 2-chloromaleylacetate (2-CMA).

FIGURE 7.8

Proposed degradation pathways of γ-HCH in *S. japonicum* UT26. (From Nagata, Y. et al., *Applied Microbiology and Biotechnology*, 76, 4, 741–752, 2007.)

a dehydrogenase (Nagata, Ohtomo et al. 1994); LinD, encoding a reductive dechlorinase (Miyauchi, Suh et al. 1998); LinE/LinEb, encoding a ring cleavage oxygenase (Endo, Kamakura et al. 2005); LinF, encoding a maleylacetate reductase (Endo, Kamakura et al. 2005); LinGH, encoding an acyl-CoA transferase (Nagata, Endo et al. 2007); and LinJ, encoding a thiolase (Nagata, Endo et al. 2007). Thus, the upper pathway included the reaction encoded by LinA to LinC, and the lower pathway included the reaction encoded by LinD to LinJ.

LinA is a protein with a 16.5 kDa molecular mass (Nagata, Hatta et al. 1993; Nagata, Imai et al. 1993) located in the periplasm of sphingomonads (Nagata, Futamura et al. 1999). LinA is a unique type of dehydrogenase with no close relatives (Nagata, Mori et al. 2001; Trantirek, Hynkova et al. 2001). Homology models of LinA suggest a catalytic mechanism similar to that of scytalone dehydratase, and the behaviours of mutants of key residues in LinA are consistent with this mechanism (Nagata, Mori et al. 2001). It is suggested that the substrate range of LinA may be limited to α-, γ- and δ-HCH (Nagata, Hatta et al. 1993). These HCHs dehydrochlorinated at the hydrogen/chlorine (H/Cl) pair stereoselectively; similarly, the intermediate PCCHs can be degraded to tetrachlorocyclohexadiene (Nagata, Hatta et al. 1993; Trantirek, Hynkova et al. 2001; Wu, Hong et al. 2007a). Due to lack of at least one adjacent transdiaxial H/Cl pair (Figure 7.1), β-HCH cannot be degraded by LinA. However, Wu, Hong et al. (2007b) reported that LinA can contribute to the dehydrochlorination of δ-PCCH, the intermediate of δ-HCH, although δ-PCCH also lacks a transdiaxial H/Cl pair.

LinB is a 32 kDa protein (Nagata, Nariya et al. 1993) located in the periplasm of the sphingomonads tested (Nagata, Futamura et al. 1999). Heterologously expressed LinB (Kmunicek, Hynkova et al. 2005) has a broad substrate preference for halogenated compounds up to an eight-carbon chain. As expected, LinB does not catalyse the hydrolytic dechlorination of γ-HCH. The major catalytic domain of LinB belongs to the large and well-characterized α/β-hydrolase fold superfamily of proteins, which characteristically carry out two-step nucleophile hydrolytic reactions (Nagata, Hatta et al. 1993; Nagata, Nariya et al. 1993). The catalytic mechanism of LinB involves an interaction of the substrate with the catalytic triad (Damborsky and Koca 1999; Hynkova, Nagata et al. 1999; Bohac, Nagata et al. 2002; Otyepka, Banas et al. 2008). It was suggested that nucleophiles attack the carbon atom and split the C–Cl bond by a displacement mechanism, resulting in the formation of an acyl-enzyme intermediate. Hydrolysis of the acyl-enzyme was followed to regenerate the active site. The –NH groups of LinB can assist the stability of the negative charge on the halide in the transition state during the reaction.

Other than Sphingomonadaceae, several other bacteria also have been found capable of HCH degradation. Manickam et al. (2007) isolated a *Xanthomonas* sp. ICH12 strain capable of γ-HCH degradation. Two intermediates, γ- PCCH and 2,5-dichlorobenzoquinone (2,5-DCBQ), were identified after the γ-HCH degradation by *Xanthomonas* sp. ICH12.

Streptomyces sp. M7 isolated from wastewater sediments was found by Cuozzo et al. (2009). It can synthesise the dechlorinase enzyme in the presence of the pesticide. This enzyme can function at alkaline pH, and dechlorinase activity of *S. japonicum* UT26 was optimal in acidic pH conditions (Phillips, Seech et al. 2001). Two metabolites, γ-PCCH and 1,3,4,6-tetrachlorocyclohexadiene, were detected after *Streptomyces* sp. M7 interacted with γ-HCH. *Streptomyces* spp. may be well suited for soil bioremediation purposes due to their mycelial growth habit, relatively rapid rates of growth, colonisation of semiselective substrates and their ability to be genetically manipulated (Shelton, Khader et al. 1996). This can be seen as adaptation to living in soil, especially to extreme environmental conditions, such as nutrient deficiency, high salt load and low pH, as well as toxic contamination (Kothe, Dimkpa et al. 2010).

7.7.3 γ-HCH Degradated by Fungi

Similar to lignin degradation, lignin-degrading fungi *Phanerochaete chrysosporium* and *Trametes hirsutus* have been shown capable of HCH degradation (Bumpus, Tien et al. 1985; Kennedy, Aust et al. 1990; Mougin, Pericaud et al. 1997). The proposed mechanism of fungal HCH degradation was similar to lignin degradation by lignin peroxidases with multiple nonspecific oxidative reactions resulting from generation of carbon-centred free radicals (Bumpus, Tien et al. 1985). The fungal inoculum might be used during soil bioremediation to enhance initial HCH dechlorination rates, and the more hydrophilic dechlorination products may become more available to indigenous microorganisms with the ability to complete the mineralisation process.

The ability of several white-rot fungi to degrade γ-HCH was tested by Arisoy (1998). The study reported that *Phanerochaete chrysosporium*, *Pleurotus sajorcaju*, *Pleurotus florida* and *Pleurotus eryngii* were all able to degrade significant (>10%) amounts of Lindane during 20 days of incubation in culture media under oxic conditions. Evidence that HCH dechlorination was extracellular was provided in other studies (Singh and Kuhad 1999). These studies provide convincing evidence that the ability of white-rot fungi to degrade complex organic pollutants might be profitably applied to the treatment soils highly contaminated by organochlorine pesticides, such as γ-HCH.

Fungi growing in symbiotic association with plant roots have unique enzymatic pathways that help organic contaminant degradation (Diez 2010). For instance, mycorrhizal fungi form symbioses with a variety of plant species and can promote plant growth and survival by reducing stresses from toxic wastes. The effects of soil HCHs on vegetation and its associated fungi was studied by Sainz et al. (2006). The authors found that a preinoculation of four plant species with an isolate of fungus obtained from the HCH-contaminated soil resulted in increased root growth and fungal colonization, indicating that the fungus increases the plants' tolerance to the toxicity of the contaminants.

7.8 Future Prospects

In consideration of chemically refractory molecules of HCHs, the evolution degradation pathway to effective detoxification of HCHs is required. Due to the different transformation pathways of the HCH isomers, the complexity and difficulty of HCH degradation are extended to a further degree. The research of aerobic HCH degradation has been studied far more intensively than anaerobic degradation, although aerobic degradation has been evidenced almost two decades later. Even for aerobic degradation, there remain several crucial gaps in our knowledge. One fundamental gap concerns an accepted scheme for the degradation pathway that may be incorrect because several disconnected observations have been reported (Raina, Hauser et al. 2007), so future work is needed to elucidate the specific metabolite pathway and detailed biochemistry of LinA and LinB enzymes. The anaerobic pathway for HCH degradation remains largely unknown. Few metabolites have been studied empirically; differences in the degradation pathway between isomers remain unresolved. Much work is needed for anaerobic degradation, including basic biochemical reactions and molecular-level gene identification. Both biostimulation and bioaugmentation technologies are still under development. Bioaugmentation is still in an early developmental stage; field study is needed to reach the full potential of this approach. For biostimulation, although certain in situ HCH remediation practice has been achieved, substantial further improvements are still needed, especially for decreasing the cost of both treatments.

7.9 Conclusions

Due to environmental persistence, the HCH group is one of the most important groups of organic environmental pollutants. Although HCHs have been banned in several countries decades ago, many countries continue using γ-HCH (lindane) for economic reasons. Studies on the fate and transformation of HCHs in the environment now provide the important body of knowledge to understand HCHs' environmental behaviour.

In this chapter, we have first discussed the role of structure on the properties, toxicity and degradation pathways of the HCH isomers. Then HCH bioremediation and the related metabolic pathways were discussed. For HCHs, chlorine substituents in equatorial positions tend to be more stable, and those in axial positions tend to be more reactive. With all chlorines in equatorial positions, β-HCH would be expected to be the most stable isomer; an increasing reactivity sequence based on HCH structure followed an order of γ-HCH > α-HCH > δ-HCH > β-HCH. Three possible elimination pathways of HCHs are substitution, nonreductive elimination and reductive elimination. HCH removal via the biodegradation under aerobic and anaerobic conditions has also been demonstrated. Aerobic degradation has been intensively studied in recent years, and sphingomonas genus bacteria are the major candidates for HCH bioremediation; however, anaerobic degradation study is still in its early stage. In in situ HCH degradation, biostimulation has been far more successful than bioaugumentaion. Rhizonsphere-enhanced microbial HCH degradation can be used as another potential remediation technology. Both laboratory and field studies suggest that the potential of microbial degradation as a remedial technology to treat HCHs is promising.

References

Abalis, I. M., M. E. Eldefrawi and A. T. Eldefrawi (1985). 'High-affinity stereospecific binding of cyclodiene insecticides and gamma-hexachlorocyclohexane to gamma-aminobutyric acid receptors of rat-brain'. *Pesticide Biochemistry and Physiology* 24(1): 95–102.

Agrawal, D., P. Sultana and G. S. D. Gupta (1991). 'Oxidative damage and changes in the glutathione redox system in erythrocytes from rats treated with hexachlorocyclohexane'. *Food and Chemical Toxicology* 29(7): 459–462.

Alamdar, A., J. H. Syed, R. N. Malik, A. Katsoyiannis, J. W. Liu, J. Li, G. Zhang and K. C. Jones (2014). 'Organochlorine pesticides in surface soils from obsolete pesticide dumping ground in Hyderabad City, Pakistan: Contamination levels and their potential for air-soil exchange'. *Science of the Total Environment* 470: 733–741.

Alvarez, A., M. L. Yanez, C. S. Benimeli and M. J. Amoroso (2012). 'Maize plants (Zea mays) root exudates enhance lindane removal by native Streptomyces strains'. *International Biodeterioration & Biodegradation* 66(1): 14–18.

Arisoy, M. (1998). 'Biodegradation of chlorinated organic compounds by white-rot fungi'. *Bulletin of Environmental Contamination and Toxicology* 60(6): 872–876.

Bachmann, A., W. Debruin, J. C. Jumelet, H. H. N. Rijnaarts and A. J. B. Zehnder (1988). 'Aerobic biomineralization of alpha-hexachlorocyclohexane in contaminated soil'. *Applied and Environmental Microbiology* 54(2): 548–554.

Barros, S. B. M., K. Simizu and V. B. C. Junqueira (1991). 'Liver lipid peroxidation-related parameters after short-term administration of hexachlorocyclohexane isomers to rats'. *Toxicology Letters* 56(1–2): 137–144.

Batten, K. M., K. M. Scow, K. F. Davies and S. P. Harrison (2006). 'Two invasive plants alter soil microbial community composition in serpentine grasslands'. *Biological Invasions* 8(2): 217–230.

Battisti, S., A. Caminiti, G. Ciotoli et al. (2013). 'A spatial, statistical approach to map the risk of milk contamination by beta-hexachlorocyclohexane in dairy farms'. *Geospatial Health* 8(1): 77–86.

Becerra-Castro, C., A. Prieto-Fernandez, P. S. Kidd et al. (2013). 'Improving performance of Cytisus striatus on substrates contaminated with hexachlorocyclohexane (HCH) isomers using bacterial inoculants: Developing a phytoremediation strategy'. *Plant and Soil* 362(1–2): 247–260.

Benimeli, C. S., M. S. Fuentes, C. M. Abate and M. J. Amoroso (2008). 'Bioremediation of lindane-contaminated soil by Streptomyces sp M7 and its effects on Zea mays growth'. *International Biodeterioration & Biodegradation* 61(3): 233–239.

Bhatt, D. K. and G. Nagda (2012). 'Modulation of acid phosphatase and lactic dehydrogenase in hexachlorocyclohexane-induced hepatocarcinogenesis in mice'. *Journal of Biochemical and Molecular Toxicology* 26(11): 439–444.

Bidlan, R., M. Afsar and H. K. Manonmami (2004). 'Bioremediation of HCH-contaminated soil: Elimination of inhibitory effects of the insecticide on radish and green gram seed germination'. *Chemosphere* 56(8): 803–811.

Bohac, M., Y. Nagata, Z. Prokop et al. (2002). 'Halide-stabilizing residues of halo-alkane dehalogenases studied by quantum mechanic calculations and site-directed mutagenesis'. *Biochemistry* 41(48): 14272–14280.

Boltner, D., S. Moreno-Morillas and J. L. Ramos (2005). '16S rDNA phylogeny and distribution of lin genes in novel hexachlorocyclohexane-degrading Sphingomonas strains'. *Environmental Microbiology* 7(9): 1329–1338.

Böltner, D., P. Godoy, J. Muñoz-Rojas, E. Duque, S. Moreno-Morillas, L. Sánchez and J. L. Ramos (2008). 'Rhizoremediation of lindane by root-colonizing *Sphingomonas*.' *Microbial Biotechnology* 1(1): 87–93.

Borga, K., G. W. Gabrielsen and J. U. Skaare (2001). 'Biomagnification of organochlorines along a Barents Sea food chain'. *Environmental Pollution* 113(2): 187–198.

Bowen, G. D. and A. D. Rovira (1999). The rhizosphere and its management to improve plant growth. *Advances in Agronomy, Vol 66*. D. L. Sparks. San Diego, Elsevier Academic Press Inc. 66: 1–102.

Bumpus, J. A., M. Tien, D. Wright and S. D. Aust (1985). 'Oxidation of persistent environmental-pollutants by a white rot fungus'. *Science* 228(4706): 1434–1436.

Buser, H.-R. and M. D. Mueller (1995). 'Isomer and enantioselective degradation of hexachlorocyclohexane isomers in sewage sludge under anaerobic conditions'. *Environmental Science & Technology* 29(3): 664–672.

Ceremonie, H., H. Boubakri, P. Mavingui, P. Simonet and T. M. Vogel (2006). 'Plasmid-encoded gamma-hexachlorocyclohexane degradation genes and insertion sequences in Sphingobium francense (ex-Sphingomonas paucimobilis Sp+)'. *FEMS Microbiology Letters* 257(2): 243–252.

Cooper, R. L., R. W. Chadwick, G. L. Rehnberg, J. M. Goldman, K. C. Booth, J. F. Hein and W. K. McElroy (1989). 'Effect of lindane on hormonal-control of reproductive function in the female rat'. *Toxicology and Applied Pharmacology* 99(3): 384–394.

Cuozzo, S. A., G. G. Rollan, C. M. Abate and M. J. Amoroso (2009). 'Specific dechlorinase activity in lindane degradation by Streptomyces sp M7'. *World Journal of Microbiology & Biotechnology* 25(9): 1539–1546.

Dadhwal, M., A. Singh, O. Prakash, S. K. Gupta, K. Kumari, P. Sharma, S. Jit, M. Verma, C. Holliger and R. Lal (2009). 'Proposal of biostimulation for hexachlorocyclohexane (HCH)-decontamination and characterization of culturable bacterial community from high-dose point HCH-contaminated soils'. *Journal of Applied Microbiology* 106(2): 381–392.

Damborsky, J. and J. Koca (1999). 'Analysis of the reaction mechanism and substrate specificity of haloalkane dehalogenases by sequential and structural comparisons'. *Protein Engineering* 12(11): 989–998.

Deeb, R. A., J. C. Spain and L. Alvarez-Cohen (1999). 'Mineralization of benzene, toluene, ethylbenzene, m-Xylene and p-Xylene by two Rhodococcus species'. *Abstracts of the General Meeting of the American Society for Microbiology* 99: 553.

Dietz, R., F. F. Riget, C. Sonne et al. (2013). 'Three decades (1983–2010) of contaminant trends in East Greenland polar bears (Ursus maritimus). Part 1: Legacy Organochlorine contaminants'. *Environment International* 59: 485–493.

Diez, M. C. (2010). 'Biological aspects involved in the degradation of organic pollutants'. *Journal of Soil Science and Plant Nutrition* 10(3): 244–267.

Dobbelaere, S., J. Vanderleyden and Y. Okon (2003). 'Plant growth-promoting effects of diazotrophs in the rhizosphere'. *Critical Reviews in Plant Sciences* 22(2): 107–149.

Duffy, B., C. Keel and G. Defago (2004). 'Potential role of pathogen signaling in multitrophic plant-microbe interactions involved in disease protection'. *Applied and Environmental Microbiology* 70(3): 1836–1842.

Endo, R., M. Kamakura, K. Miyauchi et al. (2005). 'Identification and characterization of genes involved in the downstream degradation pathway of gamma-hexachlorocyclohexane in Sphingomonas paucimobilis UT26'. *Journal of Bacteriology* 187(3): 847–853.

Endo, R., Y. Ohtsubo, M. Tsuda and Y. Nagata (2007). 'Identification and characterization of genes encoding a putative ABC-type transporter essential for utilization of gamma-hexachlorocyclohexane in Sphingobium japonicum UT26'. *Journal of Bacteriology* 189(10): 3712–3720.

Fairlee, J. R., B. L. Burback and J. J. Perry (1997). 'Biodegradation of groundwater pollutants by a combined culture of Mycobacterium vaccae and a Rhodococcus sp'. *Canadian Journal of Microbiology* 43(9): 841–846.

Fernandez, J., M. A. Arjol and C. Cacho (2013). 'POP-contaminated sites from HCH production in Sabinanigo, Spain'. *Environmental Science and Pollution Research* 20(4): 1937–1950.

Fidan, A. F., I. H. Cigerci, N. Baysu-Sozbilir, I. Kucukkurt, H. Yuksel and H. Keles (2008). 'The effects of the dose-dependent gamma-hexachlorocyclohexane (lindane) on blood and tissue antioxidant defense systems, lipid peroxidation and histopathological changes in rats'. *Journal of Animal and Veterinary Advances* 7(11): 1480–1488.

Frische, K., J. Schwarzbauer and M. Ricking (2010). 'Structural diversity of organochlorine compounds in groundwater affected by an industrial point source'. *Chemosphere* 81(4): 500–508.

Galban-Malagon, C., A. Cabrerizo, G. Caballero and J. Dachs (2013). 'Atmospheric occurrence and deposition of hexachlorobenzene and hexachlorocyclohexanes in the Southern Ocean and Antarctic Peninsula'. *Atmospheric Environment* 80: 41–49.

Ge, J., L. A. Woodward, Q. X. Li and J. Wang (2013). 'Composition, distribution and risk assessment of organochlorine pesticides in soils from the Midway Atoll, North Pacific Ocean'. *Science of the Total Environment* 452: 421–426.

Gonzalez-Mille, D. J., C. A. Ilizaliturri-Hernandez, G. Espinosa-Reyes et al. (2010). 'Exposure to persistent organic pollutants (POPs) and DNA damage as an indicator of environmental stress in fish of different feeding habits of Coatzacoalcos, Veracruz, Mexico'. *Ecotoxicology* 19(7): 1238–1248.

Grimalt, J. O., D. Carrizo, M. Gari et al. (2010). 'An evaluation of the sexual differences in the accumulation of organochlorine compounds in children at birth and at the age of 4 years'. *Environmental Research* 110(3): 244–250.

Gupta, S. K., D. Lal, P. Lata et al. (2013). 'Changes in the bacterial community and lin genes diversity during biostimulation of indigenous bacterial community of hexachlorocyclohexane (HCH) dumpsite soil'. *Microbiology* 82(2): 234–240.

Halse, A. K., M. Schlabach, A. Sweetman, K. C. Jones and K. Breivik (2012). 'Using passive air samplers to assess local sources versus long range atmospheric transport of POPs'. *Journal of Environmental Monitoring* 14(10): 2580–2590.

Heritage, A. D. and I. C. Macrae (1977). 'Degradation of lindane by cell-free preparations of clostridium-sphenoides'. *Applied and Environmental Microbiology* 34(2): 222–224.

Hinsinger, P., C. Plassard and B. Jaillard (2006). 'Rhizosphere: A new frontier for soil biogeochemistry'. *Journal of Geochemical Exploration* 88(1–3): 210–213.

Hynkova, K., Y. Nagata, M. Takagi and J. Damborsky (1999). 'Identification of the catalytic triad in the haloalkane dehalogenase from Sphingomonas paucimobilis UT26'. *FEBS Letters* 446(1): 177–181.

Imai, R., Y. Nagata, M. Fukuda, M. Takagi and K. Yano (1991). 'Molecular-cloning of a pseudomonas-paucimobilis gene encoding a 17-kilodalton polypeptide that eliminates HCl molecules from gamma-hexachlorocyclohexane'. *Journal of Bacteriology* 173(21): 6811–6819.

Ito, A., R. Yamashita, H. Takada et al. (2013). 'Contaminants in tracked seabirds showing regional patterns of marine pollution'. *Environmental Science & Technology* 47(14): 7862–7867.

Ito, M., Z. Prokop, M. Klvana et al. (2007). 'Degradation of beta-hexachlorocyclohexane by haloalkane dehalogenase LinB from gamma-hexachlorocyclohexane-utilizing bacterium Sphingobium sp MI1205'. *Archives of Microbiology* 188(4): 313–325.

Jagnow, G., K. Haider and P. C. Ellwardt (1977). 'Anaerobic dechlorination and degradation of hexachlorocyclohexane isomers by anaerobic and facultative anaerobic bacteria'. *Archives of Microbiology* 115(3): 285–292.

Jit, S., M. Dadhwal, H. Kumari et al. (2011). 'Evaluation of hexachlorocyclohexane contamination from the last lindane production plant operating in India'. *Environmental Science and Pollution Research* 18(4): 586–597.

Johri, A. K., M. Dua, D. Tuteja, R. Saxena, D. M. Saxena and R. Lal (1998). 'Degradation of alpha, beta, gamma and delta-hexachlorocyclohexanes by Sphingomonas paucimobilis'. *Biotechnology Letters* 20(9): 885–887.

Joseph, P., S. Viswanatha and M. K. Krishnakumari (1992). 'Role of vitamin-A in the haematotoxicity of hexachlorocyclohexane (HCH) in the rat'. *Journal of Environmental Science and Health Part B-Pesticides Food Contaminants and Agricultural Wastes* 27(3): 269–280.

Kannan, K., K. Ramu, N. Kajiwara, R. K. Sinha and S. Tanabe (2005). 'Organochlorine pesticides, polychlorinated biphenyls, and polybrominated diphenyl ethers in irrawaddy dolphins from India'. *Archives of Environmental Contamination and Toxicology* 49(3): 415–420.

Kennedy, D. W., S. D. Aust and J. A. Bumpus (1990). 'Comparative biodegradation of alkyl halide insecticides by the white rot fungus, phanerochaete-chrysosporium (BKM-F-1767)'. *Applied and Environmental Microbiology* 56(8): 2347–2353.

Kidd, P. S., A. Prieto-Fernandez, C. Monterroso and M. J. Acea (2008). 'Rhizosphere microbial community and hexachlorocyclohexane degradative potential in contrasting plant species'. *Plant and Soil* 302(1–2): 233–247.

Kmunicek, J., K. Hynkova, T. Jedlicka et al. (2005). 'Quantitative analysis of substrate specificity of haloalkane dehalogenase LinB from Sphingomonas paucimobilis UT26'. *Biochemistry* 44(9): 3390–3401.

Kole, R. K., H. Banerjee and A. Bhattacharyya (2001). 'Monitoring of market fish samples for endosulfan and hexachlorocyclohexane residues in and around calcutta'. *Bulletin of Environmental Contamination and Toxicology* 67(4): 554–559.

Kothe, E., C. Dimkpa, G. Haferburg, A. Schmidt, A. Schmidt and E. Schuetze (2010). Streptomycete heavy metal resistance: Extracellular and intracellular mechanisms. *Soil Heavy Metals*. I. Sherameti and A. Varma, Springer-Verlag Berlin, Heidelberger Platz 3, D-14197 Berlin, Germany. 19: 225–235.

Kowalchuk, G. A., W. H. G. Hol and J. A. Van Veen (2006). 'Rhizosphere fungal communities are influenced by Senecio jacobaea pyrrolizidine alkaloid content and composition'. *Soil Biology & Biochemistry* 38(9): 2852–2859.

Kucklick, J. R., M. M. Krahn, P. R. Becker et al. (2006). 'Persistent organic pollutants in Alaskan ringed seal (Phoca hispida) and walrus (Odobenus rosmarus) blubber'. *Journal of Environmental Monitoring* 8(8): 848–854.

Lal, R., M. Dadhwal, K. Kumari et al. (2008). 'Pseudomonas sp to Sphingobium indicum: A journey of microbial degradation and bioremediation of Hexachlorocyclohexane'. *Indian Journal of Microbiology* 48(1): 3–18.

Lal, R., C. Dogra, S. Malhotra, P. Sharma and R. Pal (2006). 'Diversity, distribution and divergence of lin genes in hexachlorocyclohexane-degrading sphingomonads'. *Trends in Biotechnology* 24(3): 121–130.

Lal, R., G. Pandey, P. Sharma et al. (2010). 'Biochemistry of microbial degradation of hexachlorocyclohexane and prospects for bioremediation'. *Microbiology and Molecular Biology Reviews* 74(1): 58–80.

Langenhoff, A. A. M., S. J. M. Staps, C. Pijls and H. H. M. Rijnaarts (2013). 'Stimulation of hexachlorocyclohexane (hch) biodegradation in a full scale in situ bioscreen'. *Environmental Science & Technology* 47(19): 11182–11188.

Liu, D., W. X. Qiu, Z. H. Xia et al. (2009). 'Oral intake of lindane by population in Taiyuan'. *Asian Journal of Ecotoxicology* 4(6): 786–792.

Macrae, I. C., K. Raghu and E. M. Bautista (1969). 'Anaerobic degradation of insecticide lindane by clostridium sp'. *Nature* 221(5183): 859–860.

Macrae, I. C., K. Raghu and T. F. Castro (1967). 'Persistence and biodegradation of 4 common isomers of benzene hexachloride in submerged soils'. *Journal of Agricultural and Food Chemistry* 15(5): 911–914.

Manickam, N., R. Misra and S. Mayilraj (2007). 'A novel pathway for the biodegradation of gamma-hexachlorocyclohexane by a Xanthomonas sp strain ICH12'. *Journal of Applied Microbiology* 102(6): 1468–1478.

Mars, A. E., T. Kasberg, S. R. Kaschabek, M. H. vanAgteren, D. B. Janssen and W. Reineke (1997). 'Microbial degradation of chloroaromatics: Use of the meta-cleavage pathway for mineralization of chlorobenzene'. *Journal of Bacteriology* 179(14): 4530–4537.

McNeish, A. S., T. Bidleman, R. T. Leah and M. S. Johnson (1999). 'Enantiomers of methylhexachlorocyclohexane and hexachlorocyclohexane in fish, shellfish, and waters of the Mersey Estuary'. *Environmental Toxicology* 14(4): 397–403.

Mertens, B., N. Boon and W. Verstraete (2006). 'Slow-release inoculation allows sustained biodegradation of gamma-hexachlorocyclohexane'. *Applied and Environmental Microbiology* 72(1): 622–627.

Middeldorp, P. J. M., M. Jaspers, A. J. B. Zehnder and G. Schraa (1996). 'Biotransformation of alpha-, beta-, gamma-, and delta-hexachlorocyclohexane under methanogenic conditions'. *Environmental Science & Technology* 30(7): 2345–2349.

Miyauchi, K., S. K. Suh, Y. Nagata and M. Takagi (1998). 'Cloning and sequencing of a 2,5-dichlorohydroquinone reductive dehalogenase gene whose product is involved in degradation of gamma-hexachlorocyclohexane by Sphingomonas paucimobilis'. *Journal of Bacteriology* 180(6): 1354–1359.

Mohn, W. W., B. Mertens, J. D. Neufeld, W. Verstraete and V. de Lorenzo (2006). 'Distribution and phylogeny of hexachlorocyclohexane-degrading bacteria in soils from Spain'. *Environmental Microbiology* 8(1): 60–68.

Morgan, J. A. W., G. D. Bending and P. J. White (2005). 'Biological costs and benefits to plant-microbe interactions in the rhizosphere'. *Journal of Experimental Botany* 56(417): 1729–1739.

Mougel, C., P. Offre, L. Ranjard, T. Corberand, E. Gamalero, C. Robin and P. Lemanceau (2006). 'Dynamic of the genetic structure of bacterial and fungal communities at different developmental stages of Medicago truncatula Gaertn. cv. Jemalong line J5'. *New Phytologist* 170(1): 165–175.

Mougin, C., C. Pericaud, J. Dubroca and M. Asther (1997). 'Enhanced mineralization of lindane in soils supplemented with the white rot basidiomycete Phanerochaete chrysosporium'. *Soil Biology & Biochemistry* 29(9–10): 1321–1324.

Mullins, L. J. (1955). 'Structure-toxicity in hexachlorocyclohexane isomers'. *Science* 122(3159): 118–119.

Nagasawa, S., R. Kikuchi, Y. Nagata, M. Takagi and M. Matsuo (1993). 'Aerobic mineralization of gamma-HCH by pseudomonas-paucimobilis UT26'. *Chemosphere* 26(9): 1719–1728.

Nagata, Y., R. Endo, M. Ito, Y. Ohtsubo and M. Tsuda (2007). 'Aerobic degradation of lindane (gamma-hexachlorocyclohexane) in bacteria and its biochemical and molecular basis'. *Applied Microbiology and Biotechnology* 76(4): 741–752.

Nagata, Y., A. Futamura, K. Miyauchi and M. Takagi (1999). 'Two different types of dehalogenases, LinA and LinB, involved in gamma-hexachlorocyclohexane degradation in Sphingomonas paucimobilis UT26 are localized in the periplasmic space without molecular processing'. *Journal of Bacteriology* 181(17): 5409–5413.

Nagata, Y., T. Hatta, R. Imai et al. (1993). 'Purification and characterization of gamma-hexachlorocyclohexane (Gamma-HCH) dehydrochlorinase (LinA) from pseudomonas-paucimobilis'. *Bioscience Biotechnology and Biochemistry* 57(9): 1582–1583.

Nagata, Y., R. Imai, A. Sakai, M. Fukuda, K. Yano and M. Takagi (1993). 'Isolation and characterization of TN5-induced mutants of pseudomonas-paucimobilis UT26 defective in gamma-hexachlorocyclohexane dehydrochlorinase (LinA)'. *Bioscience Biotechnology and Biochemistry* 57(5): 703–709.

Nagata, Y., K. Mori, M. Takagi, A. G. Murzin and J. Damborsky (2001). 'Identification of protein fold and catalytic residues of gamma-hexachlorocyclohexane dehydrochlorinase LinA'. *Proteins-Structure Function and Genetics* 45(4): 471–477.

Nagata, Y., T. Nariya, R. Ohtomo, M. Fukuda, K. Yano and M. Takagi (1993). 'Cloning and sequencing of a dehalogenase gene encoding an enzyme with hydrolase activity involved in the degradation of gamma-hexachlorocyclohexane in pseudomonas-paucimobilis'. *Journal of Bacteriology* 175(20): 6403–6410.

Nagata, Y., R. Ohtomo, K. Miyauchi, M. Fukuda, K. Yano and M. Takagi (1994). 'Cloning and sequencing of a 2,5-dichloro-2,5-cyclohexadiene-1,4-diol dehydrogenase gene involved in the degradation of gamma-hexachlorocyclohexane in pseudomonas-paucimobilis'. *Journal of Bacteriology* 176(11): 3117–3125.

Newton, S., T. Bidleman, M. Bergknut, J. Racine, H. Laudon, R. Giesler and K. Wiberg (2014). 'Atmospheric deposition of persistent organic pollutants and chemicals of emerging concern at two sites in northern Sweden'. *Environmental Science-Processes & Impacts* 16(2): 298–305.

Ngabe, B., T. F. Bidleman and R. L. Falconer (1993). 'Base hydrolysis of alpha- and gamma-hexachlorocyclohexanes'. *Environmental Science & Technology* 27(9): 1930–1933.

Orloff, H. D. (1954). 'The stereoisomerism of cyclohexane derivatives'. *Chemical Reviews* 54(3): 347–447.

Otyepka, M., P. Banas, A. Magistrato, P. Carloni and J. Damborsky (2008). 'Second step of hydrolytic dehalogenation in haloalkane dehalogenase investigated by QM/MM methods'. *Proteins-Structure Function and Bioinformatics* 70(3): 707–717.

Payne, J., M. Scholze and A. Kortenkamp (2001). 'Mixtures of four organochlorines enhance human breast cancer cell proliferation'. *Environmental Health Perspectives* 109(4): 391–397.

Phillips, T. M., H. Lee, J. T. Trevors and A. G. Seech (2006). 'Full-scale in situ bioremediation of hexachlorocyclohexane-contaminated soil'. *Journal of Chemical Technology and Biotechnology* 81(3): 289–298.

Phillips, T. M., A. G. Seech, H. Lee and J. T. Trevors (2001). 'Colorimetric assay for Lindane dechlorination by bacteria'. *Journal of Microbiological Methods* 47(2): 181–188.

Phillips, T. M., A. G. Seech, H. Lee and J. T. Trevors (2005). 'Biodegradation of hexachlorocyclohexane (HCH) by microorganisms'. *Biodegradation* 16(4): 363–392.

Porta, D., F. Fantini, E. De Felip et al. (2013). 'A biomonitoring study on blood levels of beta-hexachlorocyclohexane among people living close to an industrial area'. *Environmental Health* 12.

Quintero, J. C., M. T. Moreira, G. Feijoo and J. M. Lema (2005a). 'Anaerobic degradation of hexachlorocyclohexane isomers in liquid and soil slurry systems'. *Chemosphere* 61(4): 528–536.

Quintero, J. C., M. T. Moreira, G. Feijoo and J. M. Lema (2005b). 'Effect of surfactants on the soil desorption of hexachlorocyclohexane (HCH) isomers and their anaerobic biodegradation'. *Journal of Chemical Technology and Biotechnology* 80(9): 1005–1015.

Raina, V., A. Hauser, H. R. Buser et al. (2007). 'Hydroxylated metabolites of beta- and delta-hexachlorocyclohexane: Bacterial formation, stereochemical configuration, and occurrence in groundwater at a former production site'. *Environmental Science & Technology* 41(12): 4292–4298.

Raina, V., M. Suar, A. Singh et al. (2008). 'Enhanced biodegradation of hexachlorocyclohexane (HCH) in contaminated soils via inoculation with Sphingobium indicum B90A'. *Biodegradation* 19(1): 27–40.

Ricking, M. and J. Schwarzbauer (2008). 'HCH residues in point-source contaminated samples of the Teltow Canal in Berlin, Germany'. *Environmental Chemistry Letters* 6(2): 83–89.

Rissato, S. R., M. S. Galhiane, V. F. Ximenes et al. (2006). 'Organochlorine pesticides and polychlorinated biphenyls in soil and water samples in the northeastern part of Sao Paulo State, Brazil'. *Chemosphere* 65(11): 1949–1958.

Rodriguez-Garrido, B., T. A. Lu-Chau, G. Feijoo, F. Macias and M. C. Monterrroso (2010). 'Reductive dechlorination of alpha-, beta-, gamma-, and delta-hexachlorocyclohexane isomers with hydroxocobalamin, in soil slurry systems'. *Environmental Science & Technology* 44(18): 7063–7069.

Rooney-Varga, J. N., R. T. Anderson, J. L. Fraga, D. Ringelberg and D. R. Lovley (1999). 'Microbial communities associated with anaerobic benzene degradation in a petroleum-contaminated aquifer'. *Applied and Environmental Microbiology* 65(7): 3056–3063.

Rubinos, D., R. Villasuso, S. Muniategui, M. Barral and F. Díaz-Fierros (2007). 'Using the landfarming technique to remediate soils contaminated with hexachlorocyclohexane isomers'. *Water, Air, and Soil Pollution* 181(1–4): 385–399.

Sahu, S. K., K. K. Patnaik, S. Bhuyan, B. Sreedharan, N. Kurihara, T. K. Adhya and N. Sethunathan (1995). 'Mineralization of alpha-isomers, gamma-isomers, and beta-isomers of hexachlorocyclohexane by a soil bacterium under aerobic conditions'. *Journal of Agricultural and Food Chemistry* 43(3): 833–837.

Sainz, M. J., B. González-Penalta and A. Vilariño (2006). 'Effects of hexachlorocyclohexane on rhizosphere fungal propagules and root colonization by arbuscular mycorrhizal fungi in Plantago lanceolata'. *European Journal of Soil Science* 57(1): 83–90.

Salem, N. M., R. Ahmad and H. Estaitieh (2009). 'Organochlorine pesticide residues in dairy products in Jordan'. *Chemosphere* 77(5): 673–678.

Santos, E., R. D. Silva, H. H. C. Barretto et al. (2003). 'Levels of exposure to organochlorine pesticides in open-air dump dwellers'. *Revista De Saude Publica* 37(4): 515–522.

Sauviat, M.-P. and N. Pages (2002). 'Cardiotoxicity of lindane'. *Journal de la Societe de Biologie* 196(4): 339–348.

Sharma, P. and R. Singh (2010). 'Protective role of curcumin on lindane induced reproductive toxicity in male wistar rats'. *Bulletin of Environmental Contamination and Toxicology* 84(4): 378–384.

Shelton, D. R., S. Khader, J. S. Karns and B. M. Pogell (1996). 'Metabolism of twelve herbicides by Streptomyces'. *Biodegradation* 7(2): 129–136.

Siddique, T., B. C. Okeke, M. Arshad and W. T. Frankenberger (2002). 'Temperature and pH effects on biodegradation of hexachlorocyclohexane isomers in water and a soil slurry'. *Journal of Agricultural and Food Chemistry* 50(18): 5070–5076.

Simic, B., I. Kmetic, T. Murati and J. Kniewald (2012). 'Effects of lindane on reproductive parameters in male rats'. *Veterinarski Arhiv* 82(2): 211–220.

Singh, B. K. and R. C. Kuhad (1999). 'Biodegradation of lindane (gamma-hexachlorocyclohexane) by the white-rot fungus Trametes hirsutus'. *Letters in Applied Microbiology* 28(3): 238–241.

Singh, K., A. Malik and S. Sinha (2007). 'Persistent organochlorine pesticide residues in soil and surface water of northern indo-gangetic alluvial plains'. *Environmental Monitoring and Assessment* 125(1–3): 147–155.

Singh, P. B. and V. Singh (2008). 'Bioaccumulation of hexachlorocyclohexane, dichlorodiphenyltrichloroethane, and estradiol-17 beta in catfish and carp during the pre-monsoon season in India'. *Fish Physiology and Biochemistry* 34(1): 25–36.

Song, S. L., X. D. Ma, L. Tong et al. (2013). 'Residue levels of hexachlorocyclohexane and dichlorodiphenyl trichloroethane in human milk collected from Beijing'. *Environmental Monitoring and Assessment* 185(9): 7225–7229.

Sonne, C., R. J. Letcher, P. S. Leifsson et al. (2012). 'Temporal monitoring of liver and kidney lesions in contaminated East Greenland polar bears (Ursus maritimus) during 1999–2010'. *Environment International* 48: 143–149.

Sonne, C., S. A. B. Maehre, K. Sagerup et al. (2013). 'A screening of liver, kidney, and thyroid gland morphology in organochlorine-contaminated glaucous gulls (Larus hyperboreus) from Svalbard'. *Toxicological and Environmental Chemistry* 95(1): 172–186.

Spinelli, J. J., C. H. Ng, J. P. Weber et al. (2007). 'Organochlorines and risk of non-Hodgkin lymphoma'. *International Journal of Cancer* 121(12): 2767–2775.

Sturludottir, E., H. Gunnlaugsdottir, H. O. Jorundsdottir, E. V. Magnusdottir, K. Olafsdottir and G. Stefansson (2013). 'Spatial and temporal trends of contaminants in mussel sampled around the Icelandic coastline'. *Science of the Total Environment* 454: 500–509.

Syed, J. H., R. N. Malik, D. Liu et al. (2013). 'Organochlorine pesticides in air and soil and estimated air-soil exchange in Punjab, Pakistan'. *Science of the Total Environment* 444: 491–497.

Tian, C. G., Y. F. Li, H. L. Jia, H. L. Wu and J. M. Ma (2009). 'Modelling historical budget of alpha-hexachlorocyclohexane in Taihu Lake, China'. *Chemosphere* 77(4): 459–464.

Torres, J. P. M., C. I. R. Froes-Asmus, R. Weber and J. M. H. Vijgen (2013). 'HCH contamination from former pesticide production in Brazil-a challenge for the Stockholm Convention implementation'. *Environmental Science and Pollution Research* 20(4): 1951–1957.

Trantirek, L., K. Hynkova, Y. Nagata et al. (2001). 'Reaction mechanism and stereochemistry of gamma-hexachlorocyclohexane dehydrochlorinase LinA'. *Journal of Biological Chemistry* 276(11): 7734–7740.

van Doesburg, W., M. H. A. van Eekert, P. J. M. Middeldorp, M. Balk, G. Schraa and A. J. M. Stams (2005). 'Reductive dechlorination of beta-hexachlorocyclohexane (beta-HCH) by a Dehalobacter species in coculture with a Sedimentibacter sp'. *FEMS Microbiology Ecology* 54(1): 87–95.

Van Eekert, M. H. A., N. J. P. Van Ras, G. H. Mentink et al. (1998). 'Anaerobic transformation of beta-hexachlorocyclohexane by methanogenic granular sludge and soil microflora'. *Environmental Science & Technology* 32(21): 3299–3304.

Vijgen, J., P. C. Abhilash, Y. F. Li et al. (2011). 'Hexachlorocyclohexane (HCH) as new stockholm convention POPs-a global perspective on the management of Lindane and its waste isomers'. *Environmental Science and Pollution Research* 18(2): 152–162.

Waliszewski, S. M., R. Villalobos-Pietrini, S. Gomez-Arroyo and R. M. Infanzon (2003). 'Persistent organochlorine pesticides in Mexican butter'. *Food Additives and Contaminants* 20(4): 361–367.

Walker, K., D. A. Vallero and R. G. Lewis (1999). 'Factors influencing the distribution of lindane and other hexachlorocyclohexanes in the environment'. *Environmental Science & Technology* 33(24): 4373–4378.

Wang, H. S., Z. J. Chen, W. Wei et al. (2013). 'Concentrations of organochlorine pesticides (OCPs) in human blood plasma from Hong Kong: Markers of exposure and sources from fish'. *Environment International* 54: 18–25.

Weber, R., C. Gaus, M. Tysklind et al. (2008). 'Dioxin- and POP-contaminated sites-contemporary and future relevance and challenges'. *Environmental Science and Pollution Research* 15(5): 363–393.

Willett, K. L., E. M. Ulrich and R. A. Hites (1998). 'Differential toxicity and environmental fates of hexachlorocyclohexane isomers'. *Environmental Science & Technology* 32(15): 2197–2207.

Wong, P. S. and F. Matsumura (2007). 'Promotion of breast cancer by beta-hexachlorocyclohexane in MCF10AT1 cells and MMTV-neu mice'. *BMC Cancer* 7(130): (17 July 2007).

Wu, J., Q. Hong, P. Han, J. He and S. P. Li (2007a). 'A gene linB2 responsible for the conversion of beta-HCH and 2,3,4,5,6-pentachlorocyclohexanol in Sphingomonas sp BHC-A'. *Applied Microbiology and Biotechnology* 73(5): 1097–1105.

Wu, J., Q. Hong, Y. Sun, Y. F. Hong, Q. X. Yan and S. P. Li (2007b). 'Analysis of the role of LinA and LinB in biodegradation of delta-hexachlorocyclohexane'. *Environmental Microbiology* 9(9): 2331–2340.

Wu, W. Z., Y. Xu, K. W. Schramm and A. Kettrup (1997). 'Study of sorption, biodegradation and isomerization of HCH in stimulated sediment/water system'. *Chemosphere* 35(9): 1887–1894.

Yamamoto, S., S. Otsuka, Y. Murakami, M. Nishiyama and K. Senoo (2009). 'Genetic diversity of gamma-hexachlorocyclohexane-degrading sphingomonads isolated from a single experimental field'. *Letters in Applied Microbiology* 49(4): 472–477.

Zhang, H., Y. Luo and Q. Li (2009). 'Burden and depth distribution of organochlorine pesticides in the soil profiles of Yangtze River Delta Region, China: Implication for sources and vertical transportation'. *Geoderma* 153(1/2): 69–75.

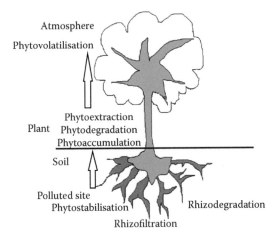

FIGURE 1.1
Concept of phytoremediation technologies.

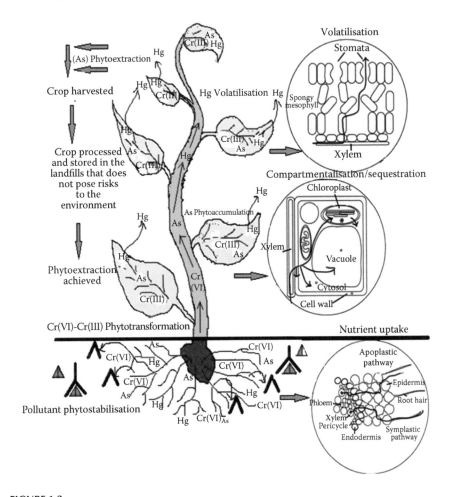

FIGURE 1.2
Different mechanism of phytotechnologies: phytoextraction of As from the soil to aerial parts of plant (leaves and stems), phytotransformation of Cr(VI) from the soil to Cr(III) in the aerial parts of the plant, phytostabilisation of metal contaminants in soil and phytovolatilisation of Hg from the soil.

FIGURE 2.2

Typical process of (a) sampling (b) isolating and (c) screening from polluted environments to determine metal tolerance. Potential isolates are subsequently (d) cultured to generate sufficient biomass for metal biosorption tests. The biomass can be prepared as (e) nonviable, (f) viable or (g) immobilised forms to determine their biosorption potential.

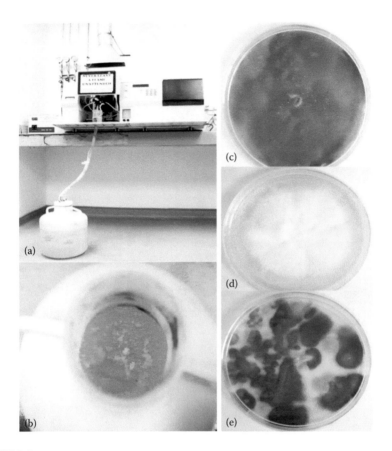

FIGURE 2.3
(a) Wastewater from AAS where (b) growth of fungi in the wastewater can be detected visually. Isolates (c) 6 (*Penicillium* sp.), (d) 9 (*Fusarium* sp.) and (e) 10 (*Aspergillus* sp.) showed the most potential in removing various metals via biosorption. (Compiled and modified from Ting ASY et al., *Proceedings of the International Congress of the Malaysian Society for Microbiology.* 8–11 December 2011, Penang, Malaysia, pp. 110–113, 2011.)

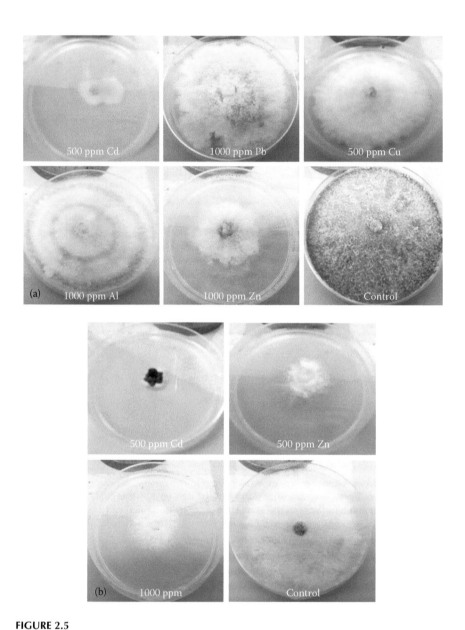

FIGURE 2.5
Metal tolerance of endophytic (a) *T. asperellum* (isolate 11) and (b) *Saccharicola bicolour* (isolate 22) from *Phragmites* toward various metals at their maximum tolerable concentrations. Changes in pigmentation and colony diameter are the two most common responses of the endophytes to high metal concentrations.

(Continued)

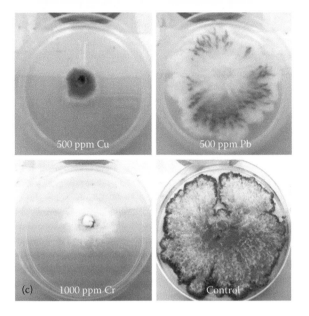

FIGURE 2.5 (CONTINUED)
Metal tolerance of endophytic (c) *Phomopsis* sp. (isolate 9) from *Phragmites* toward various metals at their maximum tolerable concentrations. Changes in pigmentation and colony diameter are the two most common responses of the endophytes to high metal concentrations. (Compiled and modified from the study by Sim, unpublished, 72; Sim CSF, Metal tolerance and biosorption potential of endophytic fungi from Phragmites. Project Report, Monash University Malaysia, p. 47, 2013.)

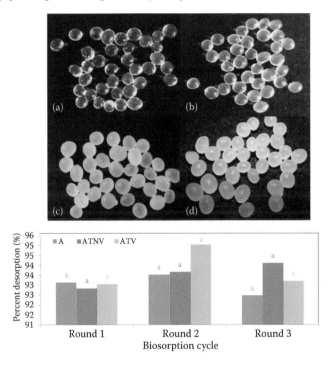

FIGURE 2.11
Appearance of (a) plain alginate beads and (b) alginate immobilised fungal cells before and after Cu biosorption (c and d). Immobilisation allows recovery of Cu and regeneration of alginate beads for three consecutive cycles. A: Plain alginate beads, ATNV: alginate immobilised dead cells, ATV: alginate immobilised live cells. Means with the same letters within treatments (A, ATNV, ATV) are not significantly different ($HSD_{(0.05)}$). (Compiled and modified from studies by Tan and Ting, unpublished, 86; Tan WS and Ting ASY, *Bioresource Technology*, 123, 290–295, 2012.)

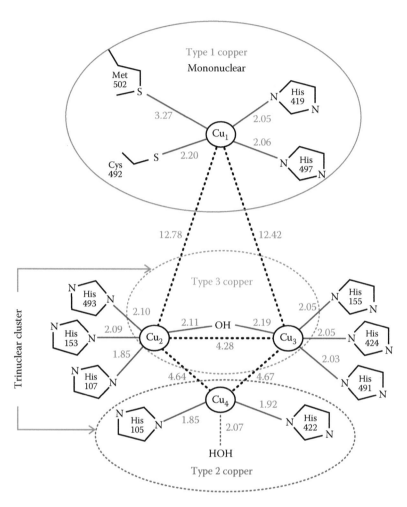

FIGURE 4.1
Structure of laccase copper centres and ligands with interatomic distances. (Reprinted from *J Mol Cat B Enz*, 68, 2, Dwivedi, U. N. et al., Singh, P., Pandey, V. P., and Kumar, A., 117–128, Copyright 2011, with permission from Elsevier.)

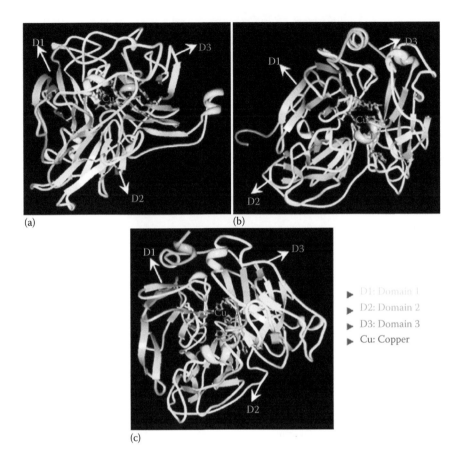

D1: Domain 1
D2: Domain 2
D3: Domain 3
Cu: Copper

FIGURE 4.2
Three-dimensional structure of (a) bacterial laccase (*Bacillus subtilis*), (b) fungal laccase (*Trametes versicolor*) and (c) plant laccase (*Populus trichocarpa*). (Reprinted from *J Mol Cat B Enz,* 68, 2, Dwivedi, U. N. et al., Singh, P., Pandey, V. P., and Kumar, A., 117–128, Copyright 2011, with permission from Elsevier.)

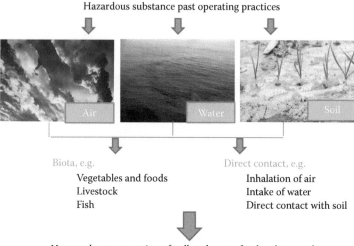

FIGURE 7.2
Hexachlorocyclohexane exposure pathways.

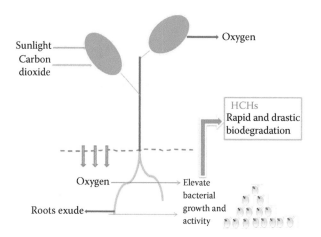

FIGURE 7.6
HCH removal by rhizosphere microorganisms.

FIGURE 9.3
Macroscopic view of a noninoculated (b) and an inoculated (a) hydrocarbon wastewater selective broth culture Erlenmeyers at the end of the bioremediation experiment. Hydrocarbon-contaminated wastewater was exposed to an indigenous consortium at 30°C and continuous shaking for 50 days. 0.1% crude oil was used as the sole source of carbon in MM media.

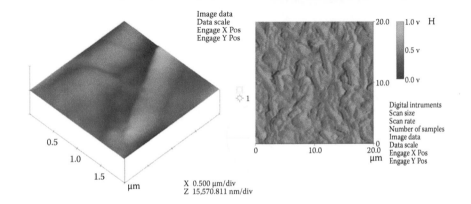

FIGURE 11.6
AFM images of *C. diolis*.

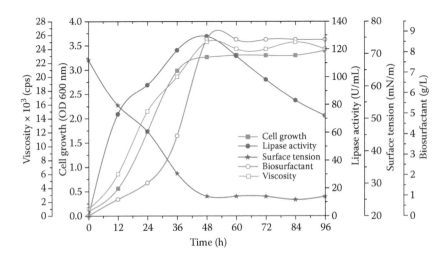

FIGURE 11.14
Profile of cell growth, lipase activity, surface tension, biosurfactant and viscosity at different incubation periods in the anaerobic fermentation of goat tallow.

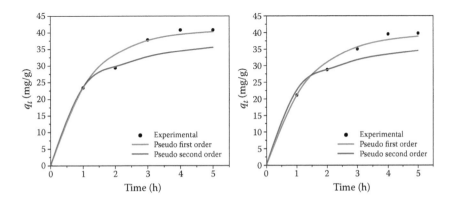

FIGURE 11.23
Pseudo first-order and pseudo second-order kinetic models for hydrolysis of edible oil waste using FAL and LMAC (conditions: concentration of edible oil waste, 4.3 g/L; pH, 3.5; temperature, 30°C; and mass of LMAC, 1 g).

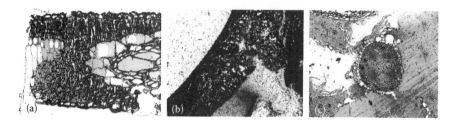

FIGURE 12.1
Cross section of *Typha* organs under TEM: (a) leaf, (b) stem and (c) root.

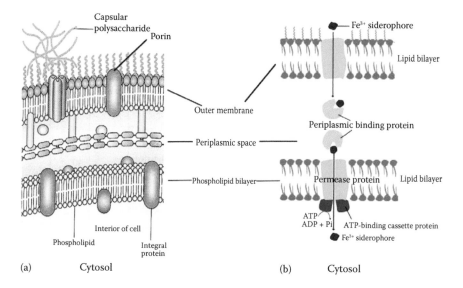

FIGURE 12.14
(a) Basic structure of Gram-negative cell wall and (b) iron siderophore transport mechanism in *E. coli*.

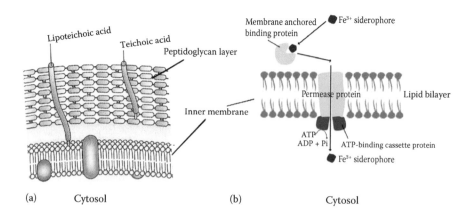

FIGURE 12.15
(a) Basic structure of Gram-positive cell wall and (b) iron transport mechanism in Gram-positive bacteria.

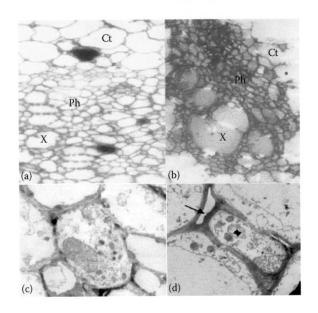

FIGURE 12.17
Light micrograph of *T. angustifolia* root shows metal deposition (dark staining) and disruption of cortex cell (b vs. a; *) and TEM micrograph shows intercellular space (→) and nucleus size reduction (◄) (d) in ST11 as compared to control (c) during metal accumulation in 60 days. Cortex (Ct), phloem (Ph) and xylem (X).

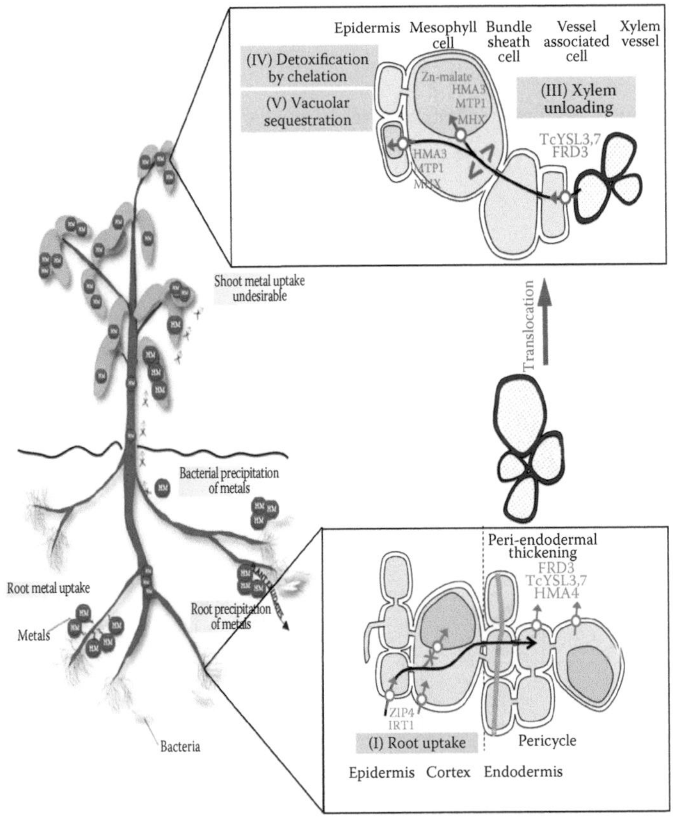

FIGURE 12.19
Major process involved in heavy metal hyperaccumulation by plants.

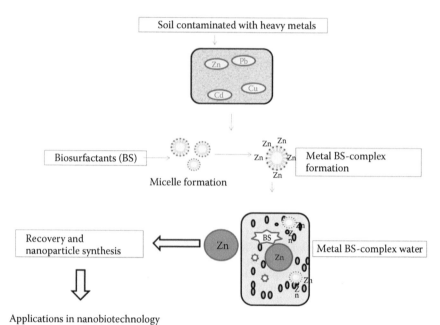

FIGURE 13.3
Biosurfactant-mediated bioremediation of heavy metals.

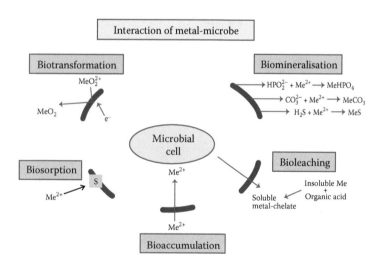

FIGURE 14.2
Interaction of metal and microbe affects bioremediation. (From Tabak, H. H. et al., *Rev Env Sci Biotech*, 4, 115–156, 2005.)

8

Biodegradation of Cellulose and Agricultural Waste Material

Nadeem Akhtar, Dinesh Goyal and Arun Goyal

CONTENTS

8.1 Introduction

Increase in global energy demand across the world has put immense pressure on utilisation of alternative energy resources, such as wind, water, sun, geothermal heat and nuclear fission as well as lignocellulosic biomass. Lignocellulosic biomass is the most abundant (worldwide 1×10^{10} million tonnes annually) plant matter that can provide alternative transportation fuel, such as bioethanol or biodiesel (Sun and Cheng 2002; Hamelinck et al. 2005; Sanchez and Cardona 2008). The prime motive in using renewable energy fuels is to reduce environmental impacts that are associated with the use of the fossil fuels (Botha and Blottnitz 2006). The key concern for the production of biofuels, such as bioethanol, biobutanol, biodiesel and biogas has gained impetus importance due to their feed and food value. Therefore, it becomes imperative to scout for alternative sources of energy that can replace conventional fossil fuels (Wan and Li 2011). Over the years, many lignocellulosic materials have been used for bioethanol production: wheat straw (Kaparaju et al. 2009); rice straw (Kyong Ko et al. 2009); sugarcane bagasse (Rabelo et al. 2008); barley and timothy grass (Naik et al. 2010); woody raw materials (Zhu et al. 2010); forest wastes, such as sawdust, wood chips and slashes and dead trees branches (Perlack et al. 2005); softwoods originating from conifers and gymnosperm (Hoadley 2000); and paper pulps (Khanna et al. 2008).

The saccharification of lignocellulosic biomass is still technically problematic because of digestibility of cellulose, which is hindered by structural and compositional factors (Mosier et al. 2005a). The primary obstacle impeding the widespread production of bioenergy from biomass feedstocks is the absence of low-cost technology for overcoming the recalcitrance of these materials mainly due to the presence of lignin (Palonen et al. 2004). Production of ethanol from lignocellulosic biomass involves hydrolysis of cellulose and hemicellulose, fermentation of sugars, separation of lignin and, finally, recovery and purification of ethanol (Lin and Tanaka 2006; Tomas-Pejo et al. 2008; Alvira et al. 2010). The most important factors to reduce the cost of ethanol production are efficient utilisation of the raw material for high productivity, greater ethanol concentration and process integration to reduce the energy demand (Galbe and Zacchi 2007; Tomas-Pejo et al. 2008). In the present scenario, pretreatment technologies are primarily targeted toward a substantial increase in the accessible surface area of cellulose for a low dosage of hydrolytic enzymes with minimum bioconversion time and inhibitory products. A large number of pretreatment techniques were studied, developed and reviewed (Carvalheiro et al. 2008; Taherzadeh and Karimi 2008; Yang and Wyman 2008; Hendriks and Zeeman 2009; Alvira et al. 2010; Geddes et al. 2011; Chaturvedi and Verma 2013) in context of lignocellulose hydrolysis. The fate of plant biomass for biofuel and bio-based value-added product generation is schematically represented in Figure 8.1.

The feasibility of a specific pretreatment process in search of promising feedstock is difficult to envision owing to the characteristics of biomasses. An effective and economical pretreatment should produce reactive cellulosic fibre for enzymatic attack, minimising the energy demand and by-products, avoiding destruction of cellulose and hemicelluloses, and possible inhibitors for hydrolytic enzymes and fermenting microorganisms (Taherzadah and Karimi 2008; Gamage et al. 2010). Several methods have been used for the pretreatment of lignocellulosic materials prior to enzymatic hydrolysis. These methods are classified into 'physical pretreatment', 'physicochemical

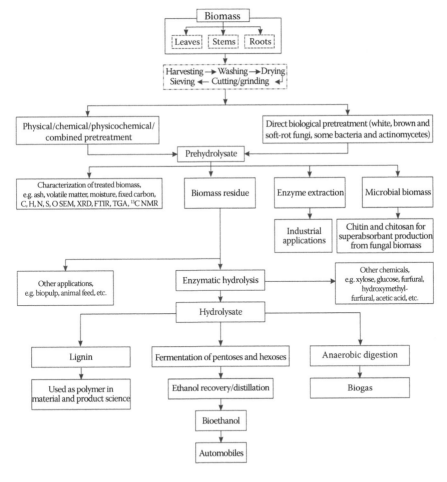

FIGURE 8.1
Schematic diagram of production of bioethanol and bio-based products from plant biomass.

pretreatment', 'chemical pretreatment' and 'biological pretreatment' (Fan et al. 1982; Wyman 1996; Berlin et al. 2006). Furthermore, combination of two or more pretreatment techniques is being used for effective digestion of biomass and recovery of selective products while minimising the formation of inhibitory compounds. Pretreatment of lignocellulosic biomass using acid (Sindhu et al. 2013), alkali (Preeti et al. 2012), ionic liquids (Weerachanchai et al. 2012), organosolvent (Sindhu et al. 2012a), organic acids (Sindhu et al. 2010), microwaves (Binod et al. 2012), ultrasound (Li et al. 2012), biologicals (Vaidya and Singh 2012) and combined pretreatment (Vani et al. 2012) has been tested to enhance the digestibility of biomass.

Pretreatment using hot water and acid is the leading strategy for degradation of cellulosic materials on industrial scales; however, this approach is expensive, slow and inefficient (Rubin 2008). In addition, the overall yield of the fermentation process decreases due to releases of inhibitors, such as weak acids, furan and phenolic compounds (Palmqvist and Hahn-Hagerdal 2000; Girio et al. 2010; Ran et al. 2014). Some of these problems could be overcome by a biological method of pretreatment involving white-, brown- and soft-rot fungi as well as bacteria and actinomycetes.

8.2 Agricultural Waste Biomass

Agricultural waste biomass is composed of cellulose $(C_6H_{10}O_5)n$, hemicellulose $(C_5H_8O_4)m$, lignin $[(C_9H_{10}O_3(OCH_3)_{0.9-1.7}]_x$, pectin, extractives, glycosylated proteins and several inorganic materials (Sjostrom 1993). The cellulose, hemicellulose and lignin contents of such biomasses fall in the range of 30%–50%, 15%–35% and 10%–20%, respectively (Petterson 1984; Mielenz 2001; Badger 2002; Knauf and Moniruzzaman 2004; Kaparaju et al. 2009; Girio et al. 2010). Cellulose and hemicelluloses are tightly linked to the lignin component through covalent and hydrogenic bonds that make the structure highly robust and resistant to any treatment (Mielenz 2001; Knauf and Moniruzzaman 2004; Edye et al. 2008).

8.2.1 Cellulose

Cellulose, the most abundant organic compound on the earth, consists of β, 1-4-polyacetal of cellobiose (4-O-β-D-glucopyranosyl-D-glucose). Cellulose is more commonly considered as a linear chain of several hundred to tens of thousands of recurring D-glucose units, linked by β(1→4) glycosidic bonds (Delmer and Amor 1995; Zhang and Lynd 2004). Cellulose, having a molecular weight of about 100,000, is the most abundant form of biologically fixed carbon and forms the structural component of a primary cell wall in green plants and algae. Lignocellulosic material consists of both crystalline and amorphous forms of cellulose, tightly bounded parallel arranged bundles of cellulosic chains formed by strong interchain hydrogen bonds, which forms a crystalline region, while this is less ordered and conspicuous in the amorphous region.

8.2.2 Hemicellulose

The second most abundant natural polymer on the earth is hemicellulose (Hendriks and Zeeman 2009; Agbor et al. 2011). Hemicellulose is the heterogenous polymer that consists of pentoses (D-xylose, D-arabinose), methyl pentoses (L-rhamnose), hexoses (D-glucose, D-mannose and D-galactose) and carboxylic acids (D-glucuronic acid, D-galacturonic acid and methyl glucuronic acid), which can be employed in the bioconversion processes for the production of ethanol and other value-added products. In hardwood, hemicelluloses are dominantly found as xylan, whereas in soft wood, glucomannan is most common (Singh et al. 2008; Hendriks and Zeeman 2009; Agbor et al. 2011). Hemicelluloses have a random, amorphous and branched structure with little resistance to hydrolysis, and they are more easily hydrolysed by acids to their monomer components (O'Dwyer 1934; Mod et al. 1981; Morohoshi 1991; Sjostrom 1993; Ademark et al. 1998). Xylan, the most common type of polysaccharide in the hemicellulose family, consists of D-xylopyranose linked together by β-1,4-linkage with less than 30,000 molecular weight and up to 200 degrees of polymerisation.

8.2.3 Lignin

Lignin, the third largest available biopolymer in nature, consists of a phenyl propane (p-coumaryl alcohol, coniferyl alcohol and sinapyl alcohol) unit linked with ester bonds that forms a complex with hemicellulose to encapsulate cellulose, making it resistant towards chemical and enzymatic hydrolysis (Hendricks and Zeeman 2009; Agbor et al. 2011). The extent of biodelignification depends upon the ratio of syringyl to guaicyl units present in different types of wood (Ruiz-Dueñas and Martínez 2009). Lignin, generally having a molecular weight less than 20,000, is found more in soft wood (pine, balsam, spruce, tamarack, fir) than hardwood (oak, walnut, maple, poplar, birch). Lignin-enriched biomass is more resistant for the depolymerisation of holocellulose (cellulose and hemicellulose) to produce fermentable sugars. In addition, during the degradation process, it may form furan (furfural and hydroxymethyl-furfural) compounds that could inhibit fermentation (Zaldivar et al. 1999; Girio et al. 2010; Ran et al. 2014).

8.2.4 Pectin

Pectin, a heteropolysaccharide of the plant cell wall, contains a backbone of α-1,4-linked D-galacturonic acid that can be methyl-esterified or substituted with acetyl groups. The polymers contain two regions: The smooth region of pectin contains D-galacturonic acids, whereas the hairy region has the back-bone of D-galacturonic acid residues, which is interrupted by α-1,2-linked L-rhamnose residues (de Souza 2013). Rhamnose residues in the hairy region may have long side chains of L-arabinose and D-galactose residues (de Vries 2003). Degrading of pectin backbone mainly requires polysaccharide lyases (PL). A large part of the fungal glycoside hydrolases involved in the degra-dation of the pectin backbone belongs to GH family 28 (Martens-Uzunova and Schaap 2009). Generally, endo- and exo-polygalacturonases of GH28 cleave the α-1,4-glycosidic bonds between the α-galacturonic acids. In addi-tion, hydrolysis of the pectin backbone also requires enzymes from other GH families: α-rhamnosidases (GH78), unsaturated glucuronyl hydrolases (GH88) and unsaturated rhamnogalacturonan hydrolases (GH105) (Brink and de Vries 2011). *Trichoderma* spp. are well known for effective degradation of cellulose (Martinez et al. 2008; Kubicek et al. 2011), and *Aspergillus* spp. produce many enzymes to degrade pectin (Martens-Uzunova and Schaap 2009).

8.3 Biological Pretreatment

Biological pretreatment mainly involves the use of white-, brown- and soft-rot fungi and is employed to release cellulose from the hemicellulose and lignin matrix. Treatment degrades lignin and hemicellulose but leaves the cellulose intact (Shi et al. 2008; Kumar et al. 2009; Sanchez et al. 2009). The biological pre-treatment requires low energy and mild operation conditions. Nevertheless, the rate of biological hydrolysis is usually very low, so this pretreatment requires a long residence time (Sun and Cheng 2002; Tengerdy and Szakacs 2003; Cardona and Sanchez 2007). Degradation of lignocellulosic biomass in terms of cellulose, hemicellulose and lignin loss is depicted in Table 8.1.

8.3.1 White-Rot Fungi

Lignolytic basidiomycetes, a physiologically distinct group of saprophytic fungi, causes white rot in wood, commonly called white-rot fungi. Over the years, white-rot fungi have been used in mineralisation of lignin (Vicuna 2000; Martinez 2002; Wong 2009; Xu et al. 2010; Suzuki et al. 2012; Tuyen et al. 2012). *Phanerochaete chrysosporium* (Shi et al. 2008), *Pycnoporus cinnabarinus* (Meza et al. 2006), *Phlebia* spp. (Arora and Sharma 2009), *Echinodontium taxodii* (Yu et al. 2010), *Irpex lacteus* (Xu et al. 2009; Dias et al. 2010; Yu et al. 2010a) and *Pycnoporus sanguineus* (Lu et al. 2010) have shown to have high lignin degradation speci-ficity, and significant levels of carbohydrate degradation have been reported for *Ceriporiopsis subvermispora* (Wan and Li 2011); *Phlebia brevispora, P. floridensis* and *P. radiata* (Sharma and Arora 2011); *Echinodontium taxodii* (Yu et al. 2010b);

TABLE 8.1

Effect of Biological Treatment on Different Types of Lignocellulosic Biomass

Microorganism	Biomass	Cellulose Loss (%)	Hemicellulose Loss (%)	Lignin Loss (%)	References
Fungi					
Dichomitus squalens	*Fagus crenata* wood chips	8.0	–	21.7	Itoh et al. (2003)
Ceriporiopsis subvermispora	*Fagus crenata* wood chips	5.0	–	6.7	Itoh et al. (2003)
	Sugarcane bagasse	9.1	21.5	38.4	Sasaki et al. (2011)
	Corn stover	<5	27	39.2	Wan and Li (2010)
Pleurotus ostreatus	*Fagus crenata* wood chips	4.1	–	10.8	Itoh et al. (2003)
Coriolus versicolor	*Fagus crenata* wood chips	13.2	–	19.5	Itoh et al. (2003)
	Rice straw	17	48	41	Taniguchi et al. (2005)
Phanerochaete chrysosporium	Rice straw	49	–	21	Taniguchi et al. (2005)
Trametes versicolor	Rice straw, corn straw	58	–	37	Taniguchi et al. (2005)
	Corn straw	47.5	–	54.6	Yu et al. (2010)
	Pinus radiata wood chips	16.8	29	31	Ferraz et al. (2001)
Echinodontium taxodii	*Salix babylonica* wood	27	51	45	Yu et al. (2009)
	Cunninghamia lanceolata wood	12	31	40	Yu et al. (2009)
	Corn straw	15.8	–	42.2	Yu et al. (2010b)
Ganoderma lucidum	Corn straw	39	–	32.7	Yu et al. (2010b)
Punctularia atropurpurascens	*Pinus radiata* wood chips	0	5	9	Ferraz et al. (2001)
Poria medula-panis	*Pinus radiata* wood chips	10	15	18	Ferraz et al. (2001)
Merulius tremellosus	*Pinus radiata* wood chips	7	18	28	Ferraz et al. (2001)
Ganoderma australe	*Pinus radiata* wood chips	26	31	35	Ferraz et al. (2001)
Ceriporia lacerata	*Pinus densiflora* wood chips	–	8	13	Lee et al. (2007)

(Continued)

TABLE 8.1 (CONTINUED)

Effect of Biological Treatment on Different Types of Lignocellulosic Biomass

Microorganism	Biomass	Cellulose Loss (%)	Hemicellulose Loss (%)	Lignin Loss (%)	References
Stereum hirsutum	*Pinus densiflora* wood chips	–	7.8	14	Lee et al. (2007)
Polyporus brumalis	*Pinus densiflora* wood chips	–	10.6	12	Lee et al. (2007)
Trametes hirsuta YJ9	Corn stover	34.06	77.84	71.99	Sun et al. (2011)
Gloephyllum trabeum KU-41	Corn stover	17	43	–	Gao et al. (2012)
Gloephylum trabeum	*Pinus radiata* wood chips	–	31	–	Monrroy et al. (2011)
	Eucalyptus globulus wood chips	–	24	–	Monrroy et al. (2011)
Laetoporus sulphureus	*Eucalyptus globulus* wood chips	–	6	–	Monrroy et al. (2011)
	Pinus radiata wood chips	21	47	3	Fissore et al. (2010)
Laetoporus sulphureus	*Pinus radiata* wood chips	–	7.5	–	Monrroy et al. (2011)
Wolfiporia cocos	*Pinus radiata* wood chips	6	23	2.6	Ferraz et al. (2001)
Laetiporus sulphureus	*Pinus radiata* wood chips	6	14.6	5.3	Ferraz et al. (2001)
Coniophora puteana	*Pinus sylvestris* wood blocks	17	–	59.7	Irbe et al. (2010)
Bacteria					
Bacillus sp.	Rice straw	3.2	16	20	Chang et al. (2014)
Cellulomonas cartea	Sugarcane trash	25.4	–	5.5	Singh et al. (2008)
Cellulomonas uda	Sugarcane trash	21.8	–	5.5	Singh et al. (2008)
Bacillus macerans	Sugarcane trash	30.4	–	5.5	Singh et al. (2008)
Zymomonas mobilis	Sugarcane trash	26.8	–	8	Singh et al. (2008)
Paenibacillus sp. AY952466	Kraft lignin	–	–	43	Chandra et al. (2007)
Aneurinibacillus aneurinilyticus	Kraft lignin	–	–	56	Chandra et al. (2007)
Bacillus sp. AY952465	Kraft lignin	–	–	37	Chandra et al. (2007)

Euc-1 (Dias et al. 2010); *Gonoderma* sp. (Bourbonnais et al. 1997; Haddadin et al. 2002; Tripathi et al. 2008; Yu et al. 2010b); *Oxysporus* sp. (Haddadin et al. 2002); *Trametes versicolor* (Yu et al. 2010b); *Pleurotus sajor-caju* (Kannan et al. 1990); and *Trichoderma reesei* (Singh et al. 2008) with 40%–60% reduction in lignin content. Biodelignification has a direct effect on fermentable sugar for ethanol production, which significantly improves the yield of biofuel production. Biodelignification has been used prior to chemical pretreatments and showed delignification up to 80% (Yu et al. 2010a,b).

8.3.2 Brown-Rot Fungi

Brown-rot fungi efficiently degrade cellulose and hemicellulose as compared to lignin, resulting in brown-rotted wood due to incomplete degradation of lignin, hence named 'brown-rot fungi'. Brown-rot fungi preferentially degrade the cellulose and hemicellulose fraction of wood but do not oxidise lignin (Hastrup et al. 2012; Schilling et al. 2012). Many of them, such as *Serpula lacrymans*, *Coniophora puteana*, *Meruliporia incrassata*, *Laetoporeus sulphureus* and *Gleophyllum trabeum*, have been used in many biodegradation studies (Rasmussen et al. 2010; Monrroy et al. 2011).

8.3.3 Soft-Rot Fungi

There are two types of soft-rot fungi: type I, consisting of biconical or cylindrical cavities that are formed within secondary walls, and type II, which refers to an erosion form of degradation (Blanchette 2000). The most efficient fungus of the type II group is *Daldinia concentric*, which primarily affects hardwood (Narayanswamy et al. 2013) and is reported to achieve 53% loss in weight of birch wood within two months (Nilsson et al. 1989). *Paecilomyces* sp. (Kluczek-Turpeinen et al. 2003) and *Cadophora* spp. (Chandel et al. 2013) are also involved in rapid biodelignification of biomass. In the early stages of classification, wood-rotting fungi *Xylariaceous ascomycetes* from genera *Daldinia*, *Hypoxylon* and *Xylaria* have often been classified under white-rot fungi, but these fungi are now considered as members of soft-rot fungi as they cause type II soft rot in many woods.

8.3.4 Bacteria and Actinomycetes

Bacteria are of less use than white-, brown- and soft-rot fungi in the biodelignification of lignocellulose materials as they lack or are a poor producer of lignolytic enzymes. Besides *Bacillus* sp. AS3 (Akhtar et al. 2012), *Bacillus circulans* and *Sphingomonas paucimobilis* (Kurakake et al. 2007), *Cellulomonas* and *Zymomonas* spp. have been been reported for biodegradation of lignocellulosic biomass (Singh et al. 2008). Bacterial delignification may lead up to 50% lignin loss, which is somewhat similar to fungi-mediated delignification. Bacteria delignification is mainly mediated by extracellular xylanases, although a synergistic effect has been observed by addition of MnP, pectinase or α-L-arabinofuranosidase (Bezalel et al. 1993; Kaur et al. 2010). However, high lignin removal might be accompanied with enhanced degradation of cellulose (31%–51%) as observed for *Bacillus macerans, Cellulomonas cartae* and *C. uda* (Singh et al. 2008). One strategy to overcome this issue is the use of cellulase-free extracts (Kaur et al. 2010) or purified enzymes (Bezalel et al. 1993) that allows up to 20% delignification levels within shorter incubation times than those observed with whole microorganism.

8.4 Enzyme Involved in Biomineralisation of Lignin

Lignin is the most complex insoluble polymer within the cell wall and is known to increase the strength as well as the recalcitrance of the plant cell wall. Biomineralisation of plant biomass is often complicated due to the large polymer nature of lignin, the oxidative environment preference of lignolytic enzymes and the irregular stereochemistry of lignin. Lignin stereochemistry makes use of less specific enzymes than the hydrolytic enzymes required for cellulose and hemicellulose degradation (Isroi et al. 2011). The most well-characterised enzymes involved in the degradation of lignin are lignin peroxidase (LiP, EC 1.11.1.14), laccase (Lac, benzenediol:oxygen oxidoreductase, EC 1.10.3.2), manganese peroxidase (MnP, Mn-dependent peroxidase, EC 1.11.1.13) and versatile peroxidase (VP, hybrid peroxidase, EC 1.11.1.16). Furthermore, some accessory enzymes, such as glyoxal oxidase (GLOX, EC 1.2.3.5) and aryl alcohol oxidase (AOO, EC 1.1.3.7), are involved in the generation of hydrogen peroxide.

8.4.1 Lignin Peroxidase (LiP)

Extracellular LiP was first reported by Tien and Kirk (1983) from *P. chrysosporium*, which was spectrally characterised by Anderson et al. (1985). The structural properties were found to be similar to those of hemoproteins (Martinez 2002). Lignin peroxidases (LiPs) generally catalyse a variety of oxidative reactions using H_2O_2 as a cofactor to degrade lignin and/or lignin-like compounds. LiPs oxidise nonphenolic units of lignin by removing an electron and creating cation radicals for chemical decomposition (Hattaka 2001). Very few fungi are known to produce extracellular LiPs (Have and Teunissen 2001); however, *P. chrysosporium, T. versicolor, Bjerkhandera* sp. and *T. cervina* are good producers of LiPs (Ghosh and Ghose 2003).

8.4.2 Manganese Peroxidase (MnP)

Manganese peroxidases (MnPs) also require H_2O_2 as an oxidant in the Mn-dependent catalysing reaction in which Mn^{2+} is converted to Mn^{3+}

(Narayanswamy et al. 2013). In addition to demethylation and depolymerisation of lignin, MnP can play an important role in the oxidation of phenolic and sulphur compounds and unsaturated fatty acids as well. *P. chrysosporium, Pleurotus ostreatus, Trametes* sp. and species that belong to Meruliaeiae, Coriolaceae and Polyporaceae are known to produce MnP (Isroi et al. 2011). *Ceriporiopsis subvermispora* efficiently degrades nonphenolic lignin structures using manganese peroxide (Costa et al. 2005). LiP is involved in direct oxidation of aromatic compounds of lignin, whereas MnP takes the help of manganese ions to oxidise the phenolic part of lignin. This gives a clue to enhance the activity of MnPs for faster bio-oxidation of lignin as Mn works as a cofactor or substrate mediator. However, *Bjerkandera* sp. BOS55 and *P. eryngii* are able to oxidise Mn^{2+} to Mn^{3+} and aromatic compounds (Have and Teunissen 2001).

8.4.3 Laccase (Lac)

Laccase is a copper-containing protein that belongs to the family oxidase. It is mainly found in bacteria, fungi and some plants. The substrate for laccase includes syringaldazine [N-N'-bis-(3,5-dimethoxy-4-hydroxybenzylidene) hydrazine], DMP (2,6-dimethoxyphenol), ABTS (2,2'-azino-bis-3-ehtylbenzothiazoline-6-sulfonicacid) and guaiacol (2-methoxyphenol). Syringaldazine is the main substrate for Lac (Baldrian 2006; Sinsabaugh 2009). Under certain conditions, it is also involved in the oxidation of 1-HBT (1-hydrobenzotriazole) (Li et al. 1998) and violuric acid (Xu et al. 2000) along with some natural mediators, such as 4-hydroxybenzoic acid, 4-hydroxybenzyl alcohol (Johannes and Majcherczyk 2000) and 3-hydroxyanthranilate (Eggert et al. 1996). High laccase titre is produced by *Pycoporous cinnabarinus*, which degrades lignin exclusively (Geng and Li 2002). Ethanol vapours have been employed to intensify the lignin-degrading capacity of this fungus, which has been used during the delignification of sugarcane bagasse (Meza et al. 2006). Out of four copper atoms present in Lac, three distinct types of copper atoms are found with different roles in the oxidation of substrate (Have and Teunissen 2001). Type I copper is involved in the reaction with the substrate, whereas type II copper and two copies of type III copper are involved in the binding, reduction of O_2 and storage of electrons originating from the reducing substrates. Almost all white-rot fungi are capable of producing a sufficient amount of Lac, which can be ameliorated by the presence of copper. Isroi et al. (2011) showed that the presence of aromatic compounds, such as VA and 2-5 xylidine, showed induction of Lac. Laccase and lignin peroxidase have been secreted by *Streptomyces cinnamomensis* (Jing and Wang 2012) and *Streptomyces lavendulae* (Jing 2010). Extracellular degradative enzymes, such as cellulase, laccase, peroxidase, xylanase and pectinase, have been known to be produced by *S. griseorubens* C-5 and were found to be effective in the degradation of rice straw (Xu and Yang 2010). Niladevi et al. (2007) reported secretion of all three types of lignin-degrading enzymes: lignin peroxidase, manganese peroxidase and laccase from *S. psammoticus*. Another study by Kunamneni et al. (2007) also found *Azospirillum lipoferum*, *Bacillus subtilis*, *Streptomyces lavendulae*, *Streptomyces cyaneus* and *Marinomonas mediterranea* to be producers of laccase.

8.4.4 Versatile Peroxidase (VP)

Versatile peroxidase (VP) acts as a licentious enzyme having LiP-like activities, such as oxidation of veratrylglycerol β-guaiacyl ether and phenolic compounds. VP also oxidises Mn^{2+} to Mn^{3+}, VA to veratraldehyde and p-dimethoxybenzene to p-benzoquinone (Isroi et al. 2011). VPs have been demonstrated in varied *Pleurotus* sp. (Camarero et al. 1999) and *Bjerkandera adusta* (Ayala Aceves et al. 2001).

8.4.5 Peroxide-Producing Enzymes

To support the biomineralisation of lignin, most of the white-rot fungi produce some accessory enzymes for H_2O_2 production, and glyoxal oxidase (GLOX) is one of them. Prior to the discovery of LiPs, it was believed that the generation of H_2O_2 and other easily diffusible activated oxygen species, such as hydroxyl radicals ($OH^{\bullet}$), superoxide anion radicals ($O^{2-\bullet}$) and singlet oxygen ($^1O_2$), might be responsible for fungal decay of lignin and lignocelluloses. Therefore, to elicit a ligninolytic oxidative reaction of LiPs and MnPs, white-rot fungi have to produce some accessory enzymes for H_2O_2 production (Narayanswamy et al. 2013). GLOX and AAO are found in some white-rot fungi, including *P. chrysosporium*. Lignin oxidation products may undergo a reduction reaction by GLOX (Narayanswamy et al. 2013); for example, arylglycerol β-aryl ether structure of lignin is oxidised by LiPs to glycolaldehyde, and this cleavage product acts as a substrate for GLOX (Hammel et al. 1994).

8.4.6 Cellobiose Dehydrogenase (CDH) in Lignin Breakdown

Cellobiose dehydrogenase (CDH; EC 1.1.99.18), a type of oxidoreductase, is produced mainly by wood-degrading fungi. *Ceriporiopsis subvermispora*, white-rot fungi, when grown on a cellulose and yeast extract–based liquid medium produced copious amounts of CDH in culture (Harreither et al. 2009). This enzyme has been isolated from many white-rot fungi, such as *P. chrysosporium*, *T. versicolor*, *P. cinnabarinus* and *Schizophyllum commune*; the brown-rot fungus *Coneophora puteana*; and the soft-rot fungi *Humicola insolens* and *Myceliophtore thermophila* (Henriksson et al. 2000). However, no significant activity of CDH has been reported to date by *C. subvermispora*, a potential biomineraliser (Harreither et al. 2009). It was suggested that CDH does not play a key role in ligninolysis (Dumonceauxa et al. 2001); the deficient CDH-mutant can still degrade or modify the lignin in a similar manner as the wild type but does not degrade cellulose (Phillips et al. 2011).

8.4.7 Low Molecular Weight Compounds or Mediators Involved in Lignin Degradation

Various low molecular weight compounds, such as veratyl alcohol (VA), manganese, oxalate and 2-chloro-1,4-dimethoxybenzene, play a role of mediator or cofactor of lignolytic enzyme systems of white-rot fungi. VA production is induced by nitrogen limitation in *P. chrysosporium*, whereas nitrogen element does not have any significant effect in *Bjerkandera* sp. VA biosynthesis (Mester et al. 1995). VA protects LiPs against H_2O_2-mediated VA and acts as a stabiliser *in vivo* and protects LiPs against H_2O_2-mediated inactivation reaction (rate-limiting step) in the LiPs catalytic cycle (Hammel et al. 1994). It has been recommended to use Mn^{2+} to enhance the bio-oxidation rate in biological pretreatment of lignocellulosic materials. However, some reports (Hames et al. 1998; Have and Teunissen 2001) showed an inhibitory effect of Mn^{2+} on the activity and production of LiPs. LiPs and MnPs are able to decompose oxalate in the presence of VA or Mn^{2+} (Hames et al. 1998; Have and Teunissen 2001), leading to formation of reactive oxygen, which participates in oxidation of lignin (Narayanswamy et al. 2013). This suggests that oxalate may be regarded as a passive sink for H_2O_2 production (Have and Teunissen 2001). A substrate for LiPs is 2-Chloro-1,4-dimethoxybenzene (2-Cl-1,4-DMB), indicating a possible active function in the wood decomposition process (Narayanswamy et al. 2013). It may also act as a redox mediator like VA (Teunissen et al. 1998; Have and Teunissen 2001) and plays an important role in protection of LiPs.

8.5 Cellulose Degradation

Cellulose, a polysaccharide consisting of linear β-1,4-linked D-glucopyranose chains, requires cellulases for its degradation. Cellulases are inducible enzymes, which are synthesised by large numbers of microorganisms, either cell-bound or extracellular during their growth on cellulosic materials (Lee 2001). Cellulases, historically, have been divided into three major groups: endoglucanase (EC 3.2.1.4), exoglucanase or cellobiohydrolase (EC 3.2.1.91) and β-glucosidase (EC 3.2.1.21), and the synergistic actions of these enzymes is a widely accepted mechanism for cellulose hydrolysis (Bhat 2000; Lynd et al. 2002; Bayer et al. 2004; Zhang and Lynd 2004; Zhang 2008). These enzymes can either be free in aerobic microorganisms or grouped in a multicomponent enzyme complex, cellulosome, in anaerobic cellulolytic bacteria (Lynd et al. 2002). Cellulase refers to a class of enzymes produced chiefly by fungi, bacteria and protozoans (Immanuel 2006), representing major costs (Kanokphorn et al. 2011) that act as biocatalyses for the conversion of various cellulosic substrates to fermentable sugars.

Cellulose has attracted attention as a renewable resource that can be converted into bioenergy and value-added bio-based products (Shahzadi et al. 2014). Enormous amounts of agricultural, industrial and municipal cellulose wastes have been accumulating or used inefficiently due to the high cost of their utilisation processes (Kim et al. 2003; Lee et al. 2008). Therefore, it has become of considerable economic interest to develop processes for effective treatment and utilisation of cellulosic wastes as a cheap carbon source. These resources are composed of leaves, stems and stalks from sources, such as corn fibre, corn stover, sugarcane bagasse, rice, rice hulls, woody crops and forest residues. Besides, there are multiple sources of lignocellulosic waste from industrial and agricultural processes, for example, citrus peel waste, coconut biomass, sawdust, paper pulp, industrial waste, municipal cellulosic solid waste and paper mill sludge (Sadhu et al. 2013).

In the last two decades, research has been aimed at developing new technologies and microbial strains to reduce the cost of cellulase production and improving the bioconversion of cellulose, particularly for the biofuel industry. Cellulases provide a key opportunity for achieving the tremendous benefits of biomass utilisation (Wen et al. 2005). But currently, two significant points of these enzyme-based bioconversion technologies are reaction conditions and the production cost of the related enzyme system (Lee et al. 2008). Therefore, much research has been aimed at obtaining new microorganisms producing celluloytic enzymes with higher specific activities and greater efficiency (Pattana et al. 2000; Subramaniyan and Prema 2000; Lee and Koo 2001; Johnvesly et al. 2002). Cellulase yields appear to depend on a complex relationship involving a variety of factors, such as inoculum size, pH, temperature, presence of inducers, medium additives, aeration and growth time (Robson and Chambliss 1984). Currently, China, India, South Korea and Taiwan have recently emerged as industrialised manufacturing centres with strong national research and development programs, and they will play a larger role in the world market of industrial cellulase production (Acharya and Chaudhary 2012).

8.6 Hemicellulose Degradation

Hemicellulose is a complex matrix composed of different residues, such as xylan, xyloglucan and mannan. The complexity of hemicellulose requires a concerted action of endo-enzymes cleaving internally the main chain, exo-enzymes releasing monomeric sugars and accessory enzymes cleaving the side chains of the polymers or associated oligosaccharides (de Souza 2013), leading to the release of various monosaccharides, disaccharides and acetic acid, depending upon the type of hemicellulose. Xylan, the main component of hemicellulose, is degraded through the action of β-1,4-endoxylanase and is responsible for the breakdown of xylan backbone to oligosaccharides, which is further cleaved to xylose by β-1,4-xylosidase. Fungal β-1,4-endoxylanases are classified as GH10 or GH11 (Polizeli et al. 2005), differing from each other in substrate specificity (Biely et al. 1997).

Family GH10 endoxylanases generally have a broad range of substrate specificity as compared to endoxylanases of family GH11 (Brink and de Vries et al. 2011). Degradation of xylan backbones, xylo-oligosaccharides and linear chains of 1,4-linked D-xylose residues are prime motives of GH10 endoxylanases. Thus, for complete degradation of substituted xylans, GH10 endoxylanases are necessary (Pollet et al. 2010). β-xylosidases are highly specific for small unsubstituted xylose oligosaccharides, and they are important for the complete degradation of xylan (de Souza 2013). In addition, β-xylosidases having transxylosylation activity have been known for synthesis of oligosaccharides, suggesting a possible application for such enzymes (Shinoyama et al. 1991; Sulistyo et al. 1995). Mannans, also referred to as galacto(gluco)mannans, consist of a backbone of β-1,4-linked D-mannose (mannans) and D-glucose (gluco-mannans) residues with D-galactose side chains (de Souza 2013). The degradation of this type of hemicellulose is performed by the action of β-endomannanases (β-mannanases) and β-mannosidases, commonly expressed by Aspergilli (de Vries 2001). In addition, hemicellulose biodegradation needs accessory enzymes, such as xylan esterases, ferulic and p-coumaric esterases, α-l-arabinofuranosidases and α-4-O-methyl glucuronosidases, acting synergistically to efficiently hydrolyse wood xylans and mannans (Perez et al. 2002).

References

Acharya, S., Chaudhary, A. 2012. Bioprospecting thermophiles for cellulose production: A review. *Brazilian Journal of Microbiology* 47: 844–856.

Ademark, P., Varga, A., Medve, J. et al. 1998. Softwood hemicellulose-degrading enzymes from Aspergillus niger: Purification and properties of a β-mannanase. *Journal of Biotechnology* 63: 199–210.

Agbor, V. B., Cicek, N., Sparling, R. et al. 2011. Biomass pretreatment: Fundamentals toward application. *Biotechnology Advances* 29: 675–685.

Akhtar, N., Sharma, A., Deka, D. et al. 2012. Characterization of cellulase produced of *Bacillus* sp. for effective degradation of leaf litter biomass. *Environmental Progress and Sustainable Energy* 32: 1195–1201.

Alvira, P., Tomas-Pejo, E., Ballesteros, M. et al. 2010. Pretreatment technologies for an efficient bioethanol production process based on enzymatic hydrolysis: A review, *Bioresource Technology* 101: 4851–4861.

Anderson, L. A., Rengnathan, V., Chiu, A. A. 1985. Spectral characterization of diarylpropane oxygenase, a novel peroxide-dependent lignin degrading heme enzyme. *Journal of Biological Chemistry* 260: 6080–6087.

Arora, D. S., Sharma, R. K. 2009. Comparative ligninolytic potential of Phlebia species and their role in improvement of in vitro digestibility of wheat straw. *Journal of Animal and Feed Science* 18: 151–161.

Ayala Aceves, M., Baratto, M. C., Basosi, R. et al. 2001. Spectroscopic characterization of a manganese-lignin peroxidase hybrid isozyme produced by Bjerkandera adusta in the absence of manganese: Evidence of a protein centred radical by hydrogen peroxide. *Journal of Molecular Catalysis B: Enzymatic* 16: 159–167.

Badger, P. C. 2002. Ethanol from cellulose: A general review. In: Janick, J., Whipkey, A. (Eds.), *Trends in new crops and new uses.* ASHS Press, Alexandria, VA, 17–21.

Baldrian, P. 2006. Fungal laccases: Occurrence and properties. *FEMS Microbiology Letters* 30: 215–242.

Bayer, E. A., Belaich, J. P., Shoham, Y. et al. 2004. The cellulosomes: Multienzyme machines for degradation of plant cell wall polysaccharides, *Annual Reviews of Microbiology*, 58: 521–554.

Berlin, A., Balakshin, M., Gilkes, N. et al. 2006. Inhibition of cellulase, xylanase and β-glucosidase activities by softwood lignin preparations. *Journal of Biotechnology* 125: 198–209.

Bezalel, L., Shoham, Y., Rosenberg, E. 1993. Characterization and delignification activity of a thermostable α-L-arabinofuranosidase from Bacillus stearother-mophilus. *Applied Microbiology and Biotechnology* 40: 57–62.

Bhat, M. K. 2000. Cellulases and related enzymes in biotechnology. *Biotechnology Advances*. 18: 355–383.

Biely, P., Vrsanska, M., Tenkanen, M. 1997. Endo-beta-1,4-xylanase families: Differences in catalytic properties. *Journal of Biotechnology* 57: 151–66.

Binod, P., Satyanagalakshmi, K., Sindhu, R. et al. 2012. Short duration microwave assisted pretreatment enhances the enzymatic saccharification and fermentable sugar yield from sugarcane bagasse. *Renewable Energy* 37: 109–116.

Blanchette, R. A. 2000. A review of microbial deterioration found in archaeological wood from different environments. *International Journal of Biodeterioration and Biodegradation* 46: 189–204.

Botha, T., Blottnitz, H. V. 2006. A comparison of the environmental benefits of bagasse derived electricity and fuel ethanol on a life-cycle basis. *Energy Policy* 34: 2654–2661.

Bourbonnais, R., Paice, M. G., Freiermuth, B. et al. 1997. Reactivities of various mediators and laccases with kraft pulp and lignin model compounds. *Applied and Environmental Microbiology* 63: 4627–4632.

Brink, J. V. D., de Vries, R. P. 2011. Fungal enzyme sets for plant polysaccharide degradation. *Applied Microbiology and Biotechnology* 91(6): 1477–1492.

Camarero, S., Sarkar, S., Ruiz-Duenas, F. J. et al. 1999. Description of a versatile peroxidase involved in the natural degradation of lignin that has both manganese peroxidase and lignin peroxidase substrate interaction sites. *Journal of Biological Chemistry* 274: 10324–10330.

Cardona, C. A., Sanchez, O. J. 2007. Fuel ethanol production: Process design trends and integration opportunities. *Bioresource Technology* 98: 2415–2457.

Carvalheiro, F., Duarte, L. C., Gírio, F. M. 2008. Hemicellulose biorefineries: A review on biomass pretreatments. *Journal of Scientific and Industrial Research* 67: 849–864.

Chandel, A. K., Goncalves, B. C. M., Strap, J. L. et al. 2013. Biodelignification of lignocellulose substrates: An intrinsic and sustainable pretreatment strategy for clean energy production. *Critical Reviews in Biotechnology* 1–13.

Chandra, R., Raj, A., Purohit, H. J. et al. 2007. Characterisation and optimisation of three potential aerobic bacterial strains for kraft lignin degradation from pulp paper waste. *Chemosphere* 67: 839–846.

Chang, Y. C., Choi, D., Takamizawa, K. et al. 2014. Isolation of *Bacillus* sp. Strains capable of decomposing alkali lignin and their application in combination with lactic acid bacteria for enhancing cellulase performance. *Bioresource Technology* 152: 429–436.

Chaturvedi, V., Verma, P. 2013. An overview of key pretreatment processes employed for bioconversion of lignocellulosic biomass into biofuels and value added products. *3Biotech* 3: 415–431.

Costa, S. M., Goncalves, A. R., Esposito, E. 2005. Ceriporiopsis subvermispora used in delignification of sugarcane bagasse prior to soda/anthraquinone pulping. *Applied Biochemistry and Biotechnology* 1–3: 695–706.

de Souza, W. R. 2013. Microbial Degradation of Lignocellulosic Biomass. Sustainable Degradation of Lignocellulosic Biomass-Techniques, Applications and Commercialization, INTECH, 207–247.

de Vries, R. P. 2003. Regulation of Aspergillus genes encoding plant cell wall polysaccharide-degrading enzymes; relevance for industrial production. *Applied Microbiology and Biotechnology* 61: 10–20.

de Vries, R. V. J. 2001. Aspergillus enzymes involved in degradation of plant cell wall polysaccharides. *Microbiology and Molecular Biology Reviews* 65: 497–522.

Delmer, D. P., Amor, Y. 1995. Cellulose biosynthesis. *The Plant Cell* 7: 987–1000.

Dias, A. A., Freitas, G.S., Marques, G. S. 2010. Enzymatic saccharification of biologically pretreated wheat straw with white-rot fungi. *Bioresource Technology* 101: 6045–6050.

Dumonceauxa, T., Bartholomewa, K., Valeanua, L. et al. 2001. Cellobiose dehydrogenase is essential for wood invasion and nonessential for kraft pulp delignification by Trametes versicolor. *Enzyme and Microbial Technology* 29: 478–489.

Edye, L. A., Doherty, W. O. S. 2008. Fractionation of a lignocellulosic material, 25, PCT International Application.

Eggert, C., Temp, U., Dean, J. F. D. et al. 1996. A fungal metabolite mediates degradation of non-lignin structures and synthetic lignin. *FEBS Letters* 391: 144–148.

Fan, L., Lee, Y., Gharpuray, M. 1982. The nature of lignocellulosics and their pretreatments for enzymatic hydrolysis. *Advances in Biochemical Engineering and Biotechnology* 23: 158–183.

Ferraz, A., Rodriguez, J., Freer, J. et al. 2001. Biodegradation of Pinus radiata softwood by white and brown-rot fungi. *World Journal of Microbiology and Biotechnology* 17: 31–34.

Fissore, A., Carrasco, L., Reyes, P. et al. 2010. Evaluation of a combined brown rot decay-chemical delignification process as a pretreatment for bioethanol production from Pinus radiata wood chips. *Journal of Industrial Microbiology and Biotechnology* 37: 893–900.

Galbe, M., Zacchi, G. 2007. Pretreatment of lignocellulosic materials for efficient bioethanol production. *Advances in Biochemical Engineering/Biotechnology* 108: 41–65.

Gamage, J., Howard, L., Zisheng, Z. 2010. Bioethanol production from lignocellulosic biomass. *Journal of Biobased Material and Bioenergy* 4: 3–11.

Gao, Z., Mori, T., Kondo, R. 2012. The pretreatment of corn stover with Gloeophyllum trabeum KU-41 for enzymatic hydrolysis. *Biotechnology for Biofuels* 5: 1–11.

Geddes, C. C., Nieves, I. U., Ingram, L. O. 2011. Advances in ethanol production. *Current Opinions in Biotechnology* 22: 312–319.

Geng, X., Li, K. 2002. Degradation of non-phenolic lignin by the white-rot fungus Pycnoporus cinnabarinus. *Applied Microbiology and Biotechnology* 60: 342–346.

Ghosh, P., and Ghose, T. K. 2003. Bioethanol in India: Recent past and emerging future. In: Scheper, T. (Ed.), *Advanced biochemical engineering/biotechnology, biotechnology in India II*, Springer, NewYork, 20: 1–27.

Girio, F. M., Fonseca, C., Carvalheiro, F. et al. 2010. Hemicelluloses for fuel ethanol: A review. *Bioresource Technology* 101: 4775–800.

Haddadin, M. S., Al-Natour, R., Al-Qsous, S. et al. 2002. Bio-degradation of lignin in olive pomace by freshly-isolated species of Basidiomycete. *Bioresource Technology* 82: 131–137.

Hamelinck, C. N., Hooijdonk, G. V., Faaij, A. P. C. 2005. Ethanol from lignocellulosic biomass: Techno-economic performance in short, middle and long-term. *Biomass and Bioenergy* 28: 384–410.

Hames, B. R., Kurek, B., Pollet, B. et al. 1998. Interaction between MnO_2 and oxalate: Formation of a natural and abiotic lignin oxidizing system. *Journal of Agricultural and Food Chemistry* 46: 5362–5367.

Hammel, K. E., Mozuch, M. D., Jensen, K. A. et al. 1994. H_2O_2 recycling during oxidation of the arylglycerol β-aryl ether lignin structure by lignin peroxidase and glyoxal oxidase. *Biochemistry* 33: 13349–13354.

Harreither, W., Sygmund, C., Dunhofen, E. et al. 2009. Cellobiose dehydrogenase from the Ligninolytic Basidiomycete Ceriporiopsis subvermispora. *Applied Environmental Microbiology* 75: 2750–2757.

Hastrup, A. C. S., Green, III. F., Lebow, P. K. et al. 2012. Enzymatic oxalic acid regulation correlated with wood degradation in four brown-rot fungi. *International Journal of Biodeterioration and Biodegradation* 75: 109–114.

Hattaka, A. 2001. Biodegradation of lignin. (eds.) MHaASc, editor: Wiley-WCH.

Have, R. T., Teunissen, P. J. M. 2001. Oxidative mechanisms involved in lignin degradation by a white-rot fungi. *Chemical Reviews* 101: 3397–3413.

Hendriks, A. T. W. M., Zeeman, G. 2009. Pretreatments to enhance the digestibility of lignocellulosic biomass. *Bioresource Technology* 100: 10–18.

Henriksson, G., Johansson, G., Pettersson, G. 2000. A critical review of cellobiose dehydrogenases. *Journal of Biotechnology* 78: 93–113.

Hoadley, R. B. 2000. *Understanding wood: A craftsman's guide to wood technology* (2nd ed.). Taunton Press, Newtown, CT.

Immanuel, G., Dhanusha, R., Prema, P. et al. 2006. Effect of different growth parameters on endoglucanase enzyme activity by bacteria isolated from coir retting effluents of estuarine environment. *International Journal of Environmental Science and Technology* 3: 25–34.

Irbe, I., Andersone, I., Andersons, B. 2011. Characterisation of the initial degradation stage of Scots pine (Pinus sylvestris L.) sapwood after attack by brown-rot fungus Coniophora puteana. *Biodegradation* 22: 719–728.

Isroi, I., Millati, R., Syamsiah, S. et al. 2011. Biological pretreatment of lignocelluloses with white-rot fungi and its applications: A review. *BioResources* 6: 5224–5259.

Itoh, H., Wada, M., Honda, Y. et al. 2003. Bioorganosolve pretreatments for simultaneous saccharification and fermentation of beech wood by ethanolysis and white rot fungi. *Journal of Biotechnology* 103: 273–280.

Jing, D. 2010. Improving the simultaneous production of laccase and lignin peroxidise from Streptomyces lavendulae by medium optimization. *Bioresource Technology* 101: 7592–7597.

Jing, D., Wang, J. 2012. Controling the simultaneous production of laccase and lignin peroxidase from Streptomyces cinnamomensis by medium formulation. *Biotechnology for Biofuels* 5: 1–15.

Johannes, C., Majcherczyk, A. 2000. Natural mediator in the oxidation of polycyclic aromatic hydrocarbon by laccase mediator system. *Applied Environmental Microbiology* 66: 524–528.

Johnvesly, B., Virupakshi, S., Patil, G. N. et al. 2002. Cellulase-free thermostable alkaline xylanase from thermophillic and alkalophillic *Bacillus* sp. JB-99. *Journal of Microbiology and Biotechnology* 12: 153–156.

Kannan, K., Oblisami, G., Loganathan, B.G. 1990. Enzymology of lignocellulose degradation by Pleurotus sajor-caju during growth on paper-mill sludge. *Biological Waste* 33: 1–8.

Kanokphorn, S., Piyaporn, V., Siripa, J. 2011. Isolation of novel cellulase from agricultural soil and application for ethanol production. *International Journal of Advanced Biotechnology and Research* 2: 230–239.

Kaparaju, P., Serrano, M., Thomsen, A. B. et al. 2009. Bioethanol, biohydrogen and biogas production from wheat straw in a biorefinery concept. *Bioresource Technology* 100: 2562–2568.

Kaur, A., Mahajan, R., Singh, A. et al. 2010. Application of cellulase-free xylanopectinolytic enzymes from the same bacterial isolate in biobleaching of kraft pulp. *Bioresource Technology* 101: 9150–9155.

Khanna, M., Onal, H., Chen, X., Huang, H. 2008. Meeting Biofuels Targets: Implications for Land Use, Greenhouse Gas Emissions and Nitrogen Use in Illinois. Paper Presented at the Transition to A Bioecnomy: Environmental and Rural Development Impact, Farm Foundation, St. Louis, MO.

Kim, K. C., Scung-soo, Y., Young, O. A. et al. 2003. Isolation and characteristics of Trichoderma harzianum FJ1 producing cellulases and xylanase. *Journal of Microbiology and Biotechnology,* 13: 1–8.

Kluczek-Turpeinen, B., Tuomela, M., Hatakka, A. et al. 2003. Lignin degradation in compost environment by the deuteromycete Paecilomycesinflatus. *Applied Microbiology and Biotechnology* 61: 374–379.

Knauf, M., Moniruzzaman, M. 2004. Lignocellulosic biomass processing: A perspective. *International Sugar Journal* 106: 147–150.

Kubicek, C. P., Herrera-Estrella, A., Seidl-Seiboth, V. et al. 2011. Comparative genome sequence analysis underscores mycoparasitism as the ancestral life style of Trichoderma. *Genome Biology* 12: R40.

Kumar, P., Barrett, D. M., Delwiche, M. J. et al. 2009. Methods for pretreatment of lignocellulosic biomass for efficient hydrolysis and biofuel production. *Industrial Engineering and Chemistry Research* 48: 3713–3729.

Kunamneni, A., Ballesteros, A., Plou, F.J., Alcalde, M. 2007. Fungal laccase-a versatile enzyme for biotechnological applications. In: Méndez-Vilas, A. (Ed.), *Communicating in current research and educational topics and trends in applied microbiology.* Formex, Spain.

Kurakake, M., Ide, N., Komaki, T. 2007. Biological pretreatment with two bacterial strains for enzymatic hydrolysis of office paper. *Current Microbiology* 54: 424–428.

Kyong Ko, J., Bak, J. S., Jung, M. W. et al. 2009. Ethanol production from rice straw using optimized aqueous ammonia soaking pretreatment and simultaneous saccharification and fermentation processes. *Bioresource Technology* 100: 4374–4380.

Lee, J. W., Gwak, K. S., Park, J. Y. et al. 2007. Biological pretreatment of softwood Pinus densiflora by three white rot fungi. *Journal of Microbiology* 45: 485–491.

Lee, S. M., Koo, M. Y. 2001. Pilot-scale production of cellulose using Trichoderma reesei Rut C-30 in fed-batch mode. *Journal of Microbiology Biotechnology* 11: 229–233.

Lee, Y. J., Kim, B. K., Lee, B. H. et al. 2008. Purification and characterization of cellulose produced by Bacillus amyoliquefaciens DL-3 utilizing rice hull. *Bioresource Technology* 99: 378–386.

Li, K. C., Helm, R. F., Eriksson, K. E. L. 1998. Mechanistic studies of the oxidation of a non-phenolic lignin model compound by the laccase/1-hydroxybenzotriazole redox system. *Biotechnology and Applied Biochemistry* 27: 239–243.

Li, Q., Ji, G., Tang, Y. et al. 2012. Ultrasound-assisted compatible in situ hydrolysis of sugarcane bagasse in cellulase-aqueouse-N-methylmorpholine-N-oxide system for improved saccharification. *Bioresource Technology* 107: 251–257.

Lin, Y., Tanaka, S. 2006. Ethanol fermentation from biomass resources: Current state and prospects. *Applied Microbiology and Biotechnology* 69: 627–642.

Lu, C., Wang, H., Luo, Y. et al. 2010. An efficient system for pre-delignification of gramineous biofuel feedstock in vitro: Application of a laccase from Pycnoporus sanguineus H275. *Process Biochemistry* 45: 1141–1147.

Lynd, L. R., Weimer, P. J., van Zyl W. H. et al. 2002. Microbial cellulose utilization: Fundamentals and biotechnology. *Microbiology and Molecular Biology Reviews* 66: 506–577.

Martens-Uzunova, E. S., Schaap, P. J. 2009. Assessment of the pectin degrading enzyme network of Aspergillus niger by functional genomics. *Fungal Genetics and Biology* 46(Suppl 1): S170–S179.

Martinez, A. T. 2002. Molecular biology and structural function of lignin degrading heme peroxidises. *Enzyme Microbial Technology* 30: 425–444.

Martinez, D., Berka, R. M., Henrissat, B. et al. 2008. Genome sequencing and analysis of the biomass-degrading fungus Trichoderma reesei (syn. Hypocrea jecorina). *Nat Biotechnol* 26(5): 553–560.

Mester, T., Jong, E., Field, J. A. 1995. Manganese regulation of veratryl alcohol in white rot fungi and its indirect effect on Lignin Peroxidase. *Applied and Environmental Microbiology* 61: 1881–1887.

Meza, J. C., Sigoillot, J. C., Lomascolo, A. et al. 2006. New process for fungal delignification of sugar-cane bagasse and simultaneous production of laccase in a vapour phase bioreactor. *Journal Agricultural and Food Chemistry* 54: 3852–3858.

Mielenz, J. R. 2001. Ethanol production from biomass: Technology and commercialization status. *Current Opinion in Microbiology* 4: 324–325.

Mod, R. R., Ory, R. L., Morris, N. M. et al. 1981. Chemical properties and interactions of rice hemicellulose with trace minerals in vitro. *Journal of Agricultural and Food Chemistry* 29: 449–454.

Monrroy, M., Ortega, I., Ramirez, M. et al. 2011. Structural change in wood by brown rot fungi and effect on enzymatic hydrolysis. *Enzyme and Microbial Technology* 49: 472–477.

Morohoshi, N. 1991. Chemical characterization of wood and its components. In Hon, D. N. S., Shiraishi, N. (Eds.), *Wood and cellulosic chemistry*. Marcel Dekker, Inc., New York, USA, 331–392.

Mosier, N., Wyman, C., Dale, B. et al. 2005a. Features of promising technologies for pretreatment of lignocellulosic biomass. *Bioresource Technology* 96: 673–686.

Naik, S., Goud, V.V., Rout, P. K. et al. 2010. Characterization of Canadian biomass for alternative renewable biofuel. *Renewable Energy* 35: 1624–1631.

Narayanaswamy, N, Dheeran, P, Verma, S. and Kumar, S. 2013. Biological pretreatment of lignocellulosic biomass for enzymatic saccharification. In: Fang, Z. (Ed.), *Pretreatment techniques for biofuels and biorefineries. Green energy and technology*, Springer Verlag Berlin Heidelberg, 3–34.

Niladevi, K. N., Sukumaran, R. K., Prema, P. 2007. Utilization of rice straw for laccase production by Streptomyces psammoticus in solid-state fermentation. *Journal of Industrial Microbiology and Biotechnology* 34: 665–674.

Nilsson, T., Daniel, G., Kirk, T. K. et al. 1989. Chemistry and microscopy of wood decay by some higher ascomycetes. *Holzfrschung* 43: 11–18.

O'Dwyer, M. H. 1934. The hemicelluloses of the wood of English oak: The composition and properties of hemicellulose A, isolated from samples of wood dried under various conditions. *Biochemical Journal* 28: 2116–2124.

Palmqvist, E., Hahn-Hägerdal, B. 2000. Fermentation of lignocellulosic hydrolysates. I: Inhibition and detoxification. *Bioresource Technology* 74: 17–24.

Palonen, H., Thomsen, A. B., Tenkanen, M. et al. 2004. Evaluation of wet oxidation pretreatment for enzymatic hydrolysis of softwood. *Applied Biochemistry and Biotechnology* 117: 1–17.

Pattana, P., Khanok, R., Khin, L. K. 2000. Isolation and properties of a cellulosome-type multienzyme complex of the thermophilic Bacteroides sp. strain P-1. *Enzyme and Microbial Technology* 26: 459–465.

Perez, J., Dorado, J. M., Rubia T. D. L. et al. 2002. Biodegradation and biological treatments of cellulose, hemicellulose and lignin: An overview. *International Microbiology* 5: 53–63.

Perlack, R. D., Wright, L., Turhollow, L. A., Graham, R. L., Stokes, B., Erbach, D. C. 2005. Biomasses feedstock for a bioenergy and bioproducts industry: The technical feasibility of a billion-ton annual supply. Oak Ridge National Laboratory Report ORNL/TM-2005/66. Oak Ridge, TN: US Dept. of Energy.

Petterson, R. C. 1984. The chemical composition of wood, in *Chemistry of Solid Wood*, edited by R. Rowell, 57–126, Adv. Chem. Ser., 207, Am. Chem. Soc., Washington, D. C.

Phillips, C. M., Beeson, W. T., Cate, J. H. et al. 2011. Cellobiose dehydrogenase and a copper-dependent polysaccharide monooxygenase potentiate cellulose by Neurospora crassa. *ACS Chemical Biology* 6: 1339–1406.

Polizeli, M. L., Rizzatti, A. C., Monti, R. et al. 2005. Xylanases from fungi: Properties and industrial applications. *Applied Microbiology and Biotechnology* 67: 577–591.

Pollet, A., Delcour, J. A., Courtin, C. M. 2010. Structural determinants of the substrate specificities of xylanases from different glycoside hydrolase families. *Critical Reviews in Biotechnology* 30: 176–191.

Preeti, V. E., Sandhya, S. V., Kuttiraja, M. et al. 2012. An evaluation of chemical pretreatment methods for improving enzymatic saccharification of chili postharvest residue. *Applied Biochemistry and Biotechnology* 167: 1489–1500.

Rabelo, S., Filho, R., Costa, A. 2008. A comparison between lime and alkaline hydrogen peroxide pretreatments of sugarcane bagasse for ethanol production. *Applied Biochemistry and Biotechnology* 144: 87–100.

Ran, H., Zhang, J., Gao, Q., Lin, Z., Bao, J. 2014. Analysis of biodegradation performance of furfural and 5-hydroxymethylfurfural by Amorphotheca resinae ZN1. *Biotechnology for Biofuels* 7: 51.

Rasmussen, M. L., Shrestha, P., Khanal, S. K. et al. 2010. Sequential saccharification of corn fiber and ethanol production by the brown-rot fungus Gloeophyllum trabeum. *Bioresource Technology* 101: 3526–3533.

Robson, L. M., Chambliss, G. H. 1984. Characterization of the cellulolytic activity of a Bacillus isolate. *Applied and Environmental Microbiology* 47: 1039.

Rubin, E. M. 2008. Genomics of cellulosic biofuels. *Nature* 454: 841–845.

Ruiz-Dueñas, F. J., Martínez, A. T. 2009. Microbial degradation of lignin: How a bulky recalcitrant polymer is efficiently recycled in nature and how we can take advantage of this. *Microbial Biotechnology* 2: 164–177.

Sadhu, S., Saha, P., Sen, S. K. et al. 2013. Production, purification and characterization of a novel thermotolerant endoglucanase (CMCase) from Bacillus strain isolated from cow dung. *Springer Plus* 2: 10.

Sanchez, M. E., Menendez, J. A., Dominguez, A. et al. 2009. Effect of pyrolysis temperature on the composition of the oils obtained from sewage sludge. *Biomass and Bioenergy* 33: 933–940.

Sanchez, O. J., Cardona, C. A. 2008. Trends in biotechnological production of fuel ethanol from different feedstocks. *Bioresource Technology* 99: 5270–5295.

Sasaki, C., Takada, R., Watanabe, T. et al. 2011. Surface carbohydrate analysis and bioethanol production of sugarcane bagasse pretreated with the white rot fungus, Ceriporiopsis subvermispora and microwave hydrothermolysis. *Bioresource Technology* 102: 9942–9946.

Schilling, J. S., Ai, J., Blanchette, R. A. et al. 2012. Lignocellulose modifications by brown rot fungi and their effects, as pretreatments on cellulolysis. *Bioresource Technology* 116: 147–154.

Shahzadi, T., Mehmood, S., Irshad, M. et al. 2014. Advances in lignocellulosic biotechnology: A brief review on lignocellulosic biomass and cellulases. *Advances in Bioscience and Biotechnology* 5: 246.

Sharma, R. K., Arora, D. S. 2011. Biodegradation of paddy straw obtained from different geographic locations by means of Phlebia spp. for animal feed. *Biodegradation* 22(1): 143–152.

Shi, J., Chinn, M. S., Sharma-Shivappa, R. R. 2008. Microbial pretreatment of cotton stalks by solid state cultivation of Phanerochaete chrysosporium. *Bioresource Technology* 99: 6556–6564.

Shinoyama, H., Ando, A., Fujii, T. et al. 1991. The possibility of enzymatic synthesis of a variety of -xylosides using the transfer reaction of Aspergillus niger xylosidase. *Agricultural and Biology Chemistry* 55: 849–850.

Sindhu, R., Binod, P., Janu, K. U. et al. 2012a. Organosolvent pretreatment and enzymatic hydrolysis of rice straw for the production of bioethanol. *World Journal of Microbiology and Biotechnology* 28: 473–483.

Sindhu, R., Binod, P., Nagalakshmi, S. et al. 2010. Formic acid as a potential pretreatment agent for conversion of sugarcane bagasse to ethanol. *Applied Biochemistry and Biotechnology* 162: 2313–2323.

Sindhu, R., Kuttiraja, M., Binod, P. et al. 2013. A novel surfactant-assisted ultrasound pretreatment of sugarcane tops for improved enzymatic release of sugars. *Bioresource Technology* 135: 67–72.

Singh, P., Suman, A., Tiwari, P. 2008. Biological pretreatment of sugarcane trash for its conversion to fermentable sugars. *World Journal of Microbiology and Biotechnology* 24: 667–673.

Sinsabaugh, R. L. 2009. Phenol oxidase, peroxidase and organic matter dynamics of soil. *Soil Biology and Biochemistry* 42: 1–14.

Sjostrom, E. 1993. *Wood chemistry: Fundamentals and applications*, (2nd ed.). San Diego: Academic Press.

Subramaniyan, S., Prema, P. 2000. Cellulase-free xylanases from Bacillus and other microorganisms. *FEMS Microbiology Letters* 183: 1–7.

Sulistyo, J., Kamiyama, Y., Yasui, T. 1995. Purification and some properties of Aspergillus pulverulentus beta-xylosidase with transxylosylation capacity. *Journal of Fermentation and Bioengineering* 79: 17–22.

Sun, F. H., Li, J., Yuan, Y. X. et al. 2011. Effect of biological pretreatment with Tramates hirsute yj9 on enzymatic hydrolysis of corn stover. *International Biodeterioration and Biodegradation* 65: 931–938.

Sun, Y., Cheng, J. 2002. Hydrolysis of lignocellulosic materials for ethanol production: A review. *Bioresource Technology* 83: 1–11.

Suzuki, H., Macdonald, J., Syed, K. et al. 2012. Comparative genomics of the white rot fungi, Phanerochaete carnosa and P. chrysosporium, to elucidate the genetic basis of the distinct wood types they colonize. *BMC Genomics* 13: 1–17.

Taherzadeh, M. J., Karimi, K. 2008. Pretreatment of lignocellulosic wastes to improve ethanol and biogas production: A review. *International Journal of Molecular Sciences* 9: 1621–1651.

Taniguchi, M., Suzuki, H., Watanabe, D. et al. 2005. Evaluation of pretreatment with Pleurotus ostreatus for enzymatic hydrolysis of rice straw. *Journal of Bioscience and Bioenergy* 100: 637–643.

Tengerdy, R. P., Szakacs, G. 2003. Bioconversion of lignocellulose in solid substrate fermentation. *Biochemical Engineering Journal* 13: 169–179.

Teunissen, P. J. M., Field, J. A. 1998. 2-Chloro-1,4-dimethoxybenzene as a novel catalytic cofactor for oxidation of anisyl alcohol by lignin peroxidase. *Applied and Environmental Microbiology* 64: 830–835.

Tien, M., Kirk, T. K. 1983. Lignin-degrading enzyme from the hymenomycete Phanerochaete chrysosporium burds. *Science* 221: 661–663.

Tomas-Pejo, E., Oliva, J. M., Ballesteros, M. et al. 2008. Comparison of SHF and SSF processes from steam-exploded wheat straw for ethanol production by xylose fermenting and robust glucose-fermenting Saccharomyces cerevisiae strains. *Biotechnology and Bioengineering* 100: 1122–1131.

Tripathi, M. K., Mishra, A. S., Misra, A. K. et al. 2008. Selection of white-rot basidiomycetes for bioconversion of mustard (Brassica compestris) straw under solid state fermentation into energy substrate for rumen micro-organism. *Letters Application Microbiology* 46: 364–370.

Tuyen, V. D., Cone, J. W., Baars, J. J. P. et al. 2012. Fungal strain and incubation period affect chemical composition and nutrient availability of wheat straw for rumen fermentation. *Bioresource Technology* 111: 336–342.

Vaidya, A., Singh, T. 2012. Pre-treatment of Pinus radiata substrates by basidiomycetes fungi to enhance enzymatic hydrolysis. *Biotechnology Letters* 34: 1263–1267.

Vani, S., Binod, P., Kuttiraja, M. et al. 2012. Energy requirement for alkali assisted microwave and high pressure reactor pretreatments of cotton plant residue and its hydrolysis for fermentable sugar production for biofuel application. *Bioresource Technology* 112: 300–307.

Vicuna, R. 2000. Ligninolysis. A very peculiar microbial process. *Molecular Biotechnology* 14: 173–176.

Wan, C., Li, Y. 2010. Microbial delignification of corn stover by Ceriporiopsis subvermispora for improving cellulose digestibility. *Enzyme Microbial Technology* 47: 31–36.

Wan, C., Li, Y. 2011. Effectiveness of microbial pretreatment by Ceriporiopsis subvermispora on different biomass feedstocks. *Bioresource Technology* 102: 7507–7512.

Weerachanchai, P., Leong, S. S. J., Chang, M. W. et al. 2012. Improvement of biomass properties by pretreatment with ionic liquids for bioconversion process. *Bioresource Technology* 111: 453–459.

Wen, Z., Liao, W., Chen, S. 2005. Production of cellulase by Trichoderma reesei from dairymanure. *Bioresource Technology* 96: 491–499.

Wong, D. 2009. Structure and action mechanism of ligninolytic enzymes. *Applied Biochemistry and Biotechnology* 157: 174–209.

Wyman, C. E. 1996. Handbook on bioethanol: Production and utilization. Washington DC: Taylor & Francis.

Xu, C., Ma, F., Zhang, X. 2009. Lignocellulose degradation among enzyme production by Irpex lacteus CD2 during solid-state fermentation of corn stover. *Journal of Bioscience and Bioengineering* 108: 372–375.

Xu, F., Kulys J. J., Duke, K. et al. 2000. Redox chemistry in laccase-catalyzed oxidation of N-hydroxy compounds. *Applied and Environmental Microbiology* 66: 2052–2056.

Xu, J., Cheng, J. J., Sharma-Shivappa, R. R. et al. 2010. Lime pretreatment of switch grass at mild temperatures for ethanol production. *Bioresource Technology* 101: 2900–2903.

Xu, J., Yang, Q. 2010. Isolation and characterization of rice straw degrading Streptomyces griseorubens C-5. *Biodegradation* 21: 107

Yang, B., Wyman, C. E. 2008. Pretreatment: The key to unlocking low cost cellulosic ethanol. *Biofuels, Bioproducts and Biorefining* 2: 26–40.

Yu, H. B., Du, W. Q., Zhang, J. et al. 2010a. Fungal treatment of cornstalks enhances the delignification and xylan loss during mild alkaline pretreatment and enzymatic digestibility of glucan. *Bioresource Technology* 101: 6728–6734.

Yu, H., Zhang, X., Song, L. et al. 2010b. Evaluation of white-rot fungi-assisted alkaline/oxidative pretreatment of corn straw undergoing enzymatic hydrolysis by cellulose. *Journal of Bioscience and Bioengineering* 110: 660–664.

Yu, J., Zhang, J., He, J. et al. 2009. Combinations of mild physical or chemical pretreatment with biological pretreatment for enzymatic hydrolysis of rice hull. *Bioresoure Technology* 100: 903–908.

Zaldivar, J., Martinez, A., Ingram, L. O. 1999. Effect of selected aldehydes on the growth and fermentation of ethanologenic Escherichia coli. *Biotechnology and Bioengineering* 65: 24–33.

Zhang, Y. H. P. and Lynd, L. R. 2005. Cellulose utilization by *Clostridium thermocellum*: Bioenergetics and hydrolysis product assimilation. Proceedings of the National Academy of Sciences USA 102(20): 7321–7325.

Zhang, Y. H. P. 2008. Reviving the carbohydrate economy via multi-product biorefineries. *Journal of Industrial Microbiolgy and Biotechnology* 35: 367.

Zhang, Y. H. P., Lynd, L. R. 2004. Toward an aggregated understanding of enzymatic hydrolysis of cellulose: Noncomplexed cellulase systems. *Biotechnology and Bioengineering* 88: 797–824.

Zhu, J., Wan, C., Li, Y. 2010. Enhanced solid-state anaerobic digestion of corn stover by alkaline pretreatment. *Bioresource Technology* 101: 7523–7528.

9

Laboratory-Scale Bioremediation Experiments on Petroleum Hydrocarbon-Contaminated Wastewater of Refinery Plants

Boutheina Gargouri

CONTENTS

9.1 Introduction

Bioremediation of contaminated aquatic and soil environments has arisen as an effective technology with a range of advantages compared to more traditional methods. Bioremediation of waste materials that contain hydrocarbons and their derivatives is based on the ability of microorganisms to increase their biomass growing on these substrates and degrading them to nontoxic products, such as H_2O and CO_2 (Leahy and Colwell 1990). Petroleum components have traditionally been divided into four fractions: saturated hydrocarbons, aromatic hydrocarbons, nitrogen–sulphur–oxygen–containing compounds (NSOs) and asphalthenes.

Biodegradation is an alternative that has been used to eliminate or minimise the effects of pollutants by using microorganisms that have biodegradation potential (Berry et al. 2006). In these environments, organic pollutants frequently occur in mixture with other synthetic as well as natural organic compounds. Various bioremediation strategies are available at present, but the use of indigenous microorganisms with adapted biochemical potential has proven to be one of the most powerful tools (Bellinaso et al. 2001).

The potentiality of the microorganisms, pointed out in literature as agents of degradation of several compounds, indicates biological treatments as the most promising alternative to reduce the environmental impact caused by oil spills. It is known that the main microorganisms consuming petroleum hydrocarbons are bacteria, yeasts and fungi (Gargouri et al. 2012; van Hamme et al. 2003; Mishra et al. 2001).

However, bacteria naturally inhabiting contaminated sites are of interest as potential agents for hydrocarbon bioremediation, and several papers have been focused on the isolation and characterisation of strains with the ability to grow using hydrocarbons as sole carbon and energy sources. Among many studies conducted on microbial biodegradation of oil-related contaminants, more than 80% are devoted to bacterial biodegradation. Bacteria are the most studied microorganisms, and the participation of bacteria during hydrocarbon mineralisation in water has been studied by many researchers (Leahy and Colwell 1990; Olivera et al. 2003). However, it is not only bacteria that are capable of hydrocarbon biodegradation but also yeasts isolated from hydrocarbon-contaminated sites that display the ability to utilise oil-related compounds (Okerentugba and Ezeronye 2003; Spencer et al. 2002).

In the present study, laboratory-scale experiments were developed in order to determine if microbes present in the refinery wastewater were able to degrade the different hydrocarbon compounds. In this context, we describe the characterisation of microbial strains with capacities to remove hydrocarbon compounds. The growth characteristics and biodegradation capacity of these bacterial and yeast strains as well as their identification by analysis of the sequence of the gene encoding 16S rDNA are reported.

9.2 Critical Importance and Advantages of Bioremediation

There is an increased interest in promoting environmental methods in the process of cleaning hydrocarbon-contaminated sites. Biological methods are less expensive and do not introduce additional chemicals to the environment. Compared to physicochemical methods, bioremediation offers a very feasible alternative for an oil spill response. This technique is considered an effective technology for treatment of hydrocarbon pollution. One reason is that the majority of the molecules in the crude oil and refined products are biodegradable.

Bioremediation through hydrocarbon biodegradation using selected microbial organisms has provided a favourable opportunity because it is environmentally friendly and cost-effective. Those microbial species or particular strains can digest hydrocarbons and utilise the resulting compound carbon as food and energy sources for growth and reproduction. Simultaneously, the hydrocarbons are hydrolysed from toxic and complicated organic compounds into nontoxic and simple inorganic compounds, such as CO_2 and H_2O, along with microbial biomass accumulation through oxidation under aerobic conditions.

Fundamentally, bioremediation uses microbes (for example, bacteria, yeast and fungi) to digest toxic organic contaminants (Sharma and Reddy 2004), such as oil, producing nontoxic products, such as water and carbon dioxide. The process of breaking down organic contaminants with microorganisms is referred to as biodegradation. This can occur in the presence of oxygen or without oxygen, known as aerobic and anaerobic conditions, respectively. This is shown diagrammatically in Figure 9.1 and, using a simplified equation as described by Sharma and Reddy (2004), in Equation 9.1.

$$\text{organic contaminant} + O_2^- \rightarrow H_2O + CO_2 + \text{cell material} + \text{energy} \qquad (9.1)$$

It has been known for 80 years that certain microorganisms are able to degrade petroleum hydrocarbons and use them as a sole source of carbon and energy for growth. The early work was summarised by Davis in 1967. Remediation of the contaminated sites can be done in many ways, which include both physicochemical and biological methods. Large numbers of methods have been developed to increase the degradation rate of petroleum products in soil and wastewater as it takes much more time than physical and chemical remediation methods. In comparison to other biological methods, bioremediation through microorganism is more efficient, but the low solubility and adsorption of high molecular weight hydrocarbons limit their availability to microorganisms. Some types of microorganism are able to degrade petroleum hydrocarbons and use them as a source of carbon and energy. The specificity of the degradation

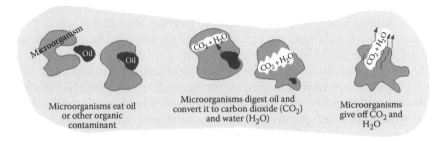

Microorganism — Oil — Oil
Microorganisms eat oil or other organic contaminant

$CO_2 + H_2O$ — $CO_2 + H_2O$
Microorganisms digest oil and convert it to carbon dioxide (CO_2) and water (H_2O)

$CO_2 + H_2O$
Microorganisms give off CO_2 and H_2O

FIGURE 9.1
Diagram representing basic concept of bioremediation. (From Devinny J, and Chang SH. 2000. Bioaugmentation for soil bioremediation. In: Wise DL, Trantolo DJ (eds.), *Bioremediation of Contaminated Soils*. Marcel Dekker, New York, pp. 465–488.)

process is related to the genetic potential of the particular microorganism to introduce molecular oxygen into hydrocarbon and to generate the intermediates that subsequently enter the general energy-yielding metabolic pathway of the cell (Millioli et al. 2009). Some bacteria are mobile and exhibit a chemotactic response, sensing the contaminant and moving toward it, while other microbes, such as fungi, grow in a filamentous form near the contaminant.

The yeast is usually encountered in environments that contain hydrophobic substrates, such as oily wastes, foods (dairy and poultry products) and hydrocarbon-contaminated sites. This feature is due to the inherent presence of several multigene families that mediate the efficient degradation of triglycerides and hydrocarbons (Fukuda 2013).

9.3 Isolation and Characterisation of Hydrocarbonoclastes Microbial Strains

Eight aerobic bacteria and two yeast strains were isolated from enrichment of petroleum hydrocarbon-contaminated water on an aerobic bioreactor (CSTR) using industrial wastewater as a sole carbon source. Basically, microbial organisms are transferred from samples collected to the above MM medium and cultured at 30°C in a rotary shaker at 180 rpm with 0.1% yeast extract until turbid growth is observed. The bacterial culture is diluted and spread on MM agar plates containing crude oil (1%) as a carbon source for selective isolation of petroleum degrader. The plates are sealed and incubated at 30°C until appearance of several colonies. Individual colonies can be purified by repeating the culture on MM agar plates containing 1% crude oil. Identification of the candidate strains will be performed based on physiological and biochemical tests or 16S rRNA sequencing.

9.3.1 Physiological and Biochemical Tests

Using phenotypic and biochemical characterisations to identify specific strains can be performed as described by Gargouri et al. (2011).

The isolated microorganism cultures were characterised by their morphological and biochemical properties. According to the data obtained using light and electron microscopy, the isolated microorganisms had the form of rods or cocci, were spore-forming or non-spore-forming, occurred as single cells or were integrated in chains and were immotile or motile with flagella. Four strains are Gram positive, and four strains are Gram negative.

9.3.2 Phylogenetic Analysis Using 16S rRNA Gene Sequence

All strains tested in this study were identified by the analysis of the sequence of the gene encoding 16S rRNA (16S rDNA). Primer fD1 and rD1 (Weisburg

TABLE 9.1

Phylogenetic Characterisation of Different Microorganisms Screened from CSTR during Bioremediation Treatment

Phylogenetic Group	Strains	Related Species	Accession No.	% Similarity
Bacteria				
Proteobacteria	HC2	*Aeromonas punctata*	DQ979324	99
	HC6	*Brucella abortus*	EU816698	99
	HC7	*Achromobacter xylosoxidans*	DQ466568	99
	HC8	*Stenotrophomonas maltophilia*	EU034540	99
Firmicutes	HC5	*Bacillus cereus*	EU855219	99
Actinobacteria	HC9	*Rhodococcus* sp.	AY927229	99
	HC12	*Micrococcus luteus*	CP001628	99
Bacteroidetes	HC10	*Myroides odoratimmus*	EU331413	99
Yeast				
	HC1	*Candida tropicalis*	FJ432611	99
	HC4	*Trichosporon asahii*	EU559346	100

et al. 1991) were used to amplify almost the full length of 16S rRNA gene from each bacteria strain. The primers ITS1 and ITS4 described by White et al. (1990) were used to amplify the 5.8S-ITS region of yeast strains. PCR programme and sequencing method was performed as described by Gargouri et al. (2011, 2013). The related sequences were collected and aligned using the MUSCLE software, and phylogenetic dendograms were constructed using the neighbour-joining method with Mega software. The topology of the distance tree was tested by resampling data with 1000 bootstraps to provide confidence estimates. Nucleotide sequences obtained in this work were deposited in the NCBI GenBank data library (Table 9.1).

9.4 Evaluation of Biodegradation Rate of Hydrocarbon Bioremediation Assays

In order to evaluate both the susceptibility of extracted total petroleum hydrocarbon (TPH) to bioremediation and the hydrocarbon biodegradation capability of indigenous microbial consortia and yeast strains, TPH biodegradation kinetics in batch cultures were determined during time course experiments.

As shown in Figure 9.2, the TPH content reached 94% on flasks containing hydrocarbonoclastes consortia and showed the greatest extent of hydrocarbon compounds of its initial value after 50 days of incubation, and this difference was evident even by visual observation (Figure 9.3). In addition, control flasks with no microbial amendments showed no significant changes in TPH during

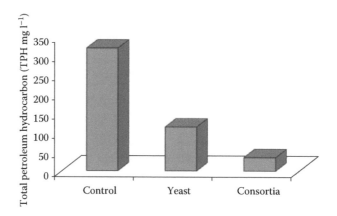

FIGURE 9.2
TPH analysis of hydrocarbon wastewater from various mesocosms incubated for 50 days.

FIGURE 9.3
(See color insert.) Macroscopic view of a noninoculated (b) and an inoculated (a) hydrocarbon wastewater selective broth culture Erlenmeyers at the end of the bioremediation experiment. Hydrocarbon-contaminated wastewater was exposed to an indigenous consortium at 30°C and continuous shaking for 50 days. 0.1% crude oil was used as the sole source of carbon in MM media.

the bioremediation process. The addition of the yeast strains only, without the presence of the microorganisms consortia, showed a limited capacity for hydrocarbon degradation (60%, TPH 115 mg l⁻¹) when compared to the mesocosmos coupled with the consortia. The addition of microorganisms (bioaugmentation) is known to enhance the extent of biodegradation by both indigenous and bioaugmented strains (Gargouri et al. 2012; Mancera-Lopez et al. 2008).

This high removal efficiency suggests that the microbial consortium from the aerobic bioreactor (CSTR) had already been exposed to contaminants. Okerentugba and Ezeronye (2003) stated that microbiological communities exposed to hydrocarbons adapt to the exposure through selective enrichment and genetic changes, resulting in an increase in the ratio of hydrocarbon-degrading versus nondegrading bacteria. To degrade hydrocarbons, it is advantageous to use native microorganisms cultured from areas with historical contamination. This approach is likely to reduce or eliminate the initial lag phase and optimise overall process time.

9.5 Potential Approaches to Improving Biodegradation of Hydrocarbons and Performances of Various Microbial Strains

Cultures on selective media were developed in order to evaluate the substrate preferences of the acclimatised microbial (bacteria and yeast) during the bioremediation assays.

Positive enrichment cultures initiated in basal medium (MM) containing 1% (v/v) crude oil as the carbon and energy source became turbid, and the dark layer of crude oil became clear, indicating that the hydrocarbons were degraded. Enrichment cultures were diluted and transferred to solid media for colony isolation. The preliminary selection of strains was based on their high capacity to degrade 1% (v/v) crude oil in a liquid medium. From these enrichments, the isolates surrounded by a clear zone on a solid medium containing crude oil were then picked up for further analysis. Six aerobic bacteria and two yeast strains were isolated from petroleum hydrocarbon–contaminated water using industrial wastewater as a sole carbon source.

Many studies are focused on the isolation and characterisation of microorganisms degrading hydrocarbon components. Numerous microorganisms, namely, bacteria, yeast and fungi, have been reported as good degraders of hydrocarbons (Katsivela et al. 2003; Rahman et al. 2002; Spencer et al. 2002). Many of these microorganisms were applied in bioremediation processes to reduce the concentration and the toxicity of various pollutants, including petroleum products (Tazaki et al. 2004).

Hydrocarbonoclastes microbial strains from this enrichment were chosen as the most superior hydrocarbon degraders and were identified to the specific level according to general principles of microbial classification. Bacterial strains belonged to the genera *Aeromonas, Bacillus, Ochrobactrum, Stenotrophomonas* and *Rhodococcus*. These bacteria have been described as the most common bacteria isolated in terrestrial as well as aquatic areas of hydrocarbon contamination. Yet, the present study revealed that crude oil–degrading microorganisms are not restricted to oil-polluted areas only, as *Ochrobactrum* and *Rhodococcus* sp. in this study have been isolated from natural habitats with no history of crude oil pollution.

Bioremediation treatment has allowed the enrichment of the microbial community with these populations capable of using contaminants. The results obtained showed that these isolates were able to degrade a wide range of *n*-alkanes (C11–C36) with good efficiency as already illustrated by Vomberg and Klinner (2000). Studies show that Actinobacteria, such as *Rhodococcus*, have the ability to degrade the molecules of C6 to C36 (van Beilen et al. 2003; van der Geize and Dijkhuizen 2004). Indeed, the strain *Rhodococcus* shows good biodegradation activity of *n*-alkanes of C10–C24 with a higher percentage of the degradation of *n*-C16 of about 79% and 87.8% for crude oil. However, for *n*-alkanes *n*-C24, the degradation is lower (Figure 9.4a).

Hydrocarbonoclastes consortium could be useful for its application in bioremediation technologies. In addition, a recent study evaluated the bioaugmentation process with an acclimatized consortium isolated from an industrial wastewater treatment plant and its performance *in situ* as a variety of factors that affect the ability of the microorganisms to degrade hydrocarbons in natural environments (Gargouri et al. 2013).

Also the results show good performance degradation by yeast grown in hydrocarbon-polluted water from the refinery industry. These results are in agreement with studies performed by Nitu and Banwari (2009). We notice that both short- and long-chain hydrocarbons of the industrial wastewater

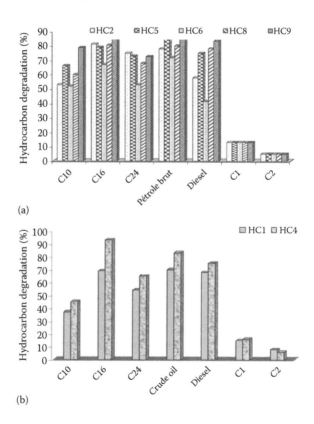

(a)

(b)

FIGURE 9.4
Biodegradation of different hydrocarbon compounds of petroleum products by selective microbial strains (a) bacteria, (b) yeast. C1 = control with no carbon source; C2 = control with no microorganism.

refinery were highly susceptible to attack by the yeast strains, probably due to the fact that they have a very efficient degradative enzyme system (Figure 9.4b). In this case, strain HC4 was better at degradation than strain HC1. The strains also showed considerable utilisation of components from aromatic fraction. Although *Candida* and *Trichosporon* sp. were capable of efficient degradation of alkanes as the sole carbon source, they were unable to completely degrade aromatic hydrocarbons as the sole carbon source. Aromatic compounds proved to be resistant to biodegradation, and we observed that *Trichosporon* can assimilate phenanthrene, anthracene and fluoranthene by 47%, 73% and 31.4%, respectively, within 7 days. The strain *Candida* sp. HC1 was also found to be effective in degrading the aromatic hydrocarbons, such as anthracene (51%) and fluoranthene (74%), after degradation for 7 days. Usually, the fractions of the petroleum containing *n*-alkanes are mostly susceptible to biodegradation, whereas saturated fractions containing branched alkanes are less vulnerable to microbial attack. The aromatic fractions are even less easily biodegraded.

The strains isolated from petroleum hydrocarbon–contaminated sites have shown promising potential in biodegradation of hydrocarbons and bioremediation of contaminated sites mostly in bench scale studies.

9.6 Monitoring and Evaluation of Biodegradation by Spectrometry Tools

To evaluate the biodegradation process of various hydrocarbons, cultured media can be extracted at certain time intervals with dichloromethane. The extract can be analysed by gas chromatography (GC), gas chromatography mass spectrometry (GC-MS), gas-liquid chromatography (GLC), high-performance liquid chromatography (HPLC) or Fourier transform infrared spectroscopy (FTIR).

With GC-MS analysis, the carrier gas is helium used at a 1 ml min^{-1} flow rate; the injector and detector temperatures are set at 250°C and 300°C, respectively, for analysis of total petroleum hydrocarbons (TPH). However, for gasoline analysis, Wongsa et al. (2004) suggested that the column temperature is first maintained at 35°C for 5 min and then increased to 220°C; for diesel oil analysis, the column temperature is set at 50°C and then ramped to 270°C with a regime of 5°C min^{-1}. In the case of lubricating oil, 320°C is needed for both injector and detector temperatures, and the column temperature can be set at 100°C initially and ramped to 320°C at a rate of 10°C min^{-1}.

GC-MS analysis was performed to identify the presence of the heavier petroleum TNA (total *n*-alkanes) compounds in the influent and effluent streams of the system and their sequential unit process removal concentrations. Resulting chromatograms can be analysed by various particular

TABLE 9.2

Principal Functional Groups Determined by FTIR Analysis in the
Hydrocarbon Wastewater Observed during Bioremediation Process

Range of Wave Numbers (cm⁻¹)	Band Assignment
3550–3230	–OH stretch
3100–3000	=C–H stretching in aromatic compound
3000–2850	–C–H stretching in aliphatic compounds
2455–2265	C–C, C≡N (nitriles)
1650–1550	C=O stretching
1639–1590	C=C stretching
1400–1000	–C–O stretching for primary alcohols
725–720	(CH2) rocking in aliphatic compounds
720–580	–C–H deformation vibration aromatic

software packages, such as Saturn Software GC/MS Workstation Version 5.52, to identify petroleum components. Detailed information has been provided elsewhere by other researchers (Gargouri et al. 2012, 2013).

FTIR spectroscopy was used to study hydrocarbon components and to compare the spectral differences between the petroleum hydrocarbon–contaminated wastewater at the beginning and at the end of bioremediation treatment. The infrared spectra of products within the wave number range from 4000 to 500 cm⁻¹, and the observed bands have been tentatively assigned (Table 9.2). A comparison of FTIR spectra of treated and untreated wastewater revealed the presence of bands pertaining to aliphatic and polycyclic aromatic hydrocarbons, including various alcohols, aldehydes and phenols.

9.7 Electrochemical Remediation Technologies for Polluted Wastewater

Electrochemical bioremediation is a relatively efficient and cost-effective technology for treating polluted wastewater. In recent years, electrochemical oxidation of refractory effluents has received a great deal of attention due to its attractive characteristics, such as versatility, energy-efficiency, amenability of automation and environmental compatibility (free chemical reagents) (Hamza et al. 2009; Méndez et al. 2012; Yavuz et al. 2010). In this context, electrochemical methods can be a promising alternative to traditional processes for the treatment of petrochemical wastewaters. Over the past years, many papers reported that electrochemical treatment has been applied successfully for the complete oxidation of various organic pollutants, including surfactants, oils, grease and gasoline residues (Martínez-Huitle et al. 2006, 2009). Moreover, a wide variety of electrode materials have been suggested, such as dimensionally stable anodes, noble metals (for example, platinum), carbon-based anodes, lead dioxide (PbO_2) and boron-doped diamond (BDD), obtaining different removal of organic matter efficiencies (Dridi Gargouri et al. 2013; Méndez et al. 2012; Zhao et al. 2010) considering that nonactive anodes, such as BDD, are useful for direct oxidation of organic material via hydroxyl radicals, and a dimension-stable anode (DSA), such as $Ti/IrO_2–Ta_2O_5$, is effective for promoting hypochlorite-mediated chemistry when chloride is present. In the case of applicability of electrochemical technology for treating petroleum hydrocarbon–contaminated wastewaters, dimensionally stable anodes, platinum and BDD have been preferentially used as electrocatalytic materials (dos Santos et al. 2014; Xing et al. 2012). However, in some cases, Pt anodes are very expensive and also subject to fouling. Recently, Rocha et al. (2012) studied the electrochemical oxidation of brine-produced water in galvanostatic conditions using platinum supported on titanium (Ti/Pt) and BDD anodes, employing a batch reactor. The results showed that complete COD removal was achieved using a BDD electrode due to the production of high amounts of hydroxyl radicals (OH) and oxidising species (Cl_2, HClO, ClO^-) (Gargouri et al. 2014). The use of these electrode materials has been proposed due to electrocatalytic features to produce *in situ* strong oxidant species.

The anodic oxidation of three classes of produced water (fresh, brine and saline) generated by the petrochemical industry using $Ti/IrO_2–Ta_2O_5$ and BDD electrodes in a flow reactor was recently studied (Cabral da Silva et al. 2013). It was found that when both electrode materials are compared under the same operating conditions, higher TOC- and COD-removal efficiencies were achieved for BDD anodes; nevertheless, the energy consumption and cost were higher when compared with the values estimated for $Ti/IrO_2–Ta_2O_5$.

Finally, we focus our attention on the electrochemical conditions that provide greater efficiency of current with lower power requirements to scale up the electrochemical treatment in order to employ it in petrochemical platforms.

9.8 Conclusion

Research using bioremediation has shown great promise to date and can result in the development of more efficient and less time-consuming technologies. Also, further research is critical to investigate its application beyond the laboratory scale and to develop the kinetics of degradation. The results proved that the acclimatised indigenous microbial population isolated from contaminated wastewater was able to feed on every hydrocarbon and that the application of a bioremediation technology could be a very useful tool for the treatment of the hydrocarbons present in the studied wastewater. However, bioremediation can be considered one of the best technologies to deal with a petroleum product–contaminated site.

9.9 Perspective

The treatment of petroleum hydrocarbon–contaminated wastewaters has undergone changes in technological approaches. Many advances have been made to improve performance, ranging from bioaugmentation to fluidised bioreactors. However, the problems of high sludge generation, low tolerance to toxic load and organic shock, coupled with a slow degradation rate, persist. The poor performance of these processes necessitates the search for more viable alternatives. An attractive wastewater treatment technique is electrochemical technologies degradation, a process that potentially mineralises all of the organic and inorganic components typically found in the refinery effluent to environmentally benign by-products. It is also an efficient and cost-effective technique that is suitable for petrochemical wastewater treatment at the advanced stage. However, available information in the literature is scarce regarding the process for the treatment of petrochemical wastewater, and this greatly limits the industrial application of the process in refinery treatment plants. Although relatively high efficiencies for electrochemical degradation have been achieved by a few reported works, optimising the process parameters would yield much better results.

References

Bellinaso ML, Henriques JAP and Gaylarde CC. 2001. Biodegradation as a biotechnological model for the teaching of biochemistry. *World Journal of Microbiology & Biotechnology* 17: 1–6.

Berry CJ, Story S, Altman DJ, Upchurch R, Whitman W, Singleton D, Płaza G and Brigmon RL. 2006. Biological treatment of petroleum in radiologically contaminated soil. In: Clayton II CJ, Lindner AS (eds.), *Remediation of Hazardous Waste in the Subsurface. Bridging Flask and Field*. American Chemical Society, Washington, DC, 87.

Cabral da Silva AJ, dos Santos EV, de Oliveira Morais CC, Martínez-Huitle CA and Leal Castroa SS. 2013. Electrochemical treatment of fresh, brine and saline produced water generated by petrochemical industry using Ti/IrO$_2$–Ta$_2$O$_5$ and BDD in flow reactor. *Chemical Engineering Journal* 233: 47–55.

Devinny J and Chang SH. 2000. Bioaugmentation for soil bioremediation. In: Wise DL, Trantolo DJ (eds.), *Bioremediation of Contaminated Soils*. Marcel Dekker, New York, pp. 465–488.

dos Santos EV, Rocha JHB, Araujo DM, Moura DC and Martínez-Huitle CA. 2014. Decontamination of produced water containing petroleum hydrocarbons by electrochemical methods: A mini-review. *Environmental Science and Pollution Research International* 21(14): 8432–8441.

Dridi Gargouri O, Gargouri B, Kallel Trabelsi S, Bouaziz M and Abdelhédi R. 2013. Synthesis of 3-O-methylgallic acid a powerful antioxidant by electrochemical conversion of syringic acid. *Biochima and Biophysica Acta* 1830: 3643–3649.

Fukuda R. 2013. Metabolism of hydrophobic carbon sources and regulation of it in n-alkane assimilating yeast Yarrowia lipolytica. *Bioscience, Biotechnology, and Biochemistry* 77: 1180–1149.

Gargouri B, Aloui F and Sayadi S. 2012. Reduction of petroleum hydrocarbons content from an engine oil refinery wastewater using a continuous stirred tank reactor monitored by spectrometry tools. *Journal of Chemical Technology and Biotechnology* 87: 238–243.

Gargouri B, Aloui F and Sayadi S. 2013. Bioremediation of petroleum hydrocarbons contaminated soil by bacterial consortium isolated from an industrial wastewater treatment plant. *Journal of Chemical Technology and Biotechnology* 89: 978–987.

Gargouri B, Dridi Gargouri O, Gargouri B, Kallel Trabelsi S, Abdelhedi R and Bouaziz M. 2014. Application of electrochemical technology for removing petroleum hydrocarbons from produced water using lead dioxide and boron-doped diamond electrodes. *Chemosphere* 117: 309–315.

Gargouri B, Karray F, Mhiri N, Aloui F and Sayadi S. 2011. Application of a continuously stirred tank bioreactor (CSTR) for bioremediation of hydrocarbon-rich industrial wastewater effluents. *Journal of Hazardous Materials* 189: 427–434.

Hamza M, Abdelhedi R, Brillas E and Sirés I. 2009. Comparative electrochemical degradation of the triphenylmethane dye Methyl Violet with boron-doped diamond and Pt anodes. *Journal of Electroanalytical Chemistry* 627: 41–50.

Katsivela E, Moore ERB and Kalogerakis N. 2003. Biodegradation of aliphatic and aromatic hydrocarbons: Specificity among bacteria isolated from refinery waste sludge. *Water Air Soil Pollution* 3: 103.

Leahy JG and Colwell RR. 1990. Microbial degradation of hydrocarbons in the environment. *Microbiology Review* 54: 305–315.

Mancera-López ME, Esparza-García F, Chávez-Gómez B, Rodríguez-Vázquez R, Saucedo-Castañeda G and Barrera-Cortés J. 2008. Bioremediation of an aged hydrocarbon-contaminated soil by a combined system of biostimulation–bioaugmentation with filamentous fungi. *International Biodeterioration & Biodegradation* 61: 151–160.

Martínez-Huitle CA and Brillas E. 2009. Decontamination of wastewaters containing synthetic organic dyes by electrochemical methods: A general review. *Applied Catalysis B: Environmental* 87: 105–145.

Martinez-Huitle CA and Ferro S. 2006. Electrochemical oxidation of organic pollutants for the wastewater treatment: Direct and indirect processes. *Chemical Society Reviews* 35: 1324–1340.

Méndez E, Pérez M, Romero O, Beltrán ED, Castro S, Corona JL, Corona A, Cuevas MC and Bustos E. 2012. Effects of electrode material on the efficiency of hydrocarbon removal by an electrokinetic remediation process. *Electrochimica Acta* 86: 148–156.

Millioli VS, Servulo ELC, Sobral LGS and de Carvalho DD. 2009. Bioremediation of crude oil-bearing soil: Evaluating the effect of Rhamnolipid addition to soil toxicity and to crude oil biodegradation efficiency. *Global NEST Journal* 11(2): 181.

Mishra S, Jyot J, Kuhad RC and Lal B. 2001. In situ bioremediation potential of an oily sludge-degrading bacterial consortium. *Current Microbiology* 43: 328–335.

Nitu S and Banwari L. 2009. Isolation of a novel yeast strain Candida digboiensis TERI ASN6 capable of degrading petroleum hydrocarbons in acidic conditions. *Journal of Environmental Management* 90: 1728–1736.

Okerentugba PO and Ezeronye OU. 2003. Petroleum degrading potentials of single and mixed microbial cultures isolated from rivers and refinery effluent in Nigeria. *African Journal of Biotechnology* 2: 288–292.

Olivera NL, Commendatore MG, Delgado O and Esteves JL. 2003. Microbial characterization and hydrocarbon biodegradation potential of natural bilge waste microflora. *Journal of Industrial Microbiology and Biotechnology* 30: 542–548.

Rahman KSM, Bana IM, Thahira J, Thayumanavan THA and Lakshmanaperumalsamy P. 2002. Towards efficient crude oil degradation by a mixed bacterial consortium. *Bioresource Technology* 85: 257.

Rocha JHB, Gomes MMS, Fernandes NS, da Silva DR and Martínez-Huitle CA. 2012. Application of electrochemical oxidation as alternative treatment of produced water generated by Brazilian petrochemical industry. *Fuel Process and Technology* 96: 80–87.

Sharma HD and Reddy KR. 2004. *Geo Environmental Engineering: Site Remediation, Waste Containment, and Emerging Waste Management Technologies.* John Wiley, Hoboken, NJ.

Spencer JFT, Ragout de Spencer AL and Laluce C. 2002. Non-conventional yeast. *Applied Microbiology and Biotechnology* 58: 147.

Tazaki CK, Asada R and Kogure K. 2004. Bioremediation of coastal areas 5 years after the Nakhodka oil spill in the Sea of Japan: Isolation and characterization of hydrocarbon-degrading bacteria. *Environmental International* 30: 911–922.

Van Beilen JB, Li Z, Duetz WA, Smits THM and Witholt B. 2003. Diversity of alkane hydroxylase systems in the environment. *Oil & Gas Science and Technology – Rev. IFP* 58(4): 427–440.

Van der Geize R and Dijkhuizen L. 2004. Harnessing the catabolic diversity of rhodococci for environmental and biotechnological applications. *Current Opinion in Microbiology* 7: 255–261.

van Hamme JD, Singh A and Ward OP. 2003. Recent advances in petroleum microbiology. *Microbiology and Molecular Biology Reviews* 503–549.

Vomberg A and Klinner U. 2000. Distribution of alkB genes within n-alkane-degrading bacteria. *Journal Applied Microbiology* 89(2): 339–348.

Weisburg WG, Barns SM, Pelletier DA and Lane DJ. 1991. 16S ribosomal DNA amplification for phylogenetic study. *Journal Bacteriology* 173: 697–703.

White TJ, Bruns T, Lee S and Taylo J. 1990. Amplification and direct sequencing of fungi ribosomal RNA genes for phylogenetics. In: Innis MA, Gelfand DH, Sninsky JJ, White TJ (eds.), *PCR Protocols. A Guide to Methods and Applications.* Academic Press, pp. 315–322.

Wongsa P, Tanaka M, Ueno A, Hasanuzzaman M, Yumoto I and Okuyama H. 2004. Isolation and characterization of novel strains of pseudomonas aeruginosa and serra-tia marcescens possessing high efficiency to degrade gasoline, kerosene, diesel oil and lubricating oil. *Current Microbiology* 49: 415–422.

Xing X, Zhu X, Li H, Jiang Y and Ni J. 2012. Electrochemical oxidation of nitrogen heterocyclic compounds at boron-doped diamond electrode. *Chemosphere* 86: 368–375.

Yavuz Y, Koparal AS and Öğütveren ÜB. 2010. Treatment of petroleum refinery wastewater by electrochemical methods. *Desalination* 258: 201–205.

Zhao G, Pang Y, Liu L, Gao J and Baoying LV. 2010. Highly efficient and energy-saving sectional treatment of landfill leachate with a synergistic system of biochemical treatment and electrochemical oxidation on a boron-doped diamond electrode. *Journal of Hazardous Materials* 179: 1078–1083.

10

Microbial Degradation of Textile Dyes for Environmental Safety

Ram Lakhan Singh, Rasna Gupta and Rajat Pratap Singh

CONTENTS

10.1 Introduction

A dye is a coloured substance that has an affinity to the substrate to which it is being applied. Most dye molecules contain a number of aromatic rings, such as those of benzene or naphthalene, linked in a fully conjugated system (Figure 10.1). The unsaturated groups that can be conjugated to make the molecule coloured are referred to as chromophores. Synthetic dyes can be classified in a number of ways, including colour, intended use, trade name, chemical constitution and basis of application (Table 10.1).

More than 10,00,000 synthetic dyes are generated worldwide with an annual production of around 7×10^5 metric tonnes (Chen et al. 2003). These dyes are widely used in the textile, paper, food, cosmetics and pharmaceutical industries, and the textile industry is the largest consumer (Franciscon et al. 2009; Singh and Singh 2011). Chemical classes of dyes employed more frequently on the industrial scale are generally azo, triphenylmethane and anthraquinone dyes. Among all the available synthetic dyes, azo dyes are the largest group used in the textile industry constituting 60%–70% of all dyestuffs produced. Azo dyes are aromatic compounds with one or more –N=N– groups and the most common synthetic dyes released into the environment (Singh et al. 2014a). All dyes do not bind to the fabric due to inefficiency in

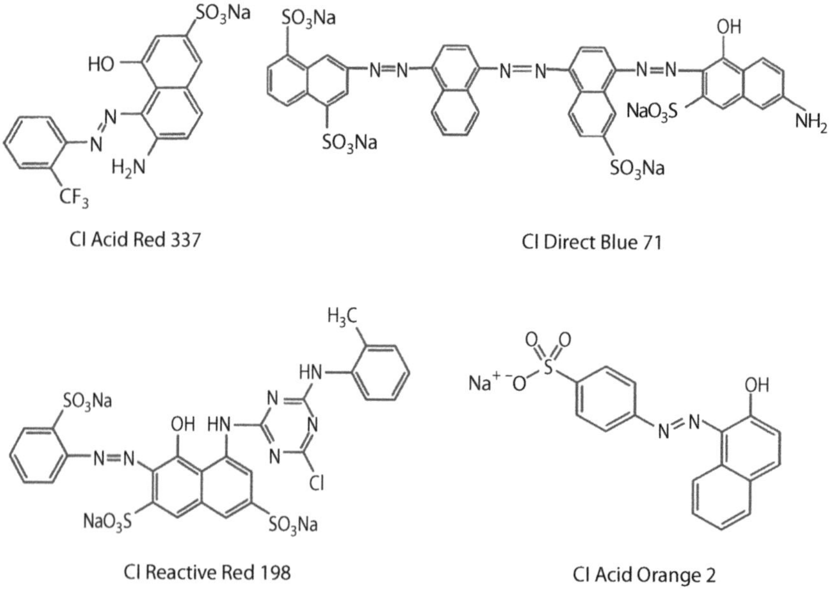

FIGURE 10.1
Structures of some textile dyes.

TABLE 10.1

Classification of Dyes on the Basis of Method of Application

Class	Chemical Types	Applications	Method of Application	Examples
Dyes Containing Anionic Functional Groups				
Acid dyes	Azo (including premetalised), anthraquinone, triphenylmethane, azine, xanthene, nitro and nitroso	Wool, silk, nylon, paper, inks and leather	Usually from neutral to acidic dye baths	CI Acid Red 337, CI Acid Blue 40
Direct dyes	Azo, phthalocyanine, stilbene and oxazine	Cotton, leather, paper and synthetics, rayon and nylon	Applied from neutral or slightly alkaline baths containing additional electrolyte	CI Direct Red 28, CI Direct Blue 71
Mordant dyes	Azo and anthraquinone	Wool, leather and anodised aluminium	Applied in conjunction with Cr salts	CI Mordant Orange 6, CI Mordant Green 17
Reactive dyes	Azo, anthraquinone, phthalocyanine, formazan, oxazine and basic	Cotton, wool, silk and nylon	A reactive site on dye reacts with functional group of fibre to bind the dye covalently under influence of heat and pH (alkaline)	CI Reactive Red 198, CI Reactive Black 5
Dyes Requiring Chemical Reaction before Application				
Vat dyes	Anthraquinone (including polycyclic quinines) and indigoids	Cotton, wool rayon and synthetics	Water-insoluble dyes solubilised by reducing with sodium hydrogen sulphite, then exhausted on fibre and reoxidised	CI Vat Blue 8, CI Vat Red 45
Azoic dyes	Azo	Cotton, rayon, cellulose acetate and polyester	Fibre impregnated with coupling component and treated with a solution of stabilised diazonium salt	CI Acid Orange 2, CI Basic Green 4

(Continued)

TABLE 10.1 (CONTINUED)

Classification of Dyes on the Basis of Method of Application

Class	Chemical Types	Applications	Method of Application	Examples
Sulphur dyes	Indeterminate structures	Cotton, rayon and synthetics	Aromatic substrate vatted with sodium sulphide and reoxidised to insoluble sulphur-containing products on fibre	CI Sulphur Black 3, CI Vat Blue 43
Dyes Containing Cationic Groups				
Basic dyes	Cyanine, hemicyanine, diazahemicyanine, diphenylmethane, triarylmethane, azo, azine, xanthene, acridine, oxazine and anthraquinone	Acrylic, paper, polyacrylonitrile, modified nylon, polyester and inks	Applied from acidic dye baths	CI Basic Blue 9, CI Basic Brown 4
Special Colourant Classes				
Disperse dyes	Azo, anthraquinone, styryl, nitro and benzodifuranone	Polyester, polyamide, acetate, acrylic and plastics	Fine aqueous dispersions often applied by high temperature/pressure or lower temperature carrier methods; dye may be padded on cloth and baked on or thermo fixed	CI Disperse Green 5, CI Disperse Yellow 23
Solvent dyes	Azo, triphenylmethane, anthraquinone and phthalocyanine	Synthetics, plastics, gasoline, varnishes, stains, inks, fats, oils and waxes	Dissolution in the substrate	CI Solvent Orange 20, CI Solvent Black 7

dyeing processes, which resulted in 10%–50% of unused dyestuff entering the wastewater directly (Singh et al. 2014b). The textile industry is one of the largest generators of wastewater contaminated with the dyestuff due to the high quantities of water used in the dyeing process (Pandey et al. 2007). Colour in textile wastewater is the first contaminant to be recognised, and the presence of very small amounts of dyes in water (less than 1 ppm for some dyes) is highly visible (Banat et al. 1996). Improper textile wastewater disposal in aqueous ecosystems affects the aesthetic merit, water transparency and gas solubility in water bodies and depicts acute toxic effects on aquatic flora and fauna, causing severe environmental problems (Singh and Singh 2012; Solis et al. 2012).

Therefore, removal of colour from textile wastewaters has been a major concern before discharging it into water bodies or onto land. Many physicochemical techniques for colour removal from dye-containing effluents have been developed, such as adsorption, coagulation, precipitation, filtration and oxidation, but these methods have many disadvantages and limitations due to their high cost, low efficiency and inapplicability to a wide variety of dyes (Lorimer et al. 2001). Biological methods, including a wide range of microorganisms, are generally considered as a viable alternative for treatment of textile effluents due to their cost effectiveness, ability to produce less sludge and environment-friendly nature (Saratale et al. 2011; Singh et al. 2013).

10.2 Microbial Decolourisation and Degradation of Textile Dyes

Different taxonomic groups of bacteria, fungi, yeasts and algae are capable of degrading azo dyes (Chen et al. 2003). The mechanism for the biodegradation of recalcitrant compounds by microbial system is generally based on the action of the biotransformation enzymes. Dye decolourisation by fungi is mainly attributed by absorption rather than degradation (Wang et al. 2009a). A variety of extracellular enzymes (lignin peroxidase, manganese peroxidase and laccase) produced by fungi may also be involved in the degradation of dyes (Joe et al. 2008). However, the longer growth cycle and long hydraulic retention time required for complete decolourisation of dyes limit the efficiency of the fungal decolourisation system. In contrast, decolourisation of dyes by a bacterial system is faster as compared to the fungal (Banat et al. 1996). It has also been reported that different trophic groups of bacteria can achieve a higher degree of degradation and even complete mineralisation of many dyes under optimum conditions (Kapdan and Erten 2007).

10.2.1 Bacterial Decolourisation and Degradation of Dyes

Diverse groups of bacteria are frequently applied for decolourisation and complete mineralisation of dyes because they are easy to cultivate and grow rapidly. The process of decolourisation by a bacterial system may be anaerobic or aerobic or involve a combination of the two. However, there are significant differences between the physiology of bacteria grown under aerobic and anaerobic conditions (Stolz 2001).

10.2.1.1 Decolourisation and Degradation of Azo Dyes by Bacteria and Its Mechanism

The initial step in the bacterial degradation of azo dyes is the reductive cleavage of the azo bond by an enzymatic biotransformation reaction under static or anaerobic conditions, which leads to the formation of colourless aromatic amines (McMullan et al. 2001). The resulting toxic aromatic amines are further degraded to simpler nontoxic forms under aerobic or anaerobic conditions (Pandey et al. 2007). Several bacterial strains have been reported that decolourise the azo dyes (Table 10.2). Decolourisation of dye under anaerobic conditions is unspecific and might be attributed to nonspecific extracellular reactions occurring between reduced compounds generated by the anaerobic biomass (Van der Zee et al. 2001). Extensive studies have been carried out using pure bacterial cultures, such as *Citrobacter* sp., *Clostridium bifermentans*, etc., for decolourisation of azo dye under anoxic/anaerobic conditions (Joe et al. 2008; Wang et al. 2009a). There are only very few bacteria that are able to grow on azo compounds as the sole carbon source. These bacteria reductively cleave $-N=N-$ bonds and utilise the resulting amines as the source of carbon and energy for their growth. Such organisms are generally specific toward their substrate. The bacterial strains *Xenophilus azovorans* KF46 and *Sphingomonas* sp. strain ICX can grow aerobically on azo dye as the sole carbon and energy source (Zimmermann et al. 1982; Coughlin et al. 1999). However, these organisms could not grow on structurally analogous sulfonated dyes.

The decolourisation of dye by bacteria is efficient and fast, but single bacterial strains usually cannot degrade azo dyes completely, and the degradation products are often toxic aromatic amines, which are more difficult to decompose (Joshi et al. 2008). The treatment systems composed of mixed microbial populations or bacterial consortia achieve a higher degree of biodegradation and mineralisation of dyes due to the synergistic or cometabolic activities of the microbial community (Khehra et al. 2005). Many researchers have reported the degradation of azo dyes by mixed cultures and bacterial consortia (Table 10.2).

The reduction of azo linkage may involve different mechanisms, such as enzymes; low molecular weight redox mediators; chemical reduction by biogenic reductants, such as sulphides; or a combination of these (Figure 10.2).

TABLE 10.2

Decolourisation of Various Azo Dyes by Bacteria

Name of Strain	Condition (pH, Temp. [°C], Agitation, Time [h])	Name of Dye, Initial Concentration and % Decolourisation	Type of Enzymes Involved	References
Decolourisation of Various Azo Dyes by Pure Bacterial Cultures				
Alishewanella sp.	7.0, 37°C, static, 6 h	Reactive Blue 59, 2500 ppm, 95%	Azo reductase, NADH-DCIP reductase	Kolekar and Kodam (2012)
Listeria sp.	37°C, aerobic, 48 h	Black B, 50 ppm each, 69%, Black HFGR, 74%, Red B5, 70%	NA	Kuberan et al. (2011)
Bacillus sp.	7.0, 40°C, aerobic, 24 h	Congo Red, 50 ppm, 85%	NA	Sawhney and Kumar (2011)
Micrococcus sp.	6.0, 35°C, aerobic, 48 h	Orange MR, 100 ppm, 93.18%	NA	Rajee and Patterson (2011)
Pseudomonas aeruginosa	7.0, 30°C, static, 24 h	Remazol Orange, 200 ppm, 94%	NA	Sarayu and Sandhya (2010)
Pseudomonas sp.	7.0, 35°C, static, 70 h	Reactive Blue 13, 200 mg/L, 83.2%	NA	Lin et al. (2010)
Aeromonas hydrophila	30°C, static, 1 h	Reactive Black 5, 300 ppm each, 67%, Reactive Red 198, 86%	NA	Hsueh et al. (2009)
Pseudomonas sp. SUK1	6.2–7.5, 30°C, static, 6 h	Reactive Red 2, 5 g/L, 96%	Oxidative and reductive	Kalyani et al. (2008)

(Continued)

TABLE 10.2 (CONTINUED)

Decolourisation of Various Azo Dyes by Bacteria

Name of Strain	Condition (pH, Temp. [°C], Agitation, Time [h])	Name of Dye, Initial Concentration and % Decolourisation	Type of Enzymes Involved	References
Decolourisation of Various Azo Dyes by Mixed Culture/Bacterial Consortium				
Bacterial consortium	7.0, 27°C, aerobic	Reactive Blue Bezaktiv 150, 96%	NA	Khouni et al. (2012)
Galactomyces geotrichum and Brevibacillus laterosporus	9.0, 30°C, aerobic-microaerophilic, 24 h	Golden Yellow, 50 ppm, 100%	Lac, Tyr, azo reductase, riboflavin reductase	Waghmode et al. (2011)
Consortium-GR (P. vulgaris and M. glutamicus)	7.0, 37°C, static, 3 h	Scarlet R and mixture of eight dyes, 50 mg/L each, 100%	Reductive	Saratale et al. (2009)
Bacillus vallismortis, Bacillus pumilus, Bacillus cereus, Bacillus subtilis, Bacillus megaterium,	37°C, aerobic, 96 h	Congo Red, 10 ppm each, 96%, Direct Red 7, 89.6%, Acid Blue 113, 81%, Direct Blue 53, 82.7%	NA	Tony et al. (2009)
Mixed microbial culture	NA, 37°C, static, 240 h	Direct Black 38, 100 mg/L, 100%	NA	Kumari et al. (2007)
Mixed cultures	6.0, 37°C, aerobic, 120–240 h	Reactive and disperse textile dyes, 0.5 g/L, 100%	NA	Asgher et al. (2007)
Mixed, mesophilic methanogenic culture	7.0, 35°C, static, 250 days	Reactive Red 2, 50–2000 mg/L, 92–87 (mg/L/day)	NA	Beydilli and Pavlostathis (2005)
Mixed culture of bacteria	30, 200 rpm, combined anaerobic–aerobic system, 24 h	Remazol Brilliant Orange 3R, Remazol Black B, and Remazol Brilliant Violet 5R; (100 mg/L), 78.9	NA	Supaka et al. (2004)

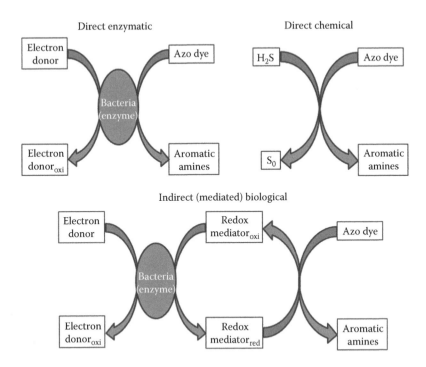

FIGURE 10.2
Different mechanisms of anaerobic azo dye reduction.

Enzymatic degradation of azo dyes requires azoreductases in the presence of reducing equivalents, NADH and NADPH (Sugumar and Thangam 2012). Azoreductases involve sequential transfer of four electrons from NADH to the azo linkage of dye via FMN. The flavin could carry out this process in two sequential steps. In each step, two electrons are transferred to the azo dye (terminal electron acceptor) by NADH or NADPH reduction with subsequent oxidation by the azo compound via a hydrazo intermediate, resulting in dye decolourisation and the formation of colourless aromatic amines (Deller et al. 2006). Anjaneya et al. (2011) studied the degradation of Metanil Yellow by *Bacillus* sp. strain AK1 and *Lysinibacillus* sp. strain AK2 in which metanilic acid and p-aminodiphenylamine are generated. Singh et al. (1991) also studied the azoreduction of Metanil Yellow using caecal microflora and detected the same metabolites. The mechanism of azoreduction of this dye may be proposed as depicted in Figure 10.3.

Azo bond degradation is oxygen labile; the presence of oxygen usually inhibits the azo bond reduction activity. Rafii et al. (1990) first reported the presence of oxygen-sensitive azoreductases in anaerobic bacteria, which mainly belonged to the genera *Clostridium* and *Eubacterium*, that decolourised sulphonated azo dyes during growth on complex media. For sulphonated azo dye, reduction could involve cytosolic flavin-dependent reductases, which

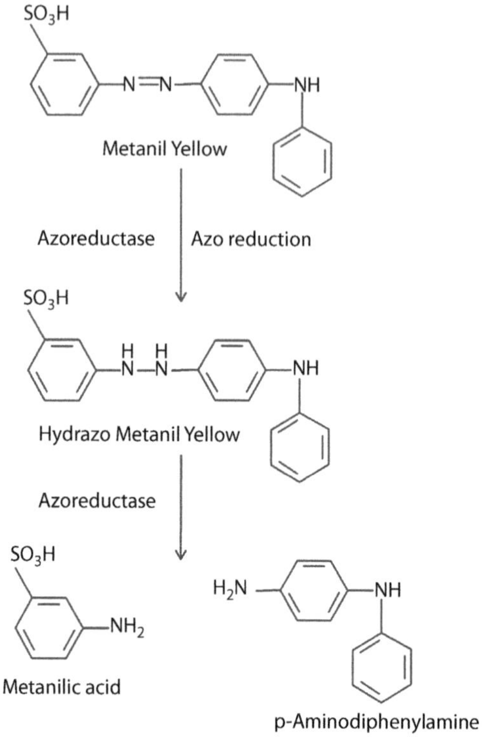

FIGURE 10.3
Proposed pathway of Metanil Yellow (azo dye) degradation.

transfer electrons via soluble flavins to azo dyes in bacterial cells with intact cell membranes (Russ et al. 2000). In bacteria that possess electron transport systems in their membranes, as in the case of aerobic or facultatively anaerobic bacteria, such as *Sphingomonas* strain BN6, the transfer of electrons from the respiratory chain to appropriate redox mediators could take place directly (Russ et al. 2000). The redox mediator acts as an electron shuttle between the dye and the reductase (Chacko and Subramaniam 2011). Some aerobic bacteria are able to reduce azo compounds with the help of oxygen-insensitive intracellular azoreductases and produce aromatic amines (Lin et al. 2010). These intracellular azoreductases showed high specificity to dye structures and required NADPH and NADH as cofactors for activity. The intracellular reduction of sulphonated azo dyes requires not only the presence of azoreductases but also a specific transport system that allows the uptake of the dye into the cells because bacterial membranes are almost impermeable to flavin-containing cofactors and, therefore, restrict the transfer of reducing equivalents by flavins from the cytoplasm to the sulphonated azo dyes (Russ et al. 2000). The ability of *X. azovorans* KF46F and *Sphingomonas* sp. strain ICX to take up Acid Orange 7 and reduce the dye in vivo shows the presence of transport as well as azoreductase enzymes in these organisms.

Reduction of sulphonated azo dyes is not dependent on their transport into the cell (Russ et al. 2000). Such mechanism involves the establishment of a link between bacterial intracellular electron transport systems and the high molecular weight azo dye molecules in the extracellular environment via a redox mediator at the bacterial cell surface. Many reports are available on the role of redox mediators in the reduction of azo bond by bacteria under anaerobic conditions (Keck et al. 1997; Brige et al. 2008). For example, Keck et al. (1997) observed that the cell suspensions of *Sphingomonas* sp. strain BN6, grown aerobically in the presence of 2-naphthyl sulfonate (NS), exhibited a 10- to 20-fold increase in decolourisation rate of an azo dye, Amaranth, under anaerobic conditions, over those grown in its absence. Even the addition of culture filtrates from these cells could enhance anaerobic decolourisation by cell suspensions grown in the absence of NS. The mechanism for redox mediator–dependent reduction of azo dyes using bacterial cell under anaerobic conditions is shown in Figure 10.4.

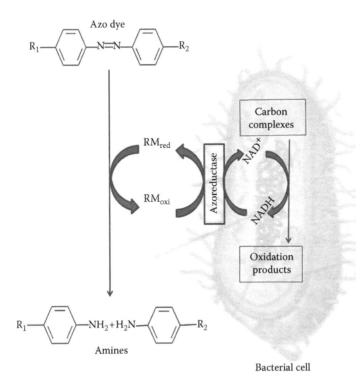

FIGURE 10.4
Mechanism for the redox mediator (RM)-dependent reduction of azo dyes by bacteria under anaerobic conditions.

10.2.1.2 Decolourisation and Degradation of Triphenylmethane Dyes by Bacteria and Its Mechanism

Many bacteria are capable of decolourising various triphenlymethane dyes under anaerobic conditions. *Shewanella decolorationis* NTOU1 could degrade Malachite Green (MG) by the way of the reduction or N-demethylation gradually under anaerobic conditions (Chen et al. 2010). Ayed et al. (2013) reported the decolourisation of triphenylmethane dye fushine, Malachite Green and Crystal Violet (CV) by *Staphylococcus epidermidis* isolated from contaminated soil. They observed that the decolourisation of these dyes involves lignin peroxidase (Lip) and laccase. A number of triphenylmethane dyes, such as Magenta, Crystal Violet, Pararosaniline, Brilliant Green and Malachite Green, were decolourised by *Kurthia* sp. under aerobic conditions (Sani and Banerjee 1999). *Pseudomonas otitidis* WL-13 showed high dye decolourisation properties under static and shaking conditions, although the decolourisation was comparatively slower under static conditions (Jing et al. 2009). Moreover, some authors reported the decolourisation of triphenylmethane dye by bacterial consortium and mixed cultures. Sharma et al. (2004) reported the decolourisation of Acid Violet dye by a consortium of five bacterial isolates. Al-Garni et al. (2013) achieved decolourisation of crystal violet by mixed bacterial culture of *P. fluorescens* and *Corynebacterium* sp. Cheriaa et al. (2012) developed a bacterial consortium CM-4 based on four bacterial strains: *Agrobacterium radiobacter*, *Bacillus* spp., *Sphingomonas paucimobilis* and *Aeromonas hydrophila*, which decolourise the Crystal Violet and Malachite Green.

The enzymes that catalyse the decolourisation of triphenylmethane dye are DCIP reductase, MG reductase, laccase, tyrosinase and lignin peroxidase. These enzymes are generally catalysing the depolymerisation, demethoxylation, decarboxylation, hydroxylation and aromatic ring opening reactions. Yatome et al. (1993) were the first to observe the degradation of Crystal Violet by *Nocardia* sp. and identify the main product as Michler's ketone. Further, Chen et al. (2007) described the biodegradation pathway of Crystal Violet by *Pseudomonas putida* and concluded that the dye is broken down by demethylation. The reduction of Malachite Green leads to formation of Leuco Malachite Green (LMG) and further into its derivative. The reductase from *Citrobacter* sp. strain KCTC 18061P could transform MG to Leuco Malachite Green (Jang et al. 2005). *Sphingomonas* sp. CM9 could use LMG as the sole source.

Wang et al. (2012) proposed a pathway for decolourisation of Malachite Green by *Exiguobacterium* sp. (Figure 10.5). The reactions of this pathway included reduction and N-demethylation and produced tridesmethyl MG or tridesmethyl Leuco Malachite Green. Benzene ring removal can be found in desmethyl-LMG's cleavage to produce (4-dimethylamino-phenyl)-phenyl-methanone and benzene, which also involves the oxidation reaction and breaking of the C–C bond. The breaking of the C–C bond results in the

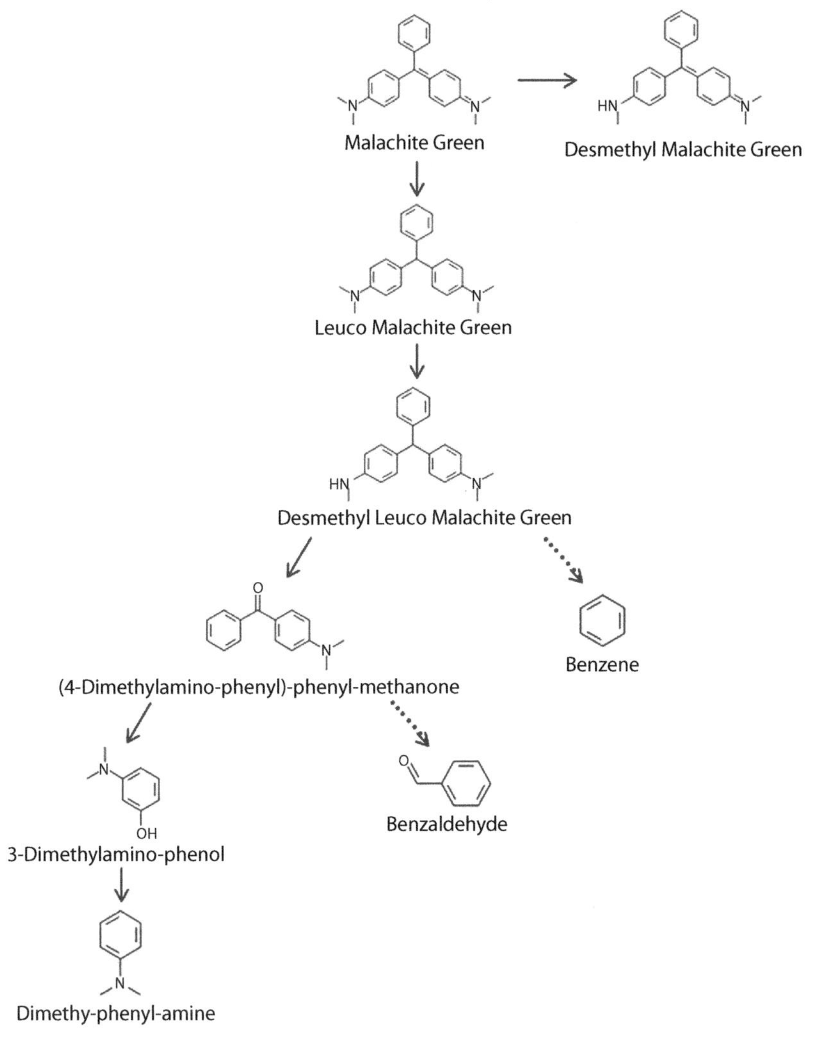

FIGURE 10.5
The proposed pathway of Malachite Green degradation in *Exiguobacterium* sp. MG2. (From Wang J et al., *PLoS ONE*, 7, 12, e51808, 2012.)

emergence of 3-dimethylamino-phenol and benzaldehyde. Finally, N, N-dimethylaniline formation requires the reaction of hydroxyl removal. In bacterial decolourisation of MG, the substrate is only known to transform into colourless LMG by an enzyme triphenylmethane reductase (TMR) (Jang et al. 2005). Jang et al. (2005) cloned a triphenylmethane reductase gene isolated from the Gram-negative *Citrobacter* sp. KCTC 18061P, which decolourise the triphenylmethane dyes. They also observed that TMR, like azoreductases, uses either NADPH or NADH as a cofactor.

10.2.1.3 Decolourisation and Degradation of Anthraquinone Dyes by Bacteria and Its Mechanism

Anthraquinone dyes have the chromophore group, $=C=O$, forming an anthraquinone complex. Many of these dyes have been reported to convert to harmful compounds, such as benzidine (Itoh et al. 1996). Several researchers reported that most anthraquinone dyes are quite resistant to biodegradation under anaerobic condition. Thongchai and Worrawit (2000) have reported that anthraquinone dyes Reactive Blue 5 and Reactive Blue 19 were not bio-degraded under anaerobic processes, but the decolourisation of these dyes was observed. It was merely a result of adsorption onto bacterial floc materials. Aksu (2001) investigated the biosorption of RB2 (anthraquinone dye with a monochlorotriazinyl reactive group) onto predried activated sludge and found that 56% of 100 mg/L dye was removed by sorption onto 500 mg/L biomass at 25°C. However, Deng et al. (2008) reported the decolourisation of Acid Blue 25 under anaerobic conditions by *Bacillus cereus* strain DC11strain via a NADH-/NADPH-dependent process. Giwa et al. (2012) also reported the bacterial decolourisation of an anthraquinone dye C. I. Reactive Blue 19 by *Bacillus cereus* under anaerobic conditions. Moreover, Lee et al. (2006) suggested that the decolourisation of reactive anthraquinone dyes is feasible only under anoxic/anaerobic conditions. Very few reports are available on decolourisation of anthraquinone dyes under aerobic conditions. Gurav et al. (2011) reported the decolourisation of Vat Red 10 dye by *Pseudomonas desmolyticum* NCIM 2112 and *Galactomyces geotrichum* MTCC 1360 under aerobic conditions.

The enzymes that catalyse the decolourisation of anthraquinone dyes are DCIP reductase, laccase, lignin peroxidase (LiP), manganese peroxidase (MnP), etc. Verma and Madamwar (2003) reported that decolourisation of Procion Brilliant Blue-H-GR by *Serratia marcescens* is catalysed by the action of MnP. Abadulla et al. (2000) used the engineered *Pseudomonas putida* cells with a bacterial laccase (WlacD) to decolourise the anthraquinone dye Acid Green 25. They found that decolourisation of Acid Green 25 dyes are Cu^{2+} and mediator-independent. This study demonstrates the methodology by which the engineered *P. putida* with surface-immobilised laccase was successfully used as a regenerable biocatalyst for biodegrading synthetic dyes, thereby opening new perspectives in the use of biocatalysts in industrial dye biotreatment. Gurav et al. (2011) reported that the degradation of Vat Red 10 by *Pseudomonas desmolyticum* NCIM 2112 and *Galactomyces geotrichum* MTCC 1360 leads to the formation of intermediates, such as diisopropylnaphthalene and naphthalene.

10.2.1.4 Factors Affecting the Rate of Decolourisation and Degradation of Dyes by Bacteria

The efficiency of the bacterial treatment process for textile dye is greatly influenced by the various physicochemical and nutritional operational parameters, such as the level of aeration, supplementation of different carbon and nitrogen sources, pH, temperature, dye structure and dye concentration.

10.2.1.4.1 *Effects of Aeration*

The presence of oxygen can either favour or inhibit the microbial degradation of azo dyes. Azoreductase-mediated degradation of azo dyes is inhibited by the presence of oxygen because oxygen was a preferable terminal electron acceptor over the azo groups in the oxidation of reduced electron carriers, such as NADH (Chang and Kuo 2000). Several researchers reported efficient dye decolourisation under static (anaerobic) conditions as compared to shaking (aerobic) conditions. Telke et al. (2008) observed 90% decolourisation of sulphonated diazo dye Reactive Red 141 under the static culture condition rather than 6% under the shaking condition by *R. radiobacter*. Wang et al. (2009b) reported more than 92% decolourisation of Reactive Black by *Enterobacter* sp. EC3 under anaerobic conditions compared to only 22% in aerobic conditions.

10.2.1.4.2 *Effects of Carbon Source Supplements*

Azo dyes are deficient in carbon sources, and microbial degradation of dyes without any supplement of carbon or nitrogen sources is very difficult (Levin et al. 2010). The decolourisation of azo dye by a microbial system generally requires complex organic sources, such as yeast extract, peptone or a combination of complex organic sources and carbohydrates (Khehra et al. 2005). These sources generate reducing equivalents, which transferred to the dye during the decolourisation process.

Several reports are available for the decolourisation of dye in the presence of additional carbon sources. Joe et al. (2008) reported that the addition of glucose to the medium enhances the decolourisation rate of Reactive Red 3B-A and Reactive Black 5 by *C. bifermentans* strains. Singh et al. (2014a) also reported that the decolourisation rate of Acid Orange by the *Staphylococcus Hominis* RMLRT03 strain was increased in the presence of glucose as a cosubstrate. The rate of decolourisation of Reactive Black 5 by *Enterobacter* sp. EC3 was found to be only 20.11%, whereas addition of glucose enhanced the decolourisation rate up to 90% (Wang et al. 2009b).

10.2.1.4.3 *Effects of Nitrogen Sources Supplements*

Organic (Peptone, yeast extract, etc.) and inorganic (ammonium sulphate, ammonium chloride, etc.) nitrogen sources as a cosubstrate are also important for bacterial decolourisation of dyes. The metabolism of organic nitrogen sources regenerate NADH, which acts as an electron donor for the reduction of azo dyes by a bacterial system (Carliell et al. 1995). In the presence of yeast extract or peptone, *A. hydrophila* efficiently decolourises the RED RBN dye (Chen et al. 2003). Telke et al. (2008) found that urea and yeast extract were effective cosubstrates for better decolourisation of Reactive Red 141 by *R. radiobacter*, whereas Moosvi et al. (2005) reported maximum decolourisation of Reactive Violet 5 by bacterial consortium RVM 11.1 in the presence of peptone and yeast extract as cosubstrates.

10.2.1.4.4 Effects of pH

pH plays a critical role for transport of nutrients across the cell membrane. The optimum pH for decolourisation of dyes is often at a neutral pH value or a slightly alkaline pH value. Generally, altering the pH within a range of 6.0 to 10.0 has very little effect on the dye reduction process.

10.2.1.4.5 Effects of Temperature

In many bacterial systems, the decolourisation rate of azo dyes increases with increasing temperature up to the optimal temperature within a defined range that depends on the system. Afterwards, there is a marginal reduction in the decolourisation activity. The reduction in decolourisation rate at a higher temperature may be attributed to thermal deactivation of azoreductase enzymes or loss of cell viability (Saratale et al. 2011).

10.2.1.4.6 Effects of Dye Concentration

The rate of dye decolourisation was gradually decreased with increasing concentration of dye due to the toxic effect of dyes on degrading microorganisms or the blockage of active sites of azoreductase enzymes by dye molecules with different structures (Tony et al. 2009).

10.2.2 Algal Decolourisation and Degradation of Dyes

Many phenolic compounds show harsh toxicity to algae as both cyanobacteria and eukaryotic microalgae are expert at biotransforming aromatic xenobiotics, including dyes and other industrial effluents. For example, *Ochromonas danica* algae have been shown to have the capacity to biodegrade xenobiotics in polluted fresh water sources.

In addition to giving the oxygen for aerobic bacterial biodegraders, microalgae can also biodegrade organic pollutants directly (Mallick 2002; Tam et al. 2002). It was accounted that more than 30 azo compounds were biodegraded and decolourised by *Chlorella pyrenoidosa*, *Chlorella vulgaris* and *Oscillatoria tenuis* in which azo dyes were decayed into simpler aromatic amines (Zhang et al. 2006). Limited growth of green microalga was accompanied by COD decline; the high percentage of COD removal for direct blue was 60.1% and 46.7% for indigo. The percentage of decolourisation for tested dyes showed that algal colour removal competence values varied from one dye to another: Indigo (89.3%), Direct Blue (79%), Remazol Brilliant Orange (75.3%) and Crystal Violet (72.5%). Results showed that isolated algae contribute to eliminating colour with a significant decrease in COD in different samples of industrial textile dyes. Several groups showed that microalgae may serve as a solution to budding environmental problems, such as the greenhouse effect and waste treatments (Craggs et al. 1997; Korner and Vermaat 1998; Parikh and Madamwar 2005). Biological remediation by unicellular green microalga was different from one dye to another; this may be credited to the adsorption or/and to the biodegradation by microalgae.

Reactive dyes cause respiratory and nasal symptoms (Docker et al. 1987), asthma and rhinitis, dermatitis (Nilsson et al. 1993), allergic contact dermatitis (Estlander 1988; Wilkinson and McGeachan 1996), mutagenicity (Przybojewska et al. 1989; Mathur et al. 2005), genotoxicity (Przybojewska et al. 1989; Dogan et al. 2005), carcinogenicity (Gonzales et al. 1988; De Roos et al. 2005) and teratogenicity (Birhanli and Ozmen 2005). Therefore, industrial effluents containing dyes should be practiced before their discharge into the environment. Jinqi and Houtian (1992) studied that azoreductase of algae is liable for degrading azo dyes into aromatic amines by breaking the azo linkage, and they have also found that azoreductase of *Chlorella vulgaris* in a provoked enzyme and the substrate itself can act as a kind of inducer. Pandey et al. (2007) reported that azoreductase can act in the breaking down of the azo bond as follows:

$$\text{Ar-N=N-Ar} \xrightarrow{\text{Azoreductase}} \text{Ar-NH2} + \text{Ar-NH2}$$

Combining the uptake removal efficiency of algae on the organic matter and the strong degradation capacity of bacteria on toxic pollutants, a composite algal–bacterial system can be formed. Recent studies have therefore shown that when proper methods for algal selection and development are used, it is possible to use microalgae to create the O_2 required by acclimatised bacteria to biodegrade hazardous pollutants, such as polycyclic aromatic hydrocarbons, phenolics and organic solvents (Muñoz and Guieysse 2006). The degradation efficacy of an algal–bacterial system was significantly higher than that of a bacteria system, suggesting that the continuation of algae can progress the growth and vitality of bacteria, and a synergistic metabolism seems to occur. In light conditions, algae absorb light energy and compile organic matter using CO_2 and assimilate anilines. At the same time, the growth of algae can produce a large amount of O_2 to provide positive conditions for bacterial degradation to anilines (Borde et al. 2003). Furthermore, bacteria can decompose organic substances secreted by algae to speed up the degradation of pollutants (Teresa and Ruben 2006). Dyes have been integrated into the algae through interaction with active functional groups. The decline rate appears to be related to the molecular structure of dyes and the species of algae used.

10.2.2.1 Mechanism of Dye Decolourisation and Degradation by Algae

The mechanisms of algae discolouration can occupy enzymatic degradation, adsorption, or both. Similar to bacteria, algae are capable of degrading azo dyes through an azoreductase to break the azo bond, resulting in the production of aromatic amines (Chacko and Subramaniam 2011). Oxidative enzymes are also concerned in the discolouration process (Priya et al. 2011).

10.2.2.2 Factors Affecting the Rate of Decolourisation and Degradation of Dyes by Algae

10.2.2.2.1 *Effects of Aeration*

In addition to providing oxygen for aerobic bacterial biodegradation, micro-algae can also biodegrade organic pollutants directly (Mallick 2002; Tam et al. 2002). Previous studies showed that more than 30 azo compounds were biodegraded and decolourised by *Chlorella Pyrenoidora*, *Chlorella vulgaris* and *Oscillatoria tenvis* in which azo dyes were decayed into simple aromatic amines (Zhang et al. 2006).

10.2.2.2.2 *Effects of Carbon Source Supplements*

Carbon is the limiting factor for growth of algae. CO_2, HCO_3^-, CO_3^- and organic compounds can all provide the source of carbon for algae dye degradation under specific conditions. Under anaerobic conditions, different end products are formed than under aerobic conditions, but many of these compounds produced under anaerobic conditions are further oxidised under aerobic conditions. In general, CO_2 is the end product for most of the organic carbon compounds. The product compounds of dye degradation, viz., CO_2, NH_3, NO_3 and PO_4, can be used again for growth of photosynthetic organisms.

10.2.2.2.3 *Effects of Nitrogen Source Supplements*

Nitrogen is a major constituent for yeast dye degradation. NO_3^- and NH_4^+ are the most important organic nitrogen sources. Urea may act as an ideal source of N for some cyanobacteria in natural aquatic systems.

10.2.2.2.4 *Effects of pH*

Alkaline pH promotes the biosorption capacity of algae. Unicellular green algae has both autotrophic and heterotrophic growth with a photosynthetic activity producing pigments (chlorophyll a). For the green algae, the optimal culture condition was pH 8.

10.2.2.2.5 *Effects of Temperature*

Temperature is the important factor for algal dye degradation. The optimum temperature for degradation of dyes by green algae is 25°C.

10.2.2.2.6 *Effects of Dye Concentration*

The decolourisation of dye was greatly influenced by the concentration of dye. Rate of dye decolourisation was gradually decreased with increasing concentration of dye due to the toxic effect of dyes on degrading microorganisms or the blockage of active sites of azoreductase enzymes by dye molecules with different structures (Tony et al. 2009). Cetin and Donmez (2006) reported that the higher the dye concentration, the longer the time required to remove the dye. They observed that at 125.6–206.3 mg/L initial dye concentrations of Remazol Blue, 100%–90% dye is decolourised at the end of a 30-h incubation period, but when concentration of dye was increased up to 462.5 mg/L, 90% decolourisation was found by mixed cultures after a 50-h incubation period. However, Saratale et al. (2009a) found that bacterial coculture reduced the increasing concentration effect of dye instead of pure culture, and this might be due to the synergistic effect of both microorganisms. Growth of green algae was measured at 665 nm wavelength. The algal growth was composed of cells activated in a medium without dyes. The percentage of chemical oxygen demand (COD) removal was calculated as follows:

$$COD = \frac{COD \text{ at } t_0 - COD \text{ at } t_1}{COD \text{ at } t_0} \times 100$$

The COD was measured after removing algal cells by centrifugation at 3000 rpm. Percentage of dye decolourisation was calculated as follows:

$$\% \text{ Decolourisation} = \frac{\text{Absorbance at } t_0 - \text{Absorbance at } t_1}{\text{Absorbance at } t_0} \times 100$$

At t_0, COD and colour absorbance were measured at the 0 time interval. At t_1, the COD and colour absorbance were measured at 5–7 days interval.

10.2.3 Fungal Decolourisation and Degradation of Dyes

Dye decolourisation by microorganism is generally performed by bacteria and basidiomycete fungi. The fungal treatment of dyes is an economical and possible alternative to the present treatment technologies (Knapp et al. 2001; Singh 2006). A particular attention was devoted to the white-rot fungi and azo dyes, the largest class of commercial dyes.

White-rot fungi are able to degrade lignin, the structural polymer of woody plants (Barr and Aust 1994). The white-rot fungi are, so far, the microorganisms most competent in degrading synthetic dyes with basidiomycetous fungi that are capable of depolymerising and mineralising lignin. Among various fungi, *Phanerochaete chrysosporium* is an efficient dye-detoxifying white-rot fungus (Knapp et al. 1995; Swamy and Ramsay 1999; Faraco et al. 2009). *P. chrysosporium* is a basidiomycete, having a capability to degrade complex compounds, such as starch, cellulose, pectin, lignin and lignocelluloses, which are characteristics of textile dye effluent. *P. chrysosporium* has emerged as a model system in textile and pulp and paper mill effluent remediation. *P. chrysosporium* has been reported to decolourise dyes using various enzymes implicated in lignin degradation, such as lignin peroxidase (Lip), manganese peroxidase (MnP) and laccase (Lac) (Swamy and Ramsay 1999; McMullan et al. 2001; Robinson et al. 2001).

In the next decade, few novel species of white-rot fungi, such as *Pleurotus ostreatus* and *Trametes versicolour* (Heinfling et al. 1997; Sukumar et al. 2009;

Pazarlioglu et al. 2010), were characterised by a capacity to degrade dyes. More powerful research was focused on *Irpex lacteus* (Novotny et al. 2009) and *Bjerkandera adusta* (Robinson et al. 2001; Eichlerova et al. 2007) in the last decade; however, in the last few years, a vast interest was generated on the decolourisation capacity of *Ceriporiopsis subvermispora* (Babic and Pavko 2007; Tanaka et al. 2009) and *Dichomites squalens* (Eichlerova et al. 2006; Pavko and Novotny 2008). *Trichoderma harzianum* has been reported to degrade textile dyes; however, *Aspergillus flavus* has been uncovered to degrade/decolourise textile dyes, including Bromophenol Blue and Congo Red. *A. flavus* shows positive results for the degradation of Bromophenol Blue and, as a result, the blue colour turns into yellow that appears around the mycelium. The positive results are also obtained for the degradation of Congo Red dye; however, it degrades more proficiently than that of Bromophenol Blue.

10.2.3.1 Mechanisms of Dye Decolourisation and Degradation by Fungi

Mechanism of fungal dye degradation involves biosorption, biodegradation and mineralisation. Biosorption is the accumulation of chemicals within the microbial biomass, which is referred to as biosorption that takes place in living or dead biomass. Waste and/or dead microbial biomass can be used as a proficient adsorbent, especially if they contain a natural polysaccharide chitin and its derivative chitosan in the cell walls. Chitosan, a cell wall component of many industrially useful fungi, has an exclusive molecular structure with a high attraction for many classes of textile dyes (Joshi et al. 2004).

Enzymatic degradation involves a family of peroxidases secreted by the fungi. The biodegradable ability of white-rot fungi to degrade and decolourise dyes is related to the capacity of fungi to degrade lignin. It is known that most of the white-rot fungi make at least two of the three highly nonspecific enzymes, such as lignin peroxidase (LiP), maganese peroxidase (MnP) and laccase, which facilitate the generation of free radicals when conducting a variety of reactions (Knapp et al. 2001; Pointing 2001). The mechanism of azo dye oxidation by peroxidases, such as LiP, probably involves the oxidation of the phenolic group to produce a radical at the carbon bearing the azo linkages. The water attacks this phenolic carbon to cleave the molecule producing phenyldiazene, and the phenyldiazene can then be oxidised by a one-electron reaction generating N_2 (Chivukula and Renganathan 1995). Synthesis and secretion of these enzymes are often stimulated by limited levels of nutrients, such as carbon and nitrogen sources. When fungal mycelium of *P. chrysosporium* is added to dye effluent containing the nutrient solution and maintained at 37°C, the solution becomes turbid as the growth of mycelium progresses in the solution. First, dye gets adsorbed on the mycelium; then the absorbed dye was decolourised by the fungi. The decolourisation was contingent on the decrease in the optical density of the dye effluent. Optical density of the solution was measured in the UV–visible spectrophotometer at the specific wavelength of the specific dye.

$$\% \text{ Decolourisation} = \frac{\text{Initial absorbance} - \text{Final absorbance}}{\text{Initial absorbance}} \times 100$$

10.2.3.2 Factors Affecting the Rate of Decolourisation and Degradation of Dyes by Fungi

The fungal growth and the enzyme production, and accordingly, decolourisation and degradation, are influenced by numerous factors, for example, carbon and nitrogen sources, pH value, agitation and aeration, temperature and initial dye concentration. Their effects are briefly presented and discussed in the following.

10.2.3.2.1 *Effects of Aeration*

Ligninolytic fungi are obligate aerobes, and therefore, they need oxygen for the growth and maintenance of their viability. In addition, lignin degradation may also require oxygen, either for the mycelial generation of H_2O_2 for peroxidases or for the direct function of oxidases. Oxygen could also act directly on lignin fragments. The oxygen demand depends on the fungus and its ligninolytic system. The oxygen supply to the culture media during the cultivation has been an attractive research approach. The major problem is its low water solubility, which is only 8 mg/L at 20°C. The aeration and agitation are essential steps in order to satisfy the microbial oxygen demand during the cultivation and to boost the oxygen gas–liquid mass transfer. This might affect the morphology of filamentous fungi, thereby decreasing the rate of enzyme production (Znidarsic and Pavko 2001).

10.2.3.2.2 *Effects of Carbon Source Supplements*

A carbon source is crucial for the fungal growth that provides supply for the oxidants that are required for the decolourisation of textile dyes. Glucose is a main carbon source; alternative sources are fructose, sucrose, maltose, xylose and glycerol, including starch and xylan that seems to be useful. Surprisingly, cellulose and its derivatives are not effective carbon sources. For the initial experiments, glucose at 5–10 g/L is a good choice. The need to add a carbon source depends on the organism and type of dye to be treated.

10.2.3.2.3 *Effects of Nitrogen Source Supplements*

The nitrogen demand differs strikingly among fungal species on the basis of fungal growth and their enzyme production. It is well known that the production of lignolytic enzymes with *P. chrysosporium* is much more successful under the conditions of nitrogen limitation. However, *B. adusta* produces more LiP and MnP in nitrogen-sufficient media. Both organic and inorganic nitrogen sources are utilised by white-rot fungi. Because the organic nitrogen sources do not appear to be advantageous, the inorganic nitrogen sources, mostly ammonium salts, have been used in fungal growth research and enzyme production.

10.2.3.2.4 Effects of pH

The decolourisation can be conducted with a whole fermentation broth (mycelium and enzymes) or with isolated enzymes. It has to be distinguished between the optimum pH for growth and enzyme production, the optimum pH for the action of isolated enzymes and the optimum pH for dye degradation. Therefore, optimum pH depends on the medium, fungus and its enzyme system as well as on the decolourisation under consideration. The majority of researchers propose that the optimum pH values are likely to be in the range of 4 to 4.5 (Knapp et al. 2001).

10.2.3.2.5 Effects of Temperature

The temperature influences the growth, enzyme production and enzymatic decolourisation rate and waste stream temperature. The optimal temperature in the range of 27°C–30°C are best for cultivation of white-rot fungi. For enzymatic reactions, the optimal temperatures are usually higher, but above 65°C, the enzyme instability and degradation take place. Various textile and dye effluents are produced in the range of temperatures 50°C–60°C. The optimal temperature for a particular decolourisation process has to be selected from case to case according to the mentioned parameters (Knapp et al. 2001; Singh 2006).

10.2.3.2.6 Effects of Dye Concentration

It is significant to optimise the initial dye concentration for colour removal. Dyes are usually toxic to microorganisms; however, their toxicity depends on the types of dye. Higher dye concentrations are always toxic. The range of initial dye concentrations studied in the literature generally varies from 50 to 1000 mg/L and depends on the investigated microorganism and types of dye.

10.2.4 Decolourisation and Degradation of Dyes by Yeast

Only a partial amount of studies about yeast decolourisation have been reported. Compared to the bacteria and filamentous algae, yeast exhibits attractive features. In recent years, there has been demanding research on dye degradation by yeast species. It is becoming a hopeful alternative to replace or supplement existing treatment processes. Compared to the bacteria and filamentous fungi, yeasts have a lot of advantages. Yeasts are an inexpensive, readily obtainable source of biomass. They not only grow quickly like bacteria, but they also have the ability to resist unfavourable environments, such as low pH, etc., by many filamentous fungi (Yu and Wen 2005). Furthermore, some yeast has been found to be capable of treating high-strength organic wastewater, such as food industry effluents, which are highly coloured (Yang 2003). Several strains of yeast have been reported that decolourise the textile dyes (Das and Charumathi 2012) (Table 10.3).

TABLE 10.3

Decolourisation and Degradation of Synthetic Dyes Using Yeasts

Yeast	Dyes	Mechanism	References
Candida krusei, *C. tropicalis*	Basic violet 3	Biodegradation, bioaccumulation, biosorption	Charumathi and Das (2010, 2011); Das et al. (2011)
Pichia fermentans MTCC 189	Basic violet 3 Acid Blue 93 Direct Red 28	Bioaccumulation	Das et al. (2010)
Trichosporon beigelii NCIM-3326	Malachite Green Crystal violet Methyl violet Navy Blue HER Red HE7B Golden Yellow 4BD Green HE4BD Orange HE2R	Biodegradation	Saratale et al. (2009)
C. albicans	Direct violet 51	Biodegradation	Vitor and Corso (2008)
Galactomyces geotrichum MTCC 1360	Methyl Red Malachite Green	Biodegradation	Jadhav et al. (2008)
Galactomyces geotrichum MTCC 1360 and *Bacillus* sp.	Brilliant Blue G	Biodegradation	Jadhav et al. (2006)
Trichoderma sp.	Vilmafix Blue RR-BB	Biodegradation	Pajot et al. (2006)
C. oleophila	Reactive Black 5	Biodegradation	Lucas et al. (2006)
Saccharomyces cerevisiae MTCC 463	Malachite Green	Biodegradation	Jadhav and Govindwar (2006)
S. Italicus CICC1201, *S. chevalieri* CICC1611, *Torulopsis candida* CICC1040	Reactive Brilliant Red	Biodegradation	Yu and Wen (2005)
Pseudozyma rugulosa Y-48	Reactive Brilliant Red Weak Acid Brilliant Red B Reactive Black KN-B Acid Mordant Yellow Acid Mordant Light Blue B Acid Mordant Red S-80 Reactive Brilliant Blue X-BR	Biodegradation	Yu and Wen (2005)
C. krusei G-1	Reactive Brilliant Red Weak Acid Brilliant Red B Acid Mordant Yellow Acid Mordant Light blue B	Biodegradation	Yu and Wen (2005)
Issatchenkia occidentalis	Dye I, II, III, IV	Biodegradation	Ramalho et al. (2002)
S. cerevisiae	Remazol Black B Remazol Blue Remazol Red RB	Bioaccumulation	Aksu (2005)

(Continued)

TABLE 10.3 (CONTINUED)

Decolourisation and Degradation of Synthetic Dyes Using Yeasts

Yeast	Dyes	Mechanism	References
Debaryomyces polymorphus	Reactive Black 5 Reactive Red M-3BE Procion Scharlach H-E3G Procion Marine H-EXL Reactive Brilliant Red K-2BP Reactive Blue KNR Reactive Yellow M-3R	Biodegradation	Yang et al. (2005)
C. tropicalis	Reactive Black 5 Reactive Red -3BE Procion Scharlach H-E3G Procion Marine H-EXL Reactive Brilliant Red K-2BP Reactive Blue KNR Reactive Yellow M-3R	Biodegradation	Yang et al. (2003)
C. tropicalis	Ramazol Blue Reactive Red Reactive Black	Bioaccumulation	Ramalho et al. (2002)
Kluyveromyces marxinus IMB3	Remazol Black B Remazol Turquoise Blue Remazol Golden Yellow Cibacron Orange	Biosorption	Bustard et al. (1998)
P. anomala	Disperse Red 15	Biodegradation	Itoh et al. (1996)
Cryptococuss heveanensis, C. rugosa, Dekkera bruxellensis, K. waltii	Reactive Black 5 Reactive Black 19	Biosorption	Polman and Breckenridge (1996)
S. cerevisiae, P. casonii	Reactive Black 19	Biosorption	Polman and Breckenridge (1996)
Candida sp.	Procyon Black Procyon Blue	Biosorption	De Angelis and Rodrigues (1987)
Rhodotorula sp.	Crystal violet	Biodegradation	Kwasiewska (1985)
R. rubra	Crystal violet	Biodegradation	Kwasiewska (1985)

Source: From Das N and Charumathi D, *Indian J. Biotechnol.* 11: 369–380, 2012.

10.2.4.1 Mechanism of Dye Decolourisation and Degradation by Yeast

The mechanism of dye degradation by various yeasts involves many processes, such as biosorption, bioaccumulation and biodegradation. If the yeast biomass appears colourless after dye removal, it suggests that biodegradation of the dye has taken place. If the yeast biomass takes the colour of the dye, then the dye removal occurs either through biosorption or bioaccumulation. The biosorption of dyes by yeast includes the nitrogen-containing group peptidomannan, peptidoglucan or protein groups that comprise the yeast biomass (Ashkenazy et al. 1997) and active groups on the cell surface, such as acidic polysaccharides, lipids, amino acids and other cellular components of the microorganism (Brady et al. 1994; Volesky and Philips 1995; Aksu 2005). Biosorption depends on a range of parameters, such as pH, initial dye concentration, biosorbent dosage and temperature. Charumathi and Das (2010) extensively studied the influence of various parameters on biosorption of Basic Violet 3 by dried biomass of *Candida tropicalis* to remove the dye in a batch system.

Kumari and Abraham (2007) screened the biosorption potential of nonviable biomass of *Saccharomyces cerevisiae* with other fungal biomasses for the removal of anionic reactive dyes, viz., Reactive Black 8, Reactive Brown 9, Reactive Green 19, Reactive Blue 38 and Reactive Blue 3. They proposed that the adsorption of anionic dye by *S. cerevisiae* might involve the nitrogen-containing groups in the peptidomannan, peptidoglucan or protein groups that occupy the yeast biomass (Wang and Hu 2008) and active groups on the cell surface, such as acidic polysaccharides, lipids, amino acids and other cellular components of microorganism (Brady et al. 1994; Aksu 2005). Bustard et al. (1998) demonstrate that nonliving biomass of *K. marxianus* IMB3 produced during ethanol-producing fermentation was capable of biosorption of textile dyes (Remazol Black B, Remazol Turquoise Blue, Remazol Red, Remazol Golden Yellow and Cibacron Orange). They proposed that the copper (Cu) atom in the Remazol Turquoise Blue played a role in the interaction between the dye and the biosorbent, resulting in maximum uptake of this dye by the yeast.

Bioaccumulation can be accomplished for the deletion of different kinds of textile dyes if the growing cells get enough instead of dyes in the growth medium. Das et al. (2010) reported the bioaccumulation of synthetic dyes, viz., Acid Blue 93, Direct Red 28 and Basic Violet 3, by growing cells of *Pichia fermentans MTCC* 189 in a growth medium prepared from sugarcane bagasse extract. Ertugrul et al. (2009) have found that *Rhodotorula mucilaginosa* could accumulate reactive dye Remazol Blue and heavy metals, viz., chromium(VI), nickel(II) and copper(II), in a medium containing molasses as a carbon and energy source (Ertugrul et al. 2009).

Biodegradation is generally considered as a phenomenon of transformation of organic compounds by living organisms, particularly microbes. It has been considered as a natural process in the microbial world. The process of biodegradation provides a carbon and energy source for microbial growth and takes an essential role in the recycling of materials in the natural ecosystem. It brings about changes in the molecular structure of a compound eventually yielding simpler (mineralisation) and comparatively harmless (nontoxic) products, such as CO_2, H_2O, NH_3, CH_4, H_2S or PO_3. When the compound is not fully broken, it is termed biotransformation (Chatterji 2005).

Charumathi and Das (2011) proposed the biodegradation of Basic Violet 3 (BV) using *C. Krusei Trichos. Trichosporum beigelii* (NCIM-3326) was reported to decolourise and degrade reactive azo dye Navy Blue HER effectively into nontoxic compounds (Saratale et al. 2009b). Induction in the enzymatic activities of NADH-DCIP reductase and azoreductase most probably degrades the Scarlet RR (100 mg/L) dye within 18 h under the shaking condition in a malt yeast medium. It was planned that the degradation by *G. geotrichum* MTCC 1360 occurred by asymmetric cleavage by peroxidases that resulted in reactive products, which then underwent demethylation and a reduction reaction to produce stable intermediates.

Biodegradation of the Direct Violet 51 azo dye by *C. albicans* isolated from industrial effluents was proposed by Vitor and Corso (2008). Jadhav et al. (2008a,b) reported the biodegradation of Methyl Red by *G. geotrichum* MTCC 1360. Pajot et al. (2006) screened 63 yeast isolates obtained from live and dead sections of the tree *Phoebe porphyria* Mez and underlying soils from Las Yungas forests (Argentina) for decolourisation of textile dyes, viz., Vilmafix Yellow 4R-HE, Vilmafix Red 7B-HE, Vilmafix Blue RR-BB and Vilmafix Green RR-4B. The isolates screening higher decolourisation efficiency were identified as *Trichosporon* spp.

C. oleophila, nonconventional ascomycetes wild yeast isolates, was identified as an outstanding degrader of diazo dye Reactive Black 5 (Lucas et al. 2006). Jadhav and Govindwar (2006) reported effective biodegradation of Malachite Green, a triphenylmethane dye carried out by *S. cerevisiae* MTCC 463. A considerable increase in the activities of NADH-DCIP reductase and MG reductase was observed in the yeast cells after decolourisation, which indicated the prominent role of reductases in the degradation process.

Two novel yeast strains, viz., *Pseudozyma rugulosa* Y-48 and *C. Krusei* G-1, were reported to be competent for decolourisation (Yu and Wen 2005). The disappearance of absorption peaks for all the dyes established that the colour removal by *P. rugulosa* Y-48 and *C. Krusei* G-1 was largely because of biodegradation rather than biosorption onto the yeast cell surface. Moreover, the cells retained their original colour as a result of biodegradation of dyes.

Yang et al. (2005) reported the decolourisation of an azo dye Reactive Black 5 by a yeast isolate *Debaryomyces polymorphus*. It could totally degrade 200 mg/L of nonhydrolysed and hydrolysed Reactive Black 5 within 24 h of cultivation. A good connection was found between dye degradation and the manganese-dependent peroxidase (MnP) enzyme production. Yang et al. (2003) reported the degradation of six azo dyes, viz., Reactive Black 5, Reactive Red M-3BE, Procion Scharlach H-E3G, Procion Marine H-EXL, Reactive Brilliant Red K-2BP and Reactive Yellow M-3R, and an anthroquinone dye, Reactive Brilliant Blue KNR, by *D. polymorphus* and *C. trophicalis*; MnP activities were induced in *D. polymorphus* cells grown in five azo dyes, viz., Reactive Black 5, Reactive Red M-3BE, *Procion Scharlach* H-E3G, *Procion Marine* H-EXL and Reactive Brilliant Red K-2BP; its activity was detected in the culture of *C. tropicalis* containing only Reactive Red M-3BE and Reactive Brilliant Red K-2BP.

Biodegradation of Crystal Violet, a triphenylmethane dye, by oxidative red yeast, viz., *Rhodotorula* sp. and *R. rubra*, was reported by Kwasiewska (1985). A linear degradation of Crystal Violet by the yeasts between the second and fourth day of incubation was investigated, which indicated the presence of an enzyme system for the degradation of Crystal Violet. It was also observed that fermentative yeast *S. cerevisiae* could not degrade Crystal Violet. A yeast-bacteria consortium containing two microorganisms, viz., *G geotrichum* MTCC 1360 and *Bacillus* sp. VUS, was found to be capable of degrading the sulphur-containing dye Brilliant Blue G, optimally at pH 9 and 50°C.

10.2.4.2 Factors Affecting the Rate of Decolourisation and Degradation of Dyes by Yeast

10.2.4.2.1 *Effects of Aeration*

Mixing is necessary to avoid sedimentation of the algal growth for dye degradation to ensure that all cells of the population are evenly exposed to the light and nutrients, to avoid thermal stratification (for example, in outdoor cultures) and to progress gas exchange between the culture medium and the air. Proper ventilation also promotes the biosorption capacity. In poor aeration conditions, the CO_2 originating from the air (containing 0.03% CO_2) bubbled through the culture, limiting the algal growth.

10.2.4.2.2 *Effects of Carbon Source Supplements*

Carbon is the limiting factor for growth of algae. CO_2, HCO_3^-, CO_3^- and organic compounds can all serve as the source of carbon for algae dye degradation under specific conditions. Under anaerobic conditions, different end products are formed than under aerobic conditions, but many of these compounds formed under anaerobic conditions are further oxidised under aerobic conditions. In general, CO_2 is the end product for most of the organic carbon compounds. The product compounds of dye degradation, viz., CO_2, NH_3, NO_3^- and PO_4, can again be used for growth of photosynthetic organisms.

10.2.4.2.3 *Effects of Nitrogen Source Supplements*

Nitrogen is a relatively major constituent for yeast dye degradation. NO_3^- and NH_4^+ are the main organic nitrogen sources. Urea may also act as a favoured source of nitrogen for some cyanobacteria in natural aquatic systems.

10.2.4.2.4 *Effects of pH*

pH is the most important parameter affecting not only the biosorption capacity but also the colour of the dye solution and the solubility of some dyes. At lower pH values, the biosorbent surface becomes protonated and acquires a net positive charge and thus increases the binding of anionic dyes to the sorbent surface. On the other hand, higher pH values increase the net negative charge on the biosorbent surface, leading to electrostatic attraction of dye cations.

10.2.4.2.5 *Effects of Temperature*

An increase in temperature decreases the viscosity of the solution containing dye, thereby enhancing the rate of the diffusion of the dye molecules across the outer boundary layer and in the internal pore of the adsorbent particles.

10.2.4.2.6 Effects of Dye Concentration

The initial concentration of dye provides an important energetic force to defeat all mass transfer resistance of the dye between the aqueous and solid phase. Hence, a higher initial concentration of the dye may increase the adsorption process (Aksu 2005). Based on many studies, it may be accomplished that yeast can be employed as a vital biological tool for developing wastewater treatment systems for decolourisation of dye-bearing effluents through processes, including biosorption, bioaccumulation and biodegradation.

10.3 Future Trends

The textile industry is now very mature, and strict environmental legislation is being imposed to control the release of dyes into the environment. The treatment of textile wastewater has become a sizable challenge over the past decades, and up to this point, there is no single and economically attractive treatment method that can effectively decolourise and detoxify the textile wastewater. Efforts will, and indeed are, already being made to reduce toxic components in dyes and produce cleaner and more environmentally safe products. As the need to reduce pollution in textile production increases, the use of microorganisms in the processing of textile is rapidly gaining recognition for its eco-friendly and nontoxic characteristics. Microorganisms are safer alternatives in a wide range of textile processes that would otherwise require harsh chemicals whose disposal would pose environmental problems. The ability of microorganisms to decolourise and metabolise dyes has long been known, and the use of bioremediation-based technologies for treating textile wastewater has attracted interest.

As regulations become more severe, companies are forced to adopt more technologically sophisticated methods. There is also a simultaneous increase in cost for waste management that many companies may not be able to handle. Thus, the achievement of effective wastewater treatment is not the only problem that must be solved. Reductions in the waste and reuse of water are also necessary. It is most important to develop effective, reachable, clean-up and environmentally friendly treatment processes. Taking into account the advances in this field, we believe that future research activities should focus on four principal areas: (i) achieving dye mineralisation in addition to decolourisation, (ii) designing novel dyes based on the introduction of substituents into the chemical structure to enhance their biodegradability, (iii) search for alternatives for dye removal from large volumes of effluents and get water into the appropriate condition so that it can be reused in the same industry and (iv) improvement and modification of the production process or implementation of new processes to reduce water use, eliminate or minimise the discharge of toxic chemicals and recycle water as many times as possible to make companies more eco-friendly at a competitive price.

10.4 Conclusions

The microbial degradation and biosorption of dyes received much attention as these are cost-effective methods for dye removal. The selection of the best treatment option for the bioremediation of a specific type of industrial wastewater is a hard task because of the complex composition of these effluents. The best option is often a combination of two or more different microorganisms or consortia, and the choice of such processes depends on the effluent composition, characteristics of the dye, cost and toxicity of the degradation products and future use of the treated water. However, the benefits of the different processes and the synergistic effect of the combination of different microorganisms must be studied carefully to create the best blend, taking advantage of the use of various strains and consortia isolated from dye-contaminated sites; the isolation of new microorganisms, such as thermo-tolerant or thermophilic microorganisms; or the adaptation of existing ones to consume dyes as their sole carbon and nitrogen sources in such a way that the effluents have low values for chemical oxygen demand, total organic content, colour and toxicity.

References

Abadulla E, Tzanov T, Costa S, Robra KH, Cavaco-Paulo A and Guebitz GM. 2000. Decolorization and detoxification of textile dyes with a laccase from *Trametes Hirsute*. *App. Environ. Microbio.* 66(8): 3357–3362.

Aksu Z. 2001. Biosorption of reactive dyes by dried activated sludge: Equilibrium and kinetic modelling. *Biochem. Eng. J.* 7(1): 79–84.

Aksu Z. 2005. Application of biosorption for the removal of organic pollutants: Review. *Process Biochem.* 40: 997–1026.

Al-Garni SM, Ghanem KM, Kabli SA and Biag AK. 2013. Decolorization of crystal violet by mono and mixed bacterial culture technique using optimized culture conditions. *Pol. J. Environ. Stud.* 22(5): 1297–1306.

Anjaneya O, Souche SY, Santoshkumar M and Karegoudar TB. 2011. Decolorization of sulfonated azo dye Metanil Yellow by newly isolated bacterial strains: *Bacillus* sp. strain AK1 and *Lysinibacillus* sp. strain AK2. *J. Hazard. Mater.* 190: 351–358.

Asgher M, Bhatti HN, Shah SAH, Javaid AM and Legge RL. 2007. Decolorization potential of mixed microbial consortia for reactive and disperse textile dye-stuffs. *Biodegradation*. 18: 311.

Ashkenazy R, Gottileb L and Yannai S. 1997. Characterization of acetone-washed yeast biomass in lead biosorption. *Biotechnol. Bioeng*. 55: 1–10.

Ayed L, Kouidhi B, Bekir K and Bakhrouf A. 2013. Biodegradation of azo and tri-phenylmethanes dyes: Cytotoxicity of dyes, slime production and enzymatic activities of *Staphylococcus epidermidis* isolated from industrial wastewater. *Afri. J. Microbiol. Res*. 7: 5550–5557.

Babič J and Pavko A. 2007. Production of ligninolytic enzymes by *Ceriporiopsis sub-vermispora* for decolourization of synthetic dyes. *Acta Chim. Slov*. 54: 730–734.

Banat IM, Nigam P, Singh D and Marchant R. 1996. Microbial decolorization of textile dye-containing effluents: A review. *Bioresour. Technol*. 58: 217–227.

Barr DP and Aust SD. 1994. Mechanisms white-rot fungi use to degrade pollutants. *Environ. Sci. Tech*. 28: 320–328.

Beydilli MI and Pavlostathis SG. 2005. Decolorization kinetics of the azo dye Reactive Red 2 under methanogenic conditions: Effect of long-term culture acclimation. *Biodegradation*. 16(2): 135–146.

Birhanli A and Ozmen M. 2005. Evaluation of toxicity and teratogenecity of six com-mercial textile dyes using the frog embryo teratogenesis assay-*Xenopus*. *Drug Chem. Toxicol*. 28: 51–65.

Borde X, Guieysse B, Delgado O, Muñoz R, Hatti-Kaul R, Nugier-Chauvin C et al. 2003. Synergistic relationships in algal–bacterial microcosms for the treatment of aromatic pollutants. *Bioresour. Technol*. 86: 293–300.

Brady D, Stoll A and Duncan JR. 1994. Biosorption of heavy metal cations by nonvi-able yeast biomass. *Environ. Technol*. 15: 429–435.

Brige A, Ba M, Borloo J, Buysschaert G, Devreese B and Van Beeumen JJ. 2008. Bacterial decolorization of textile dyes is an extracellular process requiring a multicomponent electron transfer pathway. *Microbiol. Biotechnol*. 1: 40–52.

Bustard M, McMullan G and MeHale AP. 1998. Biosorption of textile dyes by biomass derived from *Kluyveromyces marxianus* IMB3. *Bioprocess Eng*. 19: 427–430.

Carliell CM, Barclay SJ, Naidoo N, Buckley CA, Mulholland DA and Senior E. 1995. Microbial decolorization of a reactive azo dye under anaerobic conditions. *Water SA*. 21: 61–69.

Cetin D and Donmez G. 2006. Decolorization of reactive dyes by mixed cultures iso-lated from textile effluent under anaerobic conditions. *Enz. Microb. Technol*. 38: 926–930.

Chacko JT and Subramaniam K. 2011. Enzymatic degradation of azo dyes: A review. *Int. J. Environ. Sci*. 1: 1250–1260.

Chang JS and Kuo TS. 2000. Kinetics of bacterial decolorization of azo dye with *Escherichia coli* NO3. *Bioresour. Technol*. 75: 107–111.

Charumathi D and Das N. 2010. Biotechnological approach to access the performance of dried biomass of candida tropicalis for removal of Basic Violet 3 from aque-ous solutions. *Int. J. Sci. Nat*. 1: 47–52.

Charumathi D and Das N. 2011. Biodegradation of Basic Violet 3 by *Candida krusei* isolated from textile wastewater. *Biodegradation*. 22: 1169–1180.

Chatterji AK. 2005. Biotechnology and biodegradation. In *Introduction to Environmental Biotechnology*, Eastern Economy Edn., 116–134, Prentice Hall of India Private Ltd., New Delhi.

Chen CC, Liao HJ, Cheng CY, Yen CY and Chung YC. 2007. Biodegradation of crystal violet by *Pseudomonas putida*. *Biotechnol. Lett.* 29(3): 391–396.

Chen CH, Chang CF and Liu SM. 2010. Partial degradation mechanisms of malachite green and methyl violet B by *Shewanella decolorationis* NTOU1 under anaerobic conditions. *J. Hazard. Mat.* 177: 281–289.

Chen CY, Ko CW and Lee PI. 2007. Toxicity of substituted anilines to *Pseudokirchneriella subcapitata* and quantitative structure-activity relationship analysis for polar narcosis. *Environ. Toxicol. Chem.* 26: 1158–1164.

Chen KC, Wu JY, Liou DJ and Hwang SCJ. 2003. Decolorization of the textile dyes by newly isolated bacterial strains. *J. Biotechnol.* 101: 57–68.

Cheriaa J, Khaireddine M, Rouabhia M and Bakhrouf A. 2012. Removal of triphenylmethane dyes by bacterial consortium. *Sci. World J.* 512454, doi: 10.1100/2012/512454.

Chivukula M and Renganathan V. 1995. Phenolic azo dye oxidation by laccase from *Pyricularia oryzae*. *Appl. Environ. Microbiol.* 61: 4374.

Coughlin MF, Kinkle BK and Bishop PL. 1999. Degradation of azo dyes containing amino naphthol by *Sphingomonas* sp. strain ICX. *J. Industrial Microbiol. Biotechnol.* 23: 341–346.

Craggs RJ, McAuley PJ and Smith VJ. 1997. Waste-water nutrient removal by marine microalgae grown on a corrugated raceway. *Water Res.* 31: 1701–1707.

Das D, Charumathi D and Das N. 2010. Combined effect of sugarcane bagasse extract and synthetic dyes on the growth and bioaccumulation properties of *Pichia fermentans MTCC* 189. *J. Hazard. Mater.* 183: 497–505.

Das D, Charumathi D and Das N. 2011. Bioaccumulation of synthetic dye Basic violet 3 and heavy metals in single and binary systems by *Candida tropicalis* grown in sugarcane bagasse extract medium: Use of response surface methodology (RSM) and inhibition kinetics. *J. Hazard. Mater.* 186: 1541–1552.

Das N and Charumathi D. 2012. Remediation of synthetic dyes from wastewater using yeast—An overview. *Indian J. Biotechnol.* 11: 369–380.

De Angelis FE and Rodrigues GS. 1987. Azo dyes removal from industrial effluents using yeast biomass. *Arq. Biol. Technol.* 30: 301–309.

De Roos AJ, Ray RM, Gao DL, Wernli KJ, Fitzgibbons ED, Zinding F, Astrakianakis G, Thoma DB and Checkoway H. 2005. Colorectal cancer incidence among female textile workers in Shanghai, China: A case-cohort analysis of occupational exposures. *Cancer Causes and Control.* 16: 1177–1188.

Deller S, Sollner S, Trenker-El-Toukhy R, Jelesarov I, Gubitz GM and Macheroux P. 2006. Characterization of a thermostable NADPH:FMN oxidoreductase from the mesophilic bacterium *Bacillus subtilis*. *Biochemistry.* 45: 7083–7091.

Deng D, Guo J, Zeng G and Sun G. 2008. Decolorization of anthraquinone, triphenylmethane and azo dyes by a new isolated *Bacillus cereus* strain DC11. *Int. Biodeter. Biodegrad.* 62(3): 263–269.

Docker A, Wattie JM, Tpping MD, Luczynska, CM, Newman Taylor AJ, Pickering CAC et al. 1987. Clinical and immunological investigations of respiratory disease in workers using reactive dyes. *British J. Industrial Med.* 44: 534–541.

Dogan EE, Yesilada E, Ozata L and Yologlu S. 2005. Genotoxicity testing of four textile dyes in two crosses of Drosophila using wing somatic mutation and recombination test. *Drug. Chem. Toxicol.* 28 (3): 289–301.

Eichlerova I, Homolka L and Nerud F. 2007. Decolorization of high concentrations of synthetic dyes by the white rot fungus *Bjerkandera adusta* strain CCBAS 232. *Dyes and Pigments.* 75: 38–44.

Eichlerova I, Homolka L and Nerud F. 2006. Synthetic dye decolorization capacity of white rot fungus *Dichomites squalens. Biores. Technol.* 97: 2153–2159.

Ertugrul S, San NO and Donmez G. 2009. Treatment of dye (Remazol Blue) and heavy metals using yeast cells with the purpose of managing polluted textile wastewater. *Ecol. Eng.* 35: 128–134.

Estlander T. 1988. Allergic dermatoses and respiratory diseases from reactive dyes. *Contact Dermatitis.* 18: 290–297.

Faraco V, Pezzella C, Giardina P, Piscitelli A, Vanhullec S and Sannia G. 2009. Decolourization of textile dyes by the white-rot fungi *Phanerochaete chrysosporium* and Pleurotus ostreatus. *J. Chem. Technol. Biotech.* 84: 414–419.

Franciscon E, Zille A, Fantinatti-Garboggini F, Silva IS, Cavaco-Paulo A and Regina Durrant L. 2009. Microaerophilic aerobic sequential decolourization/biodegradation of textile azo dyes by a facultative *Klebsiella* sp. strain VN-31. *Process Biochemistry.* 44: 446–452.

Giwa A, Giwa FJ and Ifu BJ. 2012. Microbial decolourization of an anthraquinone dye C. I. Reactive Blue 19 using *Bacillus cereus. J. Ameri. Chem. Sci.* 2(2): 60–68.

Gonzales CA, Riboli E and Lopez Abente G. 1988. Bladder cancer among workers in the textile industry: Results of a Spanish casecontrol study. *American J. Industrial Med.* 14: 673–680.

Gurav AA, Ghosh JS and Kulkarni GS. 2011. Decolorization of anthroquinone based dye Vat Red 10 by *Pseudomonas desmolyticum* NCIM 2112 and *Galactomyces geotrichum* MTCC 1360. *Int. J. Biotech. Mol. Bio. Res.* 2(6): 93–97.

Heinfling A, Berghauer M, and Szewzyk U. 1997. Biodegradation of azo and phthalocyanine dyes by *Trametes versicolor* and *Bjerkandera adusta. Appl. Microbiol Biotechnol.* 77: 247–255.

Hsueh CC, Chen BY and Yen CY. 2009. Understanding effects of chemical structure on azo dye decolorization characteristics by *Aeromonas hydrophila. J. Hazard. Mater.* 167: 995–1001.

Itoh K, Kitade C and Yatome C. 1996. A pathway for biodegradation of an anthraquinone dye, C.I. disperse red 15, by a yeast strain *Pichia anomala. B. Environ. Conta. & Toxicol.* 56(3): 413–418.

Jadhav JP and Govindwar SP. 2006. Biotransformation of Malachite Green by Saccharomyces *cerevisiae* MTCC 463. *Yeast.* 23: 315–323.

Jadhav SU, Jadhav MU, Kagalkar AN and Govindwar SP. 2008a. Decolorization of Brilliant Blue G dye mediated by degradation of the microbial consortium of *Galactomyces geotrichum* and *Bacillus* sp. *J. Chin. Inst. Chem. Engrs.* 39(6): 563–570.

Jadhav SU, Kalme SD and Govindwar SP. 2008b. Biodegradation of Methyl Red by *Galactomyces Geotrichum* MTCC 1360. *Int. Biodeter. Biodegr.* 62: 135–142.

Jang MS, Lee YM, Kim CH, Lee JH, Kang DW et al. 2005. Triphenylmethane reductase from *Citrobacter* sp. strain KCTC18061P: Purification, characterization, gene cloning and overexpression of a functional protein in *Escherichia coli. Appl. Environ. Microbiol.* 17: 7955–7960.

Jing W, Byung-Gil J, Kyoung-Sook K, Young-Choon L and Nak-Chang S. 2009. Isolation and characterization of *Pseudomonas otitidis* WL-13 and its capacity to decolorize triphenylmethane dyes. *J. Environ. Sci.* 21: 960–964.

Jinqi L and Houtian L. 1992. Degradation of azo dyes by algae. *Environ. Poll.* 75: 273–278.

Joe MH, Lim SY, Kim DH and Lee IS. 2008. Decolorization of reactive dyes by *Clostridium bifermentans* SL186 isolated from contaminated soil. *World J. Microbiol. Biotechnol.* 24: 2221–2226.

Joshi M, Bansal R and Purwar R. 2004. Color removal from textile effluents. *Indian J. Fibre and Textile Res.* 29: 239–259.

Joshi T, Iyengar L, Singh K and Garg S. 2008. Isolation, identification and application of novel bacterial consortium TJ-1 for the decolourization of structurally different azo dyes. *Bioresour. Technol.* 99: 7115–7121.

Kalyani DC, Telke AA, Dhanve RS and Jadhav JP. 2008. Ecofriendly biodegradation and detoxification of Reactive Red 2 textile dye by newly isolated *Pseudomonas* sp. SUK1. *J. Hazard. Mater.* 163: 735–742.

Kapdan IK and Erten B. 2007. Anaerobic treatment of saline wastewater by *Halanaerobium lacusrosei*. *Process Biochemistry.* 42: 449–453.

Keck A, Klein J, Kudlich M, Stolz A, Knackmuss HJ and Mattes R. 1997. Reduction of azo dyes by redox mediators originating in the naphthalene sulfonic acid degradation pathway of *Sphingomonas* sp. strain BN6. *Appl. Environ. Microbiol.* 63: 3684–3690.

Khehra MS, Saini HS, Sharma DK, Chadha BS and Chimni SS. 2005. Comparative studies on potential of consortium and constituent pure bacterial isolates to decolorize azo dyes. *Water Res.* 39: 5135–5141.

Khouni I, Marrot B and Amar RB. 2012. Treatment of reconstituted textile wastewater containing a reactive dye in an aerobic sequencing batch reactor using a novel bacterial consortium. *Sep. Purif. Technol.* 87: 110–119.

Knapp JS, Newby PS and Reece LP. 1995. Decolourization of dyes by wood-rotting basidiomycete fungi. *Enzyme Microbiol. Technol.* 17: 664–668.

Knapp JS, Vantoch-Wood EJ and Zhang F. 2001. Use of wood–rotting fungi for the decolorization of dyes and industrial effluents. In *Fungi in Bioremediation,* GM Gadd (ed.), 253–261, Cambridge University Press, ISBN 0521781191, Cambridge.

Kolekar YM and Kodam KM. 2012. Decolorization of textile dyes by *Alishewanella* sp. KMK6. *Appl. Microbiol. Biotechnol.* 95: 521–529.

Korner S and Vermaat JE. 1998. The relative importance of *Lemna gibba* L., bacteria and algae for the nitrogen and phosohorus removal in duckweed-covered domestic wastewater. *Water Res.* 32: 3651–3661.

Kumari K and Abraham TE. 2007. Biosorption of anionic textile dyes by nonviable biomass of fungi and yeast, *Bioresour Technol.* 98: 1704–1710.

Kuberan T, Anburaj J, Sundaravadivelan C and Kumar P. 2011. Biodegradation of azo dye by *Listeria* sp. *Int. J. Environ. Sci.* 1: 1760–1770.

Kwasiewska K. 1985. Biodegradation of crystal violet (hexamethyl-p-rosaniline chloride) by oxidative red yeast, *Bull. Environ. Contam. Toxicol.* 34: 323–330.

Lee YH, Matthews RD and Pavlostathis SG. 2006. Biological decolorization of reactive anthraquinone and phthalocyanine dyes under various oxidation-reduction conditions. *Wat. Environ. Res.* 78(2): 156–169.

Levin L, Melignani E and Ramos AM. 2010. Effect of nitrogen sources and vitamins on ligninolytic enzyme production by some white-rot fungi. Dye decolorization by selected culture filtrates. *Bioresour. Technol.* 101: 4554–4563.

Lin J, Zhang X, Li Z and Lei L. 2010. Biodegradation of Reactive Blue 13 in a two-stage anaerobic/aerobic fluidized beds system with a *Pseudomonas* sp. isolate. *Bioresour. Technol.* 101(1): 34–40.

Lorimer JP, Mason TJ, Plattes M, Phull SS and Walton DJ. 2001. Degradation of dye effluent. *Pure Appl. Chem.* 73(12): 1957–1968.

Lucas MS, Amaral C, Sampaio A, Peres JA and Dias AA. 2006. Biodegradation of the diazo dye Reactive Black 5 b a wild isolate of *Candida oleophila*. *Enzyme Microb. Technol.* 39: 51–55.

Mallick N. 2002. Biotechnological potential of immobilized algae for wastewater N, P and metal removal: A review. *Biometals*. 15: 377–390.

Mathur N, Bathnagar P, Nagar P, and Bijarnia, MK. 2005. Mutagenicity assessment of effluents from textile/dye industries of Sanganer, Jaipur (India): A case study. *Ecotoxicol. Environ. Saf.* 61: 105–113.

McMullan G, Meehan C, Conneely A, Kirby N, Robinson T, Nigam P, Banat IM, Marchant R and Smyth WF. 2001. Microbial decolourisation and degradation of textile dyes. *Appl. Microbiol. Biotechnol.* 56: 81–87.

Moosvi S, Keharia H and Madamwar D. 2005. Decolourization of textile dye Reactive Violet 5 by a newly isolated bacterial consortium RVM 11.1. *World J. Microbiol. Biotechnol.* 21: 667–672.

Muñoz R. and Guieysse B. 2006. Algal–bacterial processes for the treatment of hazardous contaminants: A review, *Water Res.* 40: 2799–2815.

Nilsson R, Nordlinder R, Wass U, Meding B and Belin L. 1993. Asthma, rhinitis and dermatitis in worker exposed to reactive dyes. *B. J. Industrial Med.* 50: 65–70.

Novotny C, Cajthaml, T, Svobodova K, Susla M and Sasek V. 2009. *Irpex lacteus*, a white rot fungus with biotechnological potential – Review. *Folia Microbiol.* 54: 375–390.

Pajot HF, De Figueroa LIC and Farifia JI. 2006. Dye-decolorizing activity in isolated yeasts from the ecoregion of Las Yungas (Tucuman, Argentina). *Enzyme Microb. Technol.* 39: 51–55.

Pandey A, Singh P and Iyengar L. 2007. Bacterial decolorization and degradation of azo dyes. *Int. Biodeter. Biodegrad.* 59: 73–84.

Parikh A. and Madamwar D. 2005. Partial characterization of extracellular polysaccharides from cyanobacteria. *Bioresour. Technol.* 97: 1822–1827.

Pavko A and Novotny C. 2008. Induction of ligninolytic enzyme production by *Dichomitus squalens* on various types of immobilization support. *Acta Chim. Slov.* 55: 648–652.

Pazarlioglu NK, Akkaya A, Akdogan HA and Gungor B. 2010. Biodegradation of direct blue 15 by free and immobilized *Trametes versicolor*. *Water Environ. Res.* 82: 579–585.

Pointing SB. 2001. Feasibility of bioremediation by white-rot fungi. *Appl. Microbiol. Biotechnol.* 57: 20–33.

Polman JK and Breckenridge CR. 1996. Biomass mediated binding and recovery of textile dyes from waste effluents. *Text. Chem. Color.* 28: 31–35.

Priya B, Uma L, Ahamed AK, Subramanian G and Prabaharan D. 2011. Ability to use the diazo dye C. I. Acid Black 1 as nitrogen source by the marine cyanobacterium *Oscillatoria curviceps* BDU92191. *Bioresour. Technol.* 102: 7218–7223.

Przybojewska B, Baranski B, Spiechowicz E and Szymczak W. 1989. Mutagenic and genotoxic activity of chosen dyes and surface active compounds used in the textile industry. *Pol. J. Occup. Med.* 2: 171–185.

Rafii F, Franklin W and Cerniglia CE. 1990. Azoreductase activity of anaerobic bacteria isolated from human intestinal microflora. *Appl. Environ. Microbiol.* 56: 2146–2151.

Rajee O and Patterson J. 2011. Decolorization of azo dye (Orange MR) by an autochthonous bacterium *Micrococcus* sp. DBS 2. *Ind. J. Microbiol.* 51: 159–163.

Ramalho PA, Scholxer H, Cardoso MH, Ramalho TA and Oliveira-Campos M. 2002. Improved conditions for the aerobic reductive decolorization of azo dyes by *Candida zeylanoides*. *Enzyme Microb. Technol.* 31: 848–854.

Robinson T, McMullan G, Marchant R and Nigam P. 2001. Remediation of dyes in textile effluent: A critical review on current treatment technologies with a proposed alternative. *Bioresour. Technol.* 77: 247–255.

Russ R, Rau J and Stolz A. 2000. The function of cytoplasmic flavin reductases in the reduction of azodyes by bacteria. *Appl. Environ. Microbiol.* 66(4): 1429–1434.

Sani RK and Banerjee UC. 1999. Decolorization of triphenylmethane dyes and textile and dye-stuff effluent by *Kurthia* sp. *Enzyme Microb. Technol.* 24: 433–437.

Saratale RG, Saratale GD, Kalyani DC, Chang JS and Govindwar SP. 2009a. Enhanced decolorization and biodegradation of textile azo dye Scarlet R by using developed microbial consortium-GR. *Bioresour. Technol.* 100: 2493–2500.

Saratale RG, Saratale GD, Chang HS and Govindwar SP. 2009b. Decolorization and biodegradation of textile dyes Navy Blue HER by *Trichosporon Beigelii* NCIM-3326. *J. Hazard. Mater.* 166: 1421–1428.

Saratale RG, Saratale GD, Chang JS and Govindwar SP. 2011. Bacterial decolorization and degradation of azo dyes: A review. *J. Taiwan Inst. Chem. Eng.* 42: 138–157.

Sarayu K and Sandhya S. 2010. Aerobic biodegradation pathway for Remazol Orange by *Pseudomonas aeruginosa. Appl. Biochem. Biotechnol.* 160: 1241–1253.

Sawhney R and Kumar A. 2011. Congo red (azo dye) decolourization of local isolate VT-II inhabiting dye-effluent exposed soil. *Int. J. Environ. Sci.* 1: 1261–1267.

Sharma DK, Saini HS, Singh M, Chimni SS and Chadha BS. 2004. Biological treatment of textile dye Acid violet-17 by bacterial consortium in an up-flow immobilized cell bioreactor. *Letters in Applied Microbiology* 38: 345–350.

Singh H. 2006. *Mycoremediation-Fungal Bioremediation*: 421–471, Wiley Interscience, ISBN-13: 978-0-471-75501-2, Hoboken.

Singh RP, Singh PK and Singh RL. 2014a. Bacterial decolorization of textile azo dye acid orange by *Staphylococcus hominis* RMLRT03. Presented in International Conference on Environmental Technology and Sustainable Development: Challenges and Remedies, 107–107.

Singh PK, Singh A, Singh RP and Singh RL. 2014b. Biodecolorization of Blue GN textile dye by adapted bacterial isolate T-3. Presented in International Conference on Environmental Technology and Sustainable Development: Challenges and Remedies, 81–81.

Singh RP, Singh PK, Gupta R and Singh RL. 2013. Biodecolorization of Blue GN textile dye by adapted bacterial isolate T-6. Presented in XXXIII Annual Conference of Society of Toxicology (STOX), 123–124.

Singh RP and Singh RL. 2012. Decolorization of textile dye Red GR by bacterial culture K-2 isolated from contaminated soil. Presented in XXXII Annual Conference of Society of Toxicology (STOX), 219–219.

Singh RP and Singh RL. 2011. Effect of various parameters on decolorization of Orange II textile dye using mixed culture. Presented in XXXI Annual Conference of Society of Toxicology (STOX), 224–224.

Singh RL, Singh S, Khanna SK and Singh GB. 1991. Metabolic disposition of metanil yellow, orange II and their blend by caecal microflora. *Int. J. Toxicol. Occup. Environ. Hlth.* 1: 250.

Solis M, Solis A, Perezb HI, Manjarrezb N and Floresa M. 2012. Microbial decolouration of azo dyes: A review. *Process Biochem.* 47: 1723–1748.

Stolz A. 2001. Basic and applied aspects in the microbial degradation of azo dyes. *Appl. Microbiol. Biotechnol.* 56: 69–80.

Sugumar S and Thangam B. 2012. BiodEnz: A database of biodegrading enzymes. *Bioinformation*. 8: 40–42.

Supaka NJ, Somsak Delia KD and Pierre MLS. 2004. Microbial decolorization of reactive azo dyes in a sequential anaerobic–aerobic system. *Chem. Eng. J.* 99: 169–176.

Sukumar M, Sivasamy A and Swaminathan G. 2009. In situ biodecolorization kinetics of Acid Red 66 in aqueous solutions by *Trametes versicolor*. *J. Hazard. Mat.* 167: 660–663.

Swamy J and Ramsay JA. 1999. The evaluation of white rot fungi in the decoloration of textile dyes. *Enzyme Microb. Technol.* 24: 130–137.

Tam NF, Chong AM and Wong YS. 2002. Removal of tributyltin (TBT) by live and dead microalgal cells. *Mar. Pollut. Bull.* 45: 362–371.

Tanaka H, Koike K, Itakura S and Enoki A. 2009. Degradation of wood and enzyme production by *Ceriporiopsis subvermispora*. *Enzyme Microb. Technol.* 45: 384–390.

Telke A, Kalyani D, Jadhav J and Govindwar S. 2008. Kinetics and mechanism of reactive red 141 degradation by a bacterial isolate *Rhizobium radiobacter* MTCC 8161. *Acta Chim. Slov.* 55: 320–329.

Teresa PM and Ruben S. 2006. Differential effect of algal and soil derived dissolved organic matter on alpine lake bacterial community composition and activity. *Lminol. Oceanogr.* 51: 2527–2537.

Thongchai P and Worrawit L. 2000. Decolorization of reactive dyes with different molecular structures under different environmental conditions. *Wat. Res.* 34(17): 4177–4184.

Tony BD, Goyal D and Khanna S. 2009. Decolorization of textile azo dyes by aerobic bacterial consortium. *Int. Biodeter. Biodegr.* 63: 462–469.

Van der Zee FP, Lettinga G and Field JA. 2001. Azo dye decolorization by anaerobic granular sludge. *Chemosphere*. 44: 1169–1176.

Verma P and Madamwar D. 2003. Decolourization of synthetic dyes by a newly isolated strain of *Serratia marcescens*. *W. J. Microbio. Biotech.* 19(6): 615–618.

Vitor V and Corso CR. 2008. Decolorization of textile dye by *Candida albicans* isolated from industrial effluents. *J. Ind. Microbial. Biotechnol.* 35: 1353–1357.

Volesky B and May Philips HA. 1995. Biosorption of heavy metal cations by *Saccharomyces cerevisiae*. *Appl. Microbiol. Biotechnol.* 42: 797–806.

Waghmode TR, Kurade MB and Govindwar SP. 2011. Time dependent degradation of mixture of structurally different azo and non azo dyes by using *Galactomyces geotrichum* MTCC 1360. *Int. Biodeter. Biodegr.* 65: 479–486.

Wang BE and Hu YY. 2008. Bioaccumulation versus adsorption of reactive dye by immobilized growing *Aspergillus fumigates* beads. *J. Hazard. Mater.* 157: 1–7.

Wang J, Gao F, Liu Z, Qiao M, Niu X et al. 2012. Pathway and molecular mechanisms for Malachite Green biodegradation in *Exiguobacterium* sp. MG2. *PLoS ONE*. 7(12): e51808. doi: 10.1371/journal.pone.0051808.

Wang HJ, Su Q, Zheng XW, Tian Y, Xiong XJ and Zheng TL. 2009a. Bacterial decolorization and degradation of the reactive dye Reactive Red 180 by *Citrobacter* sp. CK3. *Int. Biodeter. Biodeg.* 63: 395–399.

Wang H, Zheng XW, Su JQ, Tian Y, Xiong XJ and Zheng TL. 2009b. Biological decolorization of the reactive dyes Reactive Black 5 by a novel isolated bacterial strain *Enterobacter* sp. EC3. *J. Hazard. Mater.* 171: 654–659.

Wilkinson SM and McGeachan K. 1996. Occupational allergic contact dermatitis from reactive dyes. *Contact Dermatitis*. 35(6): 376–378.

Yang Q, Yang M, Pritsch K, Yediler A, Hagn A, Schloter M and Kettrup A. 2003. Decolorization of synthetic dyes and production of manganese-dependent peroxidase by new fungal isolates. *Biotechnol. Lett.* 25: 709–713.

Yang Q, Yediler A, Yang M and Kettrup A. 2005. Decolorization of an azo dye, Reactive Black 5 and MnP production by yeast isolate: *Debaryomyces polymorphus. Biochem. Eng. J.* 24: 249–253.

Yatome C, Yamada S, Ogawa T and Matsui M. 1993. Degradation of crystal violet by *Nocardia coralline. Appl. Microbiol. Biotechnol.* 38: 565–569.

Yu Z and Wen X. 2005. Screening and identification of yeasts for decolouring synthetic dyes in industrial wastewater. *Int. Biodet. Biodeg.* 56: 109–114.

Zhang ZB, Liu CY, Wu ZZ, Xing L and Li PF. 2006. Detection of nitric oxide in culture media and studies on nitric oxide formation by marine microalgae. *Med. Sci. Monitor.* 12: 75–85.

Zimmermann T, Kulla H and Leisinger T. 1982. Properties of Purified Orange II Azoreductase, the enzyme initiating azo dye degradation by *Pseudomonas* KF46. *Eur. J. Biochem.* 129: 197–203.

Znidarsic P and Pavko A. 2001. The morphology of filamentous fungi in submerged cultivations as a bioprocess parameter. *Food Technol. Biotechnol.* 39: 237–252.

11

Anaerobic Biodegradation of Slaughterhouse Lipid Waste and Recovery of Bioactive Molecules for Industrial Applications

Kandasamy Ramani and Ganesan Sekaran

CONTENTS

11.1 Introduction

India has the world's largest population of livestock: 191 million cattle, 70 million buffalo, 139 million sheep and goats, 10 million pigs and more than 200 million poultry. Of them, about 36.5% of the goats, 32.5% of the sheep, 28% of the pigs, 1.9% of the buffalo and 0.9% of the cattle are being slaughtered every year in India (Nilkanth and Madhumita 2009), generating a huge amount of solid and liquid wastes. Among the solid wastes, lipid-rich solid waste, namely, tallow (Figure 11.1), is generated in considerable quantities each year from slaughterhouses (Crine et al. 2007). The tallow is a hydrophobic compound composed of saturated fatty acids (palmitic acid 26%, stearic acid 14% and myristic acid 3%), monounsaturated fatty acids (oleic acid 47%

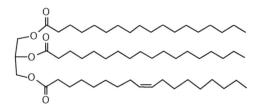

FIGURE 11.1
Structure of tallow (consists mainly of triglycerides [fat] whose major constituents are derived from stearic and oleic acids).

and palmitoleic acid 3%) and polyunsaturated fatty acids (linoleic acid 3% and linolenic acid 1%). Because of its hydrophobic nature, most of microbial strains are not efficient enough to degrade it, and thus, treatment becomes difficult. Hence, there has been no proper biological treatment/utilisation technique for the slaughterhouse tallow-laden lipid waste. A very little report suggested less efficient aerobic or anaerobic biological treatment of slaughterhouse wastewater (Kuang 2002). Hence, there has been constant research on exploration of efficient microbial systems for the degradation/utilisation of this hydrophobic tallow–slaughterhouse lipid waste (TSHL).

11.1.1 Slaughterhouse Lipid Solid Wastes for the Production of High Value–Added Biomolecules, such as Acidic Lipase and Biosurfactants

To date, lipid-rich slaughterhouse waste has been used as raw material for the production of low value–added products, such as soaps, detergents and biodiesel (Vivian et al. 2011). As the synthetic fatty esters showed prominence for quality assurance, the natural fatty esters are poorly encouraged in the soap and detergent industry. In addition to the degradation of TSHL, it may be utilised for the production of high value–added biomolecules in microbial fermentation. But the recovery of high value–added products from this waste has been a largely neglected field. Production of bioactive molecules, such as enzymes, biosurfactants, etc., using lipid solid waste as the substrate has been demonstrated as a viable technique for by-product recovery, and at the same time, the lipid solid waste could be disposed of in an environmentally sound manner. Recently, in our research articles, we reported the production of bioactive molecules, such as acidic lipases (Ramani et al. 2010a,b) and biosurfactants (Ramani et al. 2012), from the slaughterhouse lipid waste beef and goat tallow in aerobic fermentation.

11.1.2 Acidic Lipase for Industrial Applications

In recent times, lipases have been subjected to active research for their versatile industrial applications. Lipases (E.C. 3.1.1.3) are hydrolytic enzymes, which hydrolyse triacyl glycerides into free fatty acids and glycerols (Gomes et al. 2004; Pugazhenthi et al. 2004). They form an important enzyme group due to their remarkable levels of activity and stability in both aqueous and nonaqueous media, which enable several reactions, viz, catalysis, such as acidolysis, alcoholysis, aminolysis, esterification and transesterification. Lipases also show unique characteristics of chemo-, regio- and enantioselectivity. Because of these characteristics, lipases have been widely used in the medical, biotechnology, detergent, dairy, food, textile, surfactant, pharmaceutical and oleochemical industries (Maia et al. 2001; Sharma et al. 2001). The lipases used in each application are selected based on their substrate and enantio-specificity as well as their temperature and pH stability. Lipases are unique in catalysing the hydrolysis of fats and oils (triglycerides) into fatty acids and glycerol at the water–lipid interface. During hydrolysis, lipases pick up an acyl group from glycerides forming the lipase–acyl complex, which, in turn, transfers its acyl group to the O–H group of water.

Most lipases have been produced commercially from fungi (Gao et al. 2000), yeast (Dalmou et al. 2000) and bacteria (Ferreira et al. 1999). Bacterial lipases have gained importance as they are more stable at elevated temperatures and tolerant to a wide range of pH compared to the other microbial lipases. The screening of extremophiles has become an important criterion for the production of highly active and stable lipases under extreme environmental conditions.

Most of the studies have been carried out for the production of alkaline and neutral lipases (Haba et al. 2000; Shu et al. 2006; Stransky et al. 2007; Yu et al. 2007), and only a few reports are available on the production of acidic and extremely acidic lipases (Mahadik et al. 2002, 2003; Mhetras et al. 2009). The acidic lipases are defined as the lipases that can be sustained in acidic pHs (1 to 6) without denaturation, and moreover, they are active at those pHs and stable. The extremely acidic lipases (below pH 3.5) have been used considerably in the food and flavour industries (in which aroma esters, such as isoamyl acetate, are formed in acidic environments), acid bating for fur and wool and digestive aids for medical treatment (Hassan et al. 2006). Acidic lipases have been added to bread dough for the uniform production of monoglycerides, which greatly improve the resistance of the bread to staling. Also, the acidic lipases (highly active at less than pH 4.0) have been used as a substitute for pancreatic lipases in enzyme therapy (Sani 2006). Therefore, the production of acidic lipase appears to demand a huge market potential at the international level. However, the rate of production of acidic lipases in India and in an international scenario is very low. Thus, the focal theme of this chapter is on the production of acidic lipase from TSHL in anaerobic fermentation.

11.1.3 Substrates Used for the Production of Lipase

Many researchers have reported the application of vegetable oils as the substrate for the production of lipase (Haba et al. 2000; Castro-Ochoa et al. 2005; Mateos Diaz et al. 2006; Shu et al. 2006; Stransky et al. 2007; Yu et al. 2007). Papanikolaou et al. (2007) have used stearin, one of the industrial derivatives of beef tallow, for the production of lipase of low enzyme activity from *Yarrowia lipolytica*. However, reports on the utilisation of animal tallow, the predominant lipidic solid waste generated in slaughterhouses, for the production of acidic lipases are nil.

11.1.4 Biosurfactants, the Value-Added By-Product from Lipid Substrates

Synthetic surfactants are being used for many applications, but they pose environmental problems because of their toxicity and poor biodegradation (Mulligan 2005). Hence, there has been constant research on the development of biodegradable surfactants for varied applications. Biosurfactants, the microbial secondary metabolites, are produced during the fermentation of lipid substrates. Biosurfactants are multipurpose microbial amphiphilic molecules having both hydrophobic and hydrophilic domains (Figure 11.2).

The structure and configuration of hydrophobic and hydrophilic domains are related with the type of bacterial strain and carbon source considered for the production of biosurfactants. They have effective physicochemical, surface-active, and biological features applicable to several industrial and environmental processes (Benincasa et al. 2004). Various types of biosurfactants were produced by a broad spectrum of microbes during their growth on water-immiscible lipid substrates. These biosurfactants are usually less toxic and more easily biodegradable (Makkar and Cameotra 2002; Nitschke and Costa 2007). The biosurfactants are differentiated by their chemical composition, molecular weight, physicochemical properties, mode of action and microbial origin for production. Based on molecular weight, they are classified into low molecular mass biosurfactants, including glycolipids, phospholipids and lipopeptides, and into high molecular mass biosurfactants or bioemulsifiers containing amphipathic polysaccharides, proteins, lipopolysaccharides, lipoproteins or complex mixtures of these biopolymers (Figure 11.3) (Ahimou et al. 2000; Maier et al. 2000). Low molecular mass biosurfactants are efficient in lowering surface and interfacial tensions, whereas high molecular mass biosurfactants are more effective to stabilise oil-in-water emulsions. The biosurfactants accumulate at the interface between two immiscible fluids or between a fluid and a solid. By reducing surface (liquid–air) and interfacial (liquid–liquid) tensions, they lower the repulsive force between two dissimilar phases and allow these two phases to coalesce and interact more easily (Figure 11.4).

Examples of biosurfactants and their producers and applications are depicted in Table 11.1. In addition to the environmental applications of biosurfactants mentioned in Table 11.1, the biosurfactants are being used in

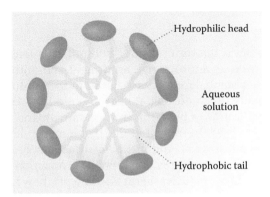

Hydrophilic head

Aqueous solution

Hydrophobic tail

FIGURE 11.2
Basic structure of biosurfactant.

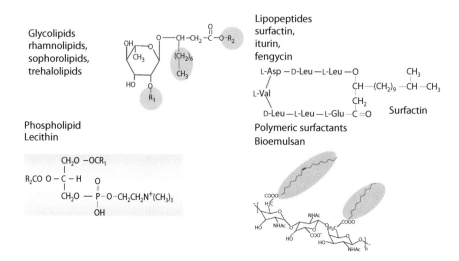

Glycolipids
rhamnolipids,
sophorolipids,
trehalolipids

Phospholipid
Lecithin

Lipopeptides
surfactin,
iturin,
fengycin

Polymeric surfactants
Bioemulsan

FIGURE 11.3
Structure of various types of biosurfactants.

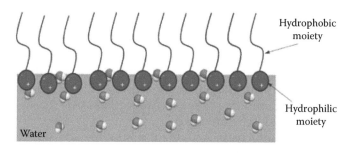

Hydrophobic
moiety

Hydrophilic
moiety

Water

FIGURE 11.4
Accumulation of biosurfactants at the interface between liquid and air.

bacterial pathogenesis, quorum sensing, biofilm formation, etc. (Singh et al. 2004). They are used as antibacterial, antifungal and antiviral agents, and they also have the potential for use as major immune-modulatory molecules and adhesive agents (Rodrigues et al. 2006). There has been an increasing interest in the application of biosurfactants on human and animal cells and cell lines (Banat et al. 2000; Singh et al. 2004). It may be viewed that the fermentation of lipid substrates could yield biosurfactants besides the production of lipases. Many reports are available on the production of biosurfactants from vegetable oils and lipidic wastes at alkaline pH; however, the literature on the production of biosurfactants using lipidic substances at acidic pH is very scanty or perhaps nil.

TABLE 11.1

Classification of Biosurfactants and Their Use in Remediation of Heavy Metal and Hydrocarbon Contaminated Sites

Group	Biosurfactant Class	Microorganism	Applications in Environmental Biotechnology
Glycolipids	Rhamnolipids	*Pseudomonas aeruginosa*, *Pseudomonas* sp.	Enhancement of the degradation and dispersion of different classes of hydrocarbons; emulsification of hydrocarbons and vegetable oils; removal of metals from soil
	Trehalolipids	*Mycobacterium tuberculosis*, *Rhodococcus erythropolis*, *Arthrobacter* sp., *Nocardia* sp., *Corynebacterium* sp.	Enhancement of the bioavailability of hydrocarbons
	Sophorolipids	*Torulopsis bombicola*, *Torulopsis petrophilum*, *Torulopsis apicola*	Recovery of hydrocarbons from dregs and muds; removal of heavy metals from sediments; enhancement of oil recovery
Fatty acids, phospolipids and neutral lipids	Corynomycolic acid	*Corynebacterium lepus*	Enhancement of bitumen recovery
	Spiculisporic acid	*Penicillium spiculisporum*	Removal of metal ions from aqueous solution; dispersion action for hydrophilic pigments; preparation of new emulsion-type organogels, superfine microcapsules (vesicles or liposomes), heavy metal sequestrants
	Phosphati-dylethanolamine	*Acinetobacter* sp. *Rhodococcus erythropolis*	Increasing the tolerance of bacteria to heavy metals
Lipopeptides	Surfactin	*Bacillus subtilis*	Enhancement of the biodegradation of hydrocarbons and chlorinated pesticides; removal of heavy metals from a contaminated soil, sediment and water; increasing the effectiveness of phytoextraction
	Lichenysin	*Bacillus licheniformis*	Enhancement of oil recovery
Polymeric biosurfactants	Emulsan	*Acinetobacter calcoaceticus* RAG-1	Stabilisation of the hydrocarbon-in-water emulsions
	Alasan	*Acinetobacter radioresistens* KA-53	
	Biodispersan	*Acinetobacter calcoaceticus* A2	Dispersion of limestone in water
	Liposan	*Candida lipolytica*	Stabilisation of hydrocarbon-in-water emulsions
	Mannoprotein	*Saccharomyces cerevisiae*	

11.2 Isolation and Identification of Anaerobic Strain *Clostridium diolis*

The conventional anaerobic fermentation of organic substrates follows these steps: hydrolysis, acidogenesis, acetogenesis and methanogenesis. The second and third steps are considered to proceed at a faster rate than the fourth step. Thus, fragmented substrates spontaneously enter the fourth step and generate CO_2 and CH_4.

Recovery of the carbon source from solid wastes in the form of value-added product may have high potential impact over biogas generation. This is possible by preventing anthropogenic methane emission and by producing enzymes. Anaerobic biofermentative processes can be used for the production of hydrolytic enzymes and volatile fatty acids (Bolzonella et al. 2005). Hence, a strategy was to be followed up to recover the value-added products, such as extracellular enzymes and long-chain fatty acids, under controlled conditions. The maximum recovery of value-added products could be obtained by carrying out the hydrolysis reaction under acidic conditions, so that the transformation to methanogenesis was curtailed. This formed the focal theme of the present investigation to isolate acidic bacterium for the fermentation of lipidic substrates.

Clostridium diolis was isolated from the excreta of flesh-eating animals (*Canis aureus*) and maintained in anaerobic broth (HiMedia) containing goat tallow at 37°C with the specific intent of finding bacterial lipases exhibiting high stability and activity. The broth was serially diluted in the prepared medium under gaseous atmosphere consisting of 80% N_2 and 20% CO_2 (300 kPa). The dilutions were plated on anaerobic agar (HiMedia) medium containing tributyrin and tallow in an anaerobic glove box and were incubated in a stainless steel anaerobic jar at 37°C and filled with gas consisting of 80% N_2 and 20% CO_2 at a pressure of 200 kPa. After 4 days of incubation, white colonies with raised centres were picked up and transferred into serum bottles.

The colony morphology of the *Clostridiun diolis* showed circular, smooth and white-coloured colonies on agar plates after 2 or 4 days of incubation. The strain C. *diolis,* isolated from the excreta of a flesh-eating animal (*Canis aureus*), could ferment sugars, such as cellobiose, galactose, glucose, trehalose, xylose, lactose, salicin, glycogen, melibiose, esculin, ribose and starch. The strain could not ferment arabinose, mannitol, sorbitol and erythritol. The Gram-staining analysis of *Clostridium diolis* showed that the strain belongs to Gram-positive organisms. The average cell size in the exponential growth phase was 1.0–2.0 μm as micrographed by SEM (Figure 11.5). The bacterium was rod- and spindle-shaped and occurred in singles. In this work, a high-resolution AFM image of *Clostridium diolis* was obtained to reveal the cell wall surface properties. The AFM deflection image (Figure 11.6) was recorded in

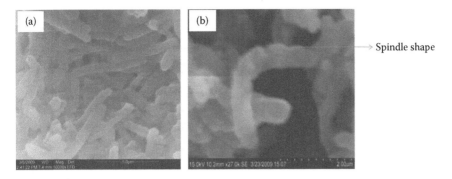

FIGURE 11.5
SEM images of *C. diolis* (a and b).

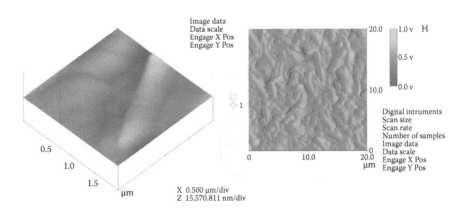

FIGURE 11.6
(See color insert.) AFM images of *C. diolis*.

noncontact mode. In the AFM image also, the surface of the *Clostridium diolis* appears to be smooth and spindle-shaped.

The 16S rDNA sequencing data indicated that the anaerobic strain was *Clostridium diolis* (Figure 11.7) with 96% similarities to those of the nearest strain *Clostridium roseum* (accession number Y18171). The nucleotide sequence reported here has been assigned an accession number FJ947160 from the NCBI Gene Bank database. This is the first report on the isolation and characterisation of lipolytic aerobic bacterium *Clostridium diolis* from excreta of a flesh-eating animal (*Canis aureus*).

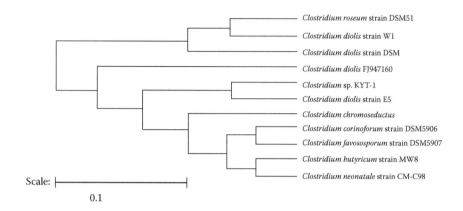

Scale: |————————————————|
     0.1

FIGURE 11.7
Phylogenetic tree showing the relationship of acidic lipase-producing strain *Clostridium diolis* FJ947160 to other *Clostridium* sp. with accession number.

11.3 Anaerobic Biodegradation of Slaughterhouse Lipid Wastes

The anaerobic digestion has been proven to be a versatile technology for the management of organic solid waste (De Baere 1999) discharged from varied industries. However, very few studies have been carried out on anaerobic digestion of slaughterhouse waste, despite lipids having higher methane potential compared to proteins or carbohydrates. This may be due to the complex nature of the substrate. Animal fat is mainly composed of esters of fatty acid and glycerol. It possesses hydrophobic characteristics, and most of the microbial communities work efficiently under a hydrophilic environment. Thus, anaerobic microbial degradation of lipids for methane recovery is one of the least investigated topics, and there has been constant research on the anaerobic digestion of lipid wastes.

In an anaerobic environment, the fats are first hydrolysed (lipolysed) into free long-chain fatty acids and glycerol. The hydrolysis process is catalysed by extracellular lipase released by acidogenic bacteria. The free LCFAs are subsequently cleaved into shorter chain fatty acids by acetogenic bacteria (Masse et al. 2002), and glycerol also mainly degraded to acetate, lactate and 1,3-propanediol (Batstone et al. 2000). During acetogenesis acetate, hydrogen and carbon dioxides are formed. The final step is the methanogenesis, which is usually the rate-limiting step in the overall anaerobic process of soluble substrates. Methanogenesis consists of acetotrophic and hydrogenotrophic processes. In acetotrophic methanogenesis, methane is produced from acetate by acetotrophic bacteria, and the hydrogenotrophic bacteria convert carbon dioxide and hydrogen into methane under hydrogenotrophic methanogenesis (Figure 11.8). The present study was confined to the hydrolysis (48 h) step alone to extract the lipase enzyme for the lipid hydrolysis.

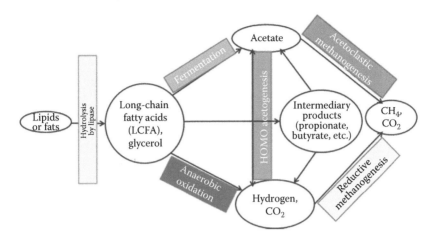

FIGURE 11.8
Overview of anaerobic digestion processes of lipids.

The anaerobic fermentation of the lipid substrates was conducted under nitrogen atmosphere in 2-L batch fermentor of composition (g/L) NaCl, 0.6; MgSO$_4$, 0.02; CaCl$_2$, 0.1; K$_2$HPO$_4$, 0.1; KH$_2$PO$_4$, 0.1; NH$_4$Cl, 0.25; and cysteine hydrochloride, 0.07. The trace element solution of 1 ml contains (g/L) CoCl$_2$, 0.2; MnCl$_2$, 0.03; FeSO$_4$, 0.02; and H$_3$BO$_3$, 0.03. The lipid substrate (goat tallow) concentration, 3.1 g/L, was added to the fermentation medium. The pH of the medium was adjusted to 6.0. The above medium was emulsified by ultrasonic sonicator (Bandelin, Germany) for 15 min at 20 Hz. The medium was autoclaved at 120°C at 15 psi for 15 min and incubated at 37°C for 96 h without agitation. The efficiency of lipid degradation by the *C. diolis* was observed to be around 92%. The yield of glycerol and free fatty acid produced was 361 and 947 mg/g of goat tallow, respectively. This shows that the *C. diolis* is an efficient strain for the degradation of slaughterhouse lipid waste goat tallow. The degradation of the goat tallow was confirmed by the following instrumental analysis.

11.4 Instrumental Analysis for the Confirmation of Hydrolysis of Goat Tallow

11.4.1 Fourier Transform Infrared (FT-IR) Spectroscopy Studies

The FT-IR study shows the functional group changes upon the anaerobic biodegradation of a lipid molecule (goat tallow). The FT-IR studies were carried out for the fermented metabolites of goat tallow (Figure 11.9) fermented at different residence times (0–96 h). Figure 11.9a shows the spectrum of unhydrolysed goat tallow at day 0; the sharp band at 1722 cm^{-1} is due to

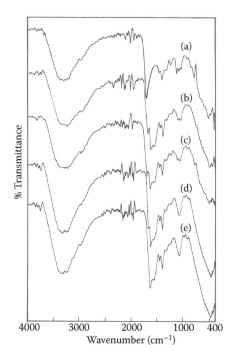

FIGURE 11.9
FT-IR spectra of fermented products of goat tallow at different fermentation periods: (a) 0 h, (b) 24 h, (c) 48 h, (d) 72 h and (e) 96 h.

the C=O stretching of esters. The $-CH_2$ scissoring vibrations were observed at 1402 cm^{-1}, and the C–O stretching vibration was found in the region 1118–1239 cm^{-1}. Figure 11.9b shows that the intensity of the C=O stretching of esters peak at 1700 cm^{-1} was steadily decreased for fermentation periods from 24 h (Figure 11.9b) to 48 h (Figure 11.9c) and then from 48 h to 72 h (Figure 11.9d) compared to the FT-IR spectrum of the zero hour. This can be attributed to the cleavage of triglyceride molecules into glycerol and fatty acids. It is supported by the following facts: The stretching band observed at 1634 cm^{-1} was due to the –C–O stretching frequency of carboxylic acid over-lapped with the hydroxyl group of glycerol. The amide II bond at 1540 cm^{-1} is due to N–H bending with a contribution from the C–N stretching vibrations, which was derived from the enzymes present in the fermented medium. The band observed at 1400 and 1067 cm^{-1} is due to the –C–O stretching vibration of carboxylic acid overlapped with –OH bending vibrations of glycerol and $-CH_2$ wagging, respectively. The FT-IR spectrum of the 96-h fermented sample (Figure 11.9e) showed the peak corresponding to –C–O stretching frequency of carboxylic acid overlapped with the hydroxyl group of glycerol at 1636 cm^{-1}. The –OH bending vibrations of glycerol were found in the region of 1399, 1401 and 1403 cm^{-1} in 48 (Figure 11.9c), 72 (Figure 11.9d) and 96 h (Figure 11.9e) of fermentation, respectively. The $-CH_2$ wagging was observed at 1065, 1074 and 1077 cm^{-1} in the 48 (Figure 11.9c), 72 (Figure 11.9d) and 96 h (Figure 11.9e) of fermentation, respectively. The amide II bond at 1540, 1524 and 1511 cm^{-1} was observed in the spectra of 48 (Figure 11.9c), 72 (Figure 11.9d), and 96 h (Figure 11.9e) of fermentation, respectively.

11.4.2 Gas Chromatography-Mass Spectrometry (GC-MS)

The GC-MS analysis suggested the intermediate product and end product formation from the goat tallow based on their molecular mass. The GC-MS of the initial goat tallow (Figure 11.10a) sample showed the parent peak and molecular ion peak at around m/z 440, and fragmentation of CH_2 groups is observed at m/z 427, 392 and 372. The ester fragmentation was observed at m/z 328 after the removal of 44 from m/z 372. The CH_3 and CH_2 groups were observed around m/z 396–372. This can be due to the presence of methyl ester, methylene dimethyl ester, 4-fluro-5-imidazole-carboxylic acid ethyl ester, benzene acetic acid and 4-(1,1-dimethylethyl)-methyl ester.

The GC-MS spectrum of fermented samples at 48 h (Figure 11.10b and c) showed that the peak corresponds to hydrolysed products, such as glycerol and fatty acids. The fragmented peak observed at m/z 83 (in which the parent peak is m/z 98) is due to the presence of the $-OH$ group (Figure 11.10c). This can be due to the presence of glycerol in the fermented sample. The molecular ion peak was observed at m/z 150, and fragmentation of acids ($-COOH$) took place; the peak was observed at m/z 91, and 78 can be due to the fragmentation of the CH_2 group present in the acid (Figure 11.10b).

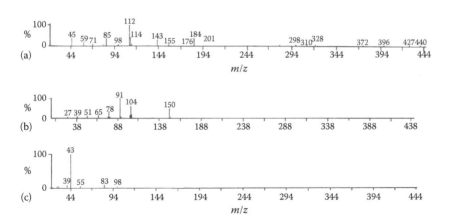

FIGURE 11.10
GC-MS of (a) unhydrolysed goat tallow, and (b and c) fermented products of hydrolysed goat tallow (48 h).

11.4.3 Nuclear Magnetic Resonance (NMR) Spectroscopy

The NMR study was carried out for the structural determination of lipid molecules in goat tallow and the intermediate products and end products formed from the goat tallow during anaerobic fermentation.

11.4.3.1 *13C-NMR Spectra*

The ^{13}C-NMR spectra were recorded for the fermentation period 0–96 h as shown in Figure 11.11. The spectrum of the day 0 sample (Figure 11.11a) showed the chemical shift (δ) of −C=O of ester present in the glycerides at δ 169 and the −CH₂ group in triglycerides at δ 29.79. The intensity of the chemical shift δ at 169 for −C=O of ester present in the glycerides and the chemical shift at δ 29.8 −CH₂ were decreased in the 24-h fermented sample (Figure 11.11b), and new peaks were observed at δ 69.52, δ 32.01 and δ 14.13; they may be attributed to the presence of −OH, CH₂ and −CH₃ groups in glycerol. This can be due to the breakage of triglycerides of the goat tallow and formation of glycerol and fatty acids. The spectra of TSHL samples fermented at 48-, 72- and 96-h residence periods (Figure 11.11c, d and e) showed the suppression of the carboxylic acid shift (δ 169) owing to the solvent effect. The presence of −OH, CH₂ and −CH₃ groups in glycerol in the fermented samples was observed at δ 69.52, δ 32.01 and δ 14.13, respectively, in the spectra of the 48-, 72- and 96-h fermentation periods (Figure 11.11c, d and e).

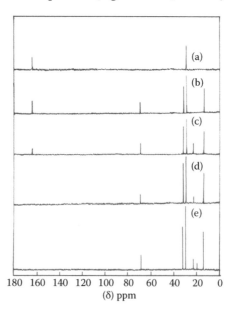

FIGURE 11.11
^{13}C-NMR analysis of fermented products of goat tallow at different fermentation periods: (a) 0 h, (b) 24 h, (c) 48 h, (d) 72 h and (e) 96 h.

11.4.3.2 ¹H NMR Spectra

The ¹H NMR spectra were recorded for the fermentation period of 0–96 h as shown in Figure 11.12. The spectrum of the day 0 sample (Figure 11.12a) showed the chemical shift (δ) of −C–O of ester present in the glycerides at δ 7.25 and −CH₂ group in triglycerides at δ 1.6 and 2.4. The intensity of the chemical shift at δ 7.25 for −C=O of ester present in the glycerides was decreased in the 24-h fermented sample (Figure 11.12b), and the presence of the new peaks observed at δ 1.7, δ 1.8 and δ 2.0 may be attributed to the presence of −OH, CH₂ and −CH₃ groups in glycerol. This can be due to the breakage of triglycerides of goat tallow and formation of glycerol and fatty acids. The spectra of 48- and 72-h fermented samples (Figure 11.12c and d) showed that the intensity of the peaks was increased from that in the 24-h fermented sample, which may be attributed to the conversion of triglycerides to glycerol and fatty acids over time. The ¹H NMR spectrum of the 96-h sample (Figure 11.12e) showed a suppression of the carboxylic acid shift at

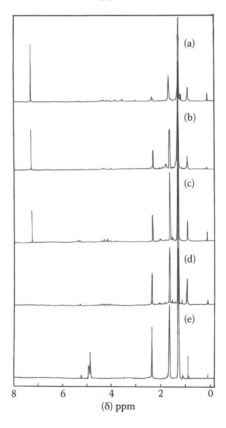

FIGURE 11.12
¹H NMR analysis of fermented products of goat tallow at different fermentation periods: (a) 0 h, (b) 24 h, (c) 48 h, (d) 72 h and (e) 96 h.

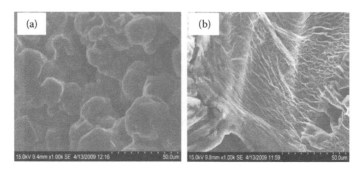

FIGURE 11.13
Surface morphology of goat tallow (a) unhydrolysed and (b) hydrolysed.

δ 7.25 owing to the solvent effect. The peak corresponds to the $-OH$, CH_2 and $-CH_3$ groups in glycerol in fermented samples observed at δ 1.7, δ 1.8 and δ 2.0 with an increase in intensity over that in the 72-h fermented sample. The new peak, which corresponds to the chemical shift of the $-OH$ group due to glycerol, was observed at δ 4.9 (Figure 11.12e).

11.4.3.3 Scanning Electron Microscopy (SEM) of the Hydrolysed and Unhydrolysed Goat Tallow

The surface morphology of the lipid molecules can be determined by the use of scanning electron microscopy. The SEM images clearly indicate the breakdown of the lipid molecule in the anaerobic fermentation. In this work, the SEM examination of the goat tallow samples confirmed the hydrolysis of goat tallow (Figure 11.13). The SEM analysis of unhydrolysed goat tallow (Figure 11.13a) showed the complex (globular) form of triglycerides. After the fermentation (Figure 11.13b), the complex (globular) form of triglyceride molecules (deformed into thread or flat, plate-like structures) is changed to an uncoupled form due to the hydrolysis by *C. diolis.* Thus, the topography of the substrate was damaged and altered completely.

11.5 Production of Bioactive Molecules from the Fermented Medium of Goat Tallow

Most of the reports on lipase production have been focused on aerobic bacteria with much less attention paid to anaerobes (Gupta et al. 2004; Petersen and Och Daniel 2006). The anaerobic lipolytic microorganisms known are very limited. However, few anaerobic lipid-hydrolysing bacteria have been reported so far, including *Anaerovibrio lipolytica, Butyrivibrio fibrisolvens* strain

S2, *Propionibacterium* (Jarvis et al. 1998) and *Selenomonas lipolytica* (Svetlitshnyi et al. 1996; Jarvis et al. 1999). Thus, there has been constant research for an anaerobic bacterial strain with the higher lipolytic activity, which enhances the hydrolysis of lipid substrates. In the present study, the maximum lipase activity produced by the *C. diolis* was around 129 U/mL at optimum culture conditions, such as time, 48 h; pH, 6.0; temperature, 37°C; and substrate concentration, 3.1 g/L; and nitrogen and metal ion sources, such as ammonium sulphate and CaCl₂, respectively, by utilising goat tallow as the substrate (Figure 11.14).

Another biomolecule, a biosurfactant (BS), produced in the same fermented medium by *C. diolis* using goat tallow was quantified and characterised. The production of biosurfactant from *C. diolis* was carried out at optimised culture conditions using TSHW as the substrate. During the period of growth in the presence of goat tallow, transformation from the solid state to liquid started from 48 h of the fermentation period. At the initial stage, the transformation of triglycerides, that is, hydrolysis by lipase to liquefy the solidus goat tallow followed by the formation of a micelle-like biosurfactant, was observed, and later, the micelle structure was retained for a period of more than 96 h. The formation of micelles and the development of vesicles are due to an increase in viscosity of the growth medium by the production of a biosurfactant in the presence of triglycerides, glycerol, fatty acids and microbial growth. The viscosity of the growth medium was enhanced with the increase in the fermentation period up to 48 h. It was retained for more than 96 h (Figure 11.14). The viscosity of the biosurfactant was 25 × 10³ cps under optimum conditions. The interface between two simple fluids is governed by the surface tension, which is isotropic when there is no surfactant; but in the presence of surfactant, the

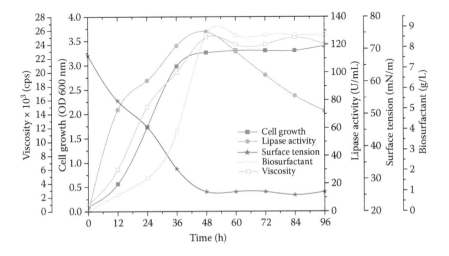

FIGURE 11.14
(See color insert.) Profile of cell growth, lipase activity, surface tension, biosurfactant and viscosity at different incubation periods in the anaerobic fermentation of goat tallow.

interfacial area is minimised, and it becomes anisotropic and increases the micellar compactness, which results in an increase in the viscosity by five-fold. The other possibilities for the formation of a micelle structure and the development of vesicles may be due to the presence of (i) fatty acids, glycerol, biosurfactants, acidic pH and unspent salts; (ii) fatty acids that include free fatty acids or mono-/diglycerides and/or unhydrolysed triglycerides; (iii) fatty acids and biosurfactants; (iv) fatty acids, glycerol and free fatty acids; (v) fatty acids, biosurfactants and free hydrogen/hydroxyl ions, etc. (Gnanamani et al. 2010). The maximum surface activity of cell-free broth showed 26 mN/m at 48 h, and it enabled the collapse of the oil drop within a fraction of a second. The critical micelle concentration (CMC) of the extracted biosurfactant from the *C. diolis* by using goat tallow was 6.6 µg/mL. Generally, the surface tension at the CMCs of various biosurfactants has been reported to be in the range of about 27 to 35 mN/m (Suk et al. 1999; Kim et al. 2000).

For practical purposes, it is important to distinguish between an efficient surfactant and an effective surfactant. Efficiency is measured by the surfactant concentration required to produce a significant reduction in the surface tension of liquid media, whereas effectiveness is measured as the minimum value to which the surface tension of liquid media can be reduced by the biosurfactant under the experimental conditions. Therefore, important characteristics of a potent surface-active agent are its ability to lower the surface tension of an aqueous solution and its low CMC (Kim et al. 2000). In general, biosurfactant activities depend on the concentration of the surface-active compounds until the CMC is obtained. At concentrations above the CMC, biosurfactant molecules associate to form micelles, bilayers and vesicles (Figure 11.15). Micelle formation enables biosurfactants to reduce the surface and interfacial tension and increase the solubility and bioavailability of hydrophobic organic compounds. The CMC is commonly used to measure the efficiency of a surfactant. Efficient biosurfactants have a low CMC, which means that less biosurfactant is required to decrease the surface tension. Micelle formation has a significant role in microemulsion formation. Microemulsions are clear and stable liquid

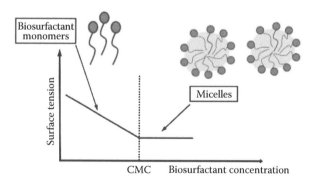

FIGURE 11.15
The relationship between biosurfactant concentration, surface tension and formation of micelles.

mixtures of water and oil domains separated by monolayer or aggregates of biosurfactants. Microemulsions are formed when one liquid phase is dispersed as droplets in another liquid phase, for example, oil dispersed in water (direct microemulsion) or water dispersed in oil (reversed microemulsion).

In the present work, the final mass of the biosurfactant extracted was 8.6 g/L (Figure 11.14) from the fermented medium at optimum culture conditions and substrate concentrations. The yield of the biosurfactant obtained from the goat tallow in the anaerobic fermentation was higher than the other reported biosurfactants from other substrates (Haba and Espuny 2000; Abouseoud et al. 2007; Ramani et al. 2012). The emulsification property of the biosurfactants was evaluated by determining the emulsifying activity with different hydrocarbons.

Emulsification is a process that forms a liquid, known as an emulsion, containing very small droplets of fat or oil suspended in a fluid, usually water. The high molecular weight biosurfactants are efficient emulsifying agents. They are often applied as an additive to stimulate bioremediation and removal of oil substances from environments. The biosurfactants exhibited different stabilisation properties for the hydrocarbons tested. Olive oil exhibited the best emulsifying activity, followed by palm oil and diesel oil (Table 11.2). This property may be exploited to use the biosurfactant in the bioremediation of hydrocarbons from their contaminated environment. They can enhance hydrocarbon bioremediation by two mechanisms. The first includes the increase in substrate bioavailability for microorganisms, and the other involves interaction with the cell surface, which increases the hydrophobicity of the surface allowing hydrophobic substrates to associate more easily with bacterial cells. By reducing surface and interfacial tensions, biosurfactants increase the surface areas of insoluble compounds leading to increased mobility and bioavailability of hydrocarbons. In consequence, biosurfactants enhance biodegradation and removal of hydrocarbons. Addition of biosurfactants can be expected to enhance hydrocarbon biodegradation by mobilisation, solubilisation or emulsification (Figure 11.16).

The mobilisation mechanism occurs at concentrations below the biosurfactant CMC. At such concentrations, biosurfactants reduce the surface and interfacial tension between air–water and soil–water systems. Due to the

TABLE 11.2

Emulsification Activity (E24) of Biosurfactants Derived from *C. diolis* Against Hydrocarbons

Hydrocarbons	E24 (%)
Olive oil	50
Palm oil	46
Crude oil	37
Kerosene	34
Diesel oil	36
Petrol	28

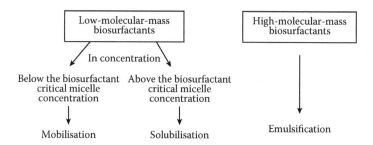

FIGURE 11.16

Mechanisms of hydrocarbon removal by biosurfactants depending on their molecular mass and concentration.

reduction of the interfacial force, contact of biosurfactants with the soil–oil system increases the contact angle and reduces the capillary force holding the oil and soil together. In turn, beyond the biosurfactant CMC, the solubilisation process takes place. At these concentrations, biosurfactant molecules associate to form micelles, which dramatically increase the solubility of oil. The hydrophobic ends of biosurfactant molecules connect together inside the micelle, while the hydrophilic ends are exposed to the aqueous phase on the exterior. Consequently, the interior of a micelle creates an environment compatible to hydrophobic organic molecules. The process of incorporation of these molecules into a micelle is known as solubilisation (Pfociniczak et al. 2011).

11.6 Purification and Characterisation of Purified *C. diolis* Biosurfactant

The purification of biosurfactants produced from *C. diolis* was carried out using a silica gel 60 chromatography column equilibrated with methanol. The fractions containing biosurfactant were identified by the oil drop collapse method, and it was pooled up and spotted on a thin layer chromatography (TLC) plate. The biosurfactant fraction yielded a single spot on the TLC plate. This fraction showed a positive reaction with ninhydrin and iodine vapour, indicating the presence of peptide–protein and lipid moieties in the molecule. The nature of the purified biosurfactants was further confirmed by FT-IR and sodium dodecyl sulphate polyacrylamide gel electrophoresis (SDS-PAGE).

The FT-IR spectrum (Figure 11.17) contained the peak corresponding to amides at 3329.21 cm^{-1}, stretching the mode of the CO–N bond at 1643.36 cm^{-1} and the N–H bond combined with the C–N stretching mode at 1532.32 cm^{-1}.

The FT-IR results suggest that the *C. diolis* biosurfactant was the lipoprotein type of biosurfactant, confirmed by SDS-PAGE. The SDS-PAGE showed that

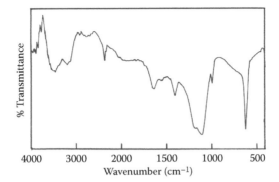

FIGURE 11.17
FT-IR spectrum of *C. diolis* biosurfactant produced from goat tallow.

the molecular weight of *C. diolis* biosurfactant produced from goat tallow during anaerobic fermentation was 13 kDa. In general, FT-IR spectra do not distinguish lipoprotein- and lipopeptide-based biosurfactants. They are differentiated in terms of their molecular mass, either >1–1.5 kDa (for lipoprotein) or <1–1.5 kDa (lipopeptide). Thus, *C. diolis* biosurfactant produced from goat tallow was categorised as a lipoprotein biosurfactant (as the molecular mass was 13 kDa). The lipid, protein and amino acid contents of the purified *C. diolis* biosurfactants were 41, 22 and 13 mg/L, respectively.

The amino acid composition of the purified *C. diolis* biosurfactant was analysed using HPLC. The dominating amino acids of the *C. diolis* biosurfactant were glutamic acid, 40 mol%; threonine, 7.14 mol%; arginine, 18.57 mol%; valine, 8.57 mol%; and lysine, 15.71 mol%, and are the predominant amino acids in the *C. diolis* biosurfactants. The presence of fatty acid was further confirmed by gas chromatography (GC) analysis (Figure 11.18). GC analysis

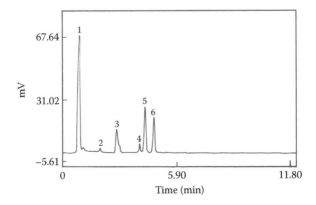

FIGURE 11.18
GC of *C. diolis* biosurfactant from goat tallow, illustrating the fatty acid composition. 1: Lauric acid, 2: myristic acid, 3: palmitic acid, 4: stearic acid, 5: oleic acid and 6: linoleic acid.

of purified lipoprotein biosurfactant showed the presence of lauric acid, 0.5%; myristic acid, 8.2%; palmitic acid, 22.3%; stearic acid, 13.6%; oleic acid, 32.1%; and linoleic acid, 13.6%.

11.7 Production and Purification of Acidic Lipases

In the present study, the maximum lipase activity produced by *C. diolis* was 129 U/mL at optimum culture conditions, such as time, 48 h; pH, 6.0; temperature, 37°C; and substrate concentration, 3.1 g/L; and nitrogen and metal ion sources, such as ammonium sulphate and $CaCl_2$, respectively, by utilising goat tallow as the substrate. The biomass in the broth was removed by centrifugation at 6500 × g for 20 min at 4°C, and the acidic lipases from *C. diolis* was purified to homogeneity using ammonium sulphate precipitation followed by chromatography column techniques, such as DEAE-cellulose column chromatography and Sephadex G-25 and G-100 column chromatography. In the anaerobic fermentation, two different acidic lipases (LipA and LipB) were produced from goat tallow with the molecular masses of 94.7 kDa (LipA) and 86 kDa (LipB) (Figure 11.19). The purified lipases produced from goat tallow possess higher molecular weight than the lipase produced from many sources. This may be due to the fact that the substrate goat tallow used in the present study was of high molecular weight consisting of 18 oleic acid molecules arranged in a chain, and it is embedded in palmitic acid residue. It is known that the essential characteristic for enzymes to cleave or solvate the high molecular weight substrates (such as tallow or vegetable oil) is that they must possess a higher number of active sites, that is, peptide linkages.

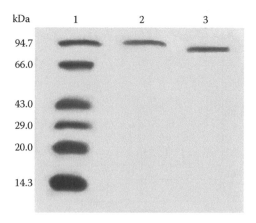

FIGURE 11.19
SDS-PAGE molecular weight profile of purified lipases. Lane 1: molecular weight marker; lane 2: lipase A (LipA); lane 3: lipase B (LipB).

Thus, enzymes with a larger number of peptide bonds are characterised by high molecular weight (Ramani and Sekaran 2010). The specific activity and yield of purified LipA were found to be 13.5 U/mg of protein and 6.9%, respectively. The specific activity and yield of purified LipB were found to be 411 U/mg of protein and 9.5%, respectively.

11.8 Characterisation of Acidic Lipases

The purified two acidic lipases (LipA and LipB) from goat tallow showed maximum lipase activity at pH 6.0. The acidic lipases, LipA and LipB, produced from goat tallow were stable in the pH ranges 3.5–8.0 and 4.0–8.0, respectively. The enzyme stability at varied temperatures confirmed that the purified acidic lipases from *C. diolis* could be classified under mesophilic enzymes that remained active at temperatures ranging from 30°C to 50°C. Both the enzymes became inactive at temperatures above 60°C, indicating the instability of conformational structures at higher temperatures. The acidic lipase activity was enhanced to 119% and 117% for LipA and LipB, respectively, with $CaCl_2$. A possible explanation of this phenomenon is that Ca^{2+} has a special enzyme-activating effect that exerts by orientation at the fat–water interface. Therefore, Ca^{2+} ions might carry out three distinct roles in lipase action: removal of fatty acids as insoluble Ca^{2+} salts in certain cases, direct enzyme activation resulting from orientation at the fat–water interface and a stabilising effect on the enzyme (Yu et al. 2007). The effect of non-ionic surfactants on lipase activity was studied by choosing the surfactants with a hydrophilic–lipophilic balance (HLB) varying in an interval of 13–40 (Tween 20, Tween 40, Tween 60, Tween 80, Triton X-100 and SDS). The bio-surfactants having higher HLB values exhibit greater dispersability in an aqueous medium. The HLB value indicates the polar ability of the molecules in arbitrary units, and the values increase with an increase in hydrophilicity (Guncheva et al. 2007). Among all the surfactants tested, the addition of 1% Triton X-100 (HLB value 13.0) to the lipase mixture enhanced the lipase activities to 206% (LipA) and 102% (LipB). The decrease in relative activities of lipases was observed in the presence of the Tween series, and SDS having an HLB value of 40 completely inhibited the relative activities.

Paramethyl sulphoxide (PMSF) and 1,10-phenanthroline inhibited the activity of lipases. The inhibitory effect of PMSF and 1,10-phenanthroline on the activity of purified acidic lipases proved that the purified acidic lipases from *C. diolis* using goat tallow are serine hydrolases (Salameh and Wiegel 2007). This suggests that the PMSF and 1,10-phenanthroline had the ability to alter the conformation of enzymes. Ethylenediaminetetraacetic acid (EDTA) and β-mercaptoethanol are neither affected by nor enhanced the activity of acidic lipases.

TABLE 11.3

Amino Acid Composition of Purified *C. diolis*
Acidic Lipases Derived from Goat Tallow

Amino Acids	Mol%	
	LipA	LipB
Asp	10.02	29.3
Glu	27.69	17.82
Ser	1.85	0.48
His	13.39	3.21
Gly	12.41	15.23
Thr	11.14	1.29
Ala	2.4	11.21
Arg	3.1	1.0
Tyr	10.78	10.29
Val	0.81	0
Met	0.51	0.04
Pala	0.4	0
Ileu	0.51	0
Lue	0.38	0
Lys	4.41	1.11

11.8.1 Amino Acid Composition Analysis by High Performance Liquid Chromatography (HPLC)

The amino acid composition of the acidic lipase showed that the acidic amino acids (glutamic acid and aspartic acid) were predominant. The percentages of polar amino acids in the purified acidic lipases from *C. diolis* using goat tallow under anaerobic conditions were 86.25% LipA and 88.73% LipB. The percentages of apolar amino acids in the acidic lipases were 18.16% LipA and 11.25% LipB (Table 11.3). The ratio of polar to apolar amino acids was in the range from 4.75 to 7.9 for the purified lipases produced from goat tallow under anaerobic conditions.

11.9 Immobilisation of Acidic Lipase in Mesoporous Activated Carbon

The low catalytic efficiency and stability of native enzymes are considered to be the main barriers for the development of their large-scale applications in many cases (Wang et al. 2006), whereas immobilised lipases generally have good physical and thermal stability and reusability. Moreover, the immobilisation of enzymes minimises the cost of product isolation and provides operational flexibility. Many carrier matrices have been evaluated, including polymers and resins (Gandhi et al. 1996; Foresti and Ferreira 2005; Wang et al. 2006), silica and silica–alumina composites (Serio et al. 2003; Blanco et al. 2004; Bai et al. 2006) and carbonaceous materials (Kovalenko et al. 2001, 2002; Zhou and Chen 2001), for immobilisation of enzymes. These systems generally have low mechanical strength and often exhibit severe diffusion limitations, leading to a relatively low enzymatic activity. To minimise the internal diffusion limitation, porous supports were mostly used in particulate form (Foresti and Ferreira 2005; Zhao et al. 2006). Most of the studies have concluded that enzymes immobilised in inorganic carrier matrices are more stable than the synthetic and natural polymers. Porous materials have certain favouring features for immobilisation compared to nonporous materials owing to their pore size, large surface area, pore volume and opened structures (Macedo et al. 2008). Mesoporous activated carbon obtained from rice husks has been reportedly used as the carrier matrix for the immobilisation of protease (Ganesh Kumar et al. 2009) and polyphenol oxidase (Kennedy et al. 2007) and lipase (Ramani et al. 2010c,d). In this work, the mesoporous activated carbon was prepared from the rice husks by a two-stage process: (1) precarbonisation at 400°C and (2) chemical activation using phosphoric acid at 800°C. The opening of the pores on the surface of the rice husk occurred due to the extraction of some volatile organic materials so as to create, upon activation, micro- and mesopores in the carbon matrix. As a result of the creation of pores, there was an increase in both the surface area and the pore volume in the activated carbon, which held a high loading capacity. The characteristics of mesoporous activated carbon prepared in the study are shown in Table 11.4.

The immobilisation of acidic lipase in MAC was studied in a batch experiment. One gram of MAC was added to 10 mL of buffer solution containing a known activity of acidic lipase, LipA (LipA was used for the immobilisation

TABLE 11.4

Characterisation of MAC

S. No	Parameters	MAC
1	Carbon (%)	45.4
2	Hydrogen (%)	0.9
3	Nitrogen (%)	0.1
4	Moisture content (%)	5.2
5	Ash content (%)	36.4
6	Apparent density (g/mL)	0.106
7	Average pore diameter (A°)	11.43
8	Point of zero charge (PZC)	6.4
9	Decolourising capacity (mg/g)	44
10	Surface area (m^2/g)	412

onto MAC and for the application studies because it showed higher activity than LipB) in an individual flask. The solutions were agitated at 100 rpm until equilibrium was reached. The optimum conditions for the immobilisation of acidic lipases in MAC were determined by varying time, pH, temperature and initial lipase activities. The maximum immobilisation capacity of the MAC was found to be 1524 U/g of MAC at optimum time, 120 min; pH, 5.5; temperature, 35°C; and initial lipase activity, 60 × 10³ U. The degree of immobilisation of acidic lipase was investigated through the magnitude of hydrolysis of olive oil as standard. The residual lipase activities in the bulk solution were determined using a lipase assay as described by Ramani et al. (2010a). The immobilised lipase activity was calculated by subtracting the final activity of lipase in solution from the initial activity of lipase. The immobilisation capacity of MAC was found to be 1834 U/g of MAC. The free acidic lipase (FAL) immobilised in MAC was named lipase-immobilised MAC (LMAC).

11.10 Adsorption Kinetic Model

In order to investigate the immobilisation of acidic lipase in MAC, the frequently used kinetic models, such as the nonlinear forms of pseudo first order (Lagergren and Svenska 1898), and pseudo second order (Ho and Mckay 1998), were employed with Equations 11.1 and 11.2.

$$q_t = q_e \left(1 - \exp^{-K_1 t}\right) \tag{11.1}$$

$$q_t = \frac{K_2 q_e^2 t}{1 + K_2 q_e t} \tag{11.2}$$

where q_e and q_t are the amount of lipase immobilised (U/g) at equilibrium, and at any time 't', and K_1 and K_2 are the first- and second-order rate constants.

The validity of the kinetic order for the immobilisation process is based on two criteria, viz., (1) the regression coefficients and (2) predicted q_e values. The first-order rate constant K_1 and the second-order rate constant K_2 were evaluated. The correlation coefficients for the first- and second-order kinetic model of immobilisation of acidic lipase in MAC were determined.

The first-order rate constant, K_1, was found to be 0.019 min⁻¹, and the second-order rate constant, K_2, was found to be 5.2 × 10⁻⁶ min⁻¹. The regression coefficients for the first- and second-order kinetic models were 0.96 and 0.988, respectively. The data indicate that the immobilisation of lipase in MAC followed pseudo second-order rate kinetics.

TABLE 11.5

Isotherm Parameters for the Immobilisation of Lipase in LMAC

Langmuir		Freundlich	
K_L (L g^{-1})	2.3	K_F (U g^{-1}) (L U^{-1})	1.36
b (L U^{-1})	0.0017	$1/n$	1.55
R^2	0.98	R^2	0.97

11.11 Equilibrium Isotherms

In the present study, the experimental data on equilibrium isotherms for immobilisation of lipases were fitted to the Langmuir (Equation 11.3) and Freundlich (Equation 11.4) models because each proposed mathematical model assumes a set of hypotheses to predict the immobilisation process.

$$q_e = \frac{K_L C_e}{1 + b C_e} \tag{11.3}$$

$$q_e = K_F C_e^{1/n} \tag{11.4}$$

where q_e is the solid phase sorbate concentration at equilibrium (U g^{-1}), C_e is the liquid phase sorbate concentration at equilibrium (U L^{-1}), K_L is the Langmuir isotherm constant, b is the Langmuir isotherm constant, K_F is the Freundlich constant and $1/n$ is the heterogeneity factor. The adsorption isotherms presented in Table 11.5 indicated that LMAC obeyed the Langmuir isotherm based on the regression coefficient (R^2).

11.12 Thermal Stability of Free and Immobilised Lipases

Thermal stability of the enzyme is one of the most important criteria for long-term and commercial applications. The immobilised enzyme is known to be more resistant to thermal energy than in the free state. The thermal stability of free and immobilised enzymes was determined by incubating them at various temperatures (10°C–50°C) for 24 h, and the residual activity was measured. The immobilised lipase in MAC exhibited a better thermal stability at all temperatures tested than the free lipase (Figure 11.20). It was observed that upon incubation for 24 h, the immobilised lipases were more stable than the free lipase. Chaubey et al. (2006) and Bayramoglu et al.

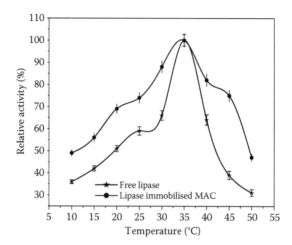

FIGURE 11.20
Thermal stability of free and immobilised lipases.

(2005) also reported that the immobilised lipase was more stable in the carrier matrix than the free enzymes. The increase in the thermal stability may be due to the presence of dangling bonds and free electrons in MAC, which act as a source for the absorption of thermal energy, and in free enzymes, the thermal energy is absorbed by the molecule itself, leading to denaturation.

11.13 Enzyme Kinetics for Free and Immobilised Lipases

Kinetic studies can yield information about the mechanism of an enzymatic reaction. Enzyme kinetics is the study of the chemical reactions that are catalysed by enzymes. In enzyme kinetics, the reaction rate is measured, and the effects of varying the conditions of the reaction are investigated. Studying an enzyme's kinetics in this way can reveal the catalytic mechanism of this enzyme. The maximum reaction rate (V_{max}) and the Michaelis–Menton constant (K_m) for purified acidic lipase and immobilised lipase in MAC were determined from the activity assay carried out for different concentrations of substrate (olive oil) (Shuler and Kargi 2005). The olive oil emulsion solutions in varied concentrations (0.06 to 0.56 mM) were prepared in distilled water containing polyvinyl alcohol, the lipase activity was measured according to the procedure of Naci and Ali (2002) and the liberated fatty acid concentration was determined. One activity unit of lipase was defined as the amount of enzyme that released 1 μmol of fatty acid per minute under assay conditions. The kinetic parameters were estimated from the Michaelis–Menton equation as shown in Equation 11.5.

$$v = \frac{[S]V_{max}}{[S] + K_m} \tag{11.5}$$

where $[S]$ is the substrate concentration (mM), and v is the initial reaction rate of the enzyme (mM/min).

The K_m values signify the extent to which the enzymes have access to the substrate molecules (Shuler and Kargi 2005). The lower the value of K_m, the higher the affinity between the enzymes and substrates. The K_m value for immobilised lipase (0.262 mM) was lower than the K_m value of free acidic lipase (0.42 mM). The lower (K_m) value of immobilised lipase indicates the higher affinity of the immobilised acidic lipase toward the substrate (enzyme–substrate complex formation). The higher K_m values for immobilised lipase could be attributed to conformational changes or lower accessibility of the substrate to the active sites of the immobilised enzyme. Hence, it can be concluded that the immobilisation of acidic lipase onto MAC did not undergo destructive conformations, and thus, loss of activity was not realised. The V_{max} value of free acidic lipase and immobilised lipase was found to be 0.19 and 0.208 mM min^{-1}, respectively. In this study, the V_{max}/K_m ratio of free acidic lipase was lower (0.46) than that of immobilised acidic lipase (0.79). In general, the greater the V_{max}/K_m values, the higher the catalytic efficiency of the immobilised enzymes for the substrates.

11.14 Characterisation of Acidic Lipase and Lipase-Immobilised MAC

11.14.1 SEM Images of MAC and LMAC

The surface morphology of MAC in a SEM image (Figure 11.21a) showed that the MAC is porous in nature with varied pore size distribution. The porous nature of MAC was generated by the chemical activation using phosphoric

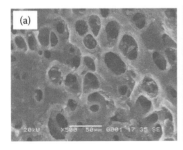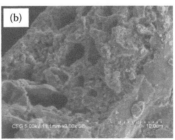

FIGURE 11.21
SEM image of (a) MAC and (b) LMAC.

acid. The generation of pores in the surface of the rice husk should be due to the destruction/decomposition of some organic components, which, upon activation, leads to creation of micro- and mesopores in the carbon matrix. The SEM image of acidic lipase immobilised in MAC (Figure 11.21b) showed that the enzyme molecules were well bound to the inner walls of the pores in the carbon matrix. On account of this, the inner surfaces of the pore wall are fully modified. The enzyme molecules entered into the deep pores of the carbon matrix, and after filling the pores, they reside at the mouth of the pores, and thereafter, they spread to the outer pore surface area. The clustering of lipase molecules observed in SEM photographs confirms that two types of interactions exist, leading to strong anchoring of lipase onto MAC contributing toward minimum drift of enzyme into the bulk solution (that is, no dislodging of enzyme, even after several washings, was observed). The interactive forces could be due to (1) enzyme–enzyme interaction or (2) interaction between the enzyme and the active sites of the MAC. This was further confirmed through FT-IR.

11.14.2 Infrared Spectra of MAC and LMAC

The infrared spectrum of MAC (Figure 11.22a) contains a wide band at 3401.75 cm^{-1} due to the O–H stretching mode of the adsorbed water molecule. The peak observed at 2922 cm^{-1} is attributed to asymmetric stretching of the CH group. The position and symmetry of this band at lower wave numbers indicate the presence of strong hydrogen bonds. The phosphate bond, which

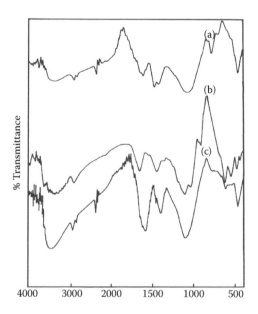

FIGURE 11.22
FT-IR of (a) MAC, (b) acidic lipase and (c) LMAC.

arises due to the phosphorous acid activation, exhibited at 1103.25 cm^{-1} along with P–O–P stretch at 795.28 cm^{-1}. Asymmetrical stretching of the carboxylate group in MAC is indicated at 1579.08 and 1459.07 cm^{-1}, respectively.

The FT-IR spectrum of acidic lipase (Figure 11.22b) from slaughterhouse lipid waste contains a broad envelope at 3431.35 cm^{-1} attributed to the overlap of –NH stretching of the amide group of proteins. The FT-IR absorption spectrum of pure lipase contains three major bands caused by peptide group vibrations in the 1800–1300 cm^{-1} spectral region (Natalello et al. 2005). The FT-IR spectrum of free acidic lipase in the present study shows that the amide I band at 1629.32 cm^{-1} is mainly due to the C–O stretching vibrations and that the amide II band with maximum absorption at 1583 cm^{-1} is due to N–H bending with a contribution from the C–N stretching vibrations. The amide III band at 1434 cm^{-1} is due to N–H bending, C–Cα and C–N stretching vibrations. The amide I and amide II bands are the most sensitive to the secondary structure of the protein, and if these bands are disturbed in the FT-IR spectrum of a protein, then it can be supposed that the protein has lost its native conformation (Vinu et al. 2004; Reshmi et al. 2007).

The FT-IR spectrum of acidic lipase immobilised MAC (Figure 11.22c) shows the N–H stretching frequency in the region of 3431.35 cm^{-1}. Complete shift of the asymmetrical stretching of the carboxylate group around 1583.95 cm^{-1} showed that there is a strong interaction between functional groups of MAC and the enzyme. This can be further confirmed by the shift of the amide I band from 1653.54 to 1629.32 cm^{-1} (sharp band). C–N stretching of amide is presented at 1458.98 cm^{-1}.

11.15 Application of Acidic Lipase in the Hydrolysis of Edible Oil Waste

Lipid-containing wastewater causes serious environmental problems due to its hazardous nature. Lipids (fats and oils) are the major organic constituent in municipal and some industrial wastewaters. Edible oil refineries, restaurants, slaughterhouses, wool scouring and the food and dairy industries are the major sources of the discharge of high concentrations of lipids (>100 mg L^{-1}) in wastewater. The conventional biological wastewater treatment system failed to treat such high concentrations of lipids in the wastewater due to the following problem: During the treatment, because of the hydrophobic nature of the lipids, they form a layer on the water surface and adsorb onto the cell wall of the microbes and hinder the diffusion of O$_2$ into the microbial cell, thereby slowing its metabolism. This may lead to the formation of a filamentous sludge and reduce the overall removal of organic matter. Because of this reason, the treatment of lipid-rich wastewater is still a challenge with conventional technologies (Crine et al. 2007). However, high fat–containing wastewater can be treated with an enzyme (lipase), and this type of enzymatic treatment system has gained much attention due to its high specificity, high catalytic efficiency and biodegradability (Prasad and Manjunath 2011).

11.15.1 Hydrolysis of Edible Oil Waste Using Acidic Lipase-Immobilised MAC

11.15.1.1 Kinetics on the Hydrolysis of Edible Oil Waste

Hydrolysis of goat tallow was carried out by using free acidic lipase (FAL) (1834 U) and LMAC (1 g [1834 U]). The optimum conditions for the maximum hydrolysis of edible oil waste by both FAL and LMAC were found to be time, 4.0 h; pH, 6.0; temperature, 30°C; and concentration of edible oil waste, 4.3 g/L (based on lipid analysis). The maximum percentage hydrolysis at the optimum lipid concentration was found to be 66% and 97% for FAL and LMAC, respectively. This indicated that the efficiency of the edible oil waste hydrolysis was enhanced by the LMAC. This confirms that there are no conformational changes in the enzymes in its immobilised state.

11.15.1.2 Determination of Hydrolysis Rate Kinetic Constants

In order to investigate the hydrolysis rate kinetic constants, the frequently used kinetic models, such as the nonlinear forms of pseudo first order (Lagergren and Svenska 1898) and pseudo second order (Ho and Mckay 1998) were employed by using Equations 11.6 and 11.7, respectively (Figure 11.23).

$$q_t = q_e(1 - \exp^{-K_1 t}) \tag{11.6}$$

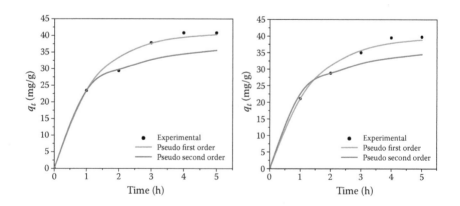

FIGURE 11.23
(See color insert.) Pseudo first-order and pseudo second-order kinetic models for hydrolysis of edible oil waste using FAL and LMAC (conditions: concentration of edible oil waste, 4.3 g/L; pH, 3.5; temperature, 30°C; and mass of LMAC, 1 g).

TABLE 11.6

Hydrolytic Rate Kinetic Constants for the Hydrolysis of Edible Oil Waste Using FAL and LMAC

	Pseudo First Order			Pseudo Second Order		
	K_1 (min^{-1})	R^2	χ^2	K_2 (g U^{-1} min^{-1})	R^2	χ^2
FAL on edible oil waste hydrolysis	0.85	0.99	0.17	0.033	0.97	0.5
LMAC for hydrolysis of edible oil waste	0.86	0.99	0.156	0.04	0.94	0.531

$$q_t = \frac{K_2 q_e^2 t}{1 + K_2 q_e t} \tag{11.7}$$

where q_e and q_t are the amount of lipid (edible oil waste) (mg/g of LMAC) at equilibrium, and at time (t), K_1 and K_2 are the first- and second-order rate constants.

The first-order rate constant K_1 and the second-order rate constant K_2 are summarised in Table 11.6. The correlation coefficients for the first- and second-order kinetic model of hydrolysis of edible oil waste using FAL and LMAC are summarised in Table 11.6. Moreover, the results confirmed that the hydrolysis of edible oil waste using free and immobilised lipases followed the first-order rate kinetic model. The results from the kinetic models confirmed that the hydrolysis of edible oil waste by the immobilised lipase under continuous mode obeyed the second-order rate kinetic model indicated by greater R^2 values (Table 11.6).

11.15.2 FT-IR Studies for the Functional Group Determination of Lipid Hydrolysates

The FT-IR spectrum of unhydrolysed edible oil waste (Figure 11.24a) shows O–H stretching vibrations at 3441 cm^{-1} and a sharp band at 1746 cm^{-1} due to the C=O stretching of esters. The CH$_2$ scissoring vibrations were observed at 1417 cm^{-1}, and the C–O stretching vibration was found in the region of 1163 cm^{-1}.

The FT-IR spectrum of hydrolysed edible oil waste (Figure 11.24b) by LMAC shows O–H stretching vibrations at 3436 cm^{-1}, and the decrease in peak corresponding to C=O stretching of esters at 1700 cm^{-1} was observed. This can be attributed to the conversion of the triglyceride form of goat tallow into glycerol and fatty acids. This was further confirmed by the presence of a band at 1627 cm^{-1} due to the –C–O stretching frequency of carboxylic acid overlapped with the hydroxyl group of glycerol. The band observed at 1450 and 1035 cm^{-1} is due to the –C–O stretching vibration of carboxylic acid overlapped with –OH bending vibrations of glycerol and –CH$_2$ wagging, respectively.

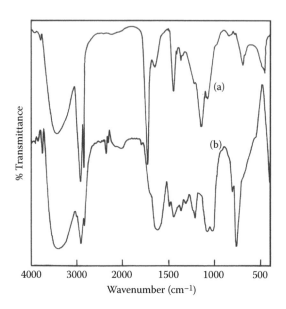

FIGURE 11.24
FT-IR spectrum of (a) unhydrolysed and (b) hydrolysed edible oil waste by LMAC.

11.16 Application of Biosurfactant for the Removal of Metals from the Aqueous Solutions Using Biosurfactant-Loaded MAC (BS-MAC)

There are numerous pollutants in the environment as a result of rapid industrial growth. Heavy metal pollution encountered worldwide is mainly due to wastewater discharge from industrial practices. In the recent past, there has been drastic interest in the application of amphiphilic biosurfactant for the heavy metal remediation process (Mulligan et al. 2005; Asha et al. 2007; Jayabarath et al. 2009; Pappalardo et al. 2010). The purified lipoprotein biosurfactant from *C. diolis* was applied for the removal of metal ions, such as calcium (Ca^{2+}), chromium (Cr^{3+}), iron (Fe^{2+}) and copper (Cu^{2+}) from aqueous solution (synthetic metal solutions) in the immobilised state (BS-MAC). The advantage of the loading of BS onto MAC is that the BS-MAC could be reused for the repeated applications of anchoring of metal ions from their contaminated environments.

The study suggested that the biosurfactant was very effective in the removal of heavy metal ions such as Cr^{3+} (89%) and Cu^{2+} (94%) when compared to other metal ions such as Ca^{2+} (68%) and Fe^{2+} (51%). The surface morphology of the BS (Figure 11.25a); BS-MAC (Figure 11.25b); and metal-adsorbed

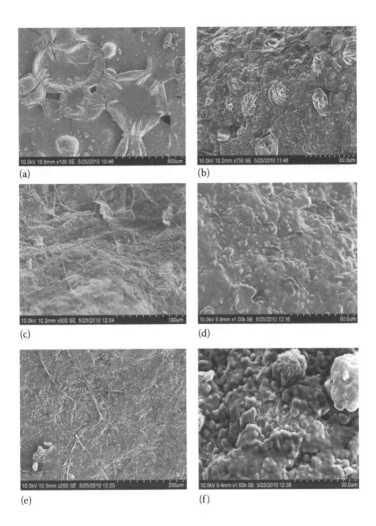

FIGURE 11.25
SEM images of (a) BS, (b) BS-MAC, (c) BS-MAC-Cr^{3+}, (d) BS-MAC-Cu^{2+}, (e) BS-MAC-Ca^{2+} and (f) BS-MAC-Fe^{3+}.

BS-MACs, such as BS-MAC-Cr^{3+} (Figure 11.25c), BS-MAC-Cu^{2+} (Figure 11.25d), BS-MAC-Ca^{2+} (Figure 11.25e) and BS-MAC-Fe^{2+} (Figure 11.25f), are shown. The SEM images clearly indicated that the BS is a potential source for the sequestration of heavy metal ions from the aqueous solutions. This property can be exploited in the bioremediation of heavy metals from their contaminated environments. Figure 11.26 shows the loading of BS onto active sites of MAC and sequestration of metal ions with the protein moieties of the BS in the BS-MAC.

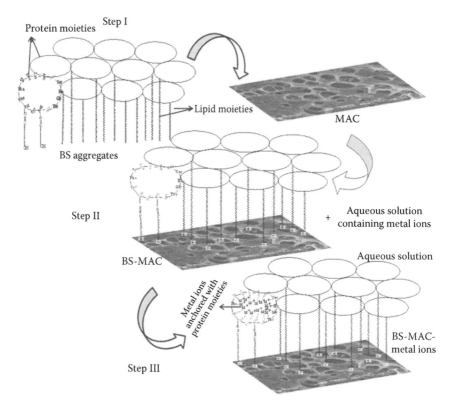

FIGURE 11.26
Pictorial representation of the loading of biosurfactant onto MAC and the sequestering of metal ions by the BS-MAC. Step I: addition of biosurfactant onto MAC; Step II: CH group in the biosurfactant binding with the active sites of the MAC; Step III: anchoring of metal ions with the protein moieties of the biosurfactant in the MAC.

11.17 Conclusions

The slaughterhouse lipid waste can be considered as a potential source for the production of high value–added products, such as acidic lipases, and microbial secondary metabolites, such as biosurfactants. Also, this chapter provides the anaerobic degradation process for the degradation of hydrophobic slaughterhouse lipid substrate goat tallow, and the degradation was confirmed by FT-IR, GC-MS, NMR and SEM analysis. The goat tallow is a potential source for the production of high molecular weight acidic lipases (two acidic lipases, LipA and LipB) through anaerobic fermentation by *C. diolis*. In addition to the high molecular weight acidic lipases, the lipoprotein biosurfactant also was produced by the strain *C. diolis* during anaerobic fermentation of goat tallow. To date, there is no report on the production of biosurfactant using goat tallow substrate in anaerobic fermentation. The *C. diolis* biosurfactant has potential application in the bioremediation (removal) of metal ions from the aqueous solutions. The acidic lipase from the anaerobic fermentation of goat tallow was immobilised in the heterogeneous carrier matrix (mesoporous activated carbon) for the hydrolysis of edible oil waste with higher efficiency, and the hydrolysis was confirmed using FT-IR spectroscopy. The study concluded that the slaughterhouse solid lipid waste, goat tallow, is a potential source for the production of high value–added bioactive molecules, such as acidic lipase and microbial secondary metabolite lipoprotein biosurfactant for industrial applications.

References

Abouseoud M, Maachi R, Amrane A. 'Biosurfactant production from olive oil by *Pseudomonas fluorescence*'. In: A. Mendez-Vilas (ed.), *Communicating Current Research and Educational Topics and Trends in Applied Microbiology*, pp. 340–347 (2007).

Ahimou F, Jacques P, Deleu M. 'Surfactin and iturin A effects on *Bacillus subtilis* surface hydrophobicity'. *Enzyme Microb Technol* 27 (2000): 749–754.

Asha AJ, Anupa N, Kirti VD et al. 'Biosurfactant technology for remediation of cadmium and lead contaminated soils'. *Chemosphere* 68 (2007): 1996–2002.

Bai YX, Li YF, Yang Y et al. 'Covalent immobilization of triacylglycerol lipase onto functionalized novel mesoporous silica supports'. *Biotechnol* 125 (2006): 574–582.

Banat IM, Makkar R, Cameotra S. 'Potential commercial applications of microbial surfactants'. *Appl Microbiol Biotechnol* 53 (2000): 495–508.

Batstone DJ, Keller J, Newell RB et al. 'Modelling anaerobic degradation of complex wastewater. I: Model development'. *Bioresour Technol* 75 (2000): 67–74.

Bayramoglu G, Kaya B, Anca MY. 'Immobilization of *Canida rugosa* lipase onto spacer-arm attached poly (GMA-HEMA-EGDMA) microspheres'. *Food Chem* 92(2) (2005): 261–268.

Benincasa M, Abalos A, Oliveira I et al. 'Chemical structure, surface properties and biological activities of the biosurfactant produced by *Pseudomonas aeruginosa* LBI from soap stock'. *Antonie Van Leeuwenhoek* 85 (2004): 1–8.

Blanco RM, Terreros P, Fernandez PM et al. 'Functionalization of mesoporous silica for lipase immobilization – Characterization of the support and the catalysts'. *J Mol Catal B: Enzym* 30 (2004): 83–93.

Bolzonella D, Pavan P, Fatone et al. 'Anaerobic fermentation of organic municipal solid wastes for the production of soluble organic compounds'. *Ind Eng Chem Res* 44(10) (2005): 3412–3418.

Castro-Ochoa LD, Rodriguez CG, Valerio GA et al. 'Screening, purification and characterization of the thermoalkalophilic lipase produced by *Bacillus thermoleovorans* CCR11'. *Enzyme Microb Technol* 37 (2005): 648–654.

Chaubey A, Parshad R, Koul S et al. '*Arthrobacter* sp. lipase immobilization for improvement in stability and enantio-selectivity'. *Appl Microbiol Biotechnol* 73 (2006): 598–606.

Crine DG, Paloumet X, Bjornsson L et al. 'Anaerobic digestion of lipid rich waste-effects of lipid concentration'. *Renew Energ* 32 (2007): 965–975.

Dalmou E, Montesinos JL, Lotti M et al. 'Effect of different carbon sources on lipase production by *Candida rugosa*'. *Enzyme Microb Technol* 26 (2000): 657–663.

De Baere L. 'Anaerobic digestion of solid waste: State-of-the art'. In: II Int. Symp. Anaerobic Dig. Solid Waste, held in Barcelona, June 15–17, J. Mata-Alvarez, A. Tilche and F. Cecchi (eds.), vol. 1, pp. 290–299, Int. Assoc. Wat. Qual. (1999).

Ferreira C, Maria A, Perolta RM. 'Production of lipase by soil fungi and partial characterization of lipase from a selected strains (*Penicillium wortmanii*)'. *J Basic Microbiol* 39 (1999): 11–15.

Foresti ML, Ferreira ML. 'Solvent-free ethyl oleate synthesis mediated by lipase from *Candida antartica* B adsorbed on polypropylene powder'. *Catal Today* 107–108 (2005): 23–30.

Gandhi NN, Vijayalakshmi V, Sawant SB et al. 'Immobilization of *Mucor miehei* lipase on ion exchange resins'. *Chem Eng J* 61 (1996): 149–156.

Ganesh Kumar A, Swarnalatha S, Kamatchi P et al. 'Immobilisation of high catalytic acid protease on functionalised mesoporous activated carbon particles'. *Biochem Eng J* 43 (2009): 185–190.

Gao XG, Cao SG, Zhang KC. 'Production, properties and application to nonaqueous enzymatic catalysis of lipase from a newly isolated *Pseudomonas* strain'. *Enzyme Microb Technol* 27 (2000): 74–82.

Gnanamani A, Kavitha V, Radhakrishnan N et al. Microbial biosurfactants and hydrolytic enzymes mediates in situ development of stable supramolecular assemblies in fatty acids released from triglycerides. *Colloids Surf B: Biointerfaces* 78 (2010): 200–207.

Gomes FM, Pereira EB, de Castro HF. 'Immobilization of lipase on chitin and its use in nonconventional biocatalysis'. *Biomacromolecules* 5 (2004): 17.

Guncheva M, Zhiryakova D, Radchenkova N et al. 'Effect of nonionic detergents on the activity of a thermostable lipase from *Bacillus stearothermophilus* MC7'. *J Mol Catal B Enzym* 49 (2007): 88–91.

Gupta R, Gupta N, Och RP. 'Bacterial lipases: An overview of production, purification and biochemical properties'. *Appl Microbiol Biotechnol* 64 (2004): 763–781.

Haba E, Bresco O, Ferrer C et al. 'Isolation of lipase-secreting bacteria by deploying used frying oil as selective substrate'. *Enzyme Microb Technol* 26 (2000): 40–44.

Haba E, Espuny MJ. 'Screening and production of rhamnolipids by *Pseudomonas aeruginosa* 47T2 NCIB 40044 from waste frying oils'. *Appl Microbiol* 88 (2000): 379–387.

Hassan F, Shah A, Hameed A. 'Industrial application of microbial lipases'. *Enzyme Microb Technol* 39 (2006): 235–251.

Ho YS, Mckay G. 'Kinetic models for the sorption of dye from aqueous solution by wood'. *Trans IChem E* 76B (1998): 183–191.

Jarvis GN, Strompl C, Moore ERB et al. 'Isolation and characterisation of obligately anaerobic, lipolytic bacteria from the rumen of red deer'. *Syst Appl Microbiol* 21 (1998): 135–143.

Jayabarath J, Shyam Sundar S, Arulmurugan R et al. 'Bioremediation of heavy metals using biosurfactants'. *Int J Biotechnol Appl* 1 (2) (2009): 50–54.

Kennedy LJ, Selvi PK, Aruna P et al. 'Immobilisation of polyphenol oxidase onto mesoporous activated carbons – Isotherm and kinetic studies'. *Chemosphere* 69 (2007): 262–270.

Kim SH, Lim EJ, Lee SO et al. 'Purification and characterization of biosurfactants from *Nocardia* sp. L-417'. *Biotechnol Appl Biochem* 31 (2000): 249–253.

Kovalenko GA, Komova OV, Simakov AV. 'Macrostructured carbonised ceramics as adsorbents for immobilization of glucoamylase'. *J Mol Catal A-Chem* 182–183 (2002): 73–80.

Kovalenko GA, Kuznetsova EV, Mogilnykh Yu, I et al. 'Catalytic filamentous carbons for immobilization of biologically active substances and non-growing bacterial cells'. *Carbon* 39 (2001): 1033–1043.

Kuang Y. 'Enhancing anaerobic degradation of lipids in wastewater by addition of co-substrate'. PhD thesis. Available at http://researchrepository.murdoch.edu .au/136/2/02Whole.pdf. (2002).

Lagergren S, Svenska BK. Zur theorie der sogenannten. 'Adsorption geloester stoffe'. *Veternskapsakad Handlingar* 24 (1898): 1–39.

Macedo JS, Otubo L, Ferreira OP et al. 'Biomorphic activated porous carbons with complex microstructures from lignocellulosic residues'. *Microporous Mesoporous Mater* 107 (2008): 276–285.

Mahadik ND, Puntambekar US, Bastawde KB et al. 'Production of acidic lipase by *Aspergillus niger* in solid state fermentation'. *Process Biochem* 38 (2002): 715–721.

Mahadik ND, Puntambekar US, Bastawde KB et al. 'Process for the preparation of acidic lipase'. *US Patent No. 06*, (2003): 534: 303.

Maia MMD, Heasley A, Camargo de Morais MM et al. 'Effect of culture conditions on lipase production by *Fusarium solani* in batch fermentation'. *Biores Technol* 76 (2001): 23–27.

Maier RM, Chavez GS. '*Pseudomonas aeruginosa* rhamnolipids: Biosynthesis and potential applications'. *Appl Microbiol Biotechnol* 54 (2000): 625–633.

Makkar RS, Cameotra SS. 'An update on the use of unconventional substrates for biosurfactant production and their new applications'. *Appl Microbiol Biotechnol* 58 (2002): 428–434.

Masse L, Masse, DI, Kennedy KJ et al. 'Neutral fat hydrolysis and long-chain fatty acid oxidation during anaerobic digestion of slaughterhouse wastewater'. *Biotechnol Bioeng* 79 (2002): 43–52.

Mateos Diaz JC, Rodríguez JA, Roussos S et al. 'Lipase from the thermotolerant fungus *Rhizopus homothallicus* is more thermostable when produced using solid-state fermentation than liquid fermentation procedures'. *Enzyme Microb Technol* 39 (2006): 1042–1050.

Mhetras NC, Bastawde KB, Gokhale DV. 'Purification and characterization of acidic lipase from *Aspergillus niger* NCIM 1207'. *Biores Technol* 100 (2009): 1486–1490.

Mulligan CN. 'Environmental applications for biosurfactants'. *Environ Pollut* 133 (2005): 183–198.

Mulligan CN, Yong N, Gibbs BF. 'Surfactant-enhanced remediation of contaminated soil: A review'. *Eng Geol* 60 (2005): 371–380.

Naci D, Ali AM. 'Partial purification of intestinal triglyceride lipase from *Cyprinion macrostomus* heckel, 1843 and effect of pH on enzyme activity'. *Turk J Biol* 26 (2002): 133–143.

Natalello A, Ami D, Brocca S et al. 'Secondary structure, conformational stability and glycosylation of a recombinant *Candida rugosa* lipase studied by Fourier transform infrared spectroscopy'. *Biochem J* 385 (2005): 511–517.

Nilkanth G, Madhumita. 'Slaughterhouse waste'. ENVIS/PPCC/NL-14, *Envis Newsletter* 4(4) (2009): 1–14.

Nitschke M, Costa SGVAO. 'Biosurfactants in food industry'. *Trends Food Sci Technol* 18 (2007): 252–259.

Papanikolaou S, Chevalot I, Panayotou MG et al. 'Industrial derivative of tallow: A promising renewable substrate for microbial lipid, single cell protein and lipase production by *Yarrowia lipolytica*'. *Electron J Biotechnol* 10 (2007): 425–435.

Pappalardo L, Jumean F, Abdo N. 'Removal of cadmium, copper, lead and nickel from aqueous solution by white, yellow and red United Arab Emirates sand'. *Am J Environ Sci* 6 (2010): 41–44.

Petersen M, Och Daniel R. 'Purification and characterization of an extracellular lipase from *Clostridium tetanomorphum*'. *World J of Microbio Biotechnol* 22 (2006): 431–435.

Pfociniczak MP, Grazyna AP, Seget ZP et al. 'Environmental applications of biosurfactants: recent advances'. *Int J Mol Sci* 12 (2011): 633–654.

Prasad MP, Manjunath K. 'Comparative study on biodegradation of lipid-rich wastewater using lipase producing bacterial species'. *Indian J Biotechnol* 10 (2011): 121–124.

Pugazhenthi G, Kumar A. 'Enzyme membrane reactor for hydrolysis of olive oil using lipase immobilized on modified PMMA composite membrane'. *J Membr Sci* 228 (2004): 187.

Ramani K, Boopathy R, Vidya C et al. 'Immobilisation of *Pseudomonas gessardii* acidic lipase derived from beef tallow onto mesoporous activated carbon and its application to hydrolysis of olive oil'. *Process Biochem* 45 (2010d): 986–992.

Ramani K, Chandan Jain S, Asit B, Mandal et al. 'Microbial induced lipoprotein biosurfactant from slaughterhouse lipid waste and its application to the removal of metal ions from aqueous solution'. *Colloids Surf B* 97 (2012): 254–263.

Ramani K, Evvie C, Sekaran G. 'Production of a novel extracellular acidic lipase from *Pseudomonas gessardii* using slaughterhouse waste as a substrate'. *J Ind Microbiol Biotechnol* 37 (2010b): 531–535.

Ramani K, John Kennedy L, Ramakrishnan M et al. 'Purification, characterization and application of acidic lipase from *Pseudomonas gessardii* using beef tallow as a substrate for fats and oil hydrolysis'. *Process Biochem* 45 (2010a): 1683–1691.

Ramani K, John Kennedy L, Vidya C et al. 'Immobilization of acidic lipase derived from *Pseudomonas gessardii* onto mesoporous activated carbon for the hydrolysis of olive oil'. *J Mol Catal B: Enzym* 62 (2010c): 59–66.

Ramani K, Sekaran G. 'Production of a novel lipase from slaughterhouse waste using *Pseudomonas gessardii*'. *(Microorganisms in Environment) Envis News Letter* 8 (2010): 2–4.

Reshmi R, Sugunan S. 'Immobilisation and characteristics of *Candida rugosa* lipase onto siliceous mesoporous molecular sieves and montmorillonite K-10 for synthesis of flavour esters'. In: Proceedings of the international conference on advanced materials and composites (ICAMC-2007). Trivandrum, India: NIIST; (2007): 819–824.

Rodrigues L, Banat IM, Teixeira J et al. 'Biosurfactants: Potential applications in medicine'. *J Antimicrob Chemother* 57 (2006): 609–618.

Salameh MA, Juergen, W. 'Purification and characterization of two highly thermophilic alkaline lipases from *Thermosyntropha lipolytica*'. *Appl Environ Microbiol* 73(23) (2007): 7725–7731.

Sani DG. 'Step up: Bacterial lipase to substitute pancreatic lipase for enzyme therapy'. Available at http://microbepundit.blogspot.com/2006/10/step-up-bacterial -lipase-to-substitute.html (accessed October 30, 2006).

Serio MD, Maturo C, De Alteriis E et al. 'Lactose hydrolysis by immobilized β-galactosidase: The effect of the supports and the kinetics'. *Catal Today* 79/80 (2003): 333–339.

Sharma R, Chisti Y, Chand BU. 'Production, purification, characterization and applications of lipases'. *Biotech Advances* 19 (2001): 627–662.

Shu CH, Xu CJ, Lin GC. 'Purification and partial characterization of a lipase from *Antrodia cinnamomea*'. *Process Biochem* 41 (2006): 734–738.

Shuler ML, Kargi F. 'Bioprocess engineering – Basic concepts'. Prentice-Hall, (2005).

Singh P, Cameotra S. 'Potential applications of microbial surfactants in biomedical sciences'. *Trends Biotechnol* 22 (2004): 142–146.

Stransky K, Zarevucka M, Kejik Z et al. 'Substrate specificity, region-selectivity and hydrolytic activity of lipases activated from *Geotrichum* sp.'. *Biochem Eng J* 34 (2007): 209–216.

Suk WS, Son HJ, Lee G et al. 'Purification and characterization of biosurfactants produced by *Pseudomonas* sp. SW1'. *J Microbiol Biotechnol* 9 (1999): 56–61.

Svetlitshnyi V, Rainey F, Wiegel J. '*Thermosyntropha lipolytica* gen. nov., sp. nov., a lipolytic, anaerobic, alkalitolerant, thermophilic bacterium utilizing short- and long-chain fatty acids in syntrophic coculture with a methanogenic archaeum'. *Int J Syst Bacteriol* 46 (1996): 1131–1137.

Vinu A, Murugesan V, Hartmann M. 'Adsorption of lysozyme over mesoporous molecular sieves MCM-41 and SBA-15: Influence of pH and aluminium incorporation'. *J Phys Chem B* 108 (2004): 7323–7330.

Vivian F, Junior AC, De Pra MC et al. 'Animal fat wastes for biodiesel production, biodiesel – Feedstocks and processing technologies'. In: M. Stoytcheva (ed.), *InTech*. Available at http://cdn.intechopen.com/pdfs-wm/22993.pdf. (accessed November 9, 2011).

Wang ZG, Wang JQ, Xu ZK. 'Immobilization of lipase from *Candida rugosa* on electrospun polysulfone membranes by adsorption'. *J Mol Catal B: Enzym* 42 (2006): 45–51.

Yu MR, Lange S, Richter S et al. 'High-level expression of extracellular lipase Lip2 from *Yarrowia lipolytica* in *Pichia pastoris* and its purification and characterization'. *Protein Expr Purif* 53 (2007): 255–263.

Zhao XS. Bao XY, Guo WP et al. 'Immobilizing catalysts on porous materials'. *Mater Today* 9 (2006): 32–39.

Zhou QZK, Chen XD. 'Effects of temperature and pH on the catalytic activity of the immobilized β-galactosidase from *Kluyveromyces lactis*'. *Biochem Eng J* 9 (2001): 33–40.

12

Mechanism of Wetland Plant Rhizosphere Bacteria for Bioremediation of Pollutants in an Aquatic Ecosystem

Ram Chandra and Vineet Kumar

CONTENTS

12.1 Introduction

The term 'rhizosphere' has been derived from the Greek word 'rhiza', meaning root, and 'sphere', meaning field of influence. The rhizosphere is the zone surrounding the roots of plants in which complex relationships exist among the plant, the soil microorganisms and the soil itself. The plant roots and the biofilm associated with them can profoundly influence the chemistry of the soil, including pH and nitrogen transformations. The term rhizosphere was coined for the first time by German agronomist and plant physiologist Lorenz Hiltner in 1904. The active reaction zone of a wetland plant is the root zone or the rhizosphere. The rhizosphere is also known to be a hot spot of microbial activities, in which physicochemical and biological processes take place that are induced by the interaction of plants, microorganisms, the soil and pollutants (Stottmeister et al. 2003). Phytoremediation efforts have largely focused on the use of plants to accelerate degradation of organic contaminants, usually in concert with root rhizosphere microorganisms, or to remove hazardous heavy metals from soils and water. Plants have been used for wastewater treatment applications over the past 300 years and began to be used for treatment of slurries and metal contamination in the mid-1970s. The wetland technology, an ecofriendly green technology, has an important influence on the biological degradation and removal mechanism of contaminants.

Wetland plants selected for wastewater treatment have to be tolerant of more extreme environmental conditions, such as elevated metal content, nutrient and organic carbon, nitrogen and sulphur. In order to grow and reproduce in an anaerobic soil environment, wetland plants must possess several morphological adaptations (aerenchyma, shallow root system, hypertrophy, pneumatophores and swollen trunk, adventitious roots, etc.) to facilitate transport of oxygen to the roots and metabolic adaptation to respire anaerobically in water-saturated soil without toxic effect.

The bioremediation of organic and inorganic pollutants by plants may occur directly through uptake, translocation into the shoot and metabolism or volatilisation indirectly through plant microbe contaminant interaction within the plant rhizosphere (Zhuang et al. 2007). Wetland plants create oxic–anoxic interfaces, thereby providing habitats for both aerobic and anaerobic microbes, facilitating nutrient recycling. The roots and rhizomes of wetland plants release a multitude of organic compounds, for example, mucilage and exudates in soil, and also provide a substrate for the attached growth of microorganisms. Microorganisms play a main role in the bio-chemical transformation of contaminants and their capability to remove toxic organic compounds added to wetlands. Wetland hosts complex micro-bial communities, including bacteria, fungi, protists and viruses (Reddy et al. 2002). The diversity of microorganisms in the wetland plant rhizosphere may be critical for the proper functioning and maintenance of the aquatic ecosystem (Ibekwe et al. 2003). In wetland soils, microbial communities may be more diverse than in upland soil. The highest portions of microorgan-isms that survive in the rhizosphere are fungi and bacteria. The bioremedia-tion of industrial pollutants by the diverse group of bacteria is associated with plant roots known as rhizosphere bacteria (rhizobacteria). These bac-teria include biodegradative bacteria, plant growth–promoting bacteria and bacteria that facilitate phytoremediation. The abundance of bacteria is 2–20 times higher in the rhizosphere than in the bulk soil (Morgan et al. 2005; Chaturvedi and Chandra 2006). Rhizobacteria play a main role in the bio-chemical transformation of contaminants and their capability in removing toxic organic and inorganic compounds added to wetland ecosystems and minimise the potential adverse effects of hazardous chemicals released into the environment.

Wetland ecosystems are essential for maintaining an ecological balance through elemental cycling and are sensitive to anthropogenic impacts. A biogeochemical cycle is the transport and transformation of chemicals in ecosystems. These are strongly influenced by the unique hydrologic condi-tions in wetlands. These processes result in changes in the chemical forms of materials and also the movement of materials within the wetland eco-system. In wetland ecosystems, the nitrogen cycle is maintained by nitri-fiers and denitrifiers, such as *Nitrosomonas, Nitrospira, Nitrosococcus* and *Nitrobacter* and *Pseudomonas, Achromobacter* and *Bacillus,* respectively. Similarly, the phosphorous and sulphur cycles are maintained by *Pseudomonas, Serratia, Pantoea, Rhizobium, Flavobacterium, Bacillus* and *Enterobacter* and *Desulfovibrio, Desulfobulbus, Desulfotomaculum* and *Desulfosporosinus,* respec-tively. Rhizobacteria have also been involved in immobilisation and accu-mulation of heavy metals in plant tissues and soil. Wetland plants and rhizobacteria have several of their own mechanisms for removal of heavy metals and metalloid contamination in the aquatic ecosystem. These mech-anisms involve production of several plant hormones, production of ACC deaminase enzyme and production of siderophores. Siderophore-mediated accumulation of heavy metals in bacteria and plants is a matter of interest. Chemically, siderophores are iron-binding proteins with molecular weights ranging from 400 to 1500 Da. The role of these compounds is to scavenge iron from the environment and to make the mineral, which is almost always essential, available to the plant and its microbial cells. Siderophores bind iron and transport it into the bacterial cell. In the transport process of the siderophore iron complex, several bacterial outer membrane proteins and receptors, periplasmic binding proteins (PBPs), permease proteins and inner membrane ATP–binding cassette (ABC) transporters play major roles. Siderophore-producing rhizobacteria enhance plant growth by increasing the bioavailability of iron near the plant root. Further, this chapter describes current knowledge about the heavy metal hyperaccumulation and detoxification mechanism in the plant. Wetland plants have been reported to have a large capacity for metal accumulation. Because of their fibrous root systems with large contact areas, aquatic plants generally have the ability to accumulate large metal concentration in the plant organ from those in the surrounding water. Histidine, nicotinamine, organic acid (citrate, malate), glutathione, heat shock protein, phytochelatin and metallothionin metabolism plays a key role in metal hyperaccumulation in plants. A hyperaccumulator constitutes an exceptional biological material for plant adaptation to survive extreme metallic environments. Plants hyperaccumulate heavy metals as a defence mechanism against natural enemies, such as herbivores.

12.2 Wetland Plants Characteristics and Their Adaptation

Wetlands are among the most important ecosystems on Earth. They are a transitional area between land and water. The term 'wetland' encompasses a broad range of wet environments, including marshes, bogs, swamps, wet meadows, tidal wetlands, floodplains and ribbon wetlands along stream channels. Conditions in wetlands vary widely, and inhabitants are often adapted to a wide range of water-quality conditions, such as salinity, temperature, tidal currents, turbidity, flooding, drought, excess pollution and low or no oxygen in the soil. Wetland habitats, with their high water levels and increased salt concentrations, are too harsh for many plants. Most organisms that thrive in these environments only do so with the help of special physiological, morphological and metabolic adaptations. The plants growing in wetlands, often called wetland plants, macrophytes or hydrophytic plants, are adapted to growing in water-saturated soils. These include aquatic vascular plants, aquatic mosses and some larger algae. Marsh plants possess specific characteristics of growth physiology that guarantee their survival even under extreme rhizosphere conditions (Stottmeister et al. 2003). Water-saturated soils become oxygen-free (anoxic or anaerobic) except for a few millimetres at the surface. However, wetland plants have several morphological and metabolic adaptations so they can grow in water-saturated soils.

12.2.1 Morphological Adaptations

Morphological adaptations include aerenchyma, hypertrophy leading to buttressed swollen trunk, pneumatophores, adventitious roots, shallow root systems, root aeration, pressurised ventilation, radial oxygen loss from roots and storage of carbohydrate reserves.

Aerenchyma and shallow root system: Wetland plants also produce shallow root systems and adventitious root to enhance oxygen uptake in the anaerobic soil. Many wetland plants have special air spaces in their roots, stems and leaves called aerenchyma through which oxygen can enter the plant and be transported to its roots as shown in Figure 12.1. In well-drained soils, the air spaces are filled with air with a high content of oxygen, and microorganisms living in the soil and roots of plants growing in the soil therefore are able to obtain oxygen directly from their surroundings. As the soil air spaces are interconnected to the atmosphere above the soil, the oxygen in the air spaces is replenished by rapid diffusion and convection from the atmosphere.

Hypertrophy: This is swelling of the stem base in woody and herbaceous plants that occurs when aerenchyma forms. In woody plants, such as bald cypress and tupelo gum, it leads to the formation of a buttressed swollen trunk.

Pneumatophores and swollen trunk: Pneumatophores and swollen trunk increase the surface area and number of lenticels that facilitate oxygen diffusion into the wetland plant. They also stabilised wetland trees in waterlogged and soft soils.

Adventitious roots: A common response of wetland plants to flooding is the formation of aquatic adventitious roots. Aquatic roots share many morphological features with sediment adventitious roots, and in addition, as aquatic roots can be exposed to light, these can form photosynthetically active chloroplasts, which we hypothesise could contribute to their O_2 and carbohydrate status.

Pressurised ventilation: Some wetland plants augment diffusion of O_2 to the roots by means of pressurised ventilation. It occurs when gradients of temperature and pressure between the atmosphere and the roots drive oxygen-rich air through the stomata of young leaves where it is expelled back into the atmosphere.

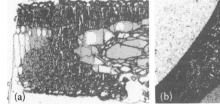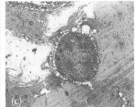

FIGURE 12.1
(See color insert.) Cross section of *Typha* organs under transmission electron microscope (TEM): (a) leaf, (b) stem and (c) root.

Radial oxygen loss: Radial oxygen loss (ROL) is another adaptation to anoxic conditions driven by diffusion. It involves leakage of oxygen from roots, oxidising the area of soil around the roots and increasing the soil's oxidation–reduction potential. ROL benefits the plants by oxidising reduced metals (Fe^{2+}, Mn^{2+}) and sulphides that, otherwise, may be toxic to the plants.

Carbohydrate reserves: One mechanism to combat short-term anoxia involves production of large carbohydrate reserves that are stored in the roots. This reserve can be used to support anaerobic metabolism when the soil is flooded or saturated for a short period of time.

12.2.2 Metabolic Adaptations

Wetland plants have several metabolic adaptations that grow in water-saturated soil. When soils are flooded and become anoxic, wetland plants respond by shifting from aerobic metabolism, with which CO_2 is the primary end product, to anaerobic metabolism, with which ethanol, lactic acid and other compounds are produced (Summers et al. 2000). Anaerobic respiration does not provide the high level of ATP needed to maintain the growth and metabolism that aerobic respiration does. It is one anaerobic pathway to maintain a high energy level, at least temporarily. However, because ethanol is toxic to plants, wetland vegetation must be able to remove excess ethanol or convert it to the nontoxic compound malate. Although some studies suggest malate production as an alternative metabolic end product (McManmon and Crawford 1971), other researchers have observed that malate levels do not always increase during anaerobic metabolism (Saglio et al. 1980; Menegus et al. 1989). Also, ethanol may not be as toxic to plants as previously believed (Cronk and Fennessy 2001) because, when flooded, ethanol diffuses out of the roots and into the rooting medium where its toxicity is diluted. Davies (1980) suggested that short-term tolerance to anaerobic conditions involves regulation of cellular pH to prevent cytoplasmic acidosis rather than accumulation of nontoxic compounds, such as malate, produced during anaerobic respiration. In plants, cytoplasmic acidosis occurs within minutes of the onset of anoxia as anaerobic metabolism kicks in and pyruvate is converted to lactic acid, leading to a decrease in cellular pH. In flood-tolerant and -intolerant plants, lactic acid is shunted to the vacuole, where it is isolated from the cytoplasm. In flood-intolerant plants, such as maize, after a short period of time (10 h), acid leaks from the vacuole and acidifies the cytoplasm (Roberts 1988). In flood-tolerant species, lactic acid and other anaerobic respiration products are stored in vacuoles. Some data suggest that isolation of lactic acid in vacuoles may not be that important because, for some species, compounds such as succinate accumulate rather than lactic acid (Summers et al. 2000).

Some important wetland plant species are listed in Table 12.1.

TABLE 12.1

Important Wetlands Plant Species

S. No.	Scientific Name	Common Name	Family
1.	*Phragmites australis*	Common reed	Poaceae
2.	*Phragmites karka*	Tall reed	Poaceae
3.	*Phragmites* sp.	Common reed	Poaceae
4.	*Juncus* spp.	Rushes	Juncaceae
5.	*Scirpus* spp.	Bulrushes	Cyperaceae
6.	*Typha angustifolia* L.	Narrow-leaved cattail	Typhaceae
7.	*Typha latifolia* L.	Broad-leaved cattail	Typhaceae
8.	*Iris pseudacorus* L.	Yellow iris	Iridaceae
9.	*Acorus calamus* L.	Sweet flag	Acoraceae
10.	*Glyceria maxima* (Hartm.) Holmb.	Reed grass	Poaceae
11.	*Carex* spp.	Sedges	Cyperaceae
12.	*Scirpus grossus*	Greater club rush	Cyperaceae
13.	*Scirpus mucronatus*	Bog bulrush	Cyperaceae
14.	*Scirpus lacustris*	Club rush	Cyperaceae
15.	*Lepironia articulata*	Tube sedge	Cyperaceae
16.	*Phylidrium lanuginosum*	Fan grass	Philyraceae
17.	*Rhynchospora corymbosa*	Golden beak sedge	Cyperaceae
18.	*Eleocharis variegata*	Spike rush	Cyperaceae
19.	*Scleria sumatrana*	Sumatran scleria	Cyperaceae
20.	*Fimbristylis globulosa*	Globular fimbristylis	Cyperaceae
21.	*Cyperus esculentus*	Yellow nutsedge	Cyperaceae
22.	*Cyperus papyrus*	Papyrus	Cyperaceae
23.	*Zinnia angustifolia*	Creeping zinnia	Cannaceae
24.	*Spartina alternifoia*	Saltmarsh cordgrass	Poaceae
25.	*Vetiveria zizaniodes*	Khus Khus grass	Gramneae
26.	*Vellisneria americana*	Tape grass	Hydrocharitacease
27.	*Peltandra virginica*	Green arrow arum	Hydrocharitacease
28.	*Canna* sp.	Canna lily	Cannaceae

12.3 Oxygen Transport Mechanism in Wetland Plant Rhizosphere

The stems and leaves of macrophytes that are submerged in the water column provide a huge surface area for biofilms. Biofilms are present on both the above- and below-ground tissue of the macrophytes. The plant tissues are colonised by dense communities of photosynthetic algae as well as by bacteria and protozoa. Likewise, the roots and rhizomes that are buried in the wetland soil provide a substrate for the attached growth of microorganisms. Roots and rhizomes of wetland plants growing in water-saturated substrates therefore must obtain oxygen from their aerial organs via transport internally in the plants (Gutknecht et al. 2006). Wetland ecosystems are characterised by hydric soils and hydrophilic plant communities and have fluctuating hydrology that gives rise to interplay between aerobic and anaerobic processes (Figure 12.2).

Wetlands contain saturated soils that are rich in organic matter. Metabolism of the organic matter by soil bacteria quickly depletes oxygen in the pore water (Mitsch and Gosselink 2000). Many wetland plants (for example, *Phragmites, Cyperus, Typha*, etc.) have an extensive oxygen transport system (aerenchyma tissue) that may exist in the roots, stems, bark, twigs and leaves (Armstrong et al. 1994) through which atmospheric oxygen passes down to the extremities of the root and to the rhizosphere, and waste gases can pass back into the air concurrently. Introduction of oxygen to the rhizosphere via diffusion through flooded soil pores is limited; therefore, plants have developed adaptations that allow them to survive in anaerobic soil. This system allows a plant to transport needed oxygen to the roots for maintaining aerobic respiration and to oxidise reducing compounds in the rhizosphere. A detailed pathway of aeration and further mechanism in *Phragmite australis* has been shown in Figure 12.3. In wetland plants, it is especially effective due to the presence of aerenchyma, and because waterlogged soils are mostly anaerobic, wetland plants are normally totally reliant upon it to support root growth. This oxygen leakage from the roots creates oxidised conditions in

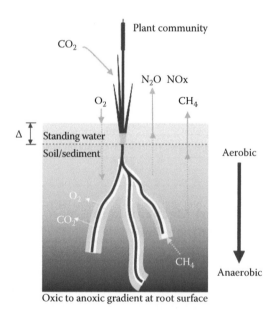

FIGURE 12.2

Wetland structure. Water table height, depth from surface and distance from plant roots create oxic to anoxic gradients. The result is a complex interplay between anaerobic and aerobic conditions that allows for a wide range of processes to occur in wetland soils.

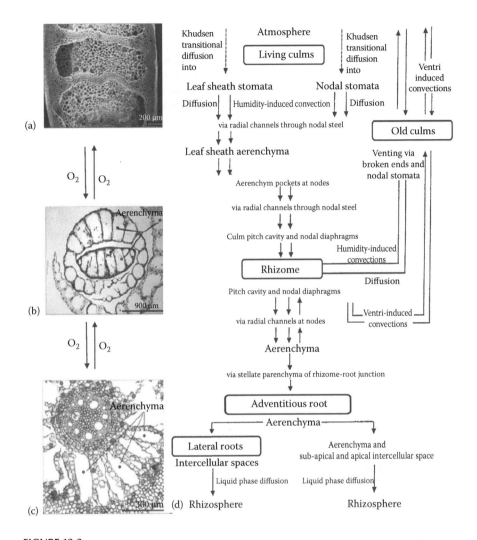

FIGURE 12.3
(a) Leaf aerenchyma, (b) shoot aerenchyma, (c) root aerenchyma and (d) pathway of diffusive and convective gas transport in wetland plant.

the otherwise anoxic substrate and stimulates both aerobic decomposition of organic matter and growth of nitrifying bacteria.

Wetland plants have a variety of mechanisms to increase the efficiency of internal oxygen diffusion in response to oxygen deficiencies associated with flooding. These mechanisms include development of intracellular gas-filled spaces or aerenchyma; development of a barrier to radial oxygen loss; and changes in root physiology, including root thickness, length of roots and arrangement of roots. It is generally accepted that aerenchyma acts as a preferential diffusion path for oxygen from shoot to root because it increases porosity and decreases tortuosity within the root. Oxygen transported through aerenchyma can ultimately diffuse radially through and out of the root into the rhizosphere, thereby aerating this zone. This phenomenon is known as radial oxygen loss (ROL). ROL can provide several benefits to the plant, including protection against phytotoxins that are present in anaerobic soils. Released oxygen oxygenates the rhizosphere and, in turn, phytotoxins, thereby reducing their toxicity while still allowing for uptake of water and nutrients. In some cases, a barrier to ROL can prevent oxygen loss; however, not all plants develop an ROL barrier. Anatomical and morphological changes, such as the formation of aerenchyma and ROL, are important adaptive mechanisms that plants have developed to overcome oxygen deficiency in the soil. In addition, the internal system of large gas spaces also reduces the internal volume of respiring tissues and oxygen consumption, thus enhancing the potential for oxygen reaching the distant underground portions of the plant. Due to such advantages, the oxygen transport system has been considered as a major mechanism critical to a plant's ability to cope with soil anaerobiosis (Armstrong et al. 1996).

12.4 Rhizosphere Bacteria

The use of wetland plants for the treatment of industrial pollutants has increased since 2005. The active reaction zone of a wetland plant is the root zone (rhizosphere). The rhizosphere is a zone between the root surface and the soil adjacent to the roots. The bacteria that live in this zone may remain in the soil that adheres to the roots after gentle shaking. The bacteria inhabiting the rhizosphere are called rhizobacteria. Plants can provide favourable conditions for microbial colonisation of the rhizosphere for symbiotic degradation and detoxification of pollutants. The microbial community structure and diversity associated with wetland plants are very important for the treatment efficiency of pollutants and ecosystem stability (Figure 12.4).

Rhizosphere bacteria play an important role for the degradation and detoxification of industrial pollutants in an aquatic ecosystem. Rhizosphere microbes utilise these pollutants as a sole carbon, nitrogen and energy source for their own growth and metabolism. Several workers have reported rhizosphere bacteria for removal of pollutants from industrial wastewater as shown in Table 12.2.

Rhizobacteria can be classified into two major groups according to their relationship with the host plants: (1) symbiotic rhizobacteria and (2) free-living rhizobacteria, which could invade the interior of cells and survive inside, which is called intracellular plant growth–promoting rhizobacteria (PGPR) (for example, nodule bacteria), or remain outside the plant cells, called

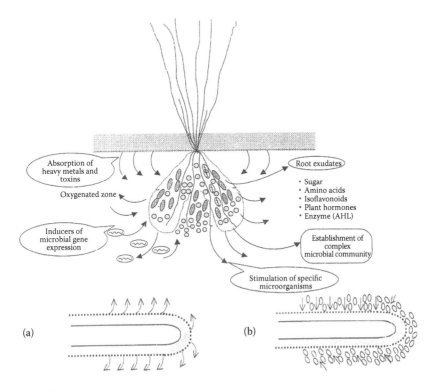

FIGURE 12.4
(a) Creation of microenvironment in wetland plants and (b) attraction of bacterial cell for attachment with root hair.

extracellular PGPR (for example, *Bacillus, Pseudomonas, Azotobacter*, etc.). PGPR are usually in contact with the root surface and improve growth of plants by several mechanisms, for example, enhanced mineral nutrition, phyto-hormone production and disease suppression. PGPR can also promote root growth. This can be caused by the ability of most rhizobacteria to produce phytohormones, for example, indole-3-acetic acid (IAA), cytokinins, gib-berellins and ethylene, which promote cell division and cell enlargement, extension of plant tissue and/or other morphological changes of the roots. The diffusion of oxygen from the roots creates aerobic, anoxic and anaer-obic zones around the roots for development of aerobic organisms in the rhizosphere. Factors such as temperature, moisture and seasonality of tem-perature and moisture act to control wetland microbial activities, resulting in changes in key biogeochemical cycles (Figure 12.5). The rhizosphere is being affected by plant root activities. The plant root releases carbon in both organic and inorganic forms; however, the organic forms are the most varied and can have the most influence on the chemical, physical and biological processes in the rhizosphere (Gutknecht et al. 2006). The composition and

TABLE 12.2

Identified Bacteria Capable of Degradation of Industrial Pollutants Growing with Wetland Plant Rhizosphere

S. No.	Wetland Plant	Rhizosphere Bacteria and Accession Number	References
1.	*Cyperus papyrus*	*Aeromonas hydrophilla, Pasteurella* sp., *Actinobacillus equuli, Aeromonas salmonicida, Kinggella* sp., *Vibrio flurialis, Propionibacterium* sp., *Corynebacterium* sp., *Staphylococcu, Enterobacter* sp., *Vibrio* sp., *Aeromonas salmonicida, Pasteurella* sp., *Actinobacillus* sp., *Micrococcus* sp., *Haemophilus* sp., *Aerobacter aerogens, Alcaligens faecalis, Bacillus cereus, Bacillus megaterium, Bacillus subtilis, Micrococcus luteus, Nocardia* sp., *Streptomyces bikinensis, Sarcina cooksoni*	Cicerone and Oremland (1988)
2.	*Phragmites cummunis*	*Bacillus odysseyi* strain 34hs1 (NR 025258), *Agrobacterium larrymooeri* strain AF3.10 (NR 026519), *Thauera selenatis* strain AX39 (NR 025212), Uncultured *Acinetobacter* sp. Clone IITR RCP27 (FJ268988), Uncultured *Acinetobacter* sp. Clone IITR RCP24 (FJ268985), Uncultured *Acinetobacter* sp. Clone IITR RCP21 (FJ268982), *Acinetobacter* sp. ATCC 31012 (AF542963), Uncultured *Acinetobacter* sp. Clone IITR RCP33 (FJ268994), Uncultured *Acinetobacter* sp. Clone IITR RCP37 (FJ268998), Uncultured *Enterobacter* sp. Clone IITR RCP19 (FJ268980), Uncultured *Klebsiella* sp. Clone IITR RCP25 (FJ268986), *Klebsiella pneumonia* subsp. Pneumonia (AF228918), Uncultured *Pantoea* sp. Clone IITR RCP34 (FJ268995), *Enterobacter aerogenes* (AB099402), Uncultured *Pantoea* sp. Clone IITR RCP26 (FJ268987), Uncultured *Klebsiella* sp. *Clone* IITR RCP28 (FJ268989), Uncultured *Enterobacter* sp. Clone IITR RCP32 (FJ268993), *Pantoea ananatis* strain 1846 *(NR 026045), Uncultured Stenotrophomonas* sp. Clone IITR RCP36 (FJ268997), Uncultured *Stenotrophomonas* sp. Clone IITR RCP23 *(FJ268984), Stenotrophomonas aciddaminiphila* strain AMX 19 (NR_025104), *Stenotrophomonas aciddaminiphila* (AF273080), Uncultured *Stenotrophomonas* sp. Clone IITR RCP22 (FJ268983), *Uncultured Stenotrophomonas* sp. Clone IITR RCP20 (FJ26898), *Uncultured Stenotrophomonas* sp. Clone IITR RCP35 (FJ268998), *Uncultured Stenotrophomonas* sp. Clone IITR RCP29 (FJ268998), Uncultured *Stenotrophomonas* sp. Clone IITR RCP30 (FJ268991), Uncultured *Stenotrophomonas* sp. Clone IITR RCP30 (FJ268998)	Chandra et al. (2011)

TABLE 12.2 (CONTINUED)

Identified Bacteria Capable of Degradation of Industrial Pollutants Growing with Wetland Plant Rhizosphere

S. No.	Wetland Plant	Rhizophere Bacteria and Accession Number	References
3.	*Phragmites australis* (L.)	*Microbacterium hydrocarbonoxydans* (AJ880397), *Achromobacter xylosoxidans* (AJ880764), *Bacillus subtilis* (AJ880760), *Bacillus megaterium* (AJ880767), *Bacillus anthracis* (AJ880766), *Bacillus licheniformis* (AJ880762), *Achromobacter xylosoxidans* (AJ880763), *Achromobacter* sp. (AJ880396), *Bacillus thuringiensis* (AJ868359), *Bacillus licheniformis* (AJ880758), *Bacillus subtilis* (AJ880761), *Staphylococcus epidermidis* (AJ880759), *Pseudomonas migulae* (AJ887999), *Alcaligens faecalis* (AJ880765), *Bacillus cereus* (AJ853737)	Chaturvedi and Chandra (2006)

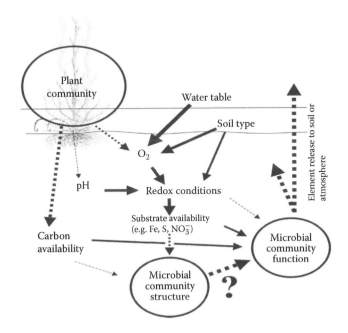

FIGURE 12.5

Relationships among controls over wetland ecosystem microbial communities and element cycling. Arrows indicate relationships, and width of arrows indicates relative importance of relationship for ecosystem functioning. Dashed arrows represent interactions that are poorly understood even though they may be important.

amount of the released compounds are influenced by many factors, including plant type, climactic conditions insect herbivore, nutrient deficiency or toxicity and the chemical, physical and biological properties of the surrounding soil. The root products imparted to the surrounding soil are generally called rhizodeposits. Rhizodeposition is partially the result of the decay of dead roots and root hairs. Plant root caps and epidermal cells secrete mucigel to supply carbohydrate sources to soil microorganisms (Jimenez et al. 2003). This carbonaceous material stimulates overall bacterial activity as well as provide substrates to support cometabolic degradation of xenobiotic hydrocarbons.

Most of the organic matter is decomposed to carbon dioxide and water in aerobic zones. Respiration and fermentation are the major mechanisms by which microorganisms break down organic pollutants into harmless substances, such as carbon dioxide (CO_2), nitrogen gas (N_2) and water (H_2O). In summary, microbial activity involved in (i) the recycling of nutrients, (ii) altering the reduction–oxidation condition of the substrate and (iii) transforming a variety of organic and inorganic compounds. Some microbial transformations are aerobic, and others are anaerobic. Microbes are capable of degrading most organic pollutants, but the rate of degradation varies considerably, depending on the chemical and structural properties of the organic compounds and the physicochemical environment in the soil. A number of processes contributing to the treatment of wastewater are the following.

Aerobic degradation: The aerobic process requires an adequate supply of molecular oxygen to decompose the organic matter and retrieve energy from it, which is needed by the bacteria to grow and multiply. The aerobic biodegradation process is rapid and more complete. Naturally occurring aerobic bacteria can decompose both natural and synthetic hazardous organic materials to harmless CO_2 and H_2O.

Methanotrophs may be promising bacteria for environmental bioremediation. Methanotrophs utilise methane and other carbon compounds, including methanol, methylated amines, halomethanes and methylated compounds containing sulphur as their sole carbon and energy source. Methanotrophs have been shown to degrade or co-oxidise diverse types of heavy metals and organic pollutants due to the presence of broad-spectrum methane monooxygenase. Broad-spectrum methane monooxygenase (MMO) enzymes are found only in methanotrophs. MMO comes in two forms, namely, the membrane-associated or particulate form (pMMO) and the soluble or cytoplasmic form (sMMO). The pMMO is found in all known methanotrophs except for the genus *Methylocella* (acidophilic), and the sMMO is present only in a few methanotroph strains.

Anaerobic degradation: Anaerobic degradation is a multistep process that occurs within constructed wetlands in the absence of dissolved oxygen, and either facultative or obligate anaerobic heterotrophic bacteria can carry out the process, called fermentation. The primary products of fermentation are acetic acid, butyric acid, lactic acid, alcohols and the gases CO_2 and H_2. The end products of fermentation can be utilised by strictly anaerobic sulphate-reducing and methane-forming bacteria. In this process, the complex 'hydrocarbons' of hazardous wastewater are converted into simpler molecules of CO_2 and CH_4 by anaerobic microbes. Anaerobic bacteria can use nitrate and sulphate as alternate electron acceptors to support their respiration. If nitrates are the source of oxygen, nitrogen is formed, and if sulphates are the source, hydrogen sulphide is formed under the anaerobic process. The anaerobic digestion consists of two distinct groups of anaerobic microorganisms: the 'acetogens' and the 'methanogens' present at several trophic levels. At the higher levels, organisms attack waste molecules through hydrolysis and fermentation, reducing them to simpler hydrocarbons CH_4 and CO_2 in the last two steps of the destructive process. The 'acetogens' convert most of the hydrocarbons in waste to acetate, CO_2, H_2 and some organic acids. However, in most freshwater wetlands, the dominant microbial activity is methanogenesis, in which 'methanogens' utilise HCO_3^- and organic substrate to produce CH_4 in the presence of the available H_2.

12.5 Removal Mechanism of Industrial Pollutants in the Wetland Ecosystem for Maintaining the Biogeochemical Cycle

The transport and transformation of chemicals in ecosystems is known as biogeochemical cycling. The diverse hydrologic conditions in wetlands have a major influence on biogeochemical cycles. Wetlands have a large number of aerobic–anaerobic zones in the water column, soil–water interface and the root zone of macrophytes. The combination of aerobic and anaerobic zones supports a wide range of microbial populations and associated processes mediated by microorganisms. Developing the root–soil interface creates a dynamic microenvironment in which microorganisms, the plant root and the soil component interact. The wetland ecosystems have abundant microflora, consisting of different heterotropic and autotropic microorganisms, including different oil-degrading bacteria and fungi (Groudeva et al. 2001; Das et al. 2003). Plants can stimulate microbe bioactivity in the root zone by the excretion of bioenhancing compounds. The plant-excreted root exudates provide a carbon and nitrogen source for soil bacteria. Root exudates from wetland plants have both positive and negative interactions among the microbes, plants and ecosystems. Environmental factors, such as temperature and light regime, affect the photosynthetic carbon fixation, which continuously influences the composition and quantity of root exudates released into the rhizosphere. Rhizosphere organisms receive these nutrients from root exudates, and nonrhizosphere organisms use organic residues in varying stages of decomposition. Specifically, it has been shown that flavonoids can support the growth of PCB-degrading bacteria. In addition, the phenols excreted by crab apple, sumac and mulberry plants can stimulate PCB-degrading bacteria and inhibit other microbes. In addition, plants can excrete surfactants, which will increase the bioavailability of the contaminant to the soil microbes. Enhanced herbicide degradation has also been found in the presence of root exudates. The different microbial compositions of bulk and rhizosphere soils reflect these nutritional modes.

The ability of wetland plants to transform and store organic matter and nutrients has resulted in a widespread use of wetland plants for wastewater treatment worldwide. Interest in the microbial biodegradation of pollutants has intensified in recent years as mankind strives to find sustainable ways to clean up contaminated environments. Microbes (bacteria and fungi) are the most important ecofriendly agents for the degradation and detoxification of industrial pollutants during the biological treatment of industrial wastewaters. The use of organisms for the removal of industrial contamination is based on the concept that all organisms could remove substances from the environment for their own growth and metabolism (Wagner et al. 2002; Gavrilescu 2005). These organisms utilise plant exudates and pollutants directly or indirectly and biotransform the products through cometabolism into harmless products, thereby supporting bioremediation. Bioremediation is an emerging technology defined as a technique that uses living organisms to manage or remediate polluted soils and as the elimination, attenuation or transformation of polluting or contaminating substances, by the use of biological processes, into their less toxic forms, and it can be applied in situ or ex situ, depending on the site at which they will be applied. As bioremediation can be effective only where environmental conditions permit microbial growth and activity, its application often involves the manipulation of environmental parameters to allow microbial growth and degradation to proceed at a faster rate. A cost-effective technique could be the use of plants to enhance microbial populations in the soil, which may stimulate degradation of organic chemicals (Schwab and Banks 1994). The elimination of pollutants and wastes from the environment is an absolute requirement to promote the sustainable development of our society with low environmental impact. But recently, it has been reported that constructed wetlands (CWs) are a low-cost and natural alternative technical method of wastewater treatment that utilises the biodegradation ability of plants and microorganisms. The first experiments on the use of wetland plants to treat wastewater were carried out in the early 1950s by Kathe Seidel in Germany. Wetland plants play a vital role in the removal and retention of pollutants and help in preventing the eutrophication of wetlands. CWs can be used for primary, secondary and tertiary treatment of municipal or domestic wastewater, storm water and agricultural and industrial wastewater, usually combined with adequate pretreatment (Kadlec et al. 2000). Recently, CWs have been widely used and accepted around the world and have become a suitable solution for wastewater treatment. CWs are effective in treating organic matter, nitrogen and phosphorus, decreasing the concentration of trace metals and organic chemicals. An important part of the treatment in a wetland treatment system is attributable to the bioaugmentive effect due to the presence and activity of plants and microorganisms. Recent research showed that the main role in the transformation and mineralisation of nutrients and organic pollutants is played not only by plants but also by microorganisms (Stottmeister et al. 2003; Vacca et al. 2005). A range of wetland plants has shown their ability to assist in the breakdown of wastewater. However, the common wetland plant *Phragmites cummunis* growing in temperate climatic conditions is the most commonly accepted wetland plant for the decolourisation and detoxification of industrial effluents (Chen et al. 2006a; Calheiros et al. 2009). CWs are designed to take advantage of the chemical and biological processes of natural wetlands to remove contaminants from wastewater. CWs have a simple operation, lower operating cost and little excess sludge production; they are environmentally friendly and offer considerable potential for conservation of wildlife. Wetland ecosystems, including constructed wetlands for wastewater treatment, are vegetated by wetland plants. Wetland plants are an important component of wetlands, and the plants have several roles in relation to the wastewater treatment processes. Plant vegetation plays an important role in the wastewater treatment process. Wetland plants species, such as *Phragmites australis, Phragmites cummunis, Typha angustifolia* L., *Typha angustala* L., *Cyperus esculentus* L., etc. can be used for the treatment of a variety of wastewaters, for example, domestic wastewater, storm water and agricultural and industrial wastewater. Thus, these species have great potential for detoxification and phytoremediation of polluted water bodies and have been used widely to treat industrial wastewater in a wetland system (Vymazal and Kropfelova 2005).

12.5.1 Process of Organic Pollutant Removal in Aquatic Ecosystem

Settleable organics (BOD, COD) are rapidly removed in wetland systems mainly by deposition and filtration. Microbial activity is the main cause for the decrease in soluble organic compounds by both aerobic and anaerobic degradation. Suspended solids passing through the root get entrapped, accumulate and finally settle by means of gravity or get metabolised by microorganisms, and particulate matter settles at the bottom of the pond. In aerobic degradation, soluble organic compounds are removed by the microbial growth on the media surfaces and attached to the roots and rhizome of the wetland plants. Organic matter contains approximately 45% to 50% carbon (C), which is utilised by a wide array of microorganisms as a source of energy (DeBusk 1999). The predominant dissolved organic matter (DOM) removal mechanism is the bio-oxidation by bacteria present in the biofilm attached to the root and in the water column (DeBusk and Reddy 1987). Additional DOM removal mechanisms include plant uptake and accumulation or degradation in the sediment after sorption onto particles. Bacteria utilise organic matter for the production of energy and synthesis of new cells. The biochemical reactions involved in energy production require the presence of electron acceptors for organic matter oxidation. The reactions, which utilise free oxygen as an electron acceptor, are predominant because they are the most energy-efficient.

12.5.2 Maintaining the Nitrogen Cycle

As we know, industrial wastewater contains a high concentration of nitrogen and causes a very serious problem of eutrophication in wastewater-receiving bodies. A simplified nitrogen cycle is shown in Figure 12.6. In CWs, nitrogen may be removed from wastewater by several processes like ammonification, nitrification–denitrification, plant uptake and physicochemical methods, such as sedimentation, ammonia stripping, breakpoint chlorination and ion exchange (Kadlec and Knight 1996).

The most important nitrogen species in a wetlands ecosystem are dissolved ammonia $\left(NH_4^+\right)$, nitrite $\left(NO_2^-\right)$ and nitrate $\left(NO_3^-\right)$. Other forms include nitrous oxide gas (N_2O), nitrogen gas (N_2), urea (organic), amino acids and amine. As it undergoes its various transformations, nitrogen is taken up by wetland plants and microflora (preferentially as NH_4^+ and NO_3^-). Organic nitrogen comprises a significant fraction of wetland biota, detritus, soils, sediments and dissolved solids. It is not readily assimilated by aquatic plants and must be converted to NH_4^+ or NO_3^- through multiple conversions, requiring long reaction times (Kadlec and Knight 1996). The ammonification, nitrification and denitrification processes are complex and the most important removal pathways for removal of nitrogen from constructed wetlands

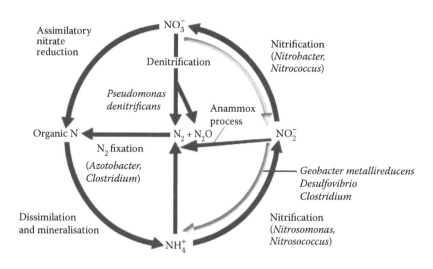

FIGURE 12.6
A simplified nitrogen cycle.

around the root zone. The major nitrogen transformations in a wetland eco-system are summarised in Table 12.3 and Figure 12.7.

Ammonification is a complex biochemical process in which organic nitro-gen is biologically converted into ammonia by several intermediate steps. This process takes place more rapidly than nitrification in the aerobic zones of the substrate. Pollutants containing nitrogen are readily degraded in both aerobic and anaerobic zones of wetland plants, releasing inorganic ammonia–nitrogen (NH_4–N). The inorganic NH_4–N is mainly removed by nitrification–denitrification processes in constructed wetlands. The rates of

TABLE 12.3

Nitrogen Transformation Influenced by Microbial Respiration in an Aquatic Macrophyte Wastewater Treatment System

Respiration		Nitrogen Transformation
Aerobic	Ammonification	$Org\text{-}N^+ \rightarrow NH_4$
	Immobilisation	$NH_4^+ \rightarrow Org\text{-}N$
	Nitrification	$NH_4^+ \rightarrow NO_3$
Anaerobic	Dissimilatory	
	NO_3^- reduction	$NO_3^- \rightarrow NH_4$
Facultative anaerobic	Denitrification	$NO_3^- \rightarrow N_2O$
	Ammonification	$Org\text{-}N \rightarrow NH_4^+$
	Immobilisation	$NH_4^+ \rightarrow Org\text{-}N$

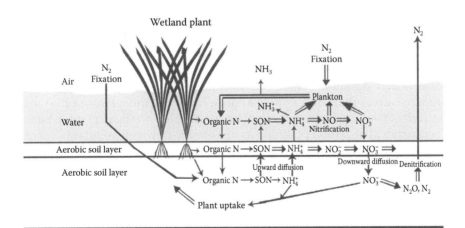

SON: Soluble organic nitrogen
$\longrightarrow$ Physical/chemical processes
$\Longrightarrow$ Bacterial/plant processes

FIGURE 12.7
Removal of nitrogen in constructed wetland.

ammonification are fastest in the oxygenated zone and then decrease as the mineralisation circuit changes from aerobic to facultative anaerobic and obligate anaerobes.

Nitrification first takes place, generally, in the rhizosphere (aerobic process). In wastewater treatment plants and CWs, the nitrification and denitrification are the major mechanisms of nitrogen removal (You et al. 2009). Nitrification is a two-step process catalysed by nitrifiers, such as *Nitrosomonas, Nitropira, Nitrosococcus* and *Nitrobacter,* followed by denitrification, and it is believed to be the major pathway for ammonia removal in both surface flow and subsurface flow of constructed wetlands (Kadlec and Knight 1996). Nitrification implies a chemolithoautotrophic oxidation of ammonia to nitrate under strict aerobic conditions and is performed in two sequential oxidative stages: ammonia to nitrite (ammonia oxidation) and nitrite to nitrate (nitrite oxidation). Ammonia oxidation is the limiting step of nitrification in several environments and is therefore critical to wastewater nitrogen removal (Choi and Hu 2008). Methane- and ammonia-oxidising bacteria play a major role in the global carbon and nitrogen cycles also. These bacteria convert most reduced carbon and nitrogen compounds (that is, CH_4 and NH_4^+) to their oxidised forms (CO_2 and NO_2^-). The contribution of methanotrophic and nitrifying bacteria to CH_4 and NH_4^+ oxidation in the rice rhizosphere was determined (Brix and Schierup 1989). Ammonia-oxidising bacteria (AOB) is predominant in wastewater treatment plants (Park and Noguera 2004). The rhizosphere of macrophytes appears to be the main site for ammonium oxidation. AOB play a major role in the nitrogen cycle in the root environment. AOB are generally known to be autotrophic and have been widely studied (Herrmann et al. 2008; Wang et al. 2010). Feray and Montuelle (2003) reported that several nitrifying strains are able to grow under mixotrophic conditions in response to environmental changes, particularly the nature of the food source. Mixotrophs grow using organic and inorganic compounds as carbon and energy sources. These bacteria may dominate aquatic environments due to their capability to use more resources than either photoautotrophic or organoheterotrophic bacteria (Eiler 2006). In Tunisia, the Joogar constructed wetland plant achieved good removal of carbon pollution as opposed to nitrogen pollutants, mainly ammonia, that remain elevated in the treated wastewater (Kouki et al. 2009). AOB have a high ecological importance and have been detected in different ecological systems, such as soil (Wang et al. 2009), the rhizosphere (Herrmann et al. 2008), freshwater (Chen et al. 2009) and wastewater treatment systems (Wang et al. 2010). In the first step, ammonia is oxidised to nitrite in an aerobic reaction catalysed by *Nitrosomonas* bacteria as shown in Equation 12.1.

$$NH_4^+ + O_2 \rightarrow NO_2 + H_2O + 2H^+ \qquad (12.1)$$

The nitrite produced is oxidised aerobically by *Nitrobacter*, forming nitrate (Equation 12.2) as follows:

$$NO_2^- + O_2 \rightarrow NO_3^- \tag{12.2}$$

The first reaction produces hydroxonium ions (acid pH), which react with natural carbonate to decrease the alkalinity. In order to perform nitrification, the *Nitrosomonas* must compete with heterotrophic bacteria for oxygen.

Denitrification is the process in which nitrate is reduced in anaerobic conditions by benthos to a gaseous form. The denitrification process contributes up to 60%–70% of the total nitrogen removal in CWs. Denitrification, caused by anaerobic bacteria, is the primary mechanism for nitrogen removal from wetland waters (Sather and Smith 1984). In this process, denitrifying bacteria (denitrifiers) decrease inorganic nitrogen, such as nitrate and nitrite, into nitrogen gas (Szekeres et al. 2002). Denitrifiers can be classified into two major species, heterotrophs and autotrophs. Heterotrophs are microbes that need organic substrates to obtain their carbon source for growth and get energy from organic matter. In contrast, autotrophs utilise inorganic substances as an energy source and CO_2 as a carbon source. The second step, denitrification, is conducted by a heterotrophic microorganism (such as *Psuedomonas, Micrococcus, Achromobactor* and *Bacillus*) under anaerobic or anoxic conditions. Denitrification can only take place in the anoxic zones of the systems as the presence of dissolved oxygen suppresses the enzyme system required for this process. In constructed wetlands, it is believed that microsites with steep oxygen gradients can be established, which allow nitrification and denitrification to occur in sequence in very close proximity to each other. Sufficient organic carbon is needed as an electron donor for nitrate reduction, which provides an energy source for denitrification microorganisms. The rate of denitrification is influenced by many factors, including nitrate concentration, microbial flora, type and quality of organic carbon source, hydroperiods, different plant species residues, the absence of O_2, redox potential, soil moisture, temperature, pH value, presence of denitrifiers, soil type, water level and the presence of overlying water (Sirivedhin and Gray 2006). Denitrification is illustrated by the following equation:

$$2NO_3^- \rightarrow 2NO_2^- \rightarrow 2NO \rightarrow N_2O \rightarrow N_2 \tag{12.3}$$

12.5.3 Maintaining the Phosphorous Cycle

Phosphorus pollution is a major concern for soil and water management. Phosphorus in water is found in the form of phosphate. Surface waters contain certain levels of phosphorus in various compounds, and it is an important constituent of living organisms. In natural conditions, the phosphate concentration in water is balanced, that is, an accessible mass of this constituent is close to the requirements of the ecological system. When the input of phosphorus to water is higher than can be assimilated by a population of living organisms, the problem of excess phosphorus content occurs. The excess content of phosphorus in the receiving water leads to extensive algae growth (eutrophication). Industrial wastewater is the major source of phosphorus contamination in the environment. A wetland ecosystem is the most widely used method for phosphorous removal from industrial wastewater. Phosphorous entering a wetland or stream is typically present in both organic and inorganic forms. In wetland soils, phosphorus occurs as soluble and insoluble complexes in both organic and inorganic forms as shown in Table 12.4.

Wetlands provide an environment for the interconversion of all forms of phosphorus. The main compartments in wetland P cycling are water, plants, microbiota and soil. The role of bacteria in solubilising inorganic phosphates in soil and making them available to plants is well known (Bhattacharya and Jain 2000). They are called phosphate-solubilising bacteria (PSB), and they convert the insoluble phosphates into soluble forms by acidification, chelation, exchange reactions and production of gluconic acid (Chen et al. 2006b). Phosphorus occurs primarily as organic phosphate esters and as inorganic forms, for example, calcium, aluminium and iron phosphates. Organic phosphates are hydrolysed by phosphatases, which liberate orthophosphate during microbial decomposition of organic material. Bacteria also liberate free orthophosphate from insoluble inorganic phosphates by producing organic or mineral acids or chelators, for example, gluconate and 2-ketogluconate, citrate, oxalate and lactate, which complex the metal resulting in dissociation or, for iron phosphates, by producing H_2S. Phosphate-solubilising activity is very important in the plant rhizosphere as shown in Figure 12.8. There are considerable populations of PSB in soil and in plant rhizospheres (Alexander 1977). These include both aerobic and anaerobic strains with a prevalence of aerobic strains in submerged soils. Emergent macrophytes have an extensive

TABLE 12.4
Major Types of Dissolved and Insoluble Phosphorus in Wetlands

Phosphorus	Soluble Forms	Insoluble Forms
Organic	Dissolved organics, for example, sugar phosphates, inositol, phosphate, phospholipids, phosphoproteins	Insoluble organic phosphorus bound in organic matter
Inorganic	Orthophosphate ($H_2PO_4^-$, $HPO_4 =$, PO_4^{3-}), polyphosphates, ferric phosphate ($FeHPO_4^-$), Calcium phosphate ($CaH_2PO_4^+$)	Clay phosphate complexes, metal hydroxide-phosphate, for example, vivianite $Fe_3(PO_4)_2$; variscite $Al(OH)_2H_2PO_4$, minerals, for example, apatite ($Ca_{10}(OH)_2(PO_4)_6$)

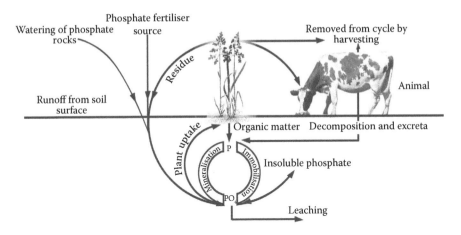

FIGURE 12.8
Phosphorous cycle in environment.

network of roots and rhizomes and have a greater potential to store P as compared to floating macrophytes as shown in Figure 12.9.

Rhizosphere microorganisms can increase or decrease the availability of phosphate (P) to plants (Marschner 2009). Rhizosphere bacteria increase P uptake by solubilising or mineralising more P than they require and by stimulating root growth. Rhizosphere microorganisms can reduce plant P availability by immobilisation of P in the microbial biomass, decomposition of P-mobilising root exudates and inhibition of root growth or mycorrhizal colonization. Many PSB belong to *Pseudomonas, Bacillus, Enterobacter, Serratia, Pantoea, Rhizobium* and *Flavobacterium* (Buch et al. 2008). Primarily, there are

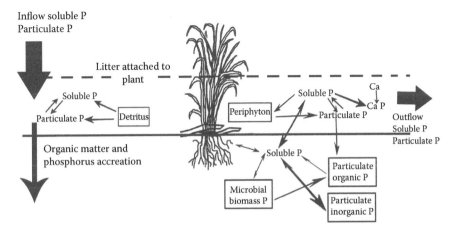

FIGURE 12.9
Scheme of P cycling as influenced by vegetation components of streams and wetlands.

two schools of thought regarding interpreting the mechanism of P solubilisation by phosphate solubiliser microorganisms (Arun 2007): (i) phosphate solubilisation by production of organic acid and (ii) phosphate solubilisation by production of phosphatase enzymes. Several reports have examined the ability of different bacterial species to solubilise insoluble inorganic phosphate compounds, such as iron-bound phosphate, tricalcium phosphate, dicalcium phosphate, hydroxyapatite and rock phosphate.

Phosphate solubilisation by production of organic acid: The major mechanism of mineral phosphate solubilisation is the action of organic acids synthesised by soil microorganisms. Organic acids are produced in the periplasmic space of some Gram-negative bacteria through a direct glucose oxidation pathway (Anthony 2004). Organic acid anions released by plant roots could potentially mobilise P but are rapidly decomposed by soil microorganisms (Van Hees et al. 2002). The organic and inorganic acids convert tricalcium phosphate to di- and monobasic phosphates with the net result of enhanced availability of the element to the plant. Other organic acids, such as glycolic, oxalic, malonic and succinic acids, have also been identified among phosphate solubilisers (Illmer and Schinner 1992). There is also experimental evidence that supports the role of organic acids in mineral phosphate solubilisation. Halder et al. (1990) showed that the organic acids isolated from a culture of *Rhizobium leguminosarum* solubilised an amount of P nearly equivalent to the amount that was solubilised by the whole culture. Inorganic phosphate–solubilising bacteria (IPSB) have been isolated from the rhizosphere of many terrestrial plants (Sundara-Rao and Sinha 1963).

Phosphate solubilisation by production of phosphatase enzymes: Mineralisation of most organic phosphorous compounds is carried out by means of enzymes, such as phosphatase (phosphohydrolase), phytase, phosphonoacetate hydrolase, D-α-glycerophosphatase and C-P lyase. Plants and microorganisms can increase the solubility of inorganic P by releasing protons; OH^- or CO_2; and organic acid anions, such as citrate, malate and oxalate, and they can mineralise organic P by the release of various phosphatase enzymes. The phosphohydrolases are clustered in acid or alkaline. The acid phosphohydrolases, unlike alkaline phosphatases, showed optimal catalytic activity at acidic to neutral pH values. Phosphatase activity is substantially increased in the rhizosphere. Soil bacteria expressing a significant level of acid phosphatases include strains from the genus *Rhizobium*, *Enterobacter, Serratia, Citrobacter, Proteus* and *Klebsiella* as well as *Pseudomonas* and *Bacillus*. The production of organic acids by PSB has been well documented. Among them, gluconic acid seems to be the most frequent agent of mineral phosphate solubilisation. It is reported as the principal organic acid produced by PSB such as *Pseudomonas* sp. (Illmer and Schinner 1992) and *Erwinia herbicola* (Liu et al. 1992). Another organic acid identified in strains with phosphate-solubilising ability is 2-ketogluconic acid, which is present in *Rhizobium leguminosarum* (Halder et al. 1990), *Rhizobium meliloti* (Halder and Chakrabartty 1993), *Bacillus firmus* (Banik and Dey 1982) and other unidentified soil bacteria (Duff and Webley 1959). Strains of *Bacillus liqueniformis* and *Bacillus amyloliquefaciens* were found to produce mixtures of lactic, isovaleric, isobutyric and acetic acids.

12.5.4 Maintaining the Sulphur Cycle

Sulphate-reducing bacteria reduce inorganic sulphates or other oxidised sulphur forms to sulphide. Sulphide is not incorporated into the organism but is released as 'free' H_2S. Dissimilatory sulphate-reducing bacteria are strict anaerobes that are severely inhibited by even small amounts of oxygen. They will, however, survive long periods of oxygen exposure and become active when anaerobic conditions are restored. Sulphate-reducing bacteria are heterotrophs and therefore require an organic carbon source. *Desulfovibrio* and *Desulfotomaculum* are the two best-known genera of sulphate-reducing bacteria. Others include *Desulfobulbus, Desulfococcus, Desulfosarcina, Desulfobacter* and *Desulfonema*, although the latter three genera are restricted to marine environments that can utilise the organic substrate (CH_2O) as a carbon source and sulphate as an electron acceptor for growth. Because sulphate-reducing bacteria can oxidise simple organic compounds and will only oxidise carbohydrates under rare circumstances, they generally rely on fermentative bacteria to break complex organic compounds into simple molecules prior to utilisation. In the bacterial conversion of sulphate to hydrogen sulphide, bicarbonate alkalinity is produced.

This is the first evidence for the existence of SRM that thrive in low-sulphate environments and were provided by enrichments from lake sediments (Ramamoorthy et al. 2006), rice paddy fields (Wind et al. 1999) and constructed wetlands (Lee et al. 2009) and included *Desulfovibrio, Desulfobulbus, Desulfotomaculum* and *Desulfosporosinus* spp. In the case of *Desulfovibrio* spp., this carbon source can be supplied by simple organic molecules, such as lactate, pyruvate and malate. These are subsequently oxidised to acetate and CO_2 with the concurrent reduction of sulphate to sulphide. Like *Desulfovibrio* spp., *Desulfotomaculum* prefer to oxidise lactate and pyruvate to acetate and CO_2, although one species, *Desulfotomaculum ruminis,* can also oxidise formate to CO_2. Several species, including *Desulfovibrio baarsii, Desulfococcus multivorans* and *Desulfotomaculum acetoxidans,* are capable of oxidising acetate to CO_2 with the concurrent reduction of oxidised sulphur species. The biological formation of methyl mercury is a caveat to sulphide production. Sulphate-reducing bacteria have been implicated in the biological methylation of mercury. Although the mechanisms of methylation are not fully understood, studies indicate that a unique coupling exists between the sulphate reduction rate (that is, sulphide production) and mercury methylation. Simultaneous enumeration of lactate-using sulphate reducers and *Thiobacilli* suggested that the most numerous sulphate reducers are in the spermosphere and the rhizosphere because of their locations. A simplified sulphur cycle is shown in Figure 12.10.

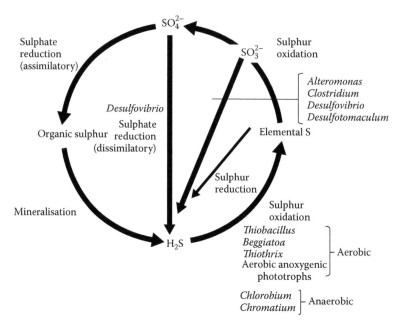

FIGURE 12.10
A simplified sulphur cycle.

12.6 Bioremediation of Heavy Metals and Metalloids

Metals play an important role in the life processes of all organisms, including bacteria, fungi and plants. Bacteria are common inhabitants of metal-contaminated sites, where they accumulate and immobilise heavy metals. Heavy metals are conventionally defined as elements with metallic properties (ductility, conductivity, stability as cations, ligand specificity, etc.) and an atomic number >20. The most common heavy metal contaminants are Cd, Cr, Cu, Hg, Pb and Zn. A metalloid is a chemical element that has properties in between those of metals and nonmetals. The six commonly recognized metalloids are boron, silicon, germanium, arsenic, antimony and tellurium. Heavy metals and metalloids are a major source of water and soil pollution, generated either through geogenic activities or industrial waste discharge. Because of their high solubility in the aquatic environments, heavy metals can be absorbed by living organisms. Once they enter the food chain, large concentrations of heavy metals may accumulate in the human body. If the metals are ingested beyond the permitted concentration, they can cause serious health disorders. Microbial interaction with small quantities of metals and metalloids does not exert a major impact on metal or metalloid distribution in the environment, whereas interaction with large quantities are required in energy metabolism, for instance, to have noticeable impact. Bacteria have a small size and a high surface area–to-volume ratio, and therefore provide a large contact area for interaction with the surrounding environment. The negative net charge of the cell envelope makes these organisms prone to accumulate metal cations from the polluted environment. There is increasing evidence that the bacteria associated with heavy metal–accumulating plants are not only adapted to but also actively involved in the accumulation process. Rhizosphere bacteria can potentially accumulate metals either by a metabolism-independent (passive) or metabolism-dependent (active) process. Passive adsorption is the dominant mechanism of metal accumulation in bacteria. Thus, overall accumulation is determined by two characteristics of the cell: the sorptivity of the cell envelope and capacity for taking up metals into the cytosol. Active uptake into the cytosol is usually slower than passive adsorption and is dependent on an element-specific transport system. The cell envelope characteristics of the bacteria determine their metal adsorption properties. The differences between Gram-positive and Gram-negative cell walls have a minor influence on the sorption behaviour of heavy metals. Heavy metal uptake can be promoted by any bacteria activity that increases the mobility and bioavailability of heavy metals. Bacterial immobilisation and sequestration of heavy metals can inhibit heavy metal uptake in the rhizosphere and reduce heavy metal toxicity in the endosphere. The functional group chemistry of the Gram-positive and Gram-negative bacteria surface is similar, but particular single constituents of the cell envelope can have great importance for metal binding. For example, the phosphoryl group of lipopoly saccharides, carboxylic group of teichoic acid and teichuronic acids or capsule-forming extracellular polymers influence the metal sorption of the cell envelope. Adverse effects of metals on the microbial cell are decreased decomposition of soil organic matter, reduced soil respiration and decreased activity of several soil enzymes. Depending on the external conditions, microbial cells have developed mechanisms to cope with high concentrations of metals.

12.6.1 Heavy Metal Accumulation in Plants

Most heavy metals have low mobility in soil and are not easily absorbed by plant roots. The bioavailability and plant uptake of heavy metals in the soils are affected by metal content, pH, Eh, water content, organic substances and other elements in the rhizosphere. The plant root and soil microbes and their interaction can improve metal bioavailability in the rhizosphere through secretion of organic acids, protein, phytochelatin (PC), amino acids and enzymes as shown in Figure 12.11. The interaction of plants and microorganisms may increase or decrease heavy metal accumulation in plants, depending on the nature of the plant–microbe interaction. Plant–microbe interaction is mediated by root exudates. Root exudates are transported across the cellular membrane and secreted into the surrounding rhizosphere. Soil organisms are attracted to these exudates as a source of food; the abundance of

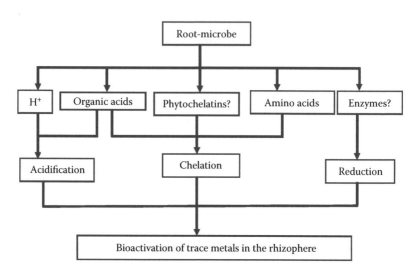

FIGURE 12.11
Process involved in heavy metal mobilisation in rhizosphere by root-microbe interaction.

soil organisms also increases close to the roots. Plant root exudates consist of a complex mixture of organic acid anions, phytosiderophores, sugars, vitamins, amino acids, purines, nucleosides, inorganic ions (for example, HCO^{3-}, OH^-, and H^+), gaseous molecules (CO_2, H_2), enzymes and root border cells, which have major direct or indirect effects on the acquisition of mineral nutrients required for plant growth. Root exudates are often divided into two classes of compounds. Low molecular weight compounds, that is, amino acids, organic acids, sugars, phenolics and other secondary metabolites, account for much of the diversity of root exudates, whereas high molecular weight exudates, that is, mucilage (polysaccharides) and proteins, are less diverse but often constitute a larger proportion of the root exudates by mass. Although the functions of most root exudates have not been determined, several compounds present in root exudates play important roles in biological processes. Some root exudates that act as metal chelators in the rhizosphere can increase the availability of metallic soil micronutrients, including iron, manganese, copper and zinc (Dakora and Phillips 2002).

Bacteria in the rhizosphere are involved in the accumulation of potentially toxic trace elements into plant tissues. Many wetland plants and bacteria have their own mechanism for removal of heavy metal and metalloid contamination in the soil. The accumulation of Zn^{2+}, Cu^{2+}, Mn^{2+} and Al^{3+} is also reported by Conard (1995). Naturally occurring rhizobacteria were found to promote selenium (Se) and mercury (Hg) bioaccumulation in plants growing in wetlands. Some rhizobacteria can exude a class of rhizobacteria secretion, such as antibiotics (including the antifungal), phosphate solubilisation, hydrocyanic acid, indole acetic acid (IAA), siderophores and 1-aminocyclopropane-1-carboxylic acid (ACC) deaminase, which increases bioavailability and facilitates root absorption of heavy metals, such as Fe (Crowley et al. 1991) and Mn (Barber and Lee 1974) as well as nonessential metals, such as Cd (Salt et al. 1995); enhances tolerance of host plants by improving the P absorption (Liu et al. 2000); and promotes plant growth (Meyer 2000). Three mechanisms of rhizosphere bacteria involved in heavy metal accumulation in plants are discussed here. These are the bacterial production of the plant hormone indole acetic acid, bacterial production of the enzyme ACC deaminase and bacterial production of siderophores.

12.6.1.1 Bacterial Production of the Plant Hormone IAA

Indole-3-acetic acid (IAA) is the main auxin hormone in plants, controlling many important processes, including cell enlargement and division, tissue differentiation and responses to light and gravity. Many plant-associated bacteria, including plant growth promotors and pathogens, are able to synthesise IAA and to influence plant development by modulating IAA levels. IAA production is a frequent feature of rhizosphere bacteria on heavy metal–contaminated sites (Dell'Amico et al. 2005).

12.6.1.2 Bacterial Production of the Enzyme ACC Deaminase

1-Aminocyclopropane-1-carboxylic acid (ACC) is a precursor molecule in the ethylene synthesis pathway of plants. Many soil- and plant-borne bacteria can synthesise ACC deaminase, an enzyme that cleaves ACC into alpha-keto-butyrate and ammonia. Plant-associated ACC deaminase producers can consume plant-borne ACC as a source for ammonia and, at the same time, inhibit ethylene synthesis in the plant (Glick 2003). Ethylene is a regulator of plant development and is involved in breaking seed dormancy, succession of early seedling stages, senescence processes and stress responses. This 'stress' ethylene causes damage to the plant organism. Treatment with ACC deaminase–producing bacteria reduced stress symptoms in plants exposed to various stress situations. The presence of toxic heavy metals can cause the formation of stress ethylene and heavy metal toxicity symptoms are partially due to the deleterious effects of ethylene (Glick 2003). ACC deaminase production has been detected in an important proportion of the rhizosphere bacteria and endophytes of Ni hyperaccumulating *Thlaspi goesingense* (Idris et al. 2004).

12.6.1.3 Bacterial Production of Siderophores

Bacterial siderophores are usually poor Fe sources for both monocot and dicot plants. On the other hand, plant-derived Fe phytosiderophore complexes appear to be a good Fe source for bacteria (Marschner and Crowley 1998). Many heavy metal accumulations by siderophores have been reported recently (Kluber et al. 1995). The siderophore formation is a common feature in wetland rhizospheres, where the accumulation of toxic metals ions, such as Fe^{3+}, Zn^{2+}, Cu^{2+}, Mn^{2+} and Al^{3+}, take place by siderophore formation. Siderophore-producing rhizosphere bacteria have therefore been hypothesised to contribute to the iron nutrition of their host plant. Moreover, many bacterial siderophores have been observed to chelate divalent heavy metal ions, including Mn, Zn, Cd, Pb and Cr(III). The complexes formed with heavy metals are less stable than those formed with iron.

Plants and microorganisms have evolved a variety of mechanisms to scavenge Fe under Fe-limiting conditions. Under physiological conditions, iron can exist in either the reduced ferrous (Fe^{2+}) form or the oxidised ferric (Fe^{3+}) form. Thus, iron is important for numerous biological processes, which include photosynthesis, respiration, the tricarboxylic acid cycle, oxygen transport, gene regulation, DNA biosynthesis, etc. Although iron is abundant in nature, it does not normally occur in its biologically relevant ferrous form. Under aerobic conditions, the ferrous ion is unstable. Via the Fenton reaction, ferric ion and reactive oxygen species are created, the latter of which can damage biological macromolecules (Touati 2000).

$$Fe^{2+} + H_2O_2 \rightarrow Fe^{3+} + OH + OH^- \qquad (12.4)$$

The ferric ion aggregates into insoluble ferric hydroxides. Because of iron's reactivity, it is sequestered into host proteins, such as transferrin, lactoferrin and ferritin. Consequently, the cellular concentration of the ferric ion is too low for microorganisms to survive by solely using free iron for survival. Microorganisms overcome this nutritional limitation in the host by procuring iron either extracellularly from transferrin, lactoferrin and precipitated ferric hydroxides or intracellularly from hemoglobin. This is accomplished by microorganisms via two general mechanisms: iron acquisition by cognate receptors using low molecular weight iron chelators, termed siderophores, and receptor-mediated iron acquisition from many microorganisms such as aerobic or facultative anaerobic bacteria and fungi release siderophores (also called siderochromes and sideramines) into the surrounding medium under Fe limiting conditions. Siderophores are low molecular weight (600–1000) organic compounds with a very high and specific affinity to chelate iron; almost 500 siderophore structures are known to date (Boukhalfa and Crumbliss 2002), which are produced by bacteria, fungi and plants as shown in Table 12.5. Siderophore-producing rhizosphere bacteria may enhance plant growth by increasing the availability of Fe near the root or by inhibiting the colonisation of roots by plant pathogens or other harmful bacteria. Bacteria produce a wide range of siderophores, for example, enterobactin, pyroverdine and ferrioxamines by bacteria. Although siderophores differ widely in their overall structure, the chemical natures of the functional groups that coordinate the iron atom are not so diverse. Siderophores incorporate either α-hydroxycarboxylic acid, catechol or hydroxamic acid moieties into their metal-binding sites and thus can be classified as either hydroxycarboxylate-, catecholate- or hydroxamate-type siderophores as shown in Figure 12.12.

TABLE 12.5

Siderophores Produced by Plant or Plant-Associated Microorganisms

S. No.	Siderophores	Producing Organisms	References
1.	Pyoverdine	*Pseudomonas fluorescens*	Cody and Gross (1987); Teintze et al. (1981)
2.	Catechols		
	Agrobactin	*Agrobacterium tumefaciens*	Ong et al. (1979)
	Enterobactin	*Enterobacteriacae*	Pollack and Neilands (1970)
	Chrysobactin	*Erwinia chrysanthemi*	Persmark et al. (1989)
3.	Hydroxamates		
	Aerobactin	*Erwinia carotovora, Enterobacter cloacae*	Ishimaru and Loper (1988); Crosa et al. (1988)
	Canadaphore	*Helminthosporium carbonum*	Letendre and Gibbons (1985)
	Cepabactin	*Pseudomonas cepacia*	Meyer et al. (1989)
	Coprogen	*Alternaria longipipes*	Jalal et al. (1988)
	Dimerum acid	*Microdochium dimerum*	Diekmann (1970)
	Ferrichrome A	*Ustilago* spp.	Budde and Leong (1989)
	Ferrirhodin	*Botrytis cinerea*	Konetschny-Rapp et al. (1988)
	Fusarinins	*Fusarium roseum*	Jalal et al. (1986);
	Fusigen	*Fusarium roseum*	Diekmann (1967)
4.	Others		
	Rhizobactin	*Rhizobium meliloti*	Schwyn and Neiland (1987)
	Mugineic acid	*Graminaceae*	Sugiura et al. (1981)

12.7 Siderophore Transport Mechanism in Bacteria

Microbial transport of Fe^{3+} generally delivers iron to cells via siderophore-mediated transport systems. Ferric-siderophore uptake in microorganisms is both receptor and energy dependent. The iron siderophore transports in Gram-positive and Gram-negative bacteria are similar. The mechanism of iron transport into the cell is best understood in *E. coli*. Siderophores bind Fe outside the cell with a very high affinity ($K_f > 10^{30}$) and are then taken up through specific receptors in the cell membrane. In *E. coli*, the ferric iron binds to the siderophore and forms an iron–siderophore complex, which crosses the outer membrane and the cytoplasmic membrane before delivering iron within the cytoplasm. In the bacterial iron uptake pathways, the pathway for the uptake of ferric siderophores is the most structurally well defined. The iron uptake siderophore pathway involves an outer membrane protein and receptors, periplasmic binding proteins (PBPs), permease protein and an inner membrane ATP-binding cassette (ABC) transporter.

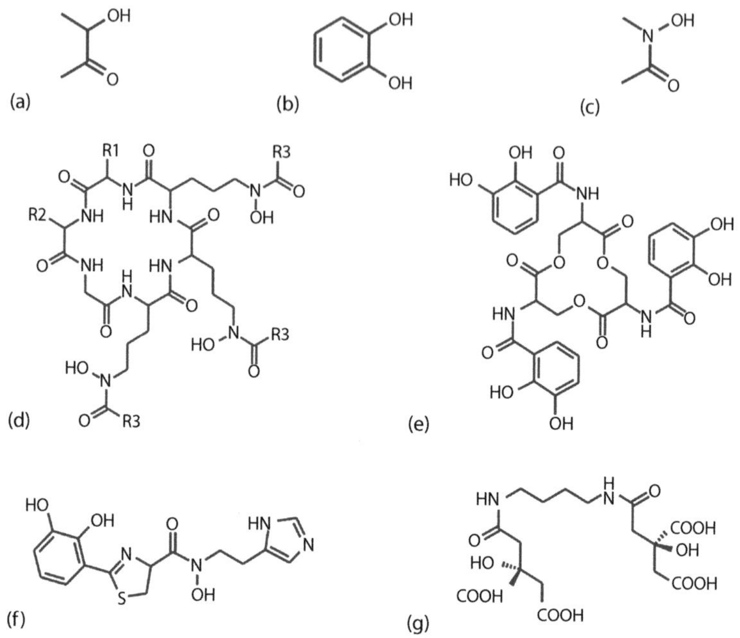

FIGURE 12.12
Functional groups found in siderophores. Although the structures of siderophores may vary, the functional groups for Fe^{3+} coordination are limited. Siderophores usually contain the following metal-chelating functional groups: (a) α-hydroxycarboxylic acid, (b) catechol or (c) hydroxamic acid, (d) hydroxamate siderophore ferrichrome, (e) the catecholate siderophore enterobactin, (f) the mixed catecholate–hydroxamate siderophore anguibactin and (g) the hydroxycarboxylate siderophore rhizoferrin.

12.7.1 Outer Membrane Protein and Receptors

The ferric complexes are too large for passive diffusion or nonspecific transport across these membranes. Translocation of iron through the bacterial outer membrane as the ferric-siderophore requires the formation of an energy-transducing complex with the proteins TonB, ExbB and ExbD, which couple the electrochemical gradient across the cytoplasmic membrane to a highly specific receptor and so promotes transport of the iron complex across the outer membrane. Translocation of ferric siderophore complex is mediated by TonB complex. ExbB is a 26-kDa cytoplasmic membrane protein. It consists of three transmembrane domains. ExbD is a 17-kDa protein that, like TonB, has only one transmembrane domain and a periplasmic domain of about 90 amino acids. Together, ExbB and ExbD couple the activity of TonB to the proton gradient of the cytoplasmic membrane.

Ferric siderophore complexes exceed the molecular weight cutoff of porins and thus require specific outer membrane receptors for their uptake into the periplasmic space: *E. coli* outer membranes hydroxamate and citrate and enterobactin receptors FhuA, FecA and FepA, respectively, as well as for the

Pseudomonas aeruginosa pyoverdine and pyochelin receptors FpvA and FptA, respectively, and the related TonB-dependent vitamin B-12 (cobalamin) receptor, BtuB.

12.7.2 Periplasmic Siderophore Binding Proteins

The periplasmic siderophore binding proteins (PBPs) are important for escorting their siderophores to the cytoplasmic membrane transporters for subsequent transport into the bacterial cell's cytoplasm. When the Fe^{3+} siderophore complex is released into the periplasm, siderophores are rapidly bound by the specific periplasmic binding proteins FhuD, FepB and FecB, respectively. In *E. coli*, the PBPs FhuD, FepB and FecB are responsible for shuttling hydroxymate, catecholate and citrate-type siderophores, respectively. The siderophore PBPs belong to cluster 8. FhuD is a ferric hydroxamate binding protein that is found in both Gram-positive and Gram-negative bacteria. FhuD exhibits a broad substrate specificity for a variety of hydroxamate siderophores, including ferrichrome, coprogen, aerobactin, ferrioxamine B and rhodoturilic acid. The structures of *E. coli* FhuD bound to gallium-bound ferrichrome (or gallichrome) and various other hydroxymate-type siderophores have all been determined. The chemically different hydroxamate siderophores are accommodated by different hydrogen bonding networks as shown in Figure 12.13.

12.7.3 ATP-Binding Cassette Transporters

Once the siderophore has been bound by a PBP, it must be transported into the cytoplasm. This is accomplished by an ABC transporter protein complex, which couples ATP hydrolysis to the transport of the siderophores. Bacterial ABC transporters are usually assembled from separate subunits rather than fused into one signal polypeptide. Such is the case for the vitamin B-12 importer from *E. coli* (BtuC2D2), the ferric-enterobactin importer from *E. coli* (FepC2D2) and the ferric-enterobactin importer from *Vibrio anguillarum* (FatC2D2). In contrast, the hydroxamate siderophore importer from *E. coli*, FhuBC, has the FhuB dimer fused into one polypeptide chain, but like the other ATP-binding domains, two copies of FhuC assemble to form a dimer (FhuBC2). The outer membrane receptor (BtuB) and the periplasmic protein (BtuF) are structurally similar to those of the siderophore uptake mechanism of bacteria. Ferric siderophores are released from the transport system at the cytoplasmic side of the cytoplasmic membrane as shown in Figure 12.14.

In Gram-positive bacteria, uptake of the iron source involves a membrane-anchored binding protein; this protein binds to Fe^{3+} and transports it into the cell cytoplasm. However, when compared to Gram-negative bacteria, there is little information on iron transport in Gram-positive bacteria. The iron transport mechanism in Gram-positive bacteria is shown in Figure 12.15. The cell wall of Gram-positive bacteria has strong metal-binding properties. Some bacteria also produce extracellular polysaccharide sheaths that bind metals.

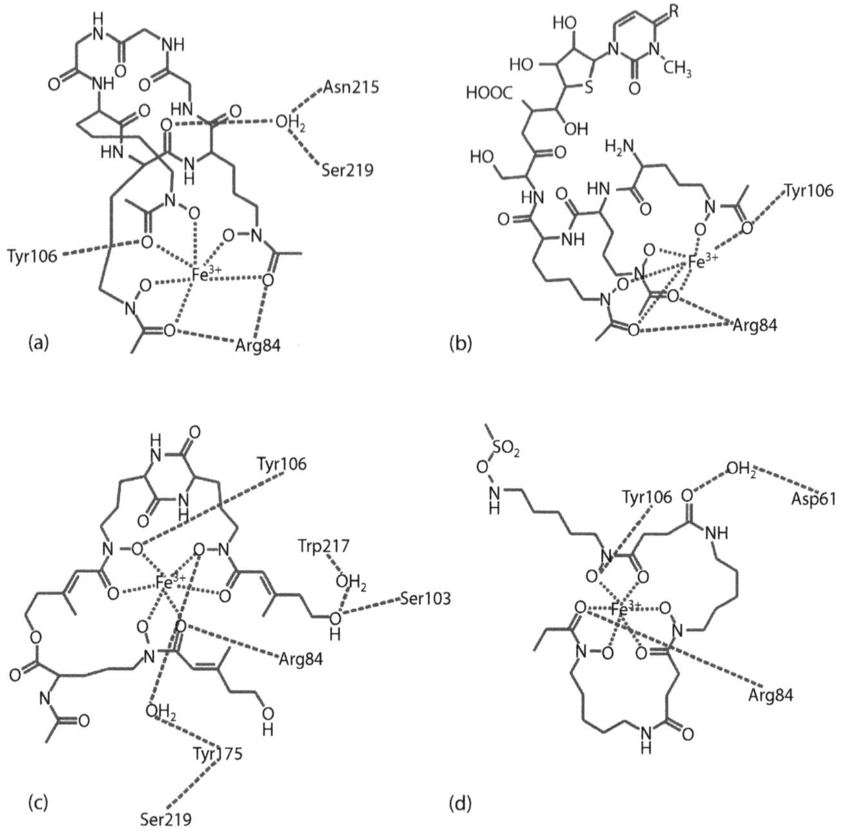

FIGURE 12.13
General structures of hydroxamate-type siderophores and their hydrogen-bonding interactions with FhuD. Comparison of binding modes of (a) ferrichrome, (b) albomycin, (c) coprogen and (d) desferal. The amino acid residues of FhuD and the water molecules that hydrogen bond with each siderophore are labelled with the hydrogen bonds represented as dashed lines.

12.8 Iron Release Mechanism from Siderophores

In the cell, cytosol iron can be released from an iron siderophore complex either by chemical breakdown or the reduction step. In this ferric form, iron is reduced to its ferrous form, and siderophores are again recycled. Once iron has been taken into the cell in the form of a ferric siderophore complex, it must be released from the carrier to become metabolically functional. Iron release from siderophore complexes in some cases is thought to be facilitated in part by the reduction of Fe^{3+} to Fe^{2+} (Dhungana and Crumbliss 2005).

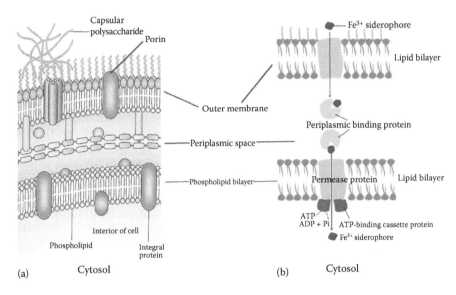

FIGURE 12.14
(See color insert.) (a) Basic structure of Gram-negative cell wall and (b) iron siderophore transport mechanism in *E. coli*.

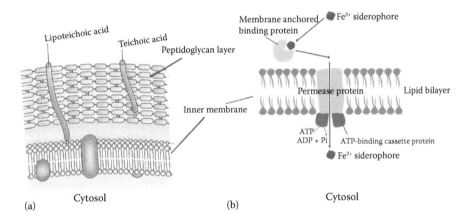

FIGURE 12.15
(See color insert.) (a) Basic structure of Gram-positive cell wall and (b) iron transport mechanism in Gram-positive bacteria.

Reduction of ferric siderophores accompanied by ligand exchange is an excellent mechanism for intracellular iron release. These ligand exchange reactions are usually slow at in vitro conditions due to the very high binding affinity and selectivity of siderophores toward Fe^{3+}, which ensures the in vivo acquisition of iron even when its concentration is very low. The first step in the dissociation of iron siderophore complexes normally involves the dissociation of a bidentate moiety (for example, hydroxamate or cate-cholate) from the fully chelated iron, which is relatively fast as a proton-driven reaction at low pH (Boukhalfa et al. 2000; Boukhalfa and Crumbliss 2000). This step represents the replacement of one binding unit from the first coordination shell of Fe^{3+} by water ligands. The lability of the fully coor-dinated iron is an important factor in iron release and exchange. Through the complex dechelation process, vacant coordination sites on iron become available, which further lowers the binding affinity of the siderophore to iron and permits the formation of ternary complexes, which may facilitate further dissociation of the siderophore complex (Boukhalfa and Crumbliss 2000; Boukhalfa et al. 2000). High $[H^+]$ is required for Fe^{3+}-siderophore dechela-tion, and even when complex dissociation occurs in acidic pH, the rate of complete ligand dissociation remains relatively slow (Boukhalfa et al. 2000). In most cases, the redox potential of iron bound in these complexes is within the physiological range for redox system (NADH, NAD^+, etc.), and so the iron is probably released by reduction (Fontecave et al. 1994). There is some evidence that suggests that iron reductases may be responsible for reducing and releasing iron from ferric siderophore complexes (Matzanke et al. 2004). In vivo, however, this reduction must coincide with additional chemical processes based on the fact that the redox potentials of many siderophores are much more negative than those of biological reducing agents, including ascorbate, glutathione and NADH. Many flavin reductases show the prop-erty to reduce a broad spectrum of ferric siderophore in vitro. Ferric sidero-phore reductases have been observed in many microorganisms. The highest activity is usually found when cell-free extracts are tested under anaerobic conditions with NADH or NADPH as a reductant. The reaction can be sum-marized as follows:

$$Fe^{3+} \text{ siderophore} + NADPH \rightarrow Fe^{2+} \text{ siderophore} + NADP^+ \qquad (12.5)$$

One possible mechanism of iron reduction utilises the presence of a strong Fe^{2+} chelator, which can act to shift the redox potential of the siderophore complex to a more positive value through either ternary complex formation or by providing a thermodynamic driving force for reduction through the formation of a highly stable Fe^{2+} complex (Boukhalfa and Crumbliss 2002). The iron release mechanisms are shown in Figure 12.16.

12.9 Mechanism of Hyperaccumulation of Heavy Metals

All plants have the ability to accumulate heavy metals, which are essential for their growth and development. Hyperaccumulators, which are often

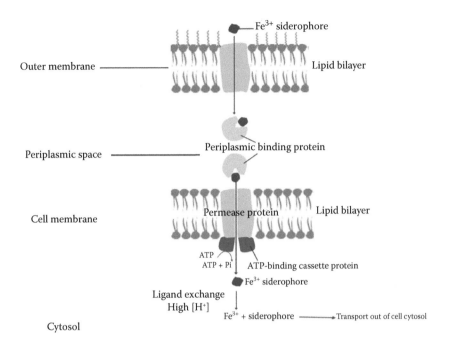

FIGURE 12.16
Iron release mechanism from siderophores in bacteria.

found growing in polluted areas, can naturally accumulate higher quantities of heavy metal in their shoots than their roots. Hyperaccumulator plants can accumulate a hundred- to a thousandfold higher levels of metals than normal plants. Threshold values used to define hyperaccumulation vary by element: Mn and Zn hyperaccumulators contain >10,000 µg/g (Reeves and Baker 2000); hyperaccumulators of As, Co, Cu, Ni, Se and Pb have >1000 µg/g (Ma et al. 2001); and hyperaccumulators of Cd have >100 µg/g (Reeves and Baker 2000). Hyperaccumulation has also been described for a few other elements, such as aluminium (Jansen et al. 2002), boron (Babaoglu et al. 2004) and iron (Rodríguez et al. 2005). Wetland macrophytes have been reported to have a larger capacity for metal accumulation. However, excessive accumulation of heavy metal can be toxic to most plants. Chandra and Yadav (2010) observed different anatomical parts of *Typha angustifolia* in the presence of heavy metal under TEM. The TEM observations of *T. angustifolia* showed physiologically and biochemically linked deformities in the plant tissue, which is shown in Figure 12.17. *T. angustifolia*, being a root accumulator, showed apparent metallic deposition and disruption of parenchyma cells in experimental pots as compared to the control under light microscopy (Figure 12.17a and b). TEM micrographs of

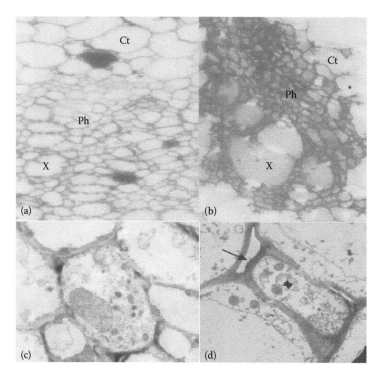

FIGURE 12.17
(See color insert.) Light micrograph of *T. angustifolia* root shows metal deposition (dark staining) and disruption of cortex cell (b vs. a; *) and TEM micrograph shows intercellular space (→)
and nucleus size reduction (◣) (d) in ST11 as compared to control (c) during metal accumulation in 60 days. Cortex (Ct), phloem (Ph) and xylem (X).

T. angustifolia root grown in an experimental pot showed shrinkage of the
cell, resulting in formation of intercellular spaces and a decrease in nucleus
size (Figure 12.17c and d).

A hyperaccumulator has the ability to both tolerate elevated levels of heavy
metals and accumulate them in a number of different plant species. The
first hyperaccumulators characterised were members of the *Brassicaceae* and
Fabaceace families. About 25% of discovered hyperaccumulator plants belong
to the family of *Brassicaceae*, in particular, to genera *Thlaspi* and *Alyssum*.
These also include the highest number of Ni hyperaccumulating taxa. More
than 400 plant species have been reported so far that hyperaccumulate metals. A relatively small group of wetland plants are capable of accumulating
heavy metals in different parts. Hyperaccumulation of heavy metal is a complex process. It involves several steps: (a) transport of heavy metal across the
plasma membrane of the root cell, (b) root-to-shoot translocation of metals
and (c) sequestration of metals in shoot vacuoles.

12.9.1 Transport of Heavy Metals Across the Plasma Membrane of Root Cells

Transport proteins and intracellular high-affinity binding sites mediate the uptake of metals across the plasma membrane of root cells. Several classes of protein have been reported in heavy metal translocation in plants. These are CPx-type heavy metal ATPase, Nramp Protein, CDF protein and ZIP Protein.

CPx-type heavy metal ATPase: CPx-type heavy metal ATPases have been identified in a wide range of organisms (for example, bacteria, archaea and eukaryotes) and have been implicated in the transport of a range of essential and also potentially toxic metals, such as Zn, Cu, Cd and Pb across the root cell membrane. Heavy metal transporters have been classified as type IB and are called CPx-ATPase because they share the common features of a conserved cystein–proline–cystein or cystein–proline–histidine or cystein–proline–serine (CPx) motif, which is through to function in heavy metal transport. CPx-ATPases are thought to be important not only in obtaining sufficient amounts of heavy metal ions for essential cell functions, but also in preventing accumulation of these ions to toxic levels.

Natural resistance-associated macrophage protein (Nramp): Natural resistance-associated macrophage protein (Nramp) is a novel family of protein that transports a wide range of metals, such as Mn^{2+}, Zn^{2+}, Cu^{2+}, Fe^{2+}, Cd^{2+}, Ni^{2+} and Co^{2+}, across membranes and have been identified in bacteria, fungi, plants and animals. In plants, Nramp transporters are expressed in roots and shoots and are involved in the transport of metal ions through the plasma membrane and the tonoplast.

Cation diffusion facilitator protein (CDFP): The cation diffusion facilitator protein (CDFP) is a ubiquitous family, members of which are found in bacteria, archaea and eukaryotes. They transport heavy metals, including cobalt, cadmium, zinc and possibly nickel, copper and mercuric ions. These proteins are considered to be efflux pumps that remove these ions from cells.

Zinc iron permease protein (ZIPP): Zinc iron permease is one of the principal metal transporter families of protein first identified in plants and is capable of transporting a variety of cations, including cadmium, iron, manganese and zinc across the plasma membrane. At this time, more than 25 ZIP family members have been identified.

12.9.2 Root-to-shoot Translocation of Metals

Efficient translocation of metal ions to the shoot requires radial symplastic passage and active loading into the xylem. Constitutively large quantities of small organic molecules are present in hyperaccumulators or roots that can operate as metal-binding ligand. The need of a ligand for all trace metals in the xylem is controversial. A key role of hyperaccumulator seems to play by free amino acids, such as histidine and nicotinamine, which form stable complexes with bivalent cations; free histidine is regarded as the most important ligand involved in Ni hyperaccumulators (Callahan et al. 2006). Physiological studies of hyperaccumulators also demonstrated higher metal concentration in the xylem sap due to enhanced xylem loading. Several types of transporters are involved in this process.

P-type ATPase-HMA: Molecular studies and mutant analysis have identified particular P-type ATPases as being responsible for the Cd and Zn loading of the xylem from the surrounding vascular tissue. The P_{1B} type ATPase is a class of protein, also known as the heavy metal transporting ATPase (HMAs). They operate roles in heavy metal transport against their electrochemical gradient using the energy provided by ATP hydrolysis and play a role in metal homeostatis and tolerance. In bacteria, P_{1B}-type ATPases are the main players in metal tolerance (Monchy et al. 2007). The overexpression of HMA4 supports a role of the HMA4 protein (which belongs to the Zn, Co, Cd, Pb HMA subclass and is localized at the xylem parenchyma plasma membrane) in Cd and Zn efflux from root symplasm into the xylem vessel, necessary for shoot hyperaccumulation.

Multidrug and toxic compound extrusion (MATE): MATE is a large family of multidrug and toxic compound extrusion (or efflux) membrane proteins that are active in heavy metal translocation in hyperaccumulator plants. Some members of the family were shown to function as drug or cation antiporters that remove toxic compounds and secondary metabolites from the cytosol by taking them out of the cell or sequestering them to the vacuole.

Oligopeptide transporters (OPT): OPT is a superfamily of oligopeptide transporters, including the yellow strip-like (YSL) subfamily. YSL mediates the loading into and unloading out of xylem at nicotinamine-metal chelates.

12.9.3 Sequestration of Metals in Shoot Vacuole

The family of cation diffusion facilitators (CDF) in plants is also called metal transporter proteins (MTPs). MTP-1 is a gene encoding in leaves of Zn/Ni hyperaccumulators (Hammond et al. 2006). It has been suggested that MTP-1, besides the role in Zn tolerance, may also play a role in enhancing Zn accumulation. The MTP protein family is involved in the transport of Zn^{2+}, Fe^{2+}, Cd^{2+}, Co^{2+} and Mn^{2+}, not only from cytoplasm to organelles or apoplasm but also from the cytoplasm to the endoplasmic reticulum. The sequestering mechanism in aerial organs of hyperaccumulators consists mainly in heavy metal removal from metabolically active cytoplasm by moving them to inactive compartments, mainly vacuole and cell walls. The vacuole is generally the amin storage site for metal in yeast and plant cells. Compartmentalisation of metals in the vacuole is also part of the tolerance mechanism of some hyperaccumulators.

Ca^{2+}/cation antiporter (CACA): In the Ca^{2+}/cation antiporter (CACA) superfamily, MHX is a vascular Mg^{+2} and Zn^{2+}/H^+ exchanger (Shaul et al. 1999). MHX protein was present in the leaves of *A. halleri* at much higher concentrations than in *Arabidopsis thaliana* and was therefore proposed to play a role in Zn vacular storage. A member of the CACA subfamily may also play a role in metal detoxification.

ATP-binding cassette (ABC): The superfamily of ABC (ATP-binding cassette) transporters are involved in vascular sequestration of various heavy metals or xenobiotics. In two subfamilies, multidrug resistant protein (MRP) and PRD, members are involved in the transport of chelated heavy metals at the organ. There is strong evidence for a role in trace metal homeostasis (Kim et al. 2006), and they may be expected to contribute to trace metal hyperaccumulator sequestration.

12.9.4 Detoxification by Chelation of Metals

A general mechanism for detoxification of heavy metals in plants is the distribution of metals to apoplast tissue, such as trichome and cell walls, and chelation of the metals by a ligand followed by sequestration of the metal ligand complex into a vacuole is shown in Figure 12.18. Small ligands, such as organic acids, may be instrumental in presenting the persistence of heavy metals as free ions in the cytoplasm, and even more organic acid chelates are primarily located. Here, we summarise some of the key ligands that seem to play a role in heavy metal hyperaccumulation in plants.

Histidine: Histidine (His) is considered to be the most important free amino acid involved in heavy metal hyperaccumulation in plants (Callahan et al. 2006). It forms stable complexes with Ni, Zn and Cd, and it is present at high concentrations in hyperaccumulator roots (Persans et al. 1999). In the Ni hyperaccumulator *Alyssum lesbiacum*, Ni exposure induced a dose-dependent increase in His in the xylem sap, which was not found in the non-hyperaccumulator congeneric species, *Asplenium montanum*. His-dependent Ni xylem loading may not be universal in *Brassicaceae*, and additional factors are required in at least *Arabidopsis thaliana*.

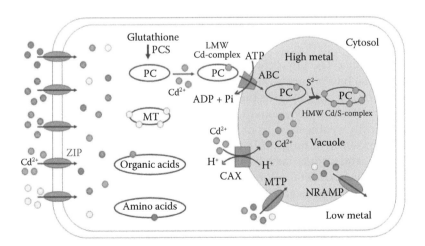

FIGURE 12.18
Role of ligand in heavy metal detoxification.

Nicotianamine: Synthesis of nicotianamine (NA) from 3 Sadenosyl-methionine (SAM) by NA synthase (NAS) is present in all plants. NA forms strong complexes with most transition metal ions. The role of NA seems to be in the movement of micronutrients throughout the plant (Stephan and Scholz 1993). The nicotianamine seems to be involved in metal hyperaccumulation both in *Arabidopsis halleri* and in *Thlaspi caerulescens*, in which several NAS genes showed higher transcript levels.

Organic acids (citrate, malate): Because of the low association constants of organic acids with metals, Callahan et al. (2006) argued against a role for organic acids in the hyperaccumulation mechanism (such as long distance transport), in spite of their constitutively elevated concentrations in hyperaccumulators (Montargès-Pelletier et al. 2008). So their role may be limited to vacuolar sequestration. The formation of metal–organic acid complexes is favoured in the acidic environment of the vacuole (Haydon and Cobbett 2007).

Glutathione: Glutathione (Glu–Cys–Gly; GSH), a tripeptide, is the most abundant low molecular weight thiol in all mitochondria-bearing eukaryotes, including plants. In plants, GSH is involved in a plethora of cellular processes, including defence against reactive oxygen species (ROS), sequestration of heavy metals and detoxification of xenobiotics. GSH does this by prior activation and conjugation with such compounds. The conjugates are subsequently transported to the vacuole and protects the plant cell from their harmful effect.

Heat shock proteins: Heat shock proteins (HSPs) act as molecular chaperones in normal protein folding and assembly but may also function in the protection and repair of protein under stress conditions. They are found in all groups of living organisms, can be classified according to molecular size and are now known to be expressed in response to a variety of stress conditions, including heavy metals. The molecular chaperones induced during heavy metal stress could prevent irreversible protein denaturation resulting from the oxidative stress linked to heavy metal exposure or help to channel their proteolytic degradation.

Phytochelatins: The phytochelatins (PCs) are a family of metal-complexing peptides that have a general structure (γ-Glu Cys)$_n$-Gly, which is generally in the range of 2–5 but can be as high as 11 (Cobbett 2000).

They are structurally related to the tripeptide glutathione (GSH) and are enzymatically synthesised from GSH. PCs are rapidly induced in cells and tissues exposed to a range of heavy metal ions, such as Cd, Ni, Cu, Zn, Ag, Hg and Pb, and anions, such as arsenate and selenite (Yang and Yang 2001).

Metallothioneins: Metallothioneins (MT) are low molecular weight proteins that bind heavy metals and are found throughout the animal and plant kingdoms. These proteins also play an important role in detoxification by sequestering metals in plant cells. Plants have been found to contain a number of genes encoding MT-like proteins having a sequence similar to animal MT proteins. A summary of heavy metal hyperaccumulation in plants is shown in Figure 12.19.

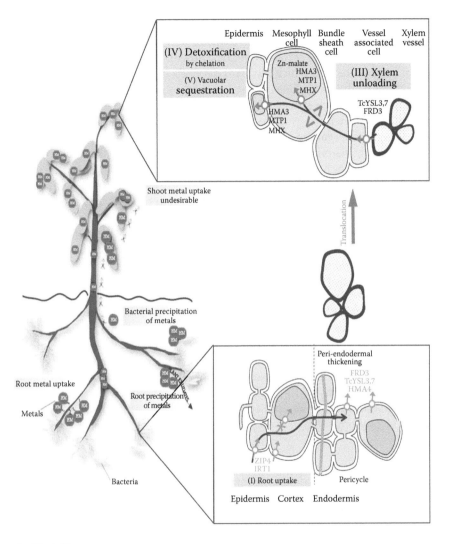

FIGURE 12.19
(See color insert.) Major process involved in heavy metal hyperaccumulation by plants.

12.10 Ecological Significance of Hyperaccumulator Plants for Their Adaptation

During the past decade, the ecological consequences of metal hyperaccumulation began to attract attention. Ecological studies of metal hyperaccumulation have been designed to provide insight into how and why metal hyperaccumulation has evolved by determining its adaptive value and to examine how the extraordinary metal concentration of hyperaccumulators impacts species relationships in the habitats in which hyperaccumulators have evolved. Several explanations have been offered to explain why hyperaccumulators take up such large quantities of certain metallic elements. Plants are locked in an evolutionary arms race in which survival depends on their ability to counter the adaptations of the other. Heavy metal accumulation in aerial tissue may function as a self-defence strategy involved in hyperaccumulator plants against some natural enemies, such as herbivores and pathogens. Plant defence against herbivores can be classified into several categories, including mechanical, chemical, visual, behavioural and associational. Most studies of plant defence chemicals have focused upon organic compounds (secondary compounds), many of which have been identified and whose roles in plant defence are well known. Besides, chemical accumulation (frequently metals or metallic compounds) in plants plays an important defensive role. Of these, the defence hypothesis has received the most supporting evidence. This hypothesis suggests that hyperaccumulation is a defensive tactic because it can protect plants from some natural enemies (for example, herbivores and pathogens). One ecological consequence of the existence of high metal herbivores is that metal may influence the herbivore's interactions with other species, such as predators or pathogens. This ability to avoid feeding on plants with high heavy metal levels might support the view that herbivores have a 'taste for metals', although no information exists on how they might do it. However, because the metal treatment will strongly affect the plant's metabolome, it might be that herbivores do not directly take metals in their food but rather metal-induced metabolites. The potential toxicity of high metal insects to predators may also raise environmental concerns about applied uses of hyperaccumulators for phytoremediation or phytomining.

References

Alexander, M. 1977. *Introduction to Soil Microbiology*. John Wiley, New York.

Anthony, C. 2004. The quinoprotein dehydrogenases for methanol and glucose. *Arch Biochem Biophys* 428: 2–9.

Armstrong, J., Armstrong, W., Beckett, P.M. et al. 1996. Pathways of aeration and the mechanisms and beneficial effects of humidity and venturi-induced convections in *Phragmites australis*. *Aquat Bot* 54: 177–197.

Armstrong, W., Brandle, R., and Jackson, M.B. 1994. Mechanisms of flood tolerance in plants. *Acta Botanica Neerlandica* 43: 307–358.

Arun, K.S. 2007. *Bio-Fertilizers for Sustainable Agriculture*. Sixth edition, Agribios publishers, Jodhpur, India.

Babaoglu, M., Gezgin, S. et al. 2004. Gypsophila spaerocephala Fenzl ex Tchihat: A boron hyperaccumulator plant species that may phytoremediate soils with toxic B levels. *Turkish Journal of Botany* 28: 273–278.

Banik, S., and Dey, B.K. 1982. Available phosphate content of an alluvial soil is influenced by inoculation of some isolated phosphate-solubilizing microorganisms. *Plant Soil* 69: 353–364.

Barber, S.A., and Lee, R.B. 1974. The effect of microorganisms on the absorption of manganese by plants. *N Phytol* 73 (1): 97–106.

Bhattacharya, P., and Jain, R.K. 2000. Phosphorus solubilizing biofertilizers in the whirlpool of rock phosphate-challenges and opportunities. *Fert News* 45: 45–52.

Boukhalfa, H., Brickman, T.J., Armstrong, S.K., and Crumbliss, A.L. 2000. Kinetics and mechanism of iron(III) dissociation from the dihydroxamate siderophores alcaligin and rhodotorulic acid. *Inorg Chem* 39: 5591–5602.

Boukhalfa, H., and Crumbliss, A.L. 2000. Multiple-path dissociation mechanism for mono- and dinuclear tris (hydroxamato)iron(III) complexes with dihydroxamic acid ligands in aqueous solution. *Inorg Chem* 39: 4318–4331.

Boukhalfa, H., and Crumbliss, A.L. 2002. Chemical aspects of siderophore mediated iron transport. *Bio-metals* 15: 325–339.

Brix, H., and Schierup, H.H. 1989. The use of aquatic macrophytes in water-pollution control. *Ambio* 18: 100–107.

Budde, A.D., and Leong, S.A. 1989. Characterisation of siderphores from Ustilago maydis. *Mycopathologia* 108: 125–133.

Buch, G., Archana, G., and Naresh, K.G. 2008. Metabolic channelling of glucose towards gluconate in phosphate solubilizing *Pseudomonas aeruginosa* P4 under phosphorus deficiency. *Res Microbiol* 159: 635–642.

Calheiros, C.S.C., Duque, A.F., Moura, A. et al. 2009. Changes in the bacterial community structure in two stage constructed wetlands with different plants for industrial wastewater treatment. *Bioresour Technol* 100: 3228–3235.

Callahan, D.L., Baker, A.J.M., Kolev, S.D., and Wedd, A.G. 2006. Metal ion ligands in hyperaccumulating plants. *J Biolog Inorg Chem* 11: 2–12.

Chandra, R., Bharagava, R.N., Kapley, A., and Purohit, H.J. 2011. Characterization of *Phragmites cummunis* rhizosphere bacterial communities and metabolic products during the two stage sequential treatment of post methanated distillery effluent by bacteria and wetland plants. *Biores Technol.* doi:10.1016/j.biortech.09.132.

Chandra, R., and Yadav, S. 2010. Potential of *Typha angustifolia* for phytoremediation of heavy metals from aqueous solution of phenol and melanoidins. *Ecol Eng* 36: 1277–1284.

Chaturvedi, S., and Chandra, R. 2006. Isolation and characterization of *Phragmites australis* (L) rhizosphere bacteria from contaminated site for bioremediation or coloured distillery effluent. *Ecol Eng* 27: 202–207.

Chen, J., Xie, P., Li, L., and Xu, J. 2009. First identification of the hepatotoxic microcystins in the serum of a chronically exposed human population together with indication of hepatocellular damage. *Toxi Sci* 108 (1): 81–89.

Chen, T.Y., Kao, C.M., Yeh, T.Y., Chien, H.Y., and Chao, A.C. 2006a. Application of a constructed wetland for industrial wastewater treatment: A pilot scale study. *Chemosphere* 64: 497–502.

Chen, Y.P., Rekha, P.D., Arun, A.B., Shen, F.T., Lai, W.A., and Young, C.C. 2006b. Phosphate solubilizing bacteria from subtropical soil and their tricalcium phosphate solubilizing abilities. *Appl Soil Ecol* 34: 33–41.

Choi, O.K., and Hu, Z.Q. 2008. Size dependent and reactive oxygen species related nanosilver toxicity to nitrifying bacteria. *Env Sci Tech* 42 (12): 4583–4588.

Cicerone, R.J., and Oremland, R.S. 1988. Biogeochemical aspects of atmospheric methane. *Global Biogeochem Cycles* 2: 299–337.

Cobbett, C.S. 2000. Phytochelatin biosynthesis and function in heavy-metal detoxification. *Curr Opin Plant Biol* 3: 211–216.

Cody, Y.S., and Gross, D.C. 1987. Characterisation Pyoverdin$_{pss}$, siderophore produced by *Pseudomonas syringae* pv. *Syringae*. *Appl Environ Microbiol* 53: 928–934.

Conard, R. 1995. Soil microbial process involved in production and consumption of atmospheric traces gases. *Adv Microb Ecol* 14: 207–250.

Cronk, J.K., and Fennessy, M.S. 2001. *Wetland Plants: Biology and Ecology*, Lewis Publishers, Boca Raton, FL.

Crosa, L.M., Wolf, M.K., Actis, L.A., Sanders-Loehr, J., and Crosa, J.H. 1988. New aerobactin mediated iron uptake system in septicemia causing strain of *Enterobacter cloacae*. *J Bacteriol* 170: 5539–5544.

Crowley, D.E., Wang, Y.C., Reid, C.P.P., and Szansiszlo, P.J. 1991. Mechanism of iron acquisition from siderophores by microorganisms and plants. *Plant Soil* 130 (1–2): 179–198.

Dakora, F.D., and Phillips, D.A. 2002. Root exudates as mediators of mineral acquisition in low-nutrient environments. *Plant Soil* 245: 35–47.

Das, A.C., Chakravarty, A., Sukul, P., and Mukherjee, D. 2003. Effect of HCH and fenvalerate on growth and distribution of microorganisms in relation to persistence of the insecticides in the rhizosphere soils of wetland rice. *Bull Environ Contam Toxicol* 70: 1059–1069.

Davies, D.D. 1980. Anaerobic metabolism and production of organic acids. In *The Biochemistry of Plants: A Comprehensive Treatise*, Vol. 2, Stumpf, P.K. and Conn, E.E. (Eds.), Academic Press: New York, pp. 581–611.

DeBusk, W.F. 1999. *Wastewater Treatment Wetlands: Applications and Treatment Efficiency*. A fact sheet of the Soil and Water Science Department, Florida Cooperative Extension Service, Institute of Food and Agricultural Sciences, University of Florida.

Debusk, W.F., and Reddy, K.R. 1987. Wastewater treatment using floating aquatic macrophytes: Contaminant removal processes and management strategies. In *Aquatic Plants for Water Treatment and Resource Recovery*, Reddy, W.K. and Smith, W.H. (Eds.), Magnolia Publications, Altamonte Springs, FL.

Dell'Amico, E., Cavalca, L., and Andreoni, V. 2005. Analysis of rhizobacterial communities in perennial Graminaceae from polluted water meadow soil, and screening of metalresistant, potentially plant growth-promoting bacteria. *FEMS Microbiol Ecol* 52: 153–162.

Dhungana, S., and Crumbliss, A.L. 2005. Coordination chemistry and redox processes in siderophore-mediated iron transport. *Geomicrobiology Journal* 22: 87–98.

Diekmann, H. 1967. Metabolic products of microorganism. 56. Fusigen: A new sideramine from fungi. *Arch Mikrobiol* 58: 1–5.

Diekmann, H. 1970. Metabolic products of microorganism. 81. Occurrence and structures of coprogen B and dimerum acid. *Arch Mikrobiol* 73: 65–67.

Duff, R.B., and Webley, D.M. 1959. 2-Ketogluconic acid as a natural chelator produced by soil bacteria. *Chem Ind* 1376–1377.

Eiler, A. 2006. Evidence for the ubiquity of mixotrophs bacteria in the upper ocean: Implications and consequences. *Appl and Enviro Microbiol* 72 (2): 7431–7437.

Feray, C., and Montuelle, B. 2003. Chemical and microbial hypotheses explaining the effect of wastewater treatment plant discharges on the nitrifying communities in freshwater sediment. *Chemosphere* 50 (7): 919–928.

Fontecave, M., Coves, J., and Pierre, J.L. 1994. Ferric reductases or flavin reductases? *Biometals* 7: 3–8.

Gavrilescu, M. 2005. Fate of pesticides in the environment and its bioremediation. *Engin Life Sciences* 5: 497–526.

Glick, B.R. 2003. Phytoremediation: Synergistic use of plants and bacteria to clean up the environment. *Biotech Advanc* 21: 383–393.

Groudeva, V.I., Groudev, S.N., and Doycheva, A.S. 2001. Bioremediation of waters contaminated with crude oil and toxic heavy metals. *Int J Miner Process* 62: 293–299.

Gutknecht, L.M.J., Goodman, R.M., and Balser, T.C. 2006. Linking soil process and microbial ecology in freshwater wetland ecosystems. *Plant Soil* 289: 17–34.

Halder, A.K., and Chakrabartty, P.K. 1993. Solubilization of inorganic phosphate by *Rhizobium*. *Folia Microbiol* 38: 325–330.

Halder, A.K., Mishra, A.K., Bhattacharyya, P., and Chakrabartty, P.K. 1990. Solubilization of rock phosphate by *Rhizobium* and *Bradyrhizobium*. *J Gen Appl Microbiol* 36: 81–92.

Hammond, J.P., Bowen, H.C., White, P.J., Mills, V., Pyke, K.A., Baker, A.J, Whiting, S.N., May, S.T., and Broadley, M.R. 2006. A comparison of the Thlaspi caerulescens and Thlaspi arvense shoot transcriptomes. *New Phytologist* 170: 239–260.

Haydon, M.J., and Cobbett, C.S. 2007. Transporters of ligands for essential metal ions in plants. *New Phytologist* 174: 499–506.

Herrmann, M., Saunders, A.M., and Schramm, A. 2008. Archaea dominate the ammonia-oxidizing community in the rhizosphere of the freshwater macrophyte *Littorella uniflora*. *Appl Environ Microbiol* 74 (10): 3279–3283.

Ibekwe, A.M., Grieve, C.M., and Lyon, S.R. 2003. Characterization of microbial communities and composition in constructed dairy wetland wastewater effluent. *Appl Environ Microbiol* 69: 5060–5069.

Idris, R., Trivonova, R., Puschenreiter, M., Wenzel, W.W., and Sessitsch, A. 2004. Bacterial communities associated with flowering plants of the Ni-hyperaccumulator *Thlaspi goesingense*. *Appl Environ Microbiol* 70: 2667–2677.

Illmer, P., and Schinner, F. 1992. Solubilization of inorganic phosphates by microorganisms isolated from forest soil. *Soil Biol Biochem* 24: 389–395.

Ishimaru, C.A., and Loper, J.E. 1988. Aerobactin production by a strain of *Erwinia carotovora* subsp. Carotovora. (Abstr) *Phytopathology* 78: 1577–1582.

Jalal, M.A.F., Love, S.K., and van der Helm, D. 1986. Diserophore mediated iron (III) uptake in Gliocladium virens. I. Properties of cis fusarinine, trans-fusarinine, dimerum acid and their ferric complexes. *J Inorg Biochem* 28: 417–430.

Jalal, M.A.F., Love, S.K., and van der Helm, D. 1988. N-Dimethylcoprogens. Three novel trihydroxamate siderophores from pathogenic fungi. *Biol Met* 1: 4–8.

Jansen, S., Broadley, M.R. et al. 2002. Aluminium hyperaccumulation in angiosperms: A review of its phylogenetic significance. *Botanical Review* 68: 235–269.

Jimenez, M.B., Flores, S.A., Zapata, E.V., Campos, E.P., Bouquel, S., and Zenteno, E. 2003. Chemical characterization of root exudates from rice (*Oryza sativa*) and their effects on the chemotactic response of endophytic bacteria. *Plant Soil* 249: 271–277.

Kadlec, R.H., and Knight, R.L. 1996. *Treatment Wetlands*. Lewis Publishers, CRC, New York.

Kadlec, R.H., Knight, R.L., Vyamazal, J., Brix, H., Cooper, P., and Haberl, R. 2000. *Constructed Wetlands for Pollution Control-Processes, Performance Design and Operation*. IWA Scientific and Technical Report No. 8. IWA, Publishing London.

Kim, D.Y., Bovet, L., Kushnir, S., Noh, E.W., Martinoia, E., and Lee, Y. 2006. At ATM3 is involved in heavy metal resistance in arabidopsis. *Plant Physiology* 140: 922–932.

Kluber, H.D., Lechner, S., and Conard R, 1995. Characterization of populations of aerobic hydrogen-oxidizing soil bacteria. *FEMS Microbiol Ecol* 16: 167.

Konetschny-Rapp, S., Jung, G., Huschka, H.G., and Winkelmann, G. 1988. Isolation and identification of the principal siderophores of the plant pathogenic fungus *Botrytis cinerea*. *Biol Met* 1: 90–98.

Kouki, S., M'Hiri, F., Saidi, N., Belaid, S., and Hassen, A. 2009. Performance of a constructed wetland treating domestic wastewaters during a macrophytes life cycle. *Desalination* 248: 131–146.

Lee, C.G., Fletcher, T.D., and Sun, G.Z. 2009. Nitrogen removal in constructed wetland systems. *Engineering Life Science* 9: 11–22.

Letendre, E.D., and Gibbons, W.A. 1985. Isolation and purification of canadaphore, a siderophore produced by *Helminthosporium arbonum*. *Biochem Biophys Res Commun* 129: 262–267.

Liu, A., Hamel, C., Hamilton, R.I., Ma, B.L., and Smith, D.L. 2000. Acquisition of Cu, Zn, Mn and Fe by mycorrhizal maize (*Zea mays* L.) grown in soil at different P and micronutrient levels. *Mycorrhiza* 9 (6): 331–336.

Liu, T.S., Lee, L.Y., Tai, C.Y., Hung, C.H., Chang, Y.S., Wolfram, J.H., Rogers, R., and Goldstein, A.H. 1992. Cloning of an Erwinia herbicola gene necessary for gluconic acid production and enhanced mineral phosphate solubilization in *Escherichia coli* HB101: Nucleotide sequence and probable involvement in biosynthesis of the co-enzyme pyrroloquinoline quinone. *J Bacteriol* 174: 14–21.

Ma, L.Q., Komar, K.M. et al. 2001. A fern that hyperaccumulates arsenic. *Nature* 209, 579.

Marschner, P. 2009. The role of rhizosphere microorganisms in relation to P uptake by plants. In *The Ecophysiology of Plant-Phosphorus Interactions*, White, P.J. and Hammond, J.P. (Eds.), Plant Ecophysiology Series, Springer, Heidelberg.

Marschner, P., and Crowley, D.E. 1998. Phytosiderophore decrease iron stress and pyoverdine production of *Pseudomonas fluorescens* Pf-5 (pvd-inaZ). *Soil Biol Biochem* 30: 1275–1280.

Matzanke, B.F., Anemuller, S., Schunemann, V., Trautwein, A.X., and Hantke, K. 2004. FhuF, part of a siderophore-reductase system. *Biochem* 43: 1386–1392.

McManmon, M., and Crawford, R.M.M. 1971. A metabolic theory of flooding tolerance: The significance of enzyme distribution and behavior. *New Phytologist* 70: 299–306.

Menegus, F., Cattaruzza, L., Chersi, A., and Fronza, G. 1989. Differences in the anaerobic lactate-succinate production and in the changes of cell sap pH for plants with high and low resistance to anoxia. *Plant Physiology* 90: 29–32.

Meyer, J.M. 2000. Pyoverdines: Pigments siderophores and potential taxonomic markers of fluorescent *Pseudomonas* species. *Arch Microbiol* 174 (3): 135–142.

Meyer, J.M., Hohnadel, D., and Halle, F. 1989. Cepabactin from *Pseudomonas cepacia*, a new type of siderphore. *J Gen Microbiol* 135: 1479–1487.

Mitsch, W.J., and Gosselink, J.G. 2000. *Wetlands*, 3rd ed. John Wiley, New York.

Monchy, S., Benotmane, M.A., Janssen, P., Vallaeys, T., Taghavi, S., van der Lelie, D., and Mergeay, M. 2007. Plasmids pMOL28 and pMOL30 of *Cupriavidus metallidurans* are specialized in the maximal viable response to heavy metals. *J Bacteriol* 189: 7417–7425.

Montargès-Pelletier, E., Chardot, V., Echevarria, G., Michot, L.J., Bauer, A., and Morel, J.L. 2008. Identification of nickel chelators in three hyperaccumulating plants: An X-ray spectroscopic study. *Phytochemistry* 69: 1695–1709.

Morgan, J.A.W., Bending, G.D., and White, P.J. 2005. Biological costs and benefits to plant–microbe interactions in the rhizosphere. *J Exp Bot* 56: 1729–1739.

Ong, S.A., Peterson, T., and Neilands, J.B. 1979. Agrobactin, a siderophore from *Agrobacterium tumefaciens*. *J Biol Chem* 254: 1860–1865.

Park, H.D., and Noguera, D.R. 2004. Evaluating the effect of dissolved oxygen on ammonia-oxidizing bacterial communities in activated sludge. *Water Res* 38: 3275–3286.

Persans, M.W., Yan, X., Patnoe, J.M., Krämer, U., and Salt, D.E. 1999. Molecular dissection of the role of histidine in nickel hyperaccumulation in Thlaspi goesingense (Hálácsy). *Plant Physiology* 121: 1117–1126.

Persmark, M., Expert, D., and Neiland, J.B. 1989. Isolation, characterization and synthesis of crysobactin, a compound with siderophore activity from *Erwinia chrysanthemi*. *J Biol Chem* 264: 3187–3193.

Pollack, J.R., and Neilands, J.B. 1970. Enterobactin, an iron transport compound from salmonella thyphimurium. *Biochem Biophys Res Comm* 5: 989–992.

Ramamoorthy, S., Sass, H., Langner, H., Schumann, P., Kroppenstedt, R.M., Spring, S., Overmann, J., and Rosenzweig, R.F. 2006. *Desulfosporosinus lacus* sp nov., a sulfate-reducing bacterium isolated from pristine freshwater lake sediments. *Int J Syst Evol Microbiol* 56: 2729–2736.

Reeves, R.D., and Baker, A.J.M. 2000. Metal-accumulating plants. In *Phytoremediation of Toxic Metals: Using Plants to Clean Up the Environment*, Raskin, I. and Ensley, B.D. (eds.), John Wiley and Sons: New York, pp. 193–229.

Reddy, K.R., Wright, A., Ogram, A., DeBusk, W.F., and Newman, S. 2002. Microbial processes regulating carbon cycling in subtropical wetlands. Symposium no. 11, Paper 982: 1–12. The 17th World Congress of Soil Science, Thailand.

Roberts, J.K.M. 1988. Cytoplasmic acidosis and flooding tolerance in crop plants. In *The Ecology and Management of Wetlands, Ecology of Wetlands*, Vol. 1, Hook, D.D. et al. (Eds.), Timber Press: Portland, Oregon, pp. 393–397.

Rodríguez, N., Menéndez, N. et al. 2005. Internal iron biomineralization in Imperata cylindrica, a perennial grass: Chemical composition, speciation and plant localization. *New Phytologist* 165: 781–789.

Saglio, P.H., Raymond, P., and Pradet, A. 1980. Metabolic activity and energy charge of excised maize root tips under anoxia. *Plant Physiology* 66: 1053–1057.

Salt, D.E., Blaylock, M., Kumar, N.P.B.A., Dushenkov, V., Ensley, B.D., Chet, I., and Raskin, I. 1995. Phytoremediation: A novel strategy for the removal of toxic metals from the environment using plants. *Biol Technol* 13 (5): 468–474.

Sather, J.H., and Smith, R.D. 1984. An overview of major wetland functions and values. Fish and Wildlife Service, U.S. Department of Interior, Washington, DC.

Schwab, A.P., and Banks, M.K. 1994. Biologically mediated dissipation of polyaromatic hydrocarbons in the root zone. In *Bioremediation Through Rhizosphere Technology*, Anderson, T.A. and Coats, J.R. (Eds.), American Chemical Society, Washington, DC.

Schwyn, B., and Neiland, J.B. 1987. Siderophores from agronomically important species of the Rhizobiacae. *Comm Agric Food Chem* 1: 95–114.

Shaul, O., Hilgemann, D.W., de-Almeida-Engler, J., Van Montagu, M., Inzé, D., and Galili, G. 1999. Cloning and characterization of a novel Mg(2+)/H(+) exchanger. *EMBO J* 18: 3973–3980.

Sirivedhin, T., and Gray, K.A. 2006. Factors affecting denitrification rates in experimental wetlands: Field and laboratory studies. *Ecol Eng* 26: 167–181.

Stephan, U.W., and Scholz, G. 1993. Nicotianamine: Mediator of transport of iron and heavy metals in the phloem? *Physiologia Plantarum* 88: 522–529.

Stottmeister, U., Wiebner, A., Kuschk, P., Kappelmeyer, U., Kastner, M., Bederski, O., Muller, R.A., and Moormann, H. 2003. Effects of plants and microorganisms in constructed wetlands for wastewater treatment. *Biotechnol Advances* 22: 93–117.

Sugiura, Y., Tanaka, H., Mino, Y., Ishida, T., Ota, N., Inoue, M., Nomoto, K., Yoshioka, H., and Takemoto, T. 1981. Structure, properties, and transport mechanism of iron (III) complex of mugieic acid, a possible phytosiderophore. *J Am Chem Soc* 103: 6979–6982.

Summers, J.E., Ratcliffe, R.G., and Jackson, M.B. 2000. Anoxia tolerance in the aquatic monocot Potamogeton pectinatus: Absence of oxygen stimulates elongation in association with an usually large Pasteur effect. *J Exp Bot* 51: 1413–1422.

Sundara Rao, W.V.B., and Sinha, M.K. 1963. Phosphate dissolving micro-organisms in the soil and rhizosphere. *Indian J Agric Sci* 33: 272–278.

Szekeres, S., Kiss, I., Kalman, M., and Soares, M. 2002. Microbial population in a hydrogen-dependent denitrification reactor. *Water Res* 36: 4088–4094.

Teintze, M., Hossain, M.B., Barnes, C.L., Leong, J., and van der Helm, D. 1981. Structure of ferric pseudobactin, a siderophore from a plant growth promoting *Pseudomonas*. *Biochemistry* 20: 6446–6457.

Touati, D. 2000. Iron and oxidative stress in bacteria. *Arch Biochem Biophys* 373 (2000): 1–6.

Vacca, G., Wand, H., Nikolausz, M., Kuschk, P., and Kastner, M. 2005. Effect of plants and filter materials on bacteria removal in pilot-scale constructed wetlands. *Water Res* 39: 1361–1373.

Van Hees, P.A.W., Jones, D.L., and Godbold, D.L. 2002. Biodegradation of low molecular weight organic acids in coniferous forest podzolic soils. *Soil Biol Biochem* 34: 1261–1272.

Vymazal, J., and Kropfelova, L. 2005. Growth of Phragmites australis and Phalaris arundinacea in constructed wetlands for wastewater treatment in the Czech Republic. *Ecol Eng* 25: 606–621.

Wagner, M., Loy, A., Nogueira, R., Purkhold, U., Lee, N., and Daims, H. 2002. Microbial community composition and function in wastewater treatment plants. *Antonie van Leeuwenhoek* 81: 665–680.

Wang, X., Wen, X., Criddle, C., Well, G., Zhang, J., and Zhao, Y. 2010. Community analysis of ammonia-oxidizing bacteria in activated sludge either wastewater treatment systems. *J Envir Sci* 22 (4): 627–634.

Wang, Y., Ke, X.B., Wu, L., and Lu, Y.H. 2009. Community composition of ammonia-oxidizing bacteria and archaea in rice field soil as affected by nitrogen fertilization. *Syst Appl Microbio* 32 (1): 27–36.

Wind, T., Stubner, S., and Conrad, R. 1999. Sulfate-reducing bacteria in rice field soil and on rice roots. *Syst Appl Microbiol* 22: 269–279.

Yang, X.E., and Yang, M.J. 2001. Some mechanisms of zinc and cadmium detoxification in a zinc and cadmium hyperaccumulating plant species (Thlaspi). In *Plant Nutrition-Food Security and Sustainability of Agro-Ecosystems*, Orst, W. et al. (Eds.), Kluwer Academic Publishers: Dordrecht, The Netherlands, pp. 444–445.

You, J., Das, A., Dolan, E.M., and Hu, Z.Q. 2009. Ammonia oxidizing Archeae involved in nitrogen removal. *Water Res* 43 (7): 1801–1809.

Zhuang, X., Chen, J., Shim, H., and Bai, Z. 2007. New advance in plant growth promoting rhizobacteria for bioremediation. *Environ Int* 33: 406–413.

13

Bioremediation of Heavy Metals Using Biosurfactants

Mohamed Yahya Khan, T.H. Swapna, Bee Hameeda and Gopal Reddy

CONTENTS

13.1 Introduction

Pollution is referred to as the presence of undesirable substances, such as inorganic (heavy metals) and organic pollutants (hydrophobic organic compounds). Heavy metals even at picomolar concentrations are extremely toxic in nature, and they are nondegradable, unlike other organic pollutants. Sources of heavy metal pollution can be (i) natural, such as geothermal activities, comets, space dust and volcanic activities, or (ii) anthropogenic sources, such as rapid industrialisation, extensive use of hydrocarbons, chlorinated hydrocarbons, chemical pesticides, etc. and (iii) e-waste. Due to the aforesaid activities, toxic metals, such as lead (Pb), chromium (Cr), nickel (Ni), cadmium (Cd), copper (Cu), zinc (Zn), arsenic (As) and mercury (Hg), are dumped into the environment. Because of their nonbiodegradable nature, they are a persistent threat to our life and environment (Singh and Cameotra 2004). In addition, these metals can enter into the food chain and can accumulate in the human body leading to birth defects, mental and physical retardation, cancer, etc. (Singh et al. 2010).

Nonbiological methods, such as ion exchange, chemical precipitation and leaching, reverse osmosis, soil replacement, thermal desorption, electrokinetic remediation and landfilling, are used for remediation of heavy metals present in water and soil (Yao et al. 2012). However, these methods need more sophisticated infrastructure and are expensive and also generate toxic sludge that affects the environment. In this scenario, bioremediation is one such eco-friendly and sustainable process to degrade or transform contaminants so that they are no longer harmful. Unlike that of hydrophobic organic pollutants, biodegradation of heavy metals into harmless CO_2 and water is not feasible. This means that the usage of microorganisms can only alter the metal contaminants and convert them into a nontoxic form and can become easily disposable. The new promising remedial strategy for treating soils contaminated with pollutants is treatment with biosurfactants. The use of biosurfactants (plant and microbial surfactants) has attracted attention as an effective and eco-friendly method for removal of heavy metals. Plant-derived surfactins include saponins, lecithins and humic acids, which are used for bioremediation of heavy metals (Soeder et al. 1996). Microbial surfactants are a diverse group of molecules, such as glycolipids, lipopeptides and lipoproteins; fatty acids; phospholipids; and polymeric structures (Ron and Rosenberg 2001). Biosurfactants act by formation of complexes with metals at the soil interface, desorption from soil particles and then removal from surface leading to their bioavailability in soil solution (Magdalena et al. 2011; Hogan et al. 2014). Over the course of the last decade, a major shift in the approach for biosurfactant-mediated bioremediation has been driven by the sustainability agenda due to its low toxic nature, better environmental compatibility and biodegradability, large-scale production using low-cost substrates and effectiveness under extreme conditions (Singh et al. 2010).

There are various strategies used by microorganisms for metal uptake; however, in this chapter, we focus on one of the novel strategies, that is, the use of biosurfactants in bioremediation of heavy metals with emphasis on the most recent results of applying microbial surfactants and their usage in enhancing heavy metal uptake and phytoremediation. Further application of these biosurfactant–metal complexes in fields of nanotechnology is discussed.

13.2 Sources of Heavy Metal Contamination

13.2.1 Natural Activities

Heavy metals have been introduced into the biosphere for thousands of years, ever since their importance and useful properties have been recognised. Heavy metals, such as Pb, As, Cd and Hg, are omnipresent in nature and cause an unfavourable result for the environment, particularly at higher concentrations. In the case of natural sources, such as the volcanic and metamorphic origin of rocks, it was observed that nickel, chromium and manganese are present in trace amounts (Nicholson et al. 2003). Weathering of sedimentary rocks containing adsorbed elements on them, forest fires, etc., add to the high load of heavy metals in the soil. Geochemical cycles and heavy metal biochemical equivalence are normal components of the Earth's crust; however, their concentration is exasperated with the advent of the industrial revolution and further resulted in a manifold rise in the usage of these metals.

13.2.2 Anthropogenic Activities

Anthropogenic sources include industrial activities such as combustion processes, continuous mining process and transportation processes that add hazardous solid wastes to the environment. Metal ions, such as Cr^{6+}, Pb^{3+} and Ni^{2+}, are added into the environment through the manufacture of chromium salts, leather tanning, industrial coolants, etc. Hexavalent chromium concentrations were found in groundwater wells near industrial areas, such as textile industries. According to standards, the specifications of drinking water quality permissible limits of chromium is ~0.05 mg/L; however, analysis of groundwater near textile industries showed that the concentration of this toxic metal ion was found to be higher than the permissible limits (Brindha et al. 2010). Arsenic (As) is found to contaminate the environment through smelting operations, thermal power plants, fuel burning, etc. Lead (Pb) contribution is found to be from lead acid batteries, paints, smelting, coal-based thermal power plants and ceramic industries. Mercury (Hg) was found from thermal power plants, hospital waste and electrical appliances (Rashmi and Pratima 2013). Nickel (Ni) from battery industries, smelting operations and thermal power plants; copper (Cu) from electroplating, smelting operations, mining, etc.; vanadium (Va) from used catalysts and sulphuric acid plants; and Zinc (Zn) from smelting and electroplating industries are being dumped into the environment (Raymond et al. 2011).

13.2.3 E-Waste Resources

Technological modernisation, rapid development of the economy and business marketing strategies resulted in the development of electronic products that are cheaper, better and more readily available than older versions due to which much electronic equipment becomes redundant more quickly than ever and is consigned to waste. Hence, a lot of e-waste with heavy metals is accumulated, and this has been a hot issue of common concern all over the world. E-waste includes metals such as Cd, Cr, Ni, Zn, Cu and Pb and is the fastest growing waste category. According to UN Environment Programme estimates, about 20–50 million tonnes of electric and electronic scrap is generated annually (Sharma Pramila et al. 2012; Karwowska et al. 2014).

13.3 Environmental and Health Hazards due to Heavy Metal Toxicity

Heavy metals entering into the environment through various sources, such as those mentioned above, are leading to deleterious effects to the biological systems in the environment. Some of these metal ions, in lower concentrations, are found to be beneficial to plants and microbes as mineral nutrients, but if the concentration exceeds the permissible limits, it leads to toxicity. According to the Agency for Toxic Substances and Disease Registry (ATSDR), lead, cadmium, arsenic and mercury are well known for showing toxicity upon exposure. Accumulation of these heavy metals generates oxidative stress in the body, causing fatal effects to important biological processes, leading to cell death (Singh et al. 2011).

The nonbiodegradable nature of heavy metals results in their prolonged persistence in the environment. Arsenic is broadly used to make insecticides, fungicides, weed killer and antifouling agents, and prolonged exposure to this metal results in arsenicosis, pigmentation and keratosis. A methylation reaction takes place in which As is converted to monomethylarsenic, dimethyl arsenic or trivalent forms. These trivalent forms tend to bind to sulphydryl groups, affecting important enzymatic reactions. It also affects metabolic reactions, such as the Krebs cycle and oxidative phosphorylation

TABLE 13.1

Health Hazards Due to the Exposure of Heavy Metals

Metal	Health Hazard	References
Arsenic	Nausea; vomiting; diarrhoea; rice water stools; haematopoetic effects, such as haemolysis, lymphopenia, and haematuria; affects liver by causing jaundice, central necrosis, etc.	Halttunen et al. (2007)
Chromium	Asthma; severely affects liver, kidney, gastrointestinal, and immune systems, carcinogenic agent may result in death	Zhitkovic (2011)
Cadmium	Affects kidney, liver, lungs; also causes cancer	Satarug et al. (2010)
Lead	Causes genotoxicity, hypertension, impaired hearing, neuropathy, fatigue, headache, abdominal pain, lethargy, encephalopathy, etc.	Bellinger et al. (1987)
Mercury	Causes acute bronchitis and death; damages central nervous system; causes tremors, erythrism, psychomotor retardation; leads to blindness, deafness and death	
Nickel	Skin, nose and throat irritation; causes shortness of breath, scarring of lungs; also affects kidneys	Geier and Geier (2007)

inhibiting ATP production. Lead toxicity depends on absorption levels in the gastrointestinal tract, lungs and skin and physiochemical conditions in which it is present. Lead has the ability to penetrate through blood vessels, brain and placenta, which leads to genotoxicity. In addition, other consequences faced are hypertension, impaired hearing, neuropathy, fatigue, headache, abdominal pain, lethargy, encephalopathy, etc. Mercury spreads throughout the body in vapour form; on inhalation, it tends to enter the bronchi and cause acute bronchitis and death. Exposure for a longer period of time can damage the central nervous system. In infants, it may lead to psychomotor retardation and, as time passes, may lead to blindness, deafness and finally death. Consumption of fish containing mercury in higher concentrations leads to minamata disease, which was reported in a few developing countries. Cadmium tends to affect kidney, liver, lungs and the skeletal system and leads to itai-itai disease. It competes with other metalloproteins to bind in their position and leads to destructive effects and is found to be one of the causes of cancer (Majumder 2013; Gaur et al. 2014). The effect of heavy metals and the related health hazards they cause are explained in Table 13.1.

13.4 Remediation Strategies of Heavy Metals

Both nonbiological (physical separation and chemical extraction) and biological methods (plants and microorganisms) can be used for remediation

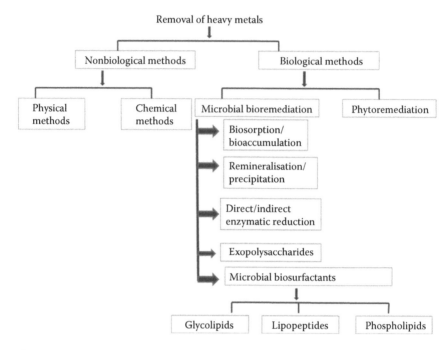

FIGURE 13.1
Strategies for remediation of toxic heavy metals in the environment.

of heavy metals. The flowchart of remediation of heavy metals is given in Figure 13.1, and the mechanisms involved are given in the following.

13.4.1 Nonbiological Methods

Increasing environmental pollution has presented a challenge in the search for technologies in the cleaning up of inorganic and organic pollution. There are various methods employed in their remediation, of which the foremost methods used are physical and chemical methods.

13.4.1.1 Physical Methods

Physical methods include soil replacement, soil spading, thermal desorption, etc. In the soil replacement process, contaminated soils are replaced with soil without contamination. In soil spading methods, contaminated soil is distributed for effective remediation with the help of flora around it (Aresta et al. 2008). Thermal desorption is another method in which metals, such as mercury and arsenic, are made volatile by using steam and microwaves, and these volatile heavy metals are recovered using a vacuum or carrier gas. Thermal desorptions are carried out at high temperatures such as 320°C–560°C. However, limitations such as expensive devices, long desorption time and non-eco-friendly manner restrict the use of these strategies, and alternatives such as biological methods are recommended.

13.4.1.2 Chemical Methods

Chemical remediation processes include chemical leaching, chemical fixation, electrokinetics, etc. Chemical leaching is a process in which reagents, fluids and gases are used to wash the contaminated soil samples and precipitation, chelation and adsorption of the pollutants using extractant solvents, such as phosphoric acid, sulphuric acid, etc. Ethylenediaminetetraacetic acid (EDTA) is observed to be an excellent extractant of heavy metals from sediments. Chemical fixation is another process in which certain reagents are added that make contaminated soil intact, thus not allowing for mobility and, in turn, restricting the spread. Electrokinetic technology is the latest technology applied in which voltage is applied to both sides of the contaminated soil, and metal ions get accumulated; an electric gradient is created, which is further processed and treated. However, these technologies are found to be complicated, have high costs and are limited to certain metals; in turn, these methods release by-products that are again contributing towards environmental pollution (Tampouris et al. 2001). Also, the procedures used are complicated and need a lot of energy, which makes it expensive and limited in application.

13.4.2 Biological Methods

Bioremediation technologies are most extensively used in comparison with physical and chemical methods due to their cost-effective and eco-friendly nature. These techniques can be used for treating various inorganic and organic pollutants. Biological methods exploit natural entities, such as plants and microorganisms, through a variety of mechanisms, such as oxidation–reduction, bioleaching, biomineralisation, biosorption, enzymatic activity and metabololites.

13.4.2.1 Phytoremediation

Phytoremediation is the process that involves the application of selected plants to degrade or assimilate, metabolise or detoxify undesirable substances, such as heavy metals, hydrocarbons, etc., from soil and water to improve their quality. Wild-type and engineered plants can be used for this process, which includes *Clerodendrum infortumatum*, *Crotan bonplandianus*, *Pistia stratiotes* and *Thlaspi caerluescens*. Phytoextraction uses the rhizosphere or root system for removal of heavy metals; however, the microbes present in the rhizosphere also play a significant role in phytoremediation. The phytovolatilisation process is another method in which *Arabidopsis thaliana* is genetically modified and has the ability to release certain root exudates, which made mercury a soil pollutant to get volatile through the transpiration process. However, the disadvantage is that the heavy metals are toxic to some extent even after volatilisation. Another important process through which bioremediation of toxic metal ions was observed is the use of a phyto-surfactant, such as saponins and humic acids, that are released from decayed roots. Saponins are nonionic biosurfactants and are used in the removal of metals from sludge samples. Pure saponins could remove heavy metals, such as Zn, Ni, Cu, Cr, Pb and Fe, from the sludge sample, and it was observed that Cr removal was improved by 24.2% by this method of repeated soil washing when compared to the crude extract (Gao et al. 2013). Phosphatidylcholine is an important group of phospholipid biosurfactants and is involved in the bioremediation of Cd and Ni (Soeder et al. 1996). The phytoremediation process is economical and cost-effective, and in addition prevents soil erosion and metal leaching.

13.4.2.2 Microorganisms for Bioremediation

Bioleaching is a process in which metal ions are removed or extracted from low-grade ores and metal concentrates using chemolithotrophic bacteria. The mechanism involved with this type of bacteria in the extraction of metals is by releasing, chelating or complexing compounds or certain organic acids. It was observed that sulphate-reducing bacteria remove metals from solutions by production of hydrogen sulphide, which precipitates metals from bioleaching solutions as sulphides (Cao et al. 2009). Uranium pollution can be remediated by the use of *Acidothiobacillus ferrooxidans* by bioleaching from uranium (Baranska and Sadowski 2013).

13.4.2.2.1 Biosorption and Bioaccumulation

Biosorption is the ability of bacteria, yeast, fungi and algae to be used for the uptake of heavy metals (Volesky 1987). Uptake of heavy metals can be metabolically mediated or can use a nonmetabolic-mediated process, such as physicochemical interactions. Apart from cells, nonliving biomass, natural residues and other biomaterials (chitin, cellulose-based materials) can also be used. Biosorption is favoured by variation in pH, and it leads to precipitate formation between negatively charged cell walls and cationic metal ions (Chen and Wang 2008). Bioaccumulation could be intracellular or extracellular where bacteria are involved in accumulation of metal ions through various ion-exchange processes.

13.4.2.2.2 Remineralisation and Precipitation of Metals

Cells utilise some metal ions as nutrients during their metabolic processes, and remineralisation takes place by complexing with metabolic by-products formed. These may even result in precipitation, which is, in turn, a metabolic-dependent or metabolic-independent process. Metabolic-dependent precipitation occurs in the presence of toxic metal ions, which generate defence mechanism compounds and precipitates metal ions. In metabolic-independent processes, precipitation occurs between cell wall components and compatible metal ions present in the vicinity (Ercole et al. 1994).

13.4.2.2.3 Direct and Indirect Enzymatic Reduction

In the direct enzymatic reduction process, metal-reducing microbes reduce the toxicity of metal ions by utilising oxidised forms of metals as electron acceptors. Chromium(VI) and uranium(VI) are reduced to nontoxic forms by certain groups of bacteria. In an indirect enzymatic reduction process, iron and sulphur-reducing bacteria form multiple complexes with pollutants, such as Cr(VI), and form insoluble precipitates, thus forming a chemically reduced reactive barrier and can reduce the contamination (Anderson et al. 2003).

13.4.2.2.4 Exo-Polysaccharides (EPS)

Microbial components that are excreted or derived from microbial biomass play an imperative role in the interaction of metal ions and the microbial cells. Exo-polysaccharides produced by bacteria is an emulsifier and is involved in metal chelation (Gutnick and Shabtai 1987). Production of EPS could be one of the methods for the biofilm-forming bacteria to grow in a contaminated environment. *Pseudomonas* sp. CU-1 isolated from an electroplating factory produced capsular EPS and showed a high Cu^{2+} adsorption ability followed by Cd^{2+}, Zn^{2+} and Ni^{2+}, which was determined using the dye displacement methods (Gooday 1982; Lau et al. 2005). EPS from cyanobacteria showed excellent chelating properties by binding onto positively charged metal ions (Roberto et al. 2011).

13.4.2.2.5 Microbial Biosurfactants

Microbial biosurfactants are amphiphilic compounds with both hydrophobic and hydrophilic portions, and these molecules decrease the surface tension of water from 72 to 27 mN/m, thus favouring bioavailability of pollutants. These microbial surfactants have many advantages over synthetically produced surfactants, such as having lower toxicity and being less harmful to the environment. Biosurfactants that play an important role in the bioremediation of inorganic and organic pollutants are rhamnolipids, trehalose lipids, sophorolipids, liposan, surfactin, etc. Structures of some of the common biosurfactants are given in Figure 13.2. These biosurfactants have the properties of emulsification, solubilisation and complex formation with metal ions during the bioremediation process, which is explained in Section 13.5.

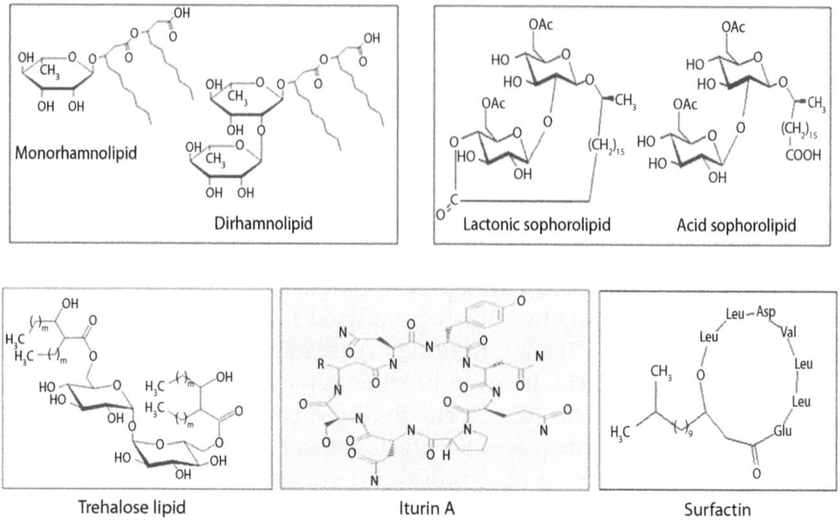

FIGURE 13.2
Chemical structure of some common biosurfactants.

13.5 Biosurfactants

13.5.1 Biosurfactant-Mediated Metal Removal

Biosurfactants can modify the surface of various metals and aggregate at interphases and favour metal separation from contaminated soils. Glycolipids (rhamnolipids, sophorolipids, trehalose lipids) and lipopeptides (surfactins) are low molecular weight surfactants and can remove metals from contaminated samples. Both cationic and anionic surfactants can be used to extract metals from aqueous solution by counterion binding without involvement of a bioleaching mechanism. Critical micelle concentration (CMC) is defined as the concentration at which maximum micelle formation takes place. Biosurfactant-mediated remediation of heavy metals was evident at concentrations below the CMC value. However, in certain studies (Mulligan and Wang 2006), bioremediation of metals below CMC was less efficient. Higher concentrations of biosurfactant can remove metals in greater amounts and vice versa (Diaz et al. 2014). Biosurfactant-mediated metal removal studies are explained below and also given in Table 13.2.

13.5.2 Rhamnolipids

Rhamnolipid surfactants are generally produced by *Pseudomonas* sp., and they are found to be amphiphilic, effective emulsifiers and foaming and dispersing agents (Marchant and Banat 2012). Due to their amphiphilic nature,

TABLE 13.2

Biosurfactant-Mediated Application for Bioremediation of Heavy Metals

Type of Biosurfactant	Microorganisms	Heavy Metal Removal	References
Glycolipids			
Rhamnolipids	*Pseudomonas. aeruginosa*	Pb, Cd, Cu, Ni, Cr	Reis et al. (2013)
Trehalose lipids	*Rhodococcus* spp.	Ni, Cr, Cu, Pb	Benedek et al. (2012)
Sophorolipids	*Candida bombicola*	Cu and Zn	Reis et al. (2013)
Phospholipids			
Spiculisporic acid	*Penicillium spiculisporum*	Cu, Ni, Zn	Hong et al. (1998)
Lipopeptide			
Surfactin	*B. subtilis*	Cu and Zn	Anil and Cameotra (2013)
Liposan	*C. lipolytica*	Pb, Cu, Zn	Raquel et al. (2012)

rhamnolipids are able to partition between interfaces, thereby decreasing interfacial tensions of contaminated solutions and hence increasing the bioavailability and mobilisation of hydrophobic substrates (Banat et al. 2010). Due to their anionic nature, rhamnolipids are applied to remove cationic metal ions, such as Zn, Cu, Pb and Cd, from contaminated soil samples (Dahrazma and Mulligan 2007). Unconventional nutrient medium, such as distillery-spent wash, was used for dirhamnolipid production by *Pseudomonas aeruginosa* BS 2, and it was used as a washing agent for removal of metals (Pb, Cu, Cd and Ni) from multimetal-contaminated soil (Juwarkar et al. 2008). It was also shown that foam produced during rhamnolipid production can also be used to wash the soil column for removing metal ions, such as Cd and Ni (Wang and Mulligan 2004). Alternate treatment of the contaminated soils with rhamnolipids from *Pseudomonas aeruginosa* CVCM 411 and bioleaching with *Acidithiobacillus* spp. could remove twice as much Zn and Fe when compared to individual applications. This study demonstrated a model of synergism by making the surfactant a facilitator for separation of metal from contaminated soil (Diaz et al. 2014). Biosurfactant trehalose lipids produced by *Rhodococcus ruber* and *Rhodococcus erythropolis* have potential activity in the degradation of polyaromatic hydrocarbons and tolerance to certain heavy metal ions (Benedek et al. 2012).

13.5.3 Sophorolipids

Sophorolipids are considered as extracellular, nonionic biosurfactants, consisting of a lactonic form and acidic form, which is generally produced by certain groups of yeasts. The lactonic form of sophorolipid is considered to have an effective biosurfactant-reducing property. Sophorolipids can efficiently remove polyaromatic hydrocarbons, aliphatic compounds and oil from contaminated sites that have limited solubility and availability of these pollutants (Kang et al. 2010). Heavy metal removal (thorium and uranium) studies from electroplate sludge using nonionic sophorolipids and saponins were done using batch and column experiments. It is observed that saponins were more efficient than sophorolipids for removal of heavy metals (Gao et al. 2013).

13.5.4 Lipopeptides

Surfactins are produced by *Bacillus* spp. and observed to have potential activity for heavy metal, polyaromatic hydrocarbon and oil removal from contaminated sites. Heavy metal–contaminated river sediments and oil-contaminated sites can be used to extract metal ions by a soil washing technique. It was observed that, after batch washing, 0.1% of surfactin could remove Cu, Zn and Cd metal ions effectively (Mulligan et al. 1999). Surfactin produced by *Bacillus subtilis* (BBK006) combined and saponin produced by soapberry could remove heavy metals from industrial soil samples. Effective removal of Pd, Cu and Zn was observed when surfactin was employed with foam fractionation and soil flushing under varied physiological conditions within 48 h (Prakash et al. 2013). Using *Bacillus* sp. MTCC 5514, reduction of Cr(VI) (toxic) to Cr(III) (nontoxic) can be explained by two different mechanisms: (i) using extracellular chromium reductase enzyme and (ii) by entrapping Cr(III) using surfactin.

In addition to terrestrial bacterial isolates, biosurfactants from marine isolates are also reported. It was observed that a biosurfactant produced by marine bacteria was capable of binding to metal ions at a concentration lower than CMC and was capable of removing the whole metal content at 5 × CMC (Das et al. 2009). Biosurfactant-mediated metal removal was explained using atomic absorption spectroscopy (AAS), Fourier transmission infrared (FTIR) and transmission electron microscopic (TEM) studies. Using AAS, it was observed that metal content in solution was lowered as metal coprecipitated with the biosurfactant. FTIR studies revealed that there was a change in the chemical nature of biosurfactant on interaction with metal, which was indicated by a shift of the peak from higher to lower frequency, indicating an ionic bond between metal ions and biosurfactants. TEM studies using biosurfactants for removal of metals revealed that the cationic metal ions bind to the anionic peptide head of the biosurfactant. Using energy dispersive x-ray spectroscopy (EDS) showed that surfactant forms a complex with the metals due to the ionic interaction between the positively charged metal and anionic biosurfactant (Das et al. 2009).

13.5.5 Liposan

Liposan extracted from yeast *Candida lipolytica* has excellent surfactant activities in treatment of metal-contaminated municipal solid waste. This lipoprotein was observed to reduce metal ions such as Pb, Cu, Cd, Fe and Zn when soil slurry was percolated with a standardised concentration of biosurfactant (Raquel et al. 2012).

13.5.6 Mechanism of Biosurfactant for Heavy Metal Removal

The following are the mechanisms by which biosurfactant-mediated metal removal from contaminated sites is reported and also depicted in Figure 13.3 for a clearer understanding.

1. Complexation of metals in solution by decreasing the solution phase activity and promoting desorption based on Le Chatelier's principle. This desorption favours sorption of free metal ions by biosurfactants, which accumulate at interfaces appearing as precipitates.

2. Biosurfactant-mediated sorption of heavy metals present in the contaminated soil particles by forming a metal–biosurfactant complex. Later on, this metal–biosurfactant complex in water helps in detachment of metal ions from biosurfactants and recovery of metals for further use in nanoparticle synthesis for medical and biotechnological applications (Singh and Cameotra 2004; Majumder 2013; Sachdev and Cameotra 2013).

3. Divalent metal ions, such as Cu^{2+}, Zn^{2+}, Ni^{2+} and Cd^{2+}, treated with the polycarboxylic acid–type biosurfactant can form metal biosurfactant micelles or can precipitate metal ions (Hong et al. 1998).

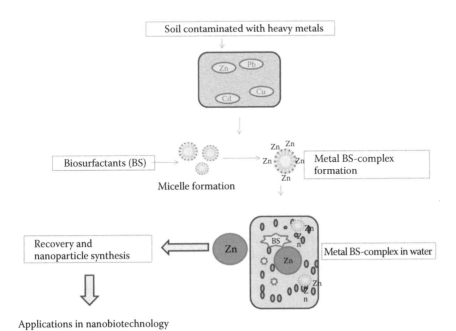

FIGURE 13.3
(See color insert.) Biosurfactant-mediated bioremediation of heavy metals.

13.5.7 Biosurfactant-Enhancing Phytoremediation

Certain chemical additives, such as EDTA, can be used to enhance phytore-mediation, but these chemicals are found to be toxic to flora around the rhi-zosphere (Wu et al. 2011). Therefore, biosurfactants released by rhizosphere microorganisms can be alternatives to chemical amendments. Biosurfactants producing *Bacillus* sp. J119 have enhanced uptake of Cd by plants, such as rape, maize and tomato. Biosurfactants released by microorganisms can interact with metal ions and, later on, desorp metal from soil particles and facilitate metal absorption by plants, enhancing the phytoremediation pro-cess (Sheng et al. 2008).

13.5.8 Biosurfactant Enhancing Cocontaminated Sites

Biosurfactants can enhance metal remediation in cocontaminated sites, such as soils containing Cd and naphthalene (Todd et al. 2000). This organic pol-lutant was utilised as a carbon substrate by the microorganisms to release biosurfactants, which, in turn, lead to remediation of both the pollutants, thus also reducing the remediation cost. It was observed that *Burkholderia* spp. could reduce Cd concentration during biodegradation of naphthalene in the presence of metal-complexing rhamnolipid, which is produced by *Pseudomonas aeruginosa*, which precipitates metal ions under laboratory con-ditions, releasing, in turn, lipopolysaccharides (Sandrin and Maier 2003). It was also observed that repeated use of rhamnolipids could reduce Cd tox-icity during the mineralisation of phenanthrene in a cocontaminated site (Maslin and Maier 2000).

13.6 Use of Metals/Biosurfactants for Nanoparticle Synthesis

Microorganisms are considered to be potential mini biofactories, which can be considered for the synthesis of nanoparticles using heavy metals. *Pseudomonas* sp. has the potential to synthesise selenium and copper nanoparticles, which could have a number of biomedical applications (Varshney et al. 2011). Hazard-ous metals, such as titanium, cadmium, mercury, manganese, selenium and the most precious palladium, have also been obtained in nanoform using common bacteria and fungi (Majumder 2013, and reference articles therein). Zn is the essential mineral element to human and plant health, and ZnO nanoparticles can be synthesised and used for various applications. ZnO are used as antifungal agents and have also been found to have biocompat-ibility to human cells (He et al. 2011). MgO nanoparticles are found to inhibit the spore germination of *Alternaria alternata*, *Fusarium oxysporum*, *Rhizopus sto-lonifer* and *Mucor plumbeus* (Wani and Shah 2012). Uranium-reducing bacteria, such as *Shewanella putrefaciens* and *Shewanella oneidensis*, are employed for UO_2 nanoparticle synthesis for further catalytic use (Baranska and Sadowski 2013).

Biosurfactants (rhamnolipids) from *Pseudomonas aeruginosa* were used to synthesise silver nanoparticles, which exhibited broad-spectrum antimicrobial activity (Ganesh et al. 2010). Sophorolipids are also used for bioleaching of metals and are used in the synthesis of silver nanoparticles as capping and reducing agents. These glyco-nanoparticles are being investigated with great vehemence these days for their widespread applications to understand protein–carbohydrate interactions and in various biomedical applications.

13.7 Conclusions and Future Perspectives

Bioremediation uses different metabolic processes to degrade or transform contaminants, so that they remain no longer in harmful form. Bioremediation has developed from the laboratory to a fully commercialised technology over the last 30 years in many industrialised countries. However, the rate and the extent of development have varied from country to country. The use of microorganisms for bioremediation is a robust strategy; however, the survival of microorganisms in the stressed environment of pollutants is a challenging fact. To overcome this, it is recommended to exploit microbial products, such as biosurfactants and extracellular polymers, which would increase the efficiency of removing the pollutants from contaminated sites. Application of biosurfactants for bioremediation of metals is an excellent option, which is eco-friendly and easily manageable. A successful bioremediation design relies on application of biosurfactants necessary for formation of a metal complex and harnessing the metal in soil solution. In addition, the harvested metals can be used for the synthesis of metal nanoparticles using microorganisms, biosurfactants and plant extracts under the green chemistry approach. The nanoparticles synthesised have profound applications in fields of diagnostics, therapeutics, medicine, drug delivery systems and agriculture.

The role of biosurfactants in the biochemical conversion of organic and inorganic contaminants has been realised, priority research needs have been identified and an effort has been made to understand the biochemical basis of contaminant degradation. However, most of the studies described were done under laboratory conditions, and efforts should be made to evaluate the role of in situ production of biosufactants. Further, there is a need for designing economic and more effective bioremediation models using biosurfactants. Nevertheless, vigilant use of these interesting surface-active molecules will surely enhance the clean-up of toxic heavy metals and provide us with a clean environment.

Acknowledgements

Author MYK thanks DST-SERB, and THS thanks DST-PURSE for fellowship and financial assistance.

References

Anderson, R. T., Vrionis, H. A., and Ortiz-Bernad, I. et al. 'Stimulating the in situ activity of Geobacter sp. to remove uranium from the ground water of a uranium-contaminated aquifer'. *Appl. Environ. Microbiol.*, no. 69 (2003): 5884–5891.

Anil, K., and Cameotra, S. S. 'Efficiency of lipopeptide biosurfactants in removal of petroleum hydrocarbons and heavy metals from contaminated soil'. *Environ. Sci. Poll. Res., 20* no. 10 (2013): 7367–7376.

Aresta, M., Dibenedetto, A., and Fragale, C. et al. 'Thermal desorption of polychlorobiphenyls from contaminated soils and their hydrodechlorination using Pd- and Rh-supported catalysts'. *Chemosph., 70* no. 6 (2008): 1052–1058.

Banat, I. M., Franzetti, A., and Gandolfi, I. 'Microbial biosurfactants production, applications and future potential'. *Appl. Microbiol. Biot., 87* (2010): 427–444.

Baranska, J. A., and Sadowski, Z. 'Bioleaching of uranium minerals and biosynthesis of UO_2 nanoparticles'. *Physiochem. Probl. Miner. Proce., 49* no. 1 (2013): 71–79.

Bellinger, D., Leviton, A., and Waternaux, C. 'Longitudinal analyses of prenatal and postnatal lead exposure and early cognitive development'. *N. Engl. J. Med.*, no. 316 (1987): 1037–1043.

Benedek, T., Mathe, I., and Salamon, R. 'Potential bacterial soil inoculant made up by Rhodococcus sp. and Pseudomonas sp. for remediation in situ of hydrocarbon and heavy metal polluted soils'. *Studia UBB Chemia, 57* no. 3 (2012): 199–211.

Brindha, K., Elango, L., and Rajesh, V. G. 'Occurence of chromium and copper in ground water around tanneries in chrompet area of Tamil nadu'. Anna University, Department of Geology, Chennai, *Int. J. Environ. Pollut., 30* no. 10 (2010): 818–822.

Cao, J., Zhang, G., and Mao, Z. et al. 'Precipitation of valuable metals from bioleaching solution by biogenic sulphides'. *Miner. Eng., 22* (2009): 289–295.

Chen, C., and Wang, J. 'Removal of Pb^{2+}, Ag^+, Cs^+, and Sr^{2+} from aqueous solution by brewery's waste biomass'. *J. Hazard. Mater.*, no. 151 (2008): 65–70.

Dahrazma, B., and Mulligan, C. N. 'Investigation of the removal of heavy metals from sediments using rhamnolipid in a continuous flow configuration'. *Chemosphere, 69* (2007): 705–711.

Das, P., Mukherjee, S., and Sen, R. 'Biosurfactant of marine origin exhibiting heavy metal remediation properties'. *Bioresour. Technol., 100* (2009): 4887–4890.

Diaz De, R. M. A., de Ranson, I. U., and Dorta, B. D. 'Metal removal from contaminated soils through bioleaching with oxidizing bacteria and rhamnolipid biosurfactant'. *Soil Sediment Contam.*, (2014).

Ercole, C., Veglio, F., and Toro, L. et al. 'Immobilisation of microbial cells for metal adsorption and desorption'. In: *Mineral Bioprocessing II*. Snow board, UT. (1994).

Ganesh, C., and Kumar Suman, K. 'Synthesis of biosurfactant-based silver nanoparticles with purified rhamnolipids isolated from *Pseudomonas aeruginosa* BS-161R'. *J. Microbiol. Biotechnol.*, 20 no. 7 (2010): 1061–1068.

Gaur, N., Gagan, F., and Mahavir, Y. 'A review with recent advancements on bioremediation-based abolition of heavy metals'. *Environ. Sci. Process. & Impact.*, 16 no. 180 (2014).

Gao, L., Naoki, K., and Hiroshi, I., 'Concentration and chemical speciation of heavy metals in sludge and removal of metals by bio-surfactants application'. *J. Chem. Chem. Eng.*, 7 (2013): 1188–1202.

Geier, D. A., and Geier, M. R. 'A prospective study of mercury toxicity biomarkers in autistic spectrum disorders'. *J. Toxicol. Environ. Health.*, no. A70 (2007): 1723–1730.

Gooday, G. W. 'Cation adsorption by lichens'. In: *Source Book of Experiments for the Teaching of Microbiology*, Primrose, S. B. and Wardlaw, A. C., Academic Press, (1982): 82–84.

Gutnick, D. L., and Shabtai, Y. 'Exopolysaccharide bioemulsifiers'. In: *Biosurfactants and Biotechnology*, Kosaric, N., Cairns, W.L. and Gray, N. C. C. (eds.), Marcel Dekker, New York (1987), pp. 211–246.

Halttunen, T., Finell, M. Sal`minen, S. 'Arsenic removal by native and chemically modified lactic acid bacteria'. *Int. J. Food Microbiol.*, 120 (2007): 173–178.

He, L., Yang, L., and Mustapha, A. 'Antifungal activity of zinc oxide nanoparticles against *Botrytis cinerea* and *Penicillium expansum*'. *Microbiol. Res.*, no. 166 (2011): 207–215.

Hogan, D. E., Tracey, A., and Veres, S. et al. 'Biosurfactant complexation of metals and application for remediation'. In: *Biosurfactants Research Trend and Applications*, C. N. Mulligan, S. K. Sharma, and A. Madhoo, (2014).

Hong, J. J., Yang, S. M., Lee, C. H., Choi, Y. K., and Kajiuchi, T. 'Ultrafiltration of divalent metal cations from aqueous solution using polycarboxylic acid type biosurfactants'. *J. Colloid Interf. Sci.*, 202 (1998): 63–73.

Juwarkar, A. A., Dubey, K. V., and Nair, A. et al. 'Bioremediation of multi-metal contaminated soil using biosurfactant – a novel approach'. *Indian J. Microbiol.*, no. 48 (2008): 142–146.

Kang, S. W., Kim, Y. B., and Shin, J. D. 'Enhanced biodegradation of hydrocarbons in soil by microbial biosurfactant, sophorolipid'. *Appl. Biochem. Biotechnol.*, no. 160 (2010): 780–790.

Karwowska, E., Andrezejewsk-Morzuch, D., and Lebkowska, M. et al. 'Bioleaching of metals from printed circuit boards supported with surfactant-producing bacteria'. *J. Hazard. Mater.*, no. 264 (2014): 203–210.

Lau, T. C., Wu, X. A., and Chua, H. 'Effect of exopolysaccharides on the adsorption of metal ions by Pseudomonas sp. CU-1'. *Water Sci. Technol.*, 52 (2005): 63–68.

Lili, H., Yang, L., and Mustapha, A. 'Antifungal activity of zinc oxide nanoparticles against Botrytis cinerea and Penicillium expansum'. *Microbiol. Res.*, no. 166 (2011): 207–215.

Magdalena, P.-P., Grażyna, A. P., and Zofia, P.-S. et al. 'Environmental applications of biosurfactants: Recent advances'. *Int. J. Mol. Sci.*, 12 (2011): 633–654.

Majumder, D. R. 'Waste to health: Bioleaching of nanoparticles from e-waste and their medical applications'. *Ind. J. Appl. Res.*, no. 3 (2013).

Marchant, R., and Banat, I. M. 'Microbial biosurfactants: Challenges and opportunities for future exploitation'. *Trends Biotechnol.*, 30 no. 11 (2012): 558–565.

Maslin, P., and Maier, R. M. 'Rhamnolipid enhanced mineralization of phenanthrene in organic metal co-contaminated soils'. *Bioremed. J.*, no. 4 (2000): 295–308.

Mulligan, C. N., and Wang, S. 'Remediation of a heavy metal-contaminated soil by a rhamnolipid foam'. *Eng. Geol.*, 85 (2006): 75–81.

Mulligan, C. N., Yong, R. N., and Gibbs, B. F. 'On the use of biosurfactants for the removal of heavy metals from oil-contaminated soil'. *Environ. Prog.*, 18 (1999): 50–54.

Nicholson, F. A., Smith, S. R., and Alloway, B. J. et al. 'An inventory of heavy metals inputs to agricultural soils in England and Wales'. *Sci. Total Environ.*, 311 no. 1–3 (2003): 205–219.

Prakash, J. M., Yuh, M. H., and Chun-Mei, H. 'Removal of Cu, Pb and Zn by foam fractionation and a soil washing process from contaminated industrial soils using soapberry-derived saponin: A comparative effectiveness assessment'. *J. Environ. Sci.*, 6 no. 25 (2013): 1180–1185.

Raquel, D. R., Juliana, M. L., Galba, M., and Campos-Takaki. 'Application of the biosurfactant produced by Candida lipolytica in the remediation of heavy metals'. *Chem. Eng. Trans.*, no. 27 (2012).

Rashmi, V., and Pratima, D. 'Heavy metal water pollution – A case study'. *Recent Res. Sci. Technol.*, 5 (2013): 98–99.

Raymond, A., Wuana, and Felix, E. O. 'Heavy metals in contaminated soils: A review of sources, chemistry, risks and best available strategies for remediation'. *International Scholarly Research Network,* 20 pages, Ecology volume (2011).

Reis, R. S., Pacheco, and Pereir, G. J. 'Biosurfactants: Production and applications, biodegradation'. *Life Sci.* (2013).

Roberto, D. P., Giovanni, C., and Emesto, M. 'Exopolysaccharide-producing cyanobacteria heavy metal removal from water: Molecular basis and practical applicability of the biosorption process'. *Appl. Microbiol. Biotechnol.*, no. 92 (2011): 697–708.

Ron, E. Z., and Rosenberg, E. 'Natural roles of biosurfactants'. *Environ. Microbiol.*, no. 3 (2001): 23229–23336.

Sachdev, D. P., and Cameotra, S. S. 'Biosurfactants in agriculture'. *Appl. Microbiol. Biotechnol.*, 97 (2013): 1005–1016.

Sandrin, T. R., and Maier, R. M. 'Impact of metals on the biodegradation of organic pollutants'. *Environ. Health Perspect.*, no. 111 (2003): 1093–1100.

Satarug, S., Garrett, S. H., Sens, M. A., and Sens, D. A. 'Cadmium, environ-mental exposure, and health outcomes'. *Environ. Health Perspect.*, 118 (2010): 182–190.

Sharma Pramila, Fulekar, M. H., and Pathak Bhawana. 'E-waste – A challenge for tomorrow'. *Res. J. Recent Sci.*, 1 no. 3 (2012): 86–93.

Sheng, X. F., He, L. Y., Wang, Q. Y., Ye, H. S., and Jiang, C. Y. 'Effects of inoculation of biosurfactant-producing Bacillus sp. J119 on plant growth and cadmium uptake in a cadmium-amended soil'. *J. Hazard. Mater.*, no. 155 (2008): 17–22.

Singh, P., and Cameotra, S. S. 'Enhancement of metal bioremediation by use of microbial surfactants'. *Biochem. Biophys. Res. Commun.*, no. 319 (2004): 291–297.

Singh, R., Gautam, N., and Mishra, A. et al. 'Heavy metal and living systems: An overview'. *Indian J. Pharmacol.*, 43 (2011): 246–253.

Singh, S. C., Randhir, S., Makkar, and Jasminder, K. 'Synthesis of biosurfactants and their advantages to microorganisms and mankind'. *Landes Bioscience and Springer,* Ramakrishna Sen (2010).

Soeder, C. J., Papaderos, A., and Kleespies, M. et al. 'Influence of pythogenic surfactants (quillaya saponin and soya lecithin) on bio-elimination off phenanthrene and fluoranthene by three bacteria'. *Appl. Microbiol. Biotechnol.*, no. 44 (1996): 654–659.

Tampouris, S., Papassiopi, N., and Paspaliaris, I. 'Removal of contaminant metals from fine grained soils, using agglomeration, chloride solutions and pile leaching techniques'. *J. Hazard. Mater., 84* no. 2–3 (2001): 297–319.

Todd, R. S., Andrea, M. C., and Maier, R. M. 'A rhamnolipid biosurfactant reduces cadmium toxicity during naphthalene biodegradation'. *Appl. Environ. Microbiol.*, no. 66 (2000): 4585–4588.

Varshney, R., Bhadauria, S., and Gaur, M. S. 'Copper nanoparticles synthesis from electroplating industry effluent'. *Nano. Biomed. Eng.*, no. 3 (2011): 115–119.

Volesky, B., 'Biosorbents for metal recovery'. *Trends Biotech., 5* no. 4 (1987): 96–101.

Wang, S., and Mulligan, C. N. 'Rhamnolipid foam enhanced remediation of cadmium and nickel contaminated soil'. *Water Air Soil Pollution, 157* (2004): 315–330.

Wani, A. H., and Shah, M. A. 'A unique and profound effect of MgO and ZnO nanoparticles on some plant pathogenic fungi'. *J. Appl. Pharm. Sci., 2* no. 3 (2012): 40–44.

Wu, P. X., Zhang, Q., and Dai, Y. P. 'Adsorption of Cu(II), Cd(II) and Cr(III) ions from aqueous solutions on humic acid modified Ca-montmorillonite'. *Geoderma, 164* (2011): 215–219.

Yao, Z., Xiec, H., and Yuc, C. et al. 'Review on remediation technologies of soil contaminated by heavy metals'. *Procedia Environ. Sci., 16* (2012): 722–729.

Zhitkovich, A. 'Chromium in drinking water: Sources, metabolism, and cancer risks'. *Chem. Res. Toxicol., 24* (2011): 1617–1629.

14

Recent Advances in Bacteria-Assisted Phytoremediation of Heavy Metals from Contaminated Soil

Jawed Iqbal and Munees Ahemad

CONTENTS

14.1 Introduction

Contrary to the existing conventional remediation technologies, phytoremediation (exploiting inherent physiological mechanisms of plants to remove pollutants or render them nontoxic) is a cost-effective, environmentally friendly and sustainable alternative to decontaminate the metal-polluted soils. Plant-based remediation can be in the form of phytostabilisation, phytoextraction, phytovolatilisation and rhizodegradation, depending upon the physical or chemical properties of contaminants present in soils (Figure 14.1). In phytoextraction, plants are used to concentrate metals from the soil into the roots and shoots of the plant; rhizodegradation is the use of plants to uptake, store and degrade contaminants within its tissue; phytostabilisation is the use of plants to reduce the mobility of heavy metals through absorption and precipitation by plants, thus reducing their bioavailability; phytovolatilisation is the uptake and release into the atmosphere of volatile materials, such as mercury or arsenic-containing compounds (Jing et al. 2007).

The major drawback that limits this approach for wide-scale application is the slow growth rate and the decreased biomass of remediating plants due to the phytotoxic effects of high concentrations of metal contaminants in soils (Table 14.1). Some of the deleterious effects of these trace metals on the physiological and biochemical processes of plants growing in metalliferous soils are the retarded growth; chlorosis; necrosis; genotoxicity through reactive oxygen species (ROS)-mediated oxidative damage; oxidation of indole-3-acetic acid (IAA); and inhibition of seed germination, water potential, transpiration rate, photosynthesis (photophosphorylation and electron transport) and catalytic efficiency of metabolic enzymes (ATPase, nitrate reductase and catalase) (Nagajyoti et al. 2010). As a result, efficiency of phytoremediation in metal-stressed soils is significantly reduced. Therefore, the foremost challenge to upgrading this promising

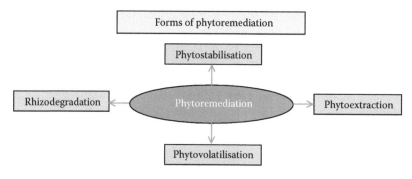

FIGURE 14.1
Plant-based phytoremediation.

TABLE 14.1

Speciation and Chemistry of Some Heavy Metals in Soils

Heavy Metals	Speciation and Chemistry
Lead	Pb occurs in 0 and +2 oxidation states. Pb (II) is the more common and reactive form of Pb. Low-solubility compounds are formed by complexation with inorganic (Cl^-, CO_3^{2-}, SO_4^{2-}, PO_4^{3-}) and organic ligands (humic and fulvic acids, EDTA, amino acids). The primary processes influencing the fate of Pb in soil include adsorption, ion exchange, precipitation and complexation with sorbed organic matter.
Chromium	Cr occurs in 0, +6 and +3 oxidation states. Cr (VI) is the dominant and toxic form of Cr at shallow aquifers. Major Cr (VI) species include chromate $\left(CrO_4^{2-}\right)$ and dichromate $\left(Cr_2O_7^-\right)$ (especially Ba^{2+}, Pb^{2+} and Ag^+). Cr (III) is the dominant form of Cr at low pH (<4). Cr (VI) can be reduced to Cr (III) by soil organic matter, S^{2-} and Fe^{2+} ions under anaerobic conditions. The leachability of Cr (VI) increases as soil pH increases.
Zinc	Zn occurs in 0 and +2 oxidation states. It forms complexes with anions, amino acids and organic acids. At high pH, Zn is bioavailable. Zn hydrolyses at pH 7.0–7.5, forming $Zn(OH)_2$. It readily precipitates under reducing conditions and may coprecipitate with hydrous oxides of Fe or manganese.
Cadmium	Cd occurs in 0 and +2 oxidation states. Hydroxide [$Cd(OH)_2$] and carbonate ($CdCO_3$) dominate at high pH, whereas Cd^{2+} and aqueous sulphate species dominate at lower pH (<8). It precipitates in the presence of phosphate, arsenate, chromate, sulphide, etc. Shows mobility at pH range 4.5–5.5.
Arsenic	As occurs in −3, 0, +3, +5 oxidation states. In aerobic environments, As (V) is dominant, usually in the form of arsenate $(AsO_4)^{3-}$. It behaves as a chelate and can coprecipitate with or adsorb into Fe oxyhydroxides under acidic conditions. Under reducing conditions, As (III) dominates, existing as arsenite $(AsO_3)^{3-}$, which is water soluble and can be adsorbed/coprecipitated with metal sulphides.
Iron	Fe occurs in 0, +2, +3 and +6 oxidation states. Organometallic compounds contain oxidation states of +1, 0, −1 and −2. Fe (IV) is a common intermediate in many biochemical oxidation reactions. Many mixed valence compounds contain both Fe (II) and Fe (III) centres, for example, magnetite and Prussian blue.
Mercury	Hg occurs in 0, +1 and +2 oxidation states. It may occur in alkylated form (methyl/ethyl mercury), depending upon the pH of the system. Hg^{2+} and Hg_2^{2+} are more stable under oxidising conditions. Sorption to soils, sediments and humic materials is pH-dependent and increases with pH.
Copper	Cu occurs in 0, +1 and +2 oxidation states. The cupric ion (Cu^{2+}) is the most toxic species of Cu, for example, $Cu(OH)^+$ and $Cu_2(OH)_2^{2+}$. In aerobic alkaline systems, $CuCO_3$ is the dominant soluble species. In anaerobic environments, CuS(s) will form in the presence of sulphur. Cu forms strong solution complexes with humic acids.

Source: Adapted from Hashim, M. A. et al., *J. Environ. Manag.*, 92, 2355–2388, 2011.

biotechnological process is to accelerate the growth of remediating plants to achieve higher biomass and concomitantly to nullify or alleviate the degree of phytotoxicity of metal pollutants. Soil is a complex ecosystem in which diverse microbial communities are considered as architects. In fact, different microbial activities and their various functional traits perform many ecosystem services, including plant production, sequestration and cycling of nutrients. Inversely, chemical and physical properties of soils (for example, soil type, soil texture, particle size, soil density, organic matter content, pH and redox conditions) greatly influence the dynamics of both structure and functions of microbial communities in the soils (Lombard et al. 2011; Schulz et al. 2013). Among the vast array of genetically diverse soil microorganisms, bacteria inhabiting in or around the rhizosphere or within the roots exhibit several important traits/activities, for example, inorganic and organic phosphate solubilisation, production of siderophores, phyto-hormones (for example, cytokinins, gibberellins and IAA), antibiotics and lytic enzymes and expression of 1-aminocyclopropane-1-carboxylate (ACC) deaminase, through which they accelerate the growth of plants by facilitating them with nutrient acquisition from soils or modulating their responses against stress factors or biological control of phytopathogens (Ahemad and Khan 2011; Glick 2012; Ahemad and Kibret 2014).

Interestingly, many of these bacterial traits have been implicated by different authors in amelioration of phytoremediation of metal-contaminated soils (Figure 14.2). For instance, phosphate solubilising and IAA producing *Bacillus weihenstephanensis* strain SM3, when used as an inoculant with *Helianthus annuus* grown in metal-spiked soil, not only promoted the overall

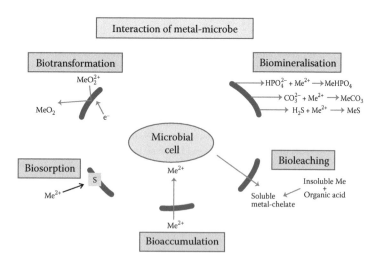

FIGURE 14.2
(See color insert.) Interaction of metal and microbe affects bioremediation. (From Tabak, H. H. et al., *Rev Env Sci Biotech*, 4, 115–156, 2005.)

plant growth but also increased the metal-extracting capacity of plants and phytoavailability of Cu and Zn in soils (Rajkumar et al. 2008). Similarly, *Pseudomonas aeruginosa* strain OSG41, exhibiting IAA and siderophore-producing capability, decreased Cr uptake and concurrently increased different growth parameters, thus phytostabilising chickpea (*Cicer arietinum*) plants (Oves et al. 2013).

The objective of this chapter is to illuminate the role of bacterial traits in phytoremediation (exclusively in phytoextraction) of metal-polluted soils, integrating knowledge from current literature and updates.

14.2 Plant Growth–Promoting Rhizobacteria (PGPR) and Its Characteristics

PGPR are valuable soil bacteria, which may facilitate plant growth and development directly and indirectly (Glick 1995). Direct stimulation may include providing plants with fixed nitrogen, phytohormones and iron, which has been sequestered by bacterial siderophores and soluble phosphate. Indirect stimulation of plant growth includes preventing phytopathogens and, consequently, promotes plant growth and development (Glick and Bashan 1997). PGPR perform some of these functions through specific enzymes, which provoke physiological changes in plants at the molecular level. Among these enzymes, bacterial ACC deaminase plays an important role in the regulation of a plant hormone, ethylene, and hence, growth and development of plants are modified (Arshad and Frankenberger 2002; Glick 2005). It has been widely studied in various species of plant growth–promoting bacteria (PGPB), such as *Agrobacterium genomovars* and *Azospirillum lipoferum* (Blaha et al. 2006), *Alcaligenes* and *Bacillus* (Belimov et al. 2001), *Burkholderia* (Pandey et al. 2005; Sessitsch et al. 2005; Blaha et al. 2006), *Enterobacter* (Penrose and Glick 2001), *Methylobacterium fujisawaense* (Madhaiyan et al. 2006), *Pseudomonas* (Belimov et al. 2001; Hontzeas et al. 2004; Blaha et al. 2006), *Ralstonia solanacearum* (Blaha et al. 2006), *Rhizobium* (Ma et al. 2004; Uchiumi et al. 2004), *Rhodococcus* (Stiens et al. 2006), *Sinorhizobium meliloti* (Belimov et al. 2005) and *Variovorax paradoxus* (Belimov et al. 2001).

14.2.1 Role of PGPR in Phytoremediation

Rhizosphere bacteria have been known to be beneficial in phytoremediation by promoting plant growth through phytohormones, mineral phosphate mobilisation and siderophore production and by immobilising heavy metals through bioaccumulation, biotransformation to less soluble or less toxic forms, chemisorption and biomineralisation (Faisal and Hasnain 2005; Khan et al. 2007; Viti and Giovannetti 2007; Haferburg and Kothe 2010). The

TABLE 14.2

Role of Plant Growth–Promoting Rhizobacteria in Phytoremediation

PGPR	Plant	PGPR and Its Function	References
Serratia marcescens BC-3 *B. amyloliquefaciens* NAR38 *Pseudomonas* sp. TLC 6-6.5-4 *Pseudomonas* sp. A3R3	*Glomus-intraradices* mangroves *Zea mays*, *Helianthus-annuus* *Alyssum-serpyllifolium*, *Brassica juncea*	Higher plant biomass and antioxidant enzyme activities. Efficiency of siderophore in bioremediation of metal-contaminated iron-deficient soils. Significantly increased Cu accumulation results in increased total biomass. Increased significant biomass and Ni content in plants grown in Ni-stressed soil.	Dong et al. (2014) Gaonkar and Bhosle (2013) Kafeng and Wusirika (2011) Ma et al. (2011c)
Psychrobacter sp. SRS8	*Ricinus-communis*, *Helianthus-annuus*	Stimulated plant growth and Ni accumulation in plant species with increased plant biomass, chlorophyll and protein content.	Ma et al. (2011a)
Bacillus species PSB10	*Cicer arietinum*	Improved growth, nodulation, chlorophyll, leghaemoglobin, seed yield and grain protein.	Wani and Khan (2010)
Bradyrhizobium sp. 750, *Pseudomonas* sp., *Ochrobactrum cytisi*	*Lupinus luteus*	Increased biomass, nitrogen content, accumulation of metals.	Dary et al. (2010)
Pseudomonas sp. SRI2, *Psychrobacter* sp. SRS8, *Bacillus* sp. SN9	*Brassica juncea*, *Brassica-oxyrrhina*	Increased biomass of the test plants and enhanced Ni accumulation in plant tissues.	Ma et al. (2009c)
Psychrobacter sp. SRA1, *Bacillus cereus* SRA10	*Brassica juncea*, *Brassica-oxyrrhina*	Enhanced metal accumulation in plant tissues by facilitating the release of Ni.	Ma et al. (2009a)
Achromobacter xylosoxidans strain Ax10	*Brassica juncea*	Significantly improved Cu uptake by plants and increased the root length, shoot length, fresh weight and dry weight of plants.	Ma et al. (2009b)
Pseudomonas aeruginosa, *Pseudomonas fluorescens*, *Ralstonia metallidurans*	*Zea mays*	Promoted plant growth, facilitated soil metal mobilisation, enhanced Cr and Pb uptake.	Braud et al. (2009)

inoculum should be composed of specific bacterial strains, which might have the following characteristics: heavy metal tolerance, phosphate solubilisation, nitrogen fixing and ability to perform biomineralisation (Haferburg and Kothe 2010). Faisal and Hasnain (2005) and Viti and Giovanetti (2007) have used bacterial biotransformation of hexavalent chrome to decrease toxic effects on plants, recording positive effects on plant growth under bacterial treatment. Previously, increased plant growth and a tendency to limit metal uptake by *Elymus repens* has been observed when using an *Azotobacter chroococcum* and *Bacillus megatherium* inoculum with dolomite amendment (Petrisor et al. 2004). Moreover, findings signifying different types of compost amendments on plant growth in heavy metal–rich substrates, which showed a reduction of heavy metal bioavailability, enhanced plant biomass, higher survival of plants and a better nutritional state (Neagoe et al. 2005; Madejon et al. 2006; Shutcha et al. 2010; Córdova et al. 2011), are described in Table 14.2. Recently, it has been observed that PGPB inoculum increases the biomass of plants grown on tailing material amended with compost and decreases the concentration of toxic metals in the biomass and in the percolating water (Nicoară et al. 2014).

14.3 Bacterial Siderophores

14.3.1 Importance of Iron on Siderophore Function

Generally, all organisms require iron [Fe (III)] as a cofactor to carry out many important metabolic functions, such as electron transfer, nitrogen fixation, DNA replication and RNA transcription. Although iron abundantly occurs in soils, its bioavailability for organisms is considerably low (10^{-7}–10^{-24} M) due to formation of oxides in aerobic habitats (Raymond et al. 2003). Under iron limitation, soil microorganisms secrete low molecular weight organic ligands, siderophores exhibiting high affinity for ferric iron (formation constant: 10^{-30} M^{-1}) (Raymond et al. 2003). Siderophores, whose concentration in soils is generally from tens of micromoles to a few millimoles per litre, release iron from minerals through extracellular solubilisation for microbial uptake (Hersman et al. 1995; Schalk et al. 2011). Soil bacteria not only utilise ferric iron bound to their own siderophores (homologous siderophores) but also scavenge iron molecules complexed with the siderophores secreted by other bacterial species (heterologous siderophores) (Geetha and Joshi 2013). Most of the studies concerning the mechanism of siderophore-mediated iron uptake have been done in Gram-negative bacteria wherein siderophores involve TonB, TBDTs (TonB-dependent transporters), ExbB, ExbD and ABC transporters or permeases to actively transport iron to the bacterial cell (Schalk et al. 2011).

14.3.2 Features of Siderophores

The chemical feature of bacterial siderophores is of great environmental significance as they not only form stable complexes with ferric iron but also bind other metal cations, for example, Ag^+, Zn^{2+}, Cu^{2+}, Co^{2+}, Cr^{2+}, Mn^{2+}, Cd^{2+}, Pb^{2+}, Ni^{2+}, Hg^{2+}, Sn^{2+}, Al^{3+}, In^{3+}, Eu^{3+}, Ga^{3+}, Tb^{3+} and Tl^+, with somewhat lower stability constants and enter the cell without affecting the cellular uptake of other nutrients (Evers et al. 1989; Chen et al. 1994; Hernlem et al. 1996; Hu and Boyer 1996; Neubauer et al. 2000; Greenwald et al. 2008; Braud et al. 2009a,b, 2010). Siderophores protect bacteria from metal toxicity as metals enter the periplasm by diffusion via porins, and siderophore–metal complexes are too large to pass through porins, consequently decreasing the concentration of free metals outside the bacterial cell (Schalk et al. 2011). Hence, siderophore-producing bacteria are more resistant to heavy metal stress than bacteria deficient with this trait (Braud et al. 2010).

14.3.3 Role of Siderophores in Metal-Stressed Plants

Plants growing in metalliferous soils face nutrient deficiency, specifically iron, owing to the adverse effects of metals on chloroplast development, photosynthetic machinery and enzymes involved; hence, their growth and development are severely hampered (Nagajyoti et al. 2010). In this case, inoculation of metal-stressed plants with siderophore-producing bacteria might be a faithful strategy to overcome the iron deficiency. Siderophore production is a widespread trait among diverse bacterial genera (Rajkumar et al. 2005; Rajkumar and Freitas 2008; Braud et al. 2009a,b; Ma et al. 2009a,b,c, 2011a,b; Tank and Saraf 2009; He et al. 2010; Ahemad and Khan 2012; Oves et al. 2013). Plant roots take up iron from the iron–siderophore complex by chelate degradation or a ligand exchange reaction, or they scavenge the whole iron–siderophore complex (Rajkumar et al. 2010). In various studies, siderophore-producing bacteria successfully augmented the chlorophyll and other growth parameters of plants in metal-stressed soils by facilitating iron uptake (Burd et al. 1998, 2000; Carrillo-Castañeda et al. 2003; Barzanti et al. 2007). Under metal stress, siderophores have been reported to promote IAA biosynthesis by chelating toxic metal species. Thus, metal chelation by siderophores does not allow metals to hinder the IAA biosynthesis; in turn, IAA displays their beneficial effects on the metal-remediating plants (Dimkpa et al. 2008).

14.3.4 Effect of Siderophores on Heavy Metal Fate

Many of the studies are concerned with the effect of siderophores on heavy metal fate in the rhizosphere, especially investigating phytoextraction techniques (Rajkumar et al. 2010, 2012; Chen et al. 2011). In some cases, siderophore-producing bacterial strains are involved in mitigating heavy metal mobility. Rajkumar et al. (2010, 2012) compiled various reports in which siderophore production is associated with increased plant biomass, and metal uptake is increased or decreased based mostly on bacterial strains, plant species and type of substrate. Moreover, another mechanism for heavy metal immobilisation is bacterial bioaccumulation and biosorption, either to the bacterial surface or bacterial biofilm. Wu et al. (2006) identified reduction in the soluble Pb in soil by *Azotobacter* and *Brevibacillus* strains. Mn and Cu have been shown to bind near the air–biofilm interface of *Pseudomonas putida* biofilms, in which bacteria survive under the protection of the biofilm (Chen et al. 2011). Bacterial bioaccumulation has also been observed to be responsible for reduced heavy metal uptake and translocation of As, Cd, Cu, Ni and Zn (Rajkumar et al. 2012). This might be a potential explanation of the distribution of Cu and Zn at the root surface observed in the inoculated treatment.

14.4 ACC Deaminase

14.4.1 Role of ACC Deaminase

Phytohormone ethylene regulates a myriad of physiological processes in plants, for example, ripening of fruits, senescence, abscission, root formation and its architecture, seed germination and responses to various biotic and abiotic stresses (inhibition of root growth, nodulation, and auxin transport) (Rodecap and Tingey 1981; Abeles et al. 1992; Arshad and Frankenberger 2002; Arshad et al. 2007). Enzymes S-adenosyl-L-methionine (SAM) synthetase, ACC synthase and ACC oxidase participate to release ethylene in plant tissues: SAM synthetase converts methionine into SAM, which is further hydrolysed by ACC synthase into 5'methylthioadenosine and ACC; through ACC oxidase, ACC produces ethylene, carbon dioxide and hydrogen cyanide (Kang et al. 2010) as described in Figure 14.3.

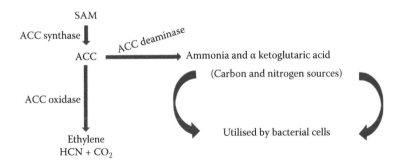

FIGURE 14.3
Representing how bacterial cells utilise carbon and nitrogen sources. (From Saleem, M. et al., *J Ind Microbiol Biotechnol*, 34, 635–648, 2007.)

14.4.2 ACC Deaminase Lowers Ethylene Level in Plants

Ethylene is an essential metabolite for plant growth and development, and its high concentration during stress conditions, including heavy metal stress, severely affects the normal performance of stressed plants. Pyridoxal 5'-phosphate-dependent and cytoplasmically located bacterial enzyme, ACC deaminase, catalyses the hydrolysis of plant-produced ACC into α-ketobutyric acid and ammonia, which are utilised as carbon and nitrogen sources by bacterial cells. Actually, ACC deaminase–containing bacteria bound to the plant roots act as a sink for root-exuded ACC, and the process of ACC exudation and its cleavage by the bacterial ACC deaminase are in equilibrium (Glick 1995). Consequently, the biosynthesis of ethylene in plant tissues is considerably reduced to lower the inhibitory level of ethylene (Glick et al. 1998; Arshad et al. 2007). Hence, ACC deaminase–containing bacteria alleviate the ill effects of the elevated levels of ethylene in plants exposed to different stresses, including heavy metal stress (Burd et al. 2000; Glick 2005; Safronova et al. 2006; Saleem et al. 2007). There is a great variability in the activity of ACC deaminases produced by different organisms. In addition, an extensive level of ACC deaminase generally binds the rhizoplane of different plants nonspecifically (Glick 2005).

14.4.3 ACC Deaminase Regulates Ethylene Level under Stress Conditions

The higher level of ethylene in response to abiotic and biotic stresses leads to inhibition of root growth and, consequently, growth of the whole plant. Ethylene synthesis is stimulated by a variety of environmental factors or stresses, which obstruct plant growth. PGPR containing ACC deaminase regulate and lower the levels of ethylene by metabolising ACC. These ACC deaminase PGPR boost plant growth, particularly under stressed conditions, by the regulation of accelerated ethylene production in response to a multitude of abiotic and biotic stresses, such as salinity, drought, water logging, temperature, pathogenicity and contaminants.

14.4.4 ACC Deaminase Enhances Phytoremediation

As better root proliferation and increased biomass are a prerequisite for the uptake and translocation of metals in hyperaccumulating plants, their inoculation with the ACC deaminase–expressing bacteria facilitates the prolific root growth and density and, thus, enhances their phytoremediation efficiency in metal-contaminated soils (Gleba et al. 1999; Agostini et al. 2003; Arshad et al. 2007). Several authors have isolated diverse genera of ACC deaminase–containing bacteria and implicated them in plant growth promotion under stressed environments (He et al. 2010; Zhang et al. 2011; George et al. 2013; Ma et al. 2013). In different reports, *Rahnella* sp. exhibiting ACC deaminase activity in conjunction with other bacterial traits improved both the metal hyperaccumulation and biomass in inoculated *Amaranthus hypochondriacus, A. mangostanus, Solanum nigrum* and *Brassica napus* plants that were grown in Cd-, Pb- and Zn-stressed soils (He et al. 2013; Yuan et al. 2013).

14.4.5 Recent Advances and Genetic Manipulation

Genetically modified bacteria could be useful for developing a better understanding of mechanisms responsible for induction of tolerance in plants inoculated with ACC deaminase bacteria against both biotic and abiotic stresses (Sheehy et al. 1991; Glick and Bashan 1997). Recent studies have demonstrated that genetic modification of PGPR expressing ACC deaminase genes helped in modulation of the root nodule in legumes and biological control of plant disease (Wang et al. 2000; Ma et al. 2004). However, plant growth promotion observed in response to inoculation with bacteria containing ACC deaminase motivated researchers to develop transgenic plants with the expression of ACC deaminase genes (Arshad and Frankenberger 2002). Recently, the focus has been now shifted to exploiting the potential of ACC deaminase genes in regulation of ethylene levels in plants exposed to various kinds of stresses, such as heavy metals, etc., along with its potential role in phytoremediation of contaminated soil environments.

14.5 Indole-3-Acetic Acid (IAA)

IAA is synthesised mainly in apical meristems of plants and affects almost every aspect of the life cycle of a plant by modulating various plant responses as reflected by the complexity of its biosynthesis, transport and signalling pathways (Santner et al. 2009). Some of the most common effects of IAA on plant physiology are promotion of stem elongation and growth, seed germination, cell division (with cytokinins), lateral bud dormancy, adventitious root formation, inducement of ethylene production, enhancement of rate of xylem and root development and inhibition of leaf abscission. Interestingly, the majority of soil microorganisms also produce this phytohormone as a secondary metabolite and modulate the plant responses by affecting the IAA pool of plants through rhizosphere interactions (Ahemad and Kibret 2014).

As mentioned previously, phytoremediation efficiency of a plant is positively linked to the root system architecture (Liphadzi et al. 2006). As metal contaminants exceeding the normal concentration in soils inhibit the growth and development of plant roots, phytoremediating plants, therefore, do not achieve sufficient root biomass in metal-stressed soils; in turn, their remediating efficiency is declined (Arshad et al. 2007; Remans et al. 2012). In this context, bacterial IAA, such as ACC deaminase, is also involved in affecting the root architecture as well as the phytoremediation process. IAA-producing bacteria increases metal extraction and the nutrient acquisition ability of plants through inducing root proliferation and increasing root surface area as well as the number of root tips (Liphadzi et al. 2006; Glick 2010; Remans et al. 2012). Moreover, a bacterial IAA trait helps to adapt the plants against inhibitory levels of heavy metals in stressed soils. For instance, Hao et al. (2012) displayed that the biomass of *Robinia pseudoacacia* plants was enhanced in a zinc-contaminated environment through inoculation with the zinc-resistant and IAA-producing *Agrobacterium tumefaciens* CCNWGS0286. In addition, they also established that the capacity of bacterial inoculum to produce IAA, rather than genes encoding metal resistance determinants, had a larger impact on the growth of host plants growing in metal-contaminated soils. Another mechanism of IAA to accelerate plant growth and development under a metal-stressed environment is the inhibition of the ethylene action by suppression of *VR-ACO1* (ACC oxidase gene) expression and induction of *VR-ACS1* (ACC synthase gene) expression (Kim et al. 2001).

14.5.1 IAA and Bioremediation of Heavy Metal

Bacteria play a critical role in the bioremediation of heavy metal pollutants in soil and wastewater. Previously, high levels of resistance to zinc, cesium, lead, arsenate and mercury in eight copper-resistant *Pseudomonas* strains have been identified (Li and Ramakrishna 2011). Moreover, these metal-resistant strains were capable of bioaccumulation of multiple metals and solubilisation of copper and produce plant growth–promoting IAA, iron-chelating siderophores and solubilised mineral phosphate and metals. The bacterial inoculation on plant growth and copper uptake by maize (*Zea mays*) was investigated using one of the isolates (*Pseudomonas* sp. TLC 6-6.5-4), and higher IAA production and phosphate as well as metal solubilisation were observed; increased copper accumulation in maize resulted as did an increased total biomass of maize (Li and Ramakrishna 2011).

14.6 Organic Acids and Biosurfactants

Application of chemical chelates (for example, EDTA) augments the bioavailability of metals in soils; such chemicals, however, have risks of metal leaching and also exert a deleterious impact on both soil fertility and soil structures (Khan et al. 2000). In contrast, environmentally important bacteria increase the mobilisation of metals in soils; in turn, their phytoavailability by lowering pH occurs through the production of low molecular weight organic acids (for example, citric acid, lactic acid, oxalic acid, succinic acid, etc.) in metalliferous soils without noxious effects on soils (Becerra-Castro et al. 2011; He et al. 2013). At low pH, metal ions are dissociated from the metal-containing compounds present in soils and, consequently, are available for plant uptake. After entering the root cells, they are accumulated in root tissues by complexation with cellular malate, citrate or histidine (Salt et al. 1999; Küpper et al. 2004). Moreover, bacteria also solubilise the metal-containing inorganic phosphates in soils by secreting different organic acids. Owing to this phosphate-solubilising property, they are also referred to as phosphate-solubilising bacteria (PSB). Different authors have reported PSB mediated solubilisation of phosphates of Fe (Song et al. 2012), Ni (Becerra-Castro et al. 2011), Cu (Li and Ramakrishna 2011), Zn (He et al. 2013) and Al (Song et al. 2012). PSB play a dual role in metal-stressed soils: First, they increase both the mobilisation of metals in soils and bioavailability to plants, and second, they help in supplying an essential nutrient, phosphorus, to the plants. Because plants exposed to the high concentrations of metal contaminants generally face a nutrient deficiency due to metal-induced oxidative stress, phosphates released from the insoluble phosphate minerals, owing to bacterially produced acid chelates, help plants to acquire greater biomass (Ahemad and Kibret 2014). For instance, Jeong et al. (2012) showed that the nonbioavailable and insoluble Cd fractions in soils were gradually solubilised by the inoculation of phosphate-solubilising *Bacillus megaterium*, and consequently, biomass and Cd hyperaccumulation in *Brassica juncea* and *Abutilon theophrasti* plants were enhanced. They inferred that the organic acids, such as IAA, exuded by PSB acidified soils, thus solubilising phosphates and sequestering Cd from the soils.

In metalliferous soils, metals are strongly bound to soil particles, and they are not separated from soils easily under normal conditions. Therefore, plants are unable to extract sufficient amount of metals from soils due to their low phytoavailability (Glick 2012). Bacterial biosurfactants are a structurally diverse group of surface-active substances whose utility in metal remediation is due to their ability to form complexes with metals entrapped in soil particles. The bond between biosurfactant and metal is stronger than that between metal and soils; therefore, the metal–biosurfactant complex is easily desorbed from the soil matrix to the soil solution due to the lowering of the interfacial tension (Pacwa-Płociniczak et al. 2011). In various studies, the producing biosurfactants have been shown to improve metal mobilisation and plant-assisted extraction in metal-contaminated soils (Braud et al. 2006; Sheng et al. 2008; Gamalero and Glick 2012; Singh and Cameotra 2013).

14.7 Association among Plants, Bacteria, Heavy Metals and Soils Enhances Phytoremediation

Phytoremediation strength depends upon the interactions among the plant, bacteria, heavy metals and the soil. The roots of plants interact with numerous microorganisms, which determine the extent of phytoremediation (Glick 1995). Functioning of associative plant–bacterial symbioses in heavy metal-polluted soil can be affected from plant-associated bacteria and the host plant. The soil microbes might play significant roles in the recycling of plant nutrients, maintenance of soil structure, detoxification of toxic chemicals and control of plant pests and plant growth (Giller et al. 1998; Elsgaard et al. 2001; Filip 2002). Hence, bacteria can enhance the remediation ability of plants or reduce the phytotoxicity of the contaminated soil. In addition, plant roots can provide root exudate as well as increase ion solubility, which may increase the remediation activity of bacteria-associated plant roots. Rhizobacteria have been shown to have several traits that can alter heavy metal bioavailability (McGrath et al. 2001; Whiting et al. 2001; Lasat 2002) through the release of chelating substances, acidification of the microenvironment and changes in redox potential (Smith and Read 1997). Abou-Shanab et al. (2003) have reported that the addition of *Sphingomonas macrogoltabidus*, *Microbacterium liquefaciens* and *Microbacterium arabinogalactanolyticum* to *Alyssum murale* grown in serpentine soil notably increased the plant uptake of Ni compared to uninoculated controls as a result of soil pH reduction. The specificity of the plant–bacteria association is dependent upon soil conditions, which can modify contaminant bioavailability, root exudate composition and nutrient levels. In addition, the metabolic requirements for heavy metal remediation may also dictate the form of the plant–bacteria interaction, either specifically or nonspecifically. Along with metal toxicity, there are often additional factors that limit plant growth in contaminated soils, including arid conditions, lack of soil structure, low water supply and nutrient deficiency (Jing et al. 2007).

14.8 Phytoremediation-Assisted Bioaugmentation

Polychlorobiphenyls (PCBs) represent 209 chlorinated molecules and are a particular threat to the environment because of their toxicity (Li et al. 2011). Tolerance of plants to PCBs has been shown to be higher for herbaceous plants, such as fescue, than for leguminous plants, in which a decrease in the root and shoot biomass was also found (Weber and Mrozek 1979; Chekol and Vough 2002).

Previously, various degradation pathways of PCBs have been described by bacteria (Bedard and Quensen 1995; Bedard et al. 1997; Wu et al. 1997). Studies have shown that among bacteria, *Burkholderia xenovorans* LB400 and *Pseudomonas pseudoalcaligenes* KF707 are able to degrade PCBs (Tremaroli et al. 2010). *B. xenovorans* LB400 has been identified as one of the most efficient among them. Contrary to other bacteria, *B. xenovorans* LB400 is known to use PCBs as a source of carbon (Parnell et al. 2010) and ably degrades PCBs into six chlorines (Furukawa and Fujihara 2008). Furthermore, genus *Burkholderia* has been well accepted in the rhizosphere of numerous plants (Suárez-Moreno et al. 2012). In bioremediation, bioaugmentation is efficient, but its application for PCBs is limited (Singer et al. 2000, 2003; Ponce et al. 2011; Sudjarid et al. 2012). In contrast, plants may play a major indirect role in the soil colonisation by microorganisms and may result in PCB degradation by *Lespedeza cuneata*, *Festuca arundinacea* and *Medicago sativa* (Chekol and Vough 2002; Chekol et al. 2004). Plant roots can assist in the spread of bacteria through soil and facilitate the penetration of impermeable soil layers (Kuiper et al. 2004), and eventually plant-assisted bioaugmentation can directly improve the remediation capacity with the attraction of water by the root system, the accumulation of the most water-soluble PCB in the rhizosphere and, directly or indirectly, the degradation or translocation of pollutants. Thus, association of the plant and bacteria for the proper bacterial colonisation of the root system thereby increases the bioremediation process (Kuiper et al. 2001).

14.9 Conclusions and Future Perspectives

In addition to accelerating the plant growth in normal soils, many bacterial traits could effectively help to expedite the phytoremediation of heavy metal–contaminated soils by protecting plants from stress factors and increasing their biomass through facilitating nutrient acquisition and metal uptake. Thus, inoculation of phytoremediating plants with bacteria exhibiting multiple traits is an excellent strategy to remediate metal-polluted soils. Previous findings suggest that there is a tremendous scope in bacteria-assisted hyperaccumulation of metals in plants growing in heavily contaminated soils, thus surpassing constraints of nutrient deficiency and reduced biomass. Moreover, exploration of new metal-mobilising attributes and unravelling the exact bacterial mechanisms assisting in metal-extracting plants would help to understand different issues pertaining to variability in results and reinforce this benign plant biotechnology for successful applicability in ecologically diverse soils.

To achieve better efficiency of bacteria-assisted phytoremediation, there is a need of vigorous screening of bacterial traits of agricultural and environmental significance under various stress conditions so that bacteria exhibiting the maximum number of beneficial activities and better root colonisation and displaying consistent and reproducible performance in agronomically diverse niches are selected to overcome the practical limitations of their application as bioinoculants with phytoremediating plants. Among different bacteria, preference should be given to endophytes for phytoremediation purposes; in this way, competition for colonisation in rhizosphere would be minimised with other soil microflora.

On the basis of recent studies, the specific strains showing good activity and colonisation potential will be useful in enhancing phytoextraction applications; however, the performance of microorganisms under natural conditions has to be investigated in detail and genetically engineered strains are likely to be superior in terms of trace element resistance and mobilisation. Furthermore, biosafety aspects have to be considered, and their release depends on the legislation. Addressing the issues of persistence and competition capacity of inoculant strains while developing their potential for plant growth promotion, stress resistance and trace element accumulation represents a promising strategy for improving current phytoremediation techniques. Recently, new biotechnological and genetic engineering tools have revolutionised the bioscience as new traits can be produced in the recipient organism by inserting or manipulating the specific genetic sequence of desired traits from the host organisms. By this approach, a bacterial strain can be genetically modified in order to increase phytoextraction or mobilisation of heavy metals.

References

Abeles FB, Morgan PW, Saltveit MEJ. 1992. *Ethylene in Plant Biology*, 2nd ed., Academic Press, New York.

Abou-Shanab RA, Angle JS, Delorme TA et al. 2003. Rhizobacterial effects on nickel extraction from soil and uptake by Alyssum murale. *N Phytol* 158(1):219–224.

Agostini E, Coniglio MS, Milrad SR, Tigier HA, Giulietti AM. 2003. Phytoremediation of 2,4-dichlorophenol by *Brassica napus* hairy root cultures. *Biotechnol Appl Biochem* 37:139–144.

Ahemad M, Khan MS. 2011. Plant growth promoting fungicide-tolerant *Rhizobium* improves growth and symbiotic characteristics of lentil (*Lens esculentus*) in fungicide-applied soil. *J Plant Growth Regul* 30:334–342.

Ahemad M, Khan MS. 2012. Effect of fungicides on plant growth promoting activities of phosphate solubilizing *Pseudomonas putida* isolated from mustard (*Brassica compestris*) rhizosphere. *Chemosphere* 86:945–950.

Ahemad M, Kibret M. 2014. Mechanisms and applications of plant growth promoting rhizobacteria: Current perspective. *J King Saud University Sci* 26:1–20.

Anjum NA, Pereira ME, Ahmad I, Duarte AC, Umar S, Khan NA (eds.) *Phytotechnologies: Remediation of Environmental Contaminants*. CRC Press, Boca Raton, FL, pp. 361–376.

Arshad M, Frankenberger WTJ. 2002. *Ethylene: Agricultural Sources and Applications*. Kluwer Academic/Plenum Publishers, New York.

Arshad M, Saleem M, Hussain S. 2007. Perspectives of bacterial ACC deaminase in phytoremediation. *Trends Biotechnol* 25:356–362.

Barzanti R, Ozino F, Bazzicalupo M et al. 2007. Isolation and characterization of endophytic bacteria from the nickel hyperaccumulator plant *Alyssum bertolonii*. *Microb Ecol* 53:306–316.

Becerra-Castro C, Prieto-Fernández A, Alvarez-Lopez V et al. 2011. Nickel solubilizing capacity and characterization of rhizobacteria isolated from hyperaccumulating and non-hyperaccumulating subspecies of *Alyssum serpyllifolium*. *Int J Phytoremediation* 1:229–244.

Bedard DL, Quensen JF III. 1995. Microbial reductive dechlorination of polychlorinated biphenyls. In: Young LY, Cerniglia CE, Cerniglia CE (eds.), *Microbial transformation and degradation of toxic organic chemicals*. Wiley-Liss Division, New York, pp. 127–216.

Bedard DL, Van Dort HM, May RJ, Smullen LA. 1997. Enrichment of microorganisms that sequentially meta, para-dechlorinate the residue of Aroclor 1260 in Housatonic River sediment. *Environ Sci Technol* 11:3308–3313.

Belimov AA, Hontzeas N, Safronova VI et al. 2005. Cadmium-tolerant plant growth-promoting bacteria associated with the roots of Indian mustard (*Brassica juncea* L. Czern.). *Soil Biol Biochem* 37:241–250.

Belimov AA, Safronova VI, Sergeyeva TA et al. 2001. Characterization of plant growth promoting rhizobacteria isolated from polluted soils and containing 1-aminocyclopropane-1-carboxylate deaminase. *Can J Microbiol* 47:242–252.

Blaha D, Combaret CP, Mirza MS, Loccoz YM. 2006. Phylogeny of the 1-aminocyclopropane-1-carboxylic acid deaminase-encoding gene acdS in phytobeneficial and pathogenic Proteobacteria and relation with strain biogeography. *FEMS Microbiol Ecol* 56:455–470.

Braud A, Geoffroy V, Hoegy F, Mislin GLA, Schalk IJ. 2010. The siderophores pyoverdine and pyochelin are involved in *Pseudomonas aeruginosa* resistence against metals: Another biological function of these two siderophores. *Environ Microbiol Report* 2:419–425.

Braud A, Hannauer M, Milsin GLA, Schalk IJ. 2009a. The *Pseudomonas aeruginosa* pyochelin-iron uptake pathway and its metal specificity. *J Bacteriol* 191:5317–5325.

Braud A, Hoegy F, Jezequel K, Lebeau T, Schalk IJ. 2009b. New insights into the metal specificity of the *Pseudomonas aeruginosa* pyoverdine-iron uptake pathway. *Environ Microbiol* 11:1079–1091.

Braud A, Jezequel K, Vieille E, Tritter A, Lebeau T. 2006. Changes in the extractability of Cr and Pb in a polycontaminated soil after bioaugmentation with microbial producers of biosurfactants, organic acids and siderophores. *Water Air Soil Pollut* 6:3–4.

Burd GI, Dixon DG, Glick BR. 1998. A plant growth promoting bacterium that decreases nickel toxicity in seedlings. *Appl Environ Microbiol* 64:3663–3668.

Burd GI, Dixon DG, Glick BR. 2000. Plant growth promoting bacteria that decrease heavy metal toxicity in plants. *Can J Microbiol* 46:237–245.

Carrillo-Castañeda G, Munoz JJ, Peralta-Videa JR, Gomez E, Gardea-Torresdey JL. 2003. Plant growth-promoting bacteria promote copper and iron translocation from root to shoot in alfalfa seedlings. *J Plant Nutr* 26:1801–1814.

Chekol T and Vough LR. 2002. Assessing the phytoremediation potential of tall fescue and Sericea Lespedeza for organic contaminants in soil. *Remed J* 3:117–128.

Chekol T, Vough LR, Chaney RL. 2004. Phytoremediation of polychlorinated biphenyl-contaminated soils: The rhizosphere effect. *Environ Int* 6:799–804.

Chen G, Chen X, Yang Y et al. 2011. Sorption and distribution of copper in unsaturated Pseudomonas putida CZ1 biofilms as determined by X-ray fluorescence microscopy. *Appl Environ Microb* 77:4719–4727.

Chen Y, Jurkewitch E, Bar-Ness E, Hadar Y. 1994. Stability constants of pseudobactin complexes with transition metals. *Soil Sci Soc Am J* 58:390–396.

Córdova S, Neaman A, González I et al. 2011. The effect of lime and compost amendments on the potential for the revegetation of metal-polluted, acidic soils. *Geoderma* 166:135–144.

Dimkpa CO, Svatos A, Dabrowska P, Schmidt A, Boland W, Kothe E. 2008. Involvement of siderophores in the reduction of metal-induced inhibition of auxin synthesis in *Streptomyces* spp. *Chemosphere* 74:19–25.

Dong R, Gu L, Guo C, Xun F, Liu J. 2014. Effect of PGPR Serratia marcescens BC-3 and AMF Glomus intraradices on phytoremediation of petroleum contaminated soil. *Ecotoxicology* 23(4):674–680.

Elsgaard L, Petersen SO, Debosz K. 2001. Effects and risk assessment of linear alkylbenzene sulfonates in agricultural soil. 1. Short-term effects on soil microbiology. *Environ Toxicol Chem* 20(8):1656–1663.

Evers A, Hancock R, Martell A, Motekaitis R. 1989. Metal ion recognition in ligands with negatively charged oxygen donor groups. Complexation of Fe (III), Ga (III), In (III), Al (III) and other highly charged metal ions. *Inorg Chem* 28:2189–2195.

Faisal M, Hasnain S. 2005. Chromate resistant *Bacillus cereus* augments sunflower growth by reducing toxicity of Cr (VI). *J Plant Biol* 48:187–194.

Filip Z. 2002. International approach to assessing soil quality by ecologically related biological parameters. *Agric Ecosyst Environ* 88(2):689–712.

Furukawa K, Fujihara H. 2008. Microbial degradation of polychlorinated biphenyls: Biochemical and molecular features. *J Biosci Bioeng* 5:433–449.

Gamalero E, Glick BR. 2012. Plant growth-promoting bacteria, mechanism and applications. Article ID 943401, 15 p.

Gaonkar T, Bhosle S. 2013. Effect of metals on a siderophore producing bacterial isolate and its implications on microbial assisted bioremediation of metal contaminated soils. *Chemosphere* 93(9):1835–1843.

Geetha SJ, Joshi SJ. 2013. Engineering rhizobial bioinoculants: A strategy to improve iron nutrition. *The Scientific World Journal 2013*: article ID 315890, 15 p.

George P, Gupta A, Gopal M, Thomas L, Thomas GV. 2013. Multifarious beneficial traits and plant growth promoting potential of *Serratia marcescens* KiSII and *Enterobacter* sp. RNF 267 isolated from the rhizosphere of coconut palms (*Cocos nucifera* L.). *World J Microbiol Biotechnol* 29:109–117.

Giller KE, Witter E, McGrath SP. 1998. Toxicity of heavy metals to microorganisms and microbial processes in agricultural soils. *Soil Biol Biochem* 30(10–11):1389–1414.

Gleba D, Borisjuk NV, Borisjuk LG et al. 1999. Use of plant roots for phytoremediation and molecular farming. *Proc Natl Acad Sci USA* 96:5973–5977.

Glick BR. 1995. The enhancement of plant growth by free-living bacteria. *Can J Microbiol* 41:109–117.

Glick BR. 2005. Modulation of plant ethylene levels by the bacterial enzyme ACC deaminase. *FEMS Microbiol Lett* 251:1–7.

Glick BR. 2010. Using soil bacteria to facilitate phytoremediation. *Biotechnol Adv* 28:367–374.

Glick BR. 2012. Plant growth-promoting bacteria: Mechanisms and applications. *Scientifica*, article ID 963401, 15 p.

Glick BR, Bashan Y. 1997. Genetic manipulation of plant growth promoting bacteria to enhance biocontrol of fungal phytopathogens. *Biotechnol Adv* 15:353–378.

Glick BR, Penrose DM, Li J. 1998. A model for the lowering of plant ethylene concentrations by plant growth-promoting bacteria. *J Theor Biol* 190:63–68.

Greenwald J, Zeder-Lutz G, Hagege A, Celia H, Pattus F. 2008. The metal dependence of pyoverdine interactions with its outer membrane receptor FpvA. *J Bacteriol* 190:6548–6558.

Haferburg G, Kothe E. 2010. Metallomics: Lessons for metalliferous soil remediation. *Appl Microbiol Biotechnol* 87:1271–1280.

Hao X, Xie P, Johnstone L, Miller SJ, Rensing C, Weia G. 2012. Genome sequence and mutational analysis of bacterium *Agrobacterium tumefaciens* CCNWGS0286 isolated from a zinc-lead mine tailing. *Appl Environ Microbiol* 78:5384–5394.

Hashim MA, Mukhopadhyay S, Sahu JN, Sengupta B. 2011. Remediation technologies for heavy metal contaminated groundwater. *J Environ Manag* 92:2355–2388.

He H, Ye Z, Yang D et al. 2013. Characterization of endophytic *Rahnella* sp. JN6 from *Polygonum pubescens* and its potential in promoting growth and Cd, Pb, Zn uptake by *Brassica napus. Chemosphere* 90:1960–1965.

He LY, Zhang YF, Ma HY et al. 2010. Characterization of copper resistant bacteria and assessment of bacterial communities in rhizosphere soils of copper-tolerant plants. *Appl Soil Ecol* 44:49–55.

Hernlem BJ, Vane LM, Sayles GD. 1996. Stability constants for complexes of the siderophore desferrioxamine B with selected heavy metal cations. *Inorganic Chimica Acta* 244:179–184.

Hersman L, Lloyd T, Sposito G. 1995. Siderophore promoted dissolution of hematite. *Geochimica Cosmochimica Acta* 59:3327–3330.

Hontzeas N, Zoidakis J, Glick BR, Abu-Omar MM. 2004. Expression and characterization of 1-aminocyclopropane-1-carboxylate deaminase from the rhizobacterium *P.* putida UW4: A key enzyme in bacterial plant growth promotion. *Biochim Biophys Acta* 1703:11–19.

Hu X, Boyer GL. 1996. Siderophore-mediated aluminium uptake by *Bacillus megaterium* ATCC19213. *Appl Environ Microbiol* 62:4044–4048.

Jeong S, Moon HS, Nama K, Kim JY, Kim TS. 2012. Application of phosphate-solubilizing bacteria for enhancing bioavailability and phytoextraction of cadmium (Cd) from polluted soil. *Chemosphere* 88:204–210.

Jing YD, He ZL, Yang XE. 2007. Role of soil rhizobacteria in phytoremediation of heavy metal contaminated soils. *J Zhejiang Univ Sci B* 8(3):192–207.

Kang BG, Kim WT, Yun HS, Chang SC. 2010. Use of plant growth-promoting rhizobacteria to control stress responses of plant roots. *Plant Biotechnol Rep* 4: 179–183.

Kefeng L, Wusirika R. 2011. Effect of multiple metal resistant bacteria from contaminated lake sediments on metal accumulation and plant growth. *J Hazard Mater* 15:189(1–2):531–539.

Khan AG, Kuek C, Chaudhry TM, Khoo CS, Hayes WJ. 2000. Role of plants, mycorrhizae and phytochelators in heavy metal contaminated land remediation. *Chemosphere* 41:197–207.

Khan MS, Zaidi A, Wani PA. 2007. Role of phosphate solubilizing microorganisms in sustainable agriculture: A review. *Agron Sustain Dev* 27:29–43.

Kim JH, Kim WT, Kang BG. 2001. IAA and N(6)-benzyladenine inhibit ethylene-regulated expression of ACC oxidase and ACC synthase genes in mungbean hypocotyls. *Plant Cell Physiol* 42:1056–1061.

Kuiper I, Bloemberg GV, Lugtenberg BJJ. 2001. Selection of a plant-bacterium pair as a novel tool for rhizostimulation of polycyclic aromatic hydrocarbon-degrading bacteria. *Mol Plant Microbe Interact* 14:1197–1205.

Kuiper I, Lagnedijk EL, Bloemberg GV et al. 2004. Rhizoremediation: A beneficial plant-microbe interaction. *Mol Plant Microbe Interact* 17:6–15.

Küpper H, Mijovilovich A, Meyer-Klaucke W, Kroneck PMH. 2004. Tissue- and age-dependent differences in the complexation of cadmium and zinc in the cadmium/zinc hyperaccumulator *Thlaspi caerulescens* (Ganges ecotype) revealed by X-ray absorption spectroscopy. *Plant Physiol* 134:748–757.

Lasat HA. 2002. Phytoextraction of toxic metals: A review of biological mechanisms. *J Environ Qual* 31(1):109–120.

Li K, Ramakrishna W. 2011. Effect of multiple metal resistant bacteria from contaminated lake sediments on metal accumulation and plant growth. *J Hazard Mater* 189:531–539.

Li Z, Kong S, Chen L et al. 2011. Concentrations, spatial distributions and congener profiles of polychlorinated biphenyls in soils from a coastal City-Tianjin, China. *Chemosphere* 85:494–501.

Liphadzi MS, Kirkham MB, Paulsen GM. 2006. Auxin-enhanced root growth for phytoremediation of sewage-sludge amended soil. *Environ Technol* 27:695–704.

Lombard N, Prestat E, van Elsas JD, Simonet P. 2011. Soil specific limitations for access and analysis of soil microbial communities by metagenomics. *FEMS Microbiol Ecol* 78:31–49.

Ma W, Charles TC, Glick BR. 2004. Expression of an exogenous 1-aminocyclopropane-1-carboxylate deaminase gene in *Sinorhizobium meliloti* increases its ability to nodulate alfalfa. *Appl Environ Microbiol* 70:5891–5897.

Ma Y, Rajkumar M, Freitas H. 2009a. Improvement of plant growth and nickel uptake by nickel resistant-plant-growth promoting bacteria. *J Hazard Mater* 166:1154–1161.

Ma Y, Rajkumar M, Freitas H. 2009b. Inoculation of plant growth promoting bacterium *Achromobacter xylosoxidans* strain Ax10 for the improvement of copper phytoextraction by *Brassica juncea*. *J Environ Manage* 90:831–837.

Ma Y, Rajkumar M, Freitas H. 2009c. Isolation and characterization of Ni mobilizing PGPB from serpentine soils and their potential in promoting plant growth and Ni accumulation by *Brassica* spp. *Chemosphere* 75:719–725.

Ma Y, Rajkumar M, Luo Y, Freitas H. 2011a. Inoculation of endophytic bacteria on host and non-host plants-effects on plant growth and Ni uptake. *J Hazard Mater* 195:230–237.

Ma Y, Rajkumar M, Luo Y, Freitas H. 2013. Phytoextraction of heavy metal polluted soils using *Sedum plumbizincicola* inoculated with metal mobilizing *Phyllobacterium myrsinacearum* RC6b. *Chemosphere* 93:1386–1392.

Ma Y, Rajkumar M, Vicente JA, Freitas H. 2011b. Inoculation of Ni-resistant plant growth promoting bacterium *Psychrobacter* sp. strain SRS8 for the improvement of nickel phytoextraction by energy crops. *Int J Phytoremediation* 13:126–139.

Madejon E, de Mora AP, Felipe E. 2006. Soil amendments reduce trace element solubility in a contaminated soil and allow regrowth of natural vegetation. *Environ Pollut* 139:40–52.

Madhaiyan M, Poonguzhali S, Ryu J, Sa T. 2006. Regulation of ethylene levels in canola (*Brassica campestris*) by 1-aminocyclopropane-1-carboxylate deaminase-containing *Methylobacterium fujisawaense*. *Planta* 224:268–278.

McGrath SP, Zhao FJ, Lombi E. 2001. Plant and rhizosphere processes involved in phytoremediation of metal-contaminated soils. *Plant Soil* 232(1–2):207–214.

Nagajyoti PC, Lee KD, Sreekanth TVM. 2010. Heavy metals, occurrence and toxicity for plants: A review. *Environ Chem Lett* 8:199–216.

Neagoe A, Ebena G, Carlsson E. 2005. The effect of soil amendments on plant performance in an area affected by acid mine drainage. *Chem Erde Geochem* 65:115–129.

Neubauer U, Nowack B, Furrer G, Schulin R. 2000. Heavy metal sorption on clay minerals affected by the siderophore desferrioxamine B. *Environ Sci Technol* 34:2749–2755.

Nicoară A, Neagoe A, Stancu P et al. 2014. Coupled pot and lysimeter experiments assessing plant performance in microbially assisted phytoremediation. *Environ Sci Pollut Res Int* 21(11):6905–6920.

Oves M, Khan MS, Zaidi A. 2013. Chromium reducing and plant growth promoting novel strain *Pseudomonas aeruginosa* OSG41 enhance chickpea growth in chromium amended soils. *Eur J Soil Biol* 56:72–83.

Pacwa-Płociniczak M, Płaza GA, Piotrowska-Seget Z, Cameotra SS. 2011. Environmental applications of biosurfactants: Recent advances. *Int J Mol Sci* 12:633–654.

Pandey P, Kang SC, Maheshwari DK. 2005. Isolation of endophytic plant growth promoting *Burkholderia* sp. MSSP from root nodules of *Mimosa pudica*. *Curr Sci* 89:170–180.

Parnell JJ, Denef VJ, Park J et al. 2010. Environmentally relevant parameters affecting PCB degradation: Carbon source- and growth phase-mitigated effects of the expression of the biphenyl pathway and associated genes in Burkholderia xenovorans LB400. *Biodegradation* 1:47–156.

Penrose DM, Glick BR. 2001. Levels of 1-aminocyclopropane-1-carboxylic acid (ACC) in exudates and extracts of canola seeds treated with plant growth-promoting bacteria. *Can J Microbiol* 47:368–372.

Petrisor IG, Dobrota S, Komnitsas K et al. 2004. Artificial inoculation- perspectives in tailings phytostabilization. *Int J Phytoremediat* 6:1–15.

Ponce BL, Latorre VK, Gonzalez M et al. 2011. Antioxidant compounds improved PCB-degradation by Burkholderia xenovorans strain LB400. *Enzyme Microbial Technol* 49:509–516.

Rajkumar M, Ae N, Prasad MNV, Freitas H. 2010. Potential of siderophore-producing bacteria for improving heavy metal phytoextraction. *Trends Biotechnol* 28:142–149.

Rajkumar M, Freitas H. 2008. Effects of inoculation of plant growth promoting bacteria on Ni uptake by Indian mustard. *Bioresour Technol* 99:3491–3498.

Rajkumar M, Lee KJ, Lee WH, Banu JR. 2005. Growth of *Brassica juncea* under chromium stress: Influence of siderophores and indole-3-acetic acid producing rhizosphere bacteria. *J Environ Biol* 26:693–699.

Rajkumar M, Ma Y, Freitas H. 2008. Characterization of metal-resistant plant-growth promoting *Bacillus weihenstephanensis* isolated from serpentine soil in Portugal. *J Basic Microbiol* 48:500–508.

Rajkumar M, Sandhya S, Prasad MNV. 2012. Perspectives of plant associated microbes in heavy metal phytoremediation. *Biotechnol Adv* 30:1562–1574.

Raymond KN, Dertz EA, Kim SS. 2003. Enterobactin: An archetype for microbial iron transport. *Proc Natl Acad Sci USA* 100:3584–3588.

Remans T, Thijs S, Truyens S et al. 2012 Understanding the development of roots exposed to contaminants and the potential of plant-associated bacteria for optimization of growth. *Ann Bot* 110:239–252.

Rodecap KD, Tingey DT. 1981. Stress ethylene: A bioassay for rhizosphere-applied phytotoxicants. *Environ Monit Assess* 1:119–127.

Safronova VI, Stepanok VV, Engqvist GL, Alekseyev YV, Belimov AA. 2006. Root-associated bacteria containing 1-aminocyclopropane-1-carboxylate deaminase improve growth and nutrient uptake by pea genotypes cultivated in cadmium supplemented soil. *Biol Fert Soils* 42:267–272.

Saleem M, Arshad M, Hussain S, Bhatti AS. 2007. Perspective of plant growth promoting rhizobacteria (PGPR) containing ACC deaminase in stress agriculture. *J Ind Microbiol Biotechnol* 34:635–648.

Salt DE, Benhamou N, Leszczyniecka M, Raskin I. 1999. A possible role for rhizobacteria in water treatment by plant roots. *Int J Phytorem* 1:67–69.

Santner A, Calderon-Villalobos LIA, Estelle M. 2009. Plant hormones are versatile chemical regulators of plant growth. *Nature Chem Biol* 5:301–307.

Schalk IJ, Hannauer M, Braud A. 2011. New roles for bacterial siderophores in metal transport and tolerance. *Environ Microbiol* 13:2844–2854.

Schulz S, Brankatschk R, Dumig A, Kogel-Knabner I, Schloter M, Zeyer J. 2013. The role of microorganisms at different stages of ecosystem development for soil formation. *Biogeosciences* 10:3983–3996.

Sessitsch A, Coenye T, Sturz AV et al. 2005. *Burkholderia phytofirmins* sp. nov., a novel plant-associated bacterium with plant beneficial properties. *Int J Syst Evol Microbiol* 55:1187–1192.

Sheehy RE, Honma M, Yamada M, Sasaki T, Martineau B, Hiatt WR. 1991. Isolation, sequence, and expression in *Escherichia coli* of the *Pseudomonas* sp. strain ACP gene encoding 1-aminocyclopropane-1-carboxylate deaminase. *J Bacteriol* 173:5260–5265.

Sheng XF, He LY, Wang QY, Ye HS, Jiang C. 2008. Effects of inoculation of biosurfactant producing *Bacillus* sp. J119 on plant growth and cadmium uptake in a cadmium amended soil. *J Hazard Mater* 155:17–22.

Shutcha MN, Mubemba MM, Faucon M et al. 2010. Phytostabilisation of copper-contaminated soil in Katanga: An experiment with three native grasses and two amendments. *Int J Phytoremediat* 12:616–632.

Singer AC, Gilbert ES, Luepromchai E et al. 2000. Bioremediation of polychlorinated biphenyl-contaminated soil using carvone and surfactant-grown bacteria. *Appl Microbiol Biotechnol* 6:838–843.

Singer AC, Smith D, Jury WA et al. 2003. Impact of the plant rhizosphere and augmentation on remediation of polychlorinated biphenyl contaminated soil. *Environ Toxicol Chem* 9:1998–2004.

Singh AK, Cameotra SS. 2013. Rhamnolipids production by multi-metal-resistant and plant-growth-promoting rhizobacteria. *Appl Biochem Biotechnol* 170:1038–1056.

Smith SE, Read DJ. 1997. *Mycorrhizal Symbiosis*. Academic Press, Inc., San Diego, CA.

Song JS, Walpola BC, Chung DY, Yoon MH. 2012. Heavy metal resistant phosphate solubilizing bacteria. *Korean J Soil Sci Fert* 45:817–821.

Stiens M, Schneiker S, Keller M, Kuhn S, Pühler A, Schlüter A. 2006. Sequence analysis of the 144-kilobase accessory plasmid psmesm11a, isolated from a dominant *Sinorhizobium meliloti* strain identiWed during a long-term Weld release experiment. *Appl Environ Microbiol* 72:3662–3672.

Suárez-Moreno ZR, Caballero-Mellado J, Coutinho BG et al. 2012. Common features of environmental and potentially beneficial plantassociated Burkholderia. *Microbial Ecol* 2:249–266.

Sudjarid W, Chen IM, Monkong W et al. 2012. Reductive dechlorination of 2,3,4-chlorobiphenyl by biostimulation and bioaugmentation. *Environment Eng Sci* 29:255–261.

Tabak HH, Lens P, van Hullebusch ED, Dejonghe W. 2005. Developments in bioremediation of soils and sediments polluted with metals and radionuclides-1. Microbial processes and mechanisms affecting bioremediation of metal contamination and influencing metal toxicity and transport. *Rev Env Sci Biotech* 4:115–156.

Tank N, Saraf M. 2009. Enhancement of plant growth and decontamination of nickel-spiked soil using PGPR. *J Basic Microbiol* 49:195–204.

Tremaroli V, Vacchi SC, Fedi S, Ceri H, Zannoni D, Turner RJ. 2010. Tolerance of Pseudomonas pseudoalcaligenes KF707 to metals, polychlorobiphenyls and chlorobenzoates: Effects on chemotaxis-, biofilm-and planktonic-grown cells. *FEMS Microbiol Ecol* 2:291–301.

Uchiumi T, Ohwada T, Itakura M et al. 2004. Expression islands clustered on the symbiosis island of the *Mesorhizobium loti* genome. *J Bacteriol* 186:2439–2448.

Viti C, Giovannetti L. 2007. Bioremediation of soils polluted with hexavalent chromium using bacteria: A challenge. In: Shree NS, Tripathi DR (eds.), *Environmental Bioremediation Technologies*. Springer, New York, pp. 57–76.

Wani PA and Khan MS. 2010. *Bacillus* species enhance growth parameters of chickpea (*Cicer arietinum* L.) in chromium stressed soils. *Food Chem Toxicol* 48:3262–3267.

Wang CKE, Glick BR, Defago G. 2000. Effect of transferring 1-aminocyclopropane-1-carboxylic acid (ACC) deaminase genes into pseudomonas Xuorescens strain CHA0 and its gacA derivative CHA96 on their growth-promoting and disease-suppressive capacities. *Can J Microbiol* 46:898–907.

Weber JB, Mrozek E Jr. 1979. Polychlorinated biphenyls: Phytotoxicity, absorption and translocation by plants, and inactivation by activated carbon. *Bull Environ Contam Toxicol* 23:412–417.

Whiting SN, de Souza MP, Terry N. 2001. Rhizosphere bacteria mobilize Zn for hyperaccumulation by *Thlaspi caerulescens*. *Environ Sci Technol* 35(15):3144–3150.

Wu Q, Bedard DL, Wiegel J. 1997. Effect of incubation temperature on the route of microbial reductive dechlorination of 2,3,4,6-tetrachlorobiphenyl in polychlorinated biphenyl (PCB)-contaminated and PCB-free freshwater sediments. *Appl Environ Microbiol* 7:2836–2843.

Wu SC, Luo YM, Cheung KC. 2006. Influence of bacteria on Pb and Zn speciation, mobility and bioavailability in soil: A laboratory study. *Environ Pollut* 144:765–773.

Yuan M, He H, Xiao L et al. 2013. Enhancement of Cd phytoextraction by two Amaranthus species with endophytic *Rahnella* sp. JN27. *Chemosphere* 103:99–104.

Zhang YF, He LY, Chen ZJ, Wang QY, Qian M, Sheng XF. 2011. Characterization of ACC deaminase-producing endophytic bacteria isolated from copper-tolerant plants and their potential in promoting the growth and copper accumulation of *Brassica napus*. *Chemosphere* 83:57–62.

Index

Page numbers followed by f and t indicate figures and tables, respectively.

T - #0214 - 111024 - C0 - 234/156/22 - PB - 9780367575830 - Gloss Lamination